THE MATHEMATICS OF FINITE ELEMENTS AND APPLICATIONS

THE MATHEMATICS OF FINITE ELEMENTS AND APPLICATIONS

Highlights 1993

Edited by

J. R. Whiteman

Brunel, The University of West London

JOHN WILEY & SONS

Chichester · New York · Brisbane · Toronto · Singapore

Copyright © 1994 by John Wiley & Sons Ltd,
                     Baffins Lane, Chichester,
                     West Sussex PO19 1UD, England
                     National   Chichester (0243) 779777
                     International   (+44) 243 779777

Other Wiley Editorial Offices

John Wiley & Sons, Inc., 605 Third Avenue,
New York, NY 10158-0012, USA

Jacaranda Wiley Ltd, 33 Park Road, Milton,
Queensland 4064, Australia

John Wiley & Sons (Canada) Ltd, 22 Worcester Road,
Rexdale, Ontario M9W 1L1, Canada

John Wiley & Sons (SEA) Pte Ltd, 37 Jalan Pemimpin #05-04,
Black B, Union Industrial Building, Singapore 2057

British Library Cataloguing in Publication Data

A catalogue record for this book is available from the British Library

ISBN 0 471 93996 X

Produced from camera-ready copy supplied by the editor
Printed and Bound in Great Britain by Bookcraft (Bath) Ltd.

CONTENTS

Preface

1 Error Estimation and Control in Computational Fluid Dynamics   1
   J.T. Oden

2 Reliability of Computational Mechanics   25
   I. Babuška

3 Numerical Techniques for Problems of Quasistatic and Dynamic Viscoelasticity   45
   S. Shaw, M.K. Warby and J.R. Whiteman.

4 Numerical Simulation of Non-Newtonian Flow through an Axisymmetric Contraction   69
   A. Baloch, P. Townsend and M.F. Webster

5 Approximate Approximations   77
   V. Maz'ya

6 A New Paradigm for Adaptive Finite Element Methods   105
   C. Johnson

7 Mesh optimality criteria for adaptive finite element computations   121
   E. Oñate and G. Bugeda

8 New Developments in Applications of hp-Adaptive BE/FE Methods to Elastic Scattering   137
   L. Demkowicz and J.T. Oden

9 A Posteriori Error Estimates for Stokes and Navier Stokes Equations   151
   M. Ainsworth

10 A Zooming-Technique Using a Hierarchical hp-Version of the Finite Element Method   159
   E. Rank

11 A Substructured Elastic-Plastic Fracture Analysis Program for Parallel Processing on Transputer Networks   171
   H. van Lengen, B. May, F.-G. Buchholz and H.A. Richard

Contents

12	An Efficient Iterative Parallel Finite Element Computational Method *K.P. Wang and J.C. Bruch, Jr.*	179
13	Application of a Variational Method for Computing Smooth Stresses, Stress Gradients, and Error Estimation in Finite Element Analysis *A. Tessler, H.R. Riggs, and S.C. Macy*	189
14	A Parallel Numerical Model for Subsurface Contaminant Transport With Biodegradation Kinetics *T. Arbogast and M.F. Wheeler*	199
15	Unstructured Grid Methods for High Speed Compressible Flows *K. Morgan, J. Peraire, J. Peiró and O. Hassan*	215
16	Simulation of Dense Non-aqueous Phase Fluid Flow in Porous Media using Collocation Finite Elements *J.F. Guarnaccia and G.F. Pinder*	243
17	A Complementary Volume Approach for Modeling Three-Dimensional Navier-Stokes Equations Using Dual Delaunay/Voronoi Tessellations *J.C. Cavendish, C.A. Hall and T.A. Porsching*	255
18	A Posteriori and A Priori Error Analysis of Finite Volume Methods *K.W. Morton and E. Süli*	267
19	Adaptive Finite Volume Methods for Hyperbolic Problems *J.A. Mackenzie, T. Sonar and E. Süli*	289
20	Reliable Finite Volume Methods for Time-Dependent Partial Differential Equations *M. Berzins and J. Ware*	299
21	Non-Reflecting Finite Elements *D. Givoli and J.B. Keller*	307
22	A Dispersion Analysis of Finite Element Methods on Triangular Grids for Maxwell's Equations *P.B. Monk, A.K. Parrott and P.J. Wesson*	315
23	Coupling of FEM and BEM for Transonic Flows *H. Berger, G. Warneke and W.L. Wendland*	323

ABSTRACTS OF OTHER PAPERS

On gradient smoothing in grids with singularity element patches   351
J. Aalto, J. Louhivirta and J.R. Whiteman

Time-dependent finite element solution of Boltzmann equation for photon transport   352
R.T. Ackroyd, C.R.E. de Oliveira and A.J.H. Goddard

Stresses computed from hierarchic plate models   352
R.L. Actis and B.A. Szabó

Boundary integral equations for the exterior acoustic problem   353
S. Amini and N.D. Maines

On the modelling errors in fracture analysis of delaminated plates   354
B. Andersson

A finite difference method of determination of local coefficients in dental enamel and other inhomogeneous permeable media   355
P. Anderson, J.C. Elliott and S.E.P. Dowker

Anisotropic finite elements and application to edge singularities   355
T. Apel

The boundary layer for the Reissner-Mindlin plate model   356
D.N. Arnold

A stablized Galerkin finite element method for solving the Boussinesq equations   356
A. Auge and G. Lube

A new elemental formulation for the dynamic analysis of structures   357
S. Azimi

A finite volume procedure to predict deformation and residual stress in castings   357
C. Bailey, M. Cross and P. Chow

Adaptive methods for grid generation and finite element solutions of time-dependent partial differential equations   358
M.J. Baines

Influence of Extensional Viscosity on Vortex Development in Complex Geometries   359
A. Baloch, P. Townsend and M.F. Webster

On the approximation of junctions between thin shells by curved C^1-finite element methods   359
    M. Bernadou

On the discretization of an optimal design problem for an unilaterally supported viscoelastic plate   360
    I. Bock and J. Lovisek

A hybrid direct-iterative method for the solution of finite element linear systems with positive definite matrices   361
    L. Brusa and F. Riccio

Hybrid Arnoldi-Newton Algorithm   362
    T. Bulenda

A FEM Approach for the Analysis of Composite Elements   362
    M. Buonsanti

Cubic finite elements in elliptic problems with interfaces and singularities   363
    P. Burda

Computation of the homogenized behaviour of a gasket   364
    P. Cartraud, C. Wielgosz and O. Debordes

Adaptive mesh refinement for the 3D magnetic code TRIFOU   364
    P. Chaussecourte, B. Métivet and G. Nicolas

Extrapolation methods for the streamline diffusion finite element method   365
    H. Chen and R. Rannacher

A new model of jointed rock masses reinforced by passive, fully-grouted bolts   366
    S.H. Chen

Structural dynamic finite element analyses of space shuttle propulsion components   367
    E.R. Christensen

Domain decomposition techniques for massively parallel machines   368
    R.K. Coomer and I.G. Graham

p-version preconditioning for the mass matrix   368
    A. Craig

Nonhomgeneous degenerate eigenvalue problem   369
    P. Drábek

Contents ix

A flux continuous approximation of the pressure equation for h-adaptive grids   370
M.G. Edwards

Parallel implementation of the hp-version of the finite element method on a shared-memory computer   371
H.C. Elman

Numerical approximation and asymptotic properties of some two dimensional plate models   371
R. Falk

The uncoupling of boundary integral and finite element methods for nonlinear boundary value problems   371
G.N. Gatica and G.C. Hsiao

New high order infinite elements with different decay types   372
L. Gavete, C. Manzano and A. Ruiz

Using explicit preconditioned schemes for solving initial/boundary value problems   373
G.A. Gravvanis and E.A. Lipitakis

Use of computer algebra systems to generate element stiffness matrices   373
D.V. Griffiths

Newton's method for obstacle problems   374
C. Grossmann

LAPACK: High performance software for linear algebra   375
S. Hammarling

Finite element methods for general second-order elliptic partial differential equations: Model problems   375
I. Harari and T.J.R. Hughes

The Fourier-finite element method for elliptic problems with axisymmetric edges   376
B. Heinrich

Nodal finite element methods for partial differential equations   376
J.P. Hennart

Finite elements for extended variational formulation in nonlinear problems   377
Z. Kestřánek

A parallel semi-discrete Galerkin splitting method for second-order hyperbolic equations   378
   A.Q.M. Khaliq

On the stability of layered circular rings   379
   B. Kovács and F.J. Szabö

Free vibrations of layered circular ring segments   379
   B. Kovács and F.J. Szabö

Finite element approximation of a nonlinear anisotropic heat conduction problem   380
   M. Křížek

A variational inequality formulation for flow theory plasticity   381
   M.S. Kuczma and J.R. Whiteman

A BE-formulation for finite strain elastoplasticity   381
   G. Kuhn

Adaptive mesh refinement in the finite element method   382
   G. Kunert

The factorization of unstructured sparse symmetric positive definite matrices on distributed memory MIMD parallel computers   382
   R. Levkovitz and G. Mitra

The application of the Finite Element Method to the solution of nonlinear partial differential equations and a comparison to the Finite Difference Method   383
   H. Maisch, H. Karl and G. Lehner

Balancing domain decomposition preconditioners and discontinuous coefficients   383
   J. Mandel

Fast iterative solvers for p-version solids, plates, and shells   384
   J. Mandel

Finite element analysis of turbulent flow through an orifice meter with a concentric obstruction of conical shape at its centre   384
   M.R. Mokhtarzadeh-Dehghan, D.J. Stephens and L. Yu

Remarks on a-posteriori error estimation for finite elasticity   385
   R. Mücke and J.R. Whiteman

Finite element analysis in nuclear safety   385
    J. Nedoma

FEM analysis of artificial substitutes of human joints and their optimal
design   386
    J. Nedoma

Numerical simulation of flow circulation in the golden horn   386
    H. Örs

Error estimates for discretized boundary element methods for three-
dimensional crack problems   387
    F. Penzel

A simple and efficient solution for the facet approximation of shells   388
    E.D. Providas

Anisotropic h-adaptive finite element method for hypersonic viscous flows
with shock-boundary layer interaction   388
    W. Rachowicz

On preconditioning for finite element equations on irregular grids   388
    A. Ramage and A.J. Wathen

A precis of developments in modelling embedded reinforcement for the finite
element analysis of reinforced concrete structures   389
    A. Ranjbaran

Improving the accuracy of finite-element solutions to two-dimensional
elliptic problems   390
    S. Rippa and B. Schiff

The numerical resolution of boundary layers and locking effects in the
Reissner-Mindlin plate model   391
    C. Schwab and M. Suri

Numerical experience with grid adjustment based on a posteriori error
estimators   391
    K. Segeth

Efficient preconditioners for incompressible flow   392
    D. Silvester and A.J. Wathen

Numerical solution of a parabolic equation with a weakly singular positive
type memory term   392
    M. Slodička

Design optimization of rubber elastic structures   393
     E. Stein and F-J. Barthold

On linear and bilinear elements for Reissner-Mindlin plates   393
     R. Stenberg

Finite element solution of unsteady-state mass transfer through a stationary
liquid   394
     J.L. Torres

A family of triangular plate bending elements, as a basis for geometrically
nonlinear shallow shell elements   395
     F. van Keulen

On an external finite element method for a second- order eigenvalue problem
on a concave 2D-domain with Dirichlet boundary conditions   395
     M. Vanmaele

The computational modelling of the thermoforming process for the creation
of axisymmetric container structures   395
     M.K. Warby and J.R. Whiteman

Element-by-element preconditioning   396
     A.J. Wathen

A posteriori error indicator and estimator for time-space FE discretization
of parabolic problems   397
     J. Weisz

Finite-element solutions for mechanical drilling methods   398
     H. Wern

Improved patch stress recovery procedure by equilibrium and boundary
conditions   398
     N-E. Wiberg and F. Abdulwahab

Matrix and image decomposition using wavelets   399
     J.R. Williams and K. Amaratunga

A general bond-slip formulation for embedded reinforcing bar elements   400
     Z.P. Wu and D. Phillips

INDEX   401

PREFACE

The increase in the range of applications of finite element methods continues unabated. Recent years have seen the expansion of the method into many areas of nonlinear fluid and solid mechanics, with an ever widening body of theory accompanying the applications. This book seeks to highlight certain aspects of the state-of-the-finite-element-art as at April, 1993, which are contributing to this expansion. All the papers result from presentations which were made at MAFELAP 1993, the Eighth Conference on the Mathematics of Finite Elements and Applications, which itself marked twenty-one years of triennial MAFELAP Conferences at Brunel University. MAFELAP 1993 took place during the period 27-30 April, 1993 and was once again organised by BICOM, the Brunel Institute of Computational Mathematics. As with all the previous conferences in the MAFELAP Series, the intention was again in 1993 to bring together at Brunel University mathematicians, engineers and other practitioners of finite elements, to discuss the theory and application of finite element techniques in as broad a context as possible.

It was very fitting that at MAFELAP 1993 the second Zienkiewicz Lecture should be presented by J Tinsley Oden. Other invited speakers were, I Babuška, M Crochet, C Johnson, V Maz'ya, K Morgan, K W Morton, E Oñate, W L Wendland, M F Wheeler, and J R Whiteman. With the exception of that of Professor Crochet, the papers resulting from the lectures by all these authors are included in this book, together with a number of additional papers which highlight other current aspects of finite element techniques. A total of over 120 presentations in lecture sessions, in mini-symposia, and in poster sessions were made at MAFELAP 1993. It has been possible to include the papers of only a small fraction of these in the current book. In order to compensate for this, and to give some details of the large range of material presented at the conference, abstracts of all other lectures and poster papers are included in this book.

At the Conference Dinner on Thursday April 29, 1993, Professor Ivo Babuška, the after-dinner speaker, emphasized that MAFELAP Conferences had been run at Brunel triennially for 21 years, with such success that they have now created for themselves an important place in the international finite element calendar. In predicting that MAFELAP Conferences would still be running in a further 21 years time, Professor Babuška pointed out that the areas of the subject which had dominated MAFELAP 1993 were error estimation, adaptivity, and reliability. He predicted that the tasks of mathematical modelling and subsequent finite element descretisation of the models would dominate MAFELAP 2014, because, by that time, the subject would have matured to the extent that robust, reliable, widely applicable adaptive finite element methods would be available and would have been incorporated into packages. The emphasis would thus have moved to the solving of more and more difficult "real world" problems.

As with all the MAFELAP Conferences, the success of MAFELAP 1993 depended on the contributions made by the conference committee, the persons who chaired the sessions, the speakers, the poster session authors and, for the first time, the organisers of the mini-symposia, as well as always on the extensive efforts of the

BICOM Research Fellows, Research Students and Visiting Research Fellows. Thanks are due to all these people, to Dr M B Reed and to the BICOM Secretary, Mrs Mary Storey, who undertook the administration of accommodation for the conference. I am pleased also to express my thanks to the United States Army Research Office, London, who contributed towards the cost of the conference. My thanks go especially to Dr Michael Warby for his wizardry with many different versions of TEX, which have allowed these proceedings to be produced in camera-ready form for the first time direct from electronic input. Finally my heartfelt thanks go to my wife, Caroline, who not only produced the index, but in reading through the entire manuscript acted as a *second editor* for the book. In addition, as always, she provided support and many ideas during the lead up period to the conference.

MAFELAP 1993 formed the focus for a nine-month period of finite element activity at BICOM. To all the colleagues who took part in this I also offer my thanks for their contributions and cheerful collaboration, and express the hope that they found their participation enjoyable, rewarding and worth while.

J R Whiteman
BICOM
November 1993

Chapter 1

Error Estimation and Control in Computational Fluid Dynamics

J. Tinsley Oden

Texas Institute for Computational Mechanics
The University of Texas at Austin
Austin, Texas 78712-1085, U.S.A.

The Second Zienkiewicz Lecture

Prologue

It is a great honour to deliver the Second Zienkiewicz Lecture at MAFELAP 1993. I am pleased to have this opportunity to pay tribute to Professor Zienkiewicz, a special friend whose work, ingenuity, integrity, and energy I have admired for three decades. His contributions to finite element theory and practice have formed an important part of the foundation of the subject as it exists today. Also, it is a pleasure to present this lecture honouring Professor Zienkiewicz at MAFELAP, which is certainly one of the most successful meetings of scholars and practitioners of finite element methods in the world, having been held regularly for nearly a quarter century. I salute Professor J. R. Whiteman for conceiving and so efficiently organizing these important gatherings, for allowing me to attend all of the MAFELAP conferences, and for giving me the honour of delivering this Zienkiewicz Lecture.

1.1 INTRODUCTION

The earliest mathematical literature on finite elements dealt with error analysis, which remains one of the main areas of focus of mathematical work on numerical methods. Does an algorithm converge? If so, at what rate? What conditions must prevail in order that it "works"? These are fundamental questions that must be asked and resolved in any careful study of permanent value in numerical mathematics. A critical component in such studies is the development of *a priori* error estimates, which give advanced notice on how a scheme may behave in the worst cases.

Nevertheless, the mathematical theory of finite element methods was slow to be embraced by practitioners. To determine if a method works, it is common to simply code it, try it on some benchmark cases, and determine its performance with heuristic

The Mathematics of Finite Elements and Applications
Edited by J. R. Whiteman © 1994 John Wiley & Sons Ltd

arguments based on the experience of the user. Thus, the utility of a mathematical theory for errors was not broadly appreciated by either the large community of scientists and engineers who use finite element methods in practice, or a large segment of the engineering research community who develop schemes for important applications.

Over the last decade, this situation has changed dramatically, primarily because of the great practicality of error estimation in significant applications of numerical methods. If the numerical error can indeed be estimated, then one can not only design schemes to control error but also schemes which generate a measure of the quality of solutions independent of the heuristic tests of yesteryear. Importantly, a reliable *a posteriori* error estimate can provide the basis for adaptive mesh strategies which can greatly improve the speed and performance of numerical schemes.

In the present study, we consider the development of effective adaptive schemes for hp finite element methods for the Navier-Stokes equations for incompressible viscous flows. The fundamental components of such schemes are, not surprisingly, theories for *a priori* error estimates, theories for *a posteriori* error estimates, and a data structure that permits h- (mesh size) and p- (spectral order) adaptation. The *a priori* estimates do not merely provide assurance that the scheme converges; they are used in a fundamental way in the design of an adaptive strategy. The *a posteriori* estimates, of course, drive the adaptive process and provide a basis for terminating a calculation when a target error level is attained. Finally, the availability of both *a priori* and *a posteriori* error measures opens new vistas for numerical mathematics, making the development of a variety of "smart" algorithms possible.

Following this introduction, a mathematical setting for the transient and steady state Navier-Stokes equations for incompressible flows is laid down together with a review of well known conditions for the existence and uniqueness of solutions to these equations. This is followed, in Section 1.3, by a review of data structures and interpolation estimates for a general class of hp finite element approximations. Here we review the data structure in use over the last several years which permits an automatic h-refinement of the mesh as well as an automatic enrichment of the spectral order p in each element. This is done in such a way that conforming finite element approximations can be obtained for second order problems, and that allows h and p to be treated as free parameters. These are, indeed, the control parameters which are manipulated to control the error. In Section 1.3, a brief description of a flow solver is described. The algorithm used in the present study is a two–step pressure based scheme in which a pressure correction is computed using the solution of a Poisson problem. The resulting algorithm is of relatively low order accuracy in the time approximation, but it is robust and it seems to work well for high Reynolds number flows. The search for a robust, efficient, high order scheme for Navier-Stokes equations remains open. In Section 1.4, new results of a rigorous *a posteriori* estimate for the steady state Navier-Stokes equations are summarized. The theory reviewed here provides a rigorous upper bound to the numerical error measured in an appropriate energy–like manner and is based on an error–residual method. In Section 1.5, a three–step hp-adaptive strategy is described. This scheme, termed here the Texas Three–Step, is based on the use of both *a priori* and *a posteriori* error estimates. The availability of an error estimate makes it possible to develop complementary parallel load balancing strategies. One such strategy is described here. In Section 1.6, the results of numerical experiments are presented. There calculations of solutions to Stokes and Navier-Stokes equations using the adaptive hp scheme are described. Also results on the performance of the *a posteriori* error estimate are presented. Finally, some major conclusions uncovered in this study and directions for future research are provided.

Acknowledgements. The work described in this paper was completed with considerable contributions and advice from a number of key collaborators. These include Dr. Mark Ainsworth from Leicester University who was a visitor at TICOM for a number of years, Mr. Weihan Wu, a Graduate Assistant at TICOM who is completing his dissertation on adaptive hp schemes for Navier-Stokes equations, and Mr. Abani Patra of TICOM who contributed to the work on parallel adaptive schemes. The work to date was completed under the support of ARPA, under Contract DABT63-92-002.

1.2 PRELIMINARIES

1.2.1 NOTATION; STATEMENT OF THE PROBLEM

We consider the flow of a viscous incompressible fluid through an open bounded region $\Omega \subset \mathbb{R}^n$, $n = 2$ or 3, with boundary $\partial\Omega$. The flow is fully characterized by a pair of fields, (u, P), $u = u(x, t)$ being the velocity at point $x \in \overline{\Omega}$ at time t and $P = P(x, t)$ being the pressure at (x, t), $t \in [0, T]$. The weak primitive–variable form of the initial-boundary-value problem for determining (u, p) is as follows:

$$
\boxed{
\begin{aligned}
&\textit{Given initial data } u_0 \in H, \textit{ body forces } f \in \\
&L^2(0, T; V^*), \textit{ surface tractions } g \in L^2\left(0, T, H_{00}^{\frac{1}{2}}(\Gamma_N)\right) \\
&\textit{for } t \in [0, T] \textit{ find } (u(t), p(t)) \textit{ such that} \\
&\quad (u_t, v) + c(u, u, v) + a(u, v) - b(P, v) \;=\; L(v) \\
&\qquad\qquad\qquad\qquad\qquad\qquad b(u, q) \;=\; 0 \\
&\qquad\qquad\qquad\qquad\qquad \forall\, v \in V, q \;\in\; Q \\
&\textit{and} \\
&\quad (u, (\cdot, 0), v) = (u_0, v) \qquad\qquad \forall\, v \in V
\end{aligned}
}
\tag{1.1}
$$

In (1.1), standard notations are used; e.g., $u_t = \partial u / \partial t$, $(u_t, v) = \int_\Omega u_t \cdot v\, dx$; $a(\cdot, \cdot)$, $b(\cdot, \cdot)$, $c(\cdot, \cdot, \cdot)$, and $L(\cdot)$ are continuous bilinear, trilinear, and linear forms on the spaces $V \times V$, $Q \times V$, $V \times V \times V$, and V respectively, given by

$$
a(u, v) = \int_\Omega \tau(u) : D(v)\, dx \;,\quad b(q, v) = \int_\Omega q \operatorname{div} v\, dx
$$

$$
c(u, v, w) = \int_\Omega u \cdot \nabla v \cdot w\, dx, \text{ and}
$$

$$
L(v) = \int_\Omega f \cdot v\, dx + \int_{\Gamma_N} g \cdot v\, ds
$$

for all $u, v, w \in V$ and $q \in Q$, where

$$V = \left\{ v \in \left(H^1(\Omega)\right)^n : \gamma v = 0 \text{ a.e. on } \Gamma_D \right\}$$

$$\tau(u) = 2\nu D(u), \quad D(u) = \left(\nabla u + \nabla u^T\right)/2$$

$$H = \{ v \in V : \operatorname{div} v = 0 \}$$

$$Q = \left\{ q \in L^2(\Omega) : \int_\Omega q \, dx = 0 \right\}$$

The forms also satisfy the conditions,

$$\left. \begin{array}{c} a(v, v) \geq \alpha |v|_1^2 \; ; \; \sup_V \dfrac{|b(q, v)|}{|v|_1} \geq \beta \|q\|_0^2 \\[1.5em] \sup_H \dfrac{|c(u, v, w)|}{|u|_1 |v|_1 |w|_1} = \gamma \end{array} \right\} \tag{1.2}$$

the first two holding for all $v \in V$ and $q \in Q$, where $\|q\|_m$, $m \geq 0$, denotes the usual Sobolev norms, and $|v|_1^2 = \int_\Omega \nabla v : \nabla v \, dx$. More generally, we demand that $u = w$ on Γ_D and that $\oint_{\partial\Omega} u_0 \cdot n \, ds = 0$ with $u_0 \cdot n = w(0) \cdot n$ on Γ_D.

We are also interested in the case of homogeneous boundary conditions. Then we seek u in $V_0 = \left(H_0^1(\Omega)\right)^n$ and $c(\cdot, \cdot, \cdot)$ has the property that

$$\left. \begin{array}{rcll} c(u, w, v) &=& -c(u, v; w) & \text{and} \\[1em] c(u, w, w) &=& 0 & \forall \, u \in H, \, v, w \in V_0 \end{array} \right\} \tag{1.3}$$

1.2.2   EXISTENCE AND UNIQUENESS

The following results are known concerning the existence and uniqueness of solutions to (1.1) for the case of homogeneous boundary conditions.

Theorem 1.1. I. The Steady State Navier-Stokes Equations. *Let (1.2) hold. Then there exists at least one solution* $(u, P) \in V_0 \cap H \times Q$ *to the steady state problem (2.1a) with homogeneous boundary conditions. Moreover, if*

$$\gamma \|f\|_{-1} < \nu^2 \tag{1.4}$$

the solution is unique and

$$|u|_1 \leq \frac{1}{\nu} \|f\|_{-1} < \frac{\nu}{\gamma} \tag{1.5}$$

II. The Transient Navier-Stokes Equations. *If (1.2) hold,* $f \in L^2\left(0, t; H^{-1}(\Omega)\right)$ *and* $u_0 \in H$*, then there exists a solution*

$$(u, P) \in L^2\left(0, T; V_0 \cap H\right) \times L^2(0, T; Q)$$

to the transient problem (2.1a). The solution is unique for $n = 2$ *and is also unique for* $n = 3$ *under additional conditions on* f*,* u_0*, and* ν*.*

A proof of these results, together with specific conditions of the regularity and magnitude of the data for uniqueness in the three–dimensional case, can be found in the book of Temam (1985).

1.3 $h\text{-}p$ -FINITE ELEMENT APPROXIMATIONS

Our general approach to hp-finite element data structures and to constructing hp-meshes in which the mesh size h and the local spectral order p of shape functions can be varied as independent parameters is well documented in earlier papers (see Oden, et al. (1989), Demkowicz, et al. (1989)), and only a brief review is needed here. The principal features are listed as follows:

- A family $\mathcal{P}$ of partitions of Ω is considered such that

$$\overline{\Omega} = \bigcup_K^{N(\mathcal{P})} \{\overline{\Omega}_K \, ; \, \Omega_K \in \mathcal{P}\} \, , \ \Omega_K \cap \Omega_L = \emptyset, \ K \neq L$$

each element Ω_K being the image of a master element $\overline{\widehat{\Omega}} = [-1,1]^n$ under an invertible map F_K.
- On $\widehat{\Omega}$, 3^n nodes are defined; these are classified as vertex nodes, edge nodes, and interior nodes if $n = 2$ and vertex, edge, face, and interior nodes if $n = 3$.
- Standard bilinear shape functions are assigned the vertex nodes (for $n = 2$), e.g.,

$$\widehat{\chi}^i(\xi, \eta) = \frac{1}{4}(1 \pm \xi)(1 \pm \eta) \qquad i = 1, 2, 3, 4$$

and analogous trilinear functions for $n = 3$. For $n = 2$, at the midpoints of each edge of $\widehat{\Omega}$, edge functions of the type

$$\widehat{\chi}_{\partial\widehat{\Omega}}^{\alpha,k}(\xi, \eta) = \begin{cases} \dfrac{1}{2}(1 \pm \eta)\varphi_k(\xi) \\[2ex] \dfrac{1}{2}(1 \pm \xi)\varphi_k(\eta) \end{cases}$$

are defined, with $\alpha = 1, 2, 3, 4$, $k = 2, 3, \cdots, p_\alpha$, and

$$\varphi_k(s) = \sqrt{\frac{2k-1}{2}} \int_{-1}^{s} P_{k-1}(t)dt$$

where P_{k-1} is the Legendre polynomial of degree $k - 1$. At the interior node $(0,0) \in \widehat{\Omega}$, we use bubble functions

$$\widehat{b}_{ij}(\xi, \eta) = \varphi_i(\xi)\varphi_j(\eta) \quad 2 \leq i \leq p_{01}, \ 2 \leq j \leq p_{02}$$

Analogous functions are used for faces and interior nodes for the three–dimensional case. The element shape functions define a space

$$\widehat{M}\left(\widehat{\Omega}; p_1, p_2, p_3, p_4; p_{01}, p_{02}\right) = \text{span}\Big\{ \widehat{\chi}^i, \widehat{\chi}_{\partial\Omega}^{\alpha,k}, \widehat{b}_{rs} : \begin{array}{l} 1 \leq i \leq 4, \\ 2 \leq k \leq \max_i p_i, \\ 2 \leq r, s \leq \max(p_{01}, p_{02}) \end{array} \Big\}$$

etc.

- An h-refinement, derefinement strategy is used which permits 1–irregular refinements of the mesh. In two–dimensional problems, elements can be bisected into four sons by the introduction of constrained or irregular nodes. Values of the hp functions at the constrained nodes take m interpolated values of the degrees of freedom at remaining active nodes (see Demkowicz, et al. (1989)). For higher degree shape functions, more elaborate constraint conditions are imposed to maintain continuity across interelement boundaries.
- For the p-adaptivity, continuity of the global basis functions is maintained by enriching the edge functions of Ω_K to match the highest–degree polynomial used on the common interelement boundary. When adjoining elements interface with polynomials of differing degree, the edge functions on the common boundary are enriched by the addition of edge functions of a degree equal to the largest polynomial degree of any element sharing that edge.

This data structure provides for the construction of general hp-spaces of the type

$$S^{hp}(\Omega) = \left\{ v^{hp} = v^{hp}(\boldsymbol{x}) \in C^0\left(\overline{\Omega}\right) : v_K^{hp} = v^{hp}\Big|_{\Omega_K} \in M_K \right\}$$

Locally, we have the interpolation property,

$$\left\| u - \widetilde{u}^{hp} \right\|_{s,\Omega_K} \leq C \, \frac{h_K^{\min(p_K+1,r)}}{p_K^{r-2}} \|u\|_{s,\Omega_K}$$

where $\widetilde{u}^{hp}$ is an appropriate hp-interpolant of u and $M_K = \{v = v(\boldsymbol{x}) = \hat{v} \circ F_K^{-1}, \hat{v} \in \widehat{M}(\widehat{\Omega}; p_1^K, p_2^K, p_2^K; p_{01}^K, p_{02}^K)\}$ and F_K is an invertible map from the master element $\widehat{\Omega}$ to element Ω_K.

Turning now to hp-finite element approximations of (1.1), we introduce the spaces

$$\begin{aligned}
V^{hp}(\Omega &= \left\{ v^{hp} \in \left(S^{hp}(\Omega)\right)^n \cap V \right\} \\
Q^{hp}(\Omega &= S^{hp}(\Omega) \cap Q
\end{aligned}$$

Alternatively, there is no need to demand that the approximate pressures q^{hp} be continuous across interelement boundaries and we could also use $\widetilde{Q}^{hp}(\Omega) = \{q^{hp} \in Q : q_K^{hp} \in M_K\}$.

A finite element approximation of (1.1) is of the form

$$
\boxed{
\begin{aligned}
&\text{Given initial data } \boldsymbol{u}_0 \in H^h, \text{ body forces } \boldsymbol{f} \in L^2(0,T;V^*) \\
&\text{and surface tractions } \boldsymbol{g} \in L^2(0,T; H_{00}^{\frac{1}{2}}(\Gamma_N)), \\
&\text{for } t \in [0,T] \text{ find} \\
&(\boldsymbol{u}^{hp}(t), P^{hp}(t)) \in V^{hp}(\Omega) \times Q^{hp}(\Omega) \\
&\text{such that} \\[4pt]
&\left(\boldsymbol{u}_t^{hp}, \boldsymbol{v}\right) + c\left(\boldsymbol{u}^{hp}, \boldsymbol{u}^{hp}, \boldsymbol{v}\right) + a\left(\boldsymbol{u}^{hp}, \boldsymbol{v}\right) - b\left(P^{hp}, \boldsymbol{v}\right) = L(\boldsymbol{v}) \\
&\hspace{6cm} b\left(q, \boldsymbol{u}^{hp}\right) = 0 \\
&\hspace{3cm} \forall\, \boldsymbol{v} \in V^{hp}(\Omega),\ q \in Q^{hp}(\Omega) \\
&\text{and } \left(\boldsymbol{u}^{hp}(0), \boldsymbol{v}\right) = (\boldsymbol{u}_0, \boldsymbol{v}) \quad \forall\, \boldsymbol{v} \in V^{hp}(\Omega)
\end{aligned}
}
\tag{1.6}
$$

Here

$$
H^h = \left\{ \boldsymbol{v} \in V^{hp}(\Omega) : b(q, \boldsymbol{v}) = 0 \quad \forall\, q \in Q^{hp}(\Omega) \right\}
$$

To compute $(\boldsymbol{u}^{hp}, P^{hp})$, we must also introduce an appropriate flow solver. Ideally, a scheme which is high order accurate in the time approximation is desirable so as to balance the high–order spatial approximations. At the same time, the LBB condition establishes compatibility between velocity and pressure approximations for numerical stability. Unfortunately, a robust and efficient high–order scheme for Navier-Stokes calculations does not seem to exist (despite claims in the literature to the contrary). Since we are also interested in steady state cases, we adopt as a compromise in the present study a pressure–based scheme which, while low–order in temporal accuracy, is robust and functions well on nonuniform hp meshes. A brief outline of the scheme follows.

0. For a partition, $0 = t_0 < t_1 < \cdots < t_N = T$, $t_{n+1}^n - t_n = \Delta t$, denote $\boldsymbol{u}^n(\boldsymbol{x}) = \boldsymbol{u}^{hp}(\boldsymbol{x}, n\Delta t)$ and $P^n(\boldsymbol{x}) = P^{hp}(\boldsymbol{x}, n\Delta t)$ and suppose that $b(q, \boldsymbol{u}^n) = 0$ $\forall\, q \in Q^{hp}(\Omega)$ and $\boldsymbol{u}^n = \boldsymbol{w}^n$ on Γ_D.
1. We advance the solution in time via a momentum step, formally given by

$$
\begin{aligned}
\frac{d\widetilde{\boldsymbol{u}}^{n+1}}{dt} &= \boldsymbol{F}\left(\widetilde{\boldsymbol{u}}^{n+1}, t\right) \\
&= \boldsymbol{f}(t) - \left(\widetilde{\boldsymbol{u}}^{n+1} \cdot \boldsymbol{\nabla}\widetilde{\boldsymbol{u}}^{n+1}\right) + \boldsymbol{\nabla} \cdot 2\nu \boldsymbol{D}\left(\widetilde{\boldsymbol{u}}^{n+1}\right)
\end{aligned}
$$

subject to boundary conditions, $\widetilde{\boldsymbol{u}}^{n+1} = \boldsymbol{w}^{n+1}$ on Γ_D, and, e.g.,

$$
\left.
\begin{aligned}
2\nu\left(\partial\widetilde{u}_n^{n+1}/\partial n\right) &= P^n + \tau_n^{n+1} \\
\nu\left(\partial\widetilde{u}_\tau^{n+1} + \partial\widetilde{u}^{n+1}/\partial n\right) &= \tau_\tau^{n+1}
\end{aligned}
\right\} \text{ on } \Gamma_N
$$

where τ_n, τ_τ denote normal and tangential components of the viscous stress tensors. In the present analysis, this step is accomplished using an implicit singly–diagonal

Runge-Kutta scheme in which the nonlinearities are resolved using the Newton-Raphson algorithm.

2. Next comes the pressure correction: we calculate a function φ^{n+1} from the Poisson problem

$$-\Delta\varphi^{n+1} \;=\; -\operatorname{div}\widetilde{\boldsymbol{u}}^{n+1}$$

$$\frac{\partial\varphi^{n+1}}{\partial n} \;=\; 0 \text{ on } \Gamma_D$$

$$\varphi^{n+1} \;=\; \Delta t\, P^n \text{ on } \Gamma_N$$

3. Velocity and pressure corrections are then

$$P^{n+1} \;=\; \Delta t^{-1}\varphi^{n+1} \text{ in } \overline{\Omega}$$

$$\boldsymbol{u}^{n+1} \;=\; \widetilde{\boldsymbol{u}}^{n+1} - \boldsymbol{\nabla}\varphi^{n+1} \text{ in } \overline{\Omega}$$

$$\boldsymbol{u}^{n+1} \;=\; \boldsymbol{w}^{n+1} \text{ on } \Gamma_D$$

4. Set $n := n+1$, $\widetilde{\boldsymbol{u}}^n = \boldsymbol{u}^n$, and go to Step 1.

Such algorithms have been discussed by many investigators and are variants of the classical Chorin scheme (1968); for a review, see Gresho (1990).

1.4 *A POSTERIORI* ERROR ESTIMATION

We are now ready to consider the issue of *a posteriori* error estimation for the Navier-Stokes equations. The aim is to produce rigorous, computable upper bounds to the approximation errors

$$e = \boldsymbol{u} - \boldsymbol{u}^{hp} \text{ and } E = P - P^{hp}$$

in appropriate norms. Complete details of this theory are given in a forthcoming paper (Oden, *et al.* (1993)) and are reviewed in the paper by Ainsworth and Wu in this volume.

We shall focus on the steady state case involving homogeneous boundary conditions, although our applications are considerably more general. The following problem class and assumptions are considered:

1. $(\boldsymbol{u}, P) \in V_0 \times Q$ are unique solutions of the problem

$$c(\boldsymbol{u},\boldsymbol{u},\boldsymbol{v}) + a(\boldsymbol{u},\boldsymbol{v}) - b(P,\boldsymbol{v}) + b(q,\boldsymbol{u})$$

$$= L(\boldsymbol{v}) \qquad \forall\ (\boldsymbol{v},q) \in V_0 \times Q \tag{1.7}$$

where now

$$L(\boldsymbol{v}) = \int_\Omega \boldsymbol{f} \cdot \boldsymbol{v}\, d\boldsymbol{x}, \qquad \boldsymbol{f} \in L^2(\Omega)$$

$\|\boldsymbol{f}\|_{H^{-1}}\gamma < \nu^2$, with Ω and open bounded Lipschitz domain in $\mathbb{R}^n$, $n \le 3$.

2. $(\boldsymbol{u}^{hp}, P^{hp}) \in V^{hp}(\Omega) \times Q^{hp}(\Omega)$ are hp finite element approximations of $(\boldsymbol{u}, P)$, uniquely determined for a fixed hp-mesh as solutions of

$$c\left(\boldsymbol{u}^{hp}, \boldsymbol{u}^{hp}, \boldsymbol{v}\right) + a\left(\boldsymbol{u}^{hp}, \boldsymbol{v}\right) - b\left(P^{hp}, \boldsymbol{v}\right) + b\left(q, \boldsymbol{u}^{hp}\right) = L(\boldsymbol{v})$$

$$\forall\, (\boldsymbol{v}, q) \in V^{hp}(\Omega) \cap V_0 \times Q^{hp}(\Omega) \qquad (1.8)$$

3. There exist h_0, p_0 such that for all $h < h_0$, $p > p_0$, $p = \max_K p_K$, $p_K = \max\{p_i^K, p_{01}^K, p_{02}^K, 1 \le i \le 4\}$ a family $\mathcal{P}^{hp}$ of hp meshes can be generated such that the scheme is convergent. A stronger sufficient condition is that an *a priori* estimate of the following type holds:

$$\|(e, E)\| \le C \sum_{K=1}^{N(\mathcal{P}^{hp})} \frac{h_K^{\mu_K}}{p_K^{\nu_K}}$$

where

$$\|(e, E)\|^2 = |e|_{1,\Omega}^2 + \|E\|_0^2$$

$$h_K = \mathrm{dia}\,(\Omega_K), \quad p_K = \text{maximum complete polynomial defined on } \Omega_K$$

$$\mu_K = \min(r, p_K), \quad \nu_K = r - 1$$

$$\text{for } \boldsymbol{u} \in V_0 \cap (H^r(\Omega))^n, P \in Q \cap \widehat{H}^{r-1}(\Omega)$$

These conditions merely assert that we can produce convergent hp-approximations of (1.7) and that the rate of convergence is affected by the regularity (r) of $\boldsymbol{u}$ and P.

Next we select an appropriate norm in which the error is to be estimated. Toward this end, we introduce the inner products,

$$\left.\begin{aligned}
A(\boldsymbol{v}, \boldsymbol{w}) &= a(\boldsymbol{v}, \boldsymbol{w}) \\[2mm]
&= \sum_{K=1}^{N} A_K(\boldsymbol{v}, \boldsymbol{w}) \quad \boldsymbol{v}, \boldsymbol{w} \in V_0 \\[4mm]
B(q, s) &= (q, s) \\[2mm]
&= \sum_{K=1}^{N} (q, s)_K \qquad q, s \in Q
\end{aligned}\right\} \qquad (1.9)$$

where $A_K(\boldsymbol{v}, \boldsymbol{w}) = \int_{\Omega_K} 2\nu D(\boldsymbol{v}) : D(\boldsymbol{w}) dx$ and $(q, s)_K = \int_{\Omega_K} qs\, dx$. It is worth noting that alternative inner products could be used here (e.g., $\widetilde{A}_K(\boldsymbol{v}, \boldsymbol{w}) = \int_{\Omega_K} \nabla \boldsymbol{v} : \nabla \boldsymbol{w}\, dx$). Corresponding to these choices, we have the energy–like norm,

$$|||(\boldsymbol{v}, q)|||^2 = A(\boldsymbol{v}, \boldsymbol{v}) + B(q, q) \qquad (1.10)$$

The importance of this choice is seen in the following result.

Theorem 1.2. *Let (e, E) be the approximation errors (1.7) and let conditions 1–3 above hold. Further, let L denote a constant such that $|u|_1 \leq L < \nu/\gamma$. Further, let $(\varphi, \psi) \in V_0 \times Q$ denote the unique solution of the problem*

$$A(\varphi, v) + B(\psi, q) = a(e, u) - b(E, v) - b(q, e)$$

$$+ c(u, u, v) - c\left(u^{hp}, u^{hp}, v\right)$$

$$\forall \, (v, q) \in V \times Q \tag{1.11}$$

Then constants k_1 and k_2 exists such that as, $\frac{h}{p} \to 0$,

$$k_1 |||(e, E)||| \leq |e|_1^2 + \|E\|_0^2 \leq k_2 |||(e, E)||| \tag{1.12}$$

where

$$|||(e, E)||| = A(\varphi, \varphi) + B(\psi, \psi) \tag{1.13}$$

To obtain an upper bound to the error in this norm, we return to the mesh $\mathcal{P}^{hp}$ and construct local elementwise boundary–value problems on each element "loaded" by the FEM residuals.

$$
\begin{aligned}
\text{Find } (\varphi_K, \psi_K) \quad &\in \quad V_K \times Q_K \text{ such that} \\[2mm]
A_K(\varphi_K, v_K) \quad &= \quad f_K(v_K) - a_K\left(u_K^{hp}, v_K\right) \\[2mm]
&+ \quad b_K\left(P_K^{hp}, v_K\right) - c_K\left(u_K^{hp}, u_K^{hp}, v_K\right) \\[2mm]
&+ \quad \oint_{\partial\Omega_K} \langle n_K \cdot \sigma\left(u^{hp}, P^{hp}\right)\rangle \cdot v_K \, ds \\[2mm]
B\left(\psi_K, q_K\right) \quad &= \quad b_K\left(q_K, u_K^{hp}\right) \\[2mm]
&\qquad \forall \, (v_K, q_K) \in V_K \times Q_K, \\[2mm]
&\qquad \forall \, \Omega_K,, \; 1 \leq K \leq N\left(\mathcal{P}^{hp}\right)
\end{aligned}
\tag{1.14}
$$

Here, subscripts K denote restrictions to element Ω_K (e.g., $V_K(Q_K)$ are spaces of restrictions of functions in $V_0(Q)$ to Ω_K, $v_K = v\big|_{\overline{\Omega}_K}$, etc.).

The boundary term in (1.14) is of critical importance and is computed using a *flux balancing* or *equilibration* technique (see Ainsworth and Oden (1992, 1993(i), 1993(ii), and Wu et al. (1993)). In particular, for each pair of neighbouring elements Ω_K and Ω_L, we introduce piecewise linear functions $\alpha_{KL}(s)$ defined on the interelement boundaries

$\Gamma_{KL} = \partial\Omega_K \cap \partial\Omega_L$ such that $\alpha_{KL}(s) + \alpha_{LK}(s) = 1$. The averaged approximate flux on Γ_{KL} is then given by,

$$\langle n_K \cdot \sigma \left(u^{hp}, P^{hp} \right) \rangle = n_K \left\{ \left(1 - \alpha_{KL}(s) \right) \sigma_K \left(u^{hp}, P^{hp} \right) + \alpha_{KL}(s) \sigma_L \left(u^{hp}, P^{hp} \right) \right\} \tag{1.15}$$

where $s \in \Gamma_{KL}$ and n_K is a unit vector exterior and normal to $\partial\Omega_K$. These functions α_{KL} are then selected so that the element residual and the boundary residuals (on the right hand side of (1.14)) are balanced, rendering the local problem solvable. Details on the equilibration process used to calculate the α_{KL}'s are given in Ainsworth and Oden (1992).

Finally, we have

Theorem 1.3. *Let the assumptions of Theorem 1.2 hold and let* (φ_K, ψ_K) *be the solutions of the element residual problems (1.14). Then the error* (e, E) *of the finite element approximation of the Navier-Stokes equations satisfies the following bound:*

$$|||(e, E)|||^2 \leq \sum_{K=1}^{N} |||(\varphi_K, \operatorname{div} u_K^{hp})|||_K^2 \tag{1.16}$$

where

$$|||(\varphi_K, \operatorname{div} u_K^{hp})|||_K^2 = A_K(\varphi_K, \varphi_K) + \int_{\Omega_K} \left(\operatorname{div} u_K^{hp} \right)^2 dx \tag{1.17}$$

A proof of this result is given in a forthcoming paper by Oden, Ainsworth, and Wu (1993).To implement this estimate in practical calculations, (1.14) is solved approximately over each element Ω_K using polynomial approximations of φ_K and v which are, in general, of higher degree than those used in obtaining u^{hp} and P^{hp}. The resulting indicator

$$\theta^2 = \sum_{K=1}^{N} \theta_K^2, \quad \theta_K = |||(\varphi_K, \operatorname{div} u_K^{hp})|||_K \tag{1.18}$$

provides a rigorous upper bound to the error if these local problems are resolved sufficiently accurately.

1.5 PARALLEL *HP*-ADAPTIVE STRATEGIES: THE TEXAS 3–STEP

1.5.1 A THREE–STEP *HP* ADAPTIVE STRATEGY

With an effective means for *a posteriori* error estimation in hand, various adaptive strategies can be developed to control the numerical error by appropriate refinements and enrichments of the *hp* mesh. Here we outline a three–step scheme that attempts to minimize the computational effort required to achieve a target error. To simplify the discussion, we temporarily confine our attentions to a model scalar elliptic problem on two–dimensional polygonal domains:

$$\left.\begin{array}{rcl} -\Delta u = f & \text{in} & \Omega \subset \mathbb{R}^2 \\[2mm] u = 0 & \text{on} & \partial\Omega \end{array}\right\} \tag{1.19}$$

An *a priori* error estimate for an hp-approximation of (1.19) is

$$\left\| u - u^{hp} \right\|_1 \leq \sum_{K=1}^{N} \frac{h_K^{2\mu_K}}{p_K^{2\nu_K}} \Lambda_K$$

where $\Lambda_K = C\|f\|_s$ if $f \in H^s(\Omega)$, in which case $u \in H^r(\Omega)$ and $r \leq \min(s+2, 1+\alpha-\varepsilon)$ where $\alpha = \pi/\max(\alpha_A)$, α_A = interior angle at vertex A on $\partial\Omega$. The exponents μ_K, ν_K are generally unknown and it is necessary to estimate them in some way. In this model problem, $\mu_K = \min(p_K, r-1)$, $\nu_K = r-1$.

The three–step scheme is based on two ideas: the asymptotic estimate (1.19) is treated as an equality and the actual error is available to sufficient accuracy through an *a posteriori* error estimate. Then we have

$$\sum_{K=1}^{N} \eta_K^2 \approx \|e\|_1^2 = \sum_{K=1}^{N} \Lambda_K^2 \frac{h_K^{2\mu_K}}{p_K^{2\nu_K}}$$

The basic steps of the scheme now follow:

1.5.2   NOTATION AND INITIALIZATION

We introduce some preliminary notation which is used in describing the algorithm

$$\begin{aligned} \eta \;\; &= \;\; \text{error index } \|e\|_1/\|u\|_1 \\[2mm] \theta_K \;\; &= \;\; \text{local error indicator provided by an appropriate} \\ & \quad\;\; \textit{a posteriori} \text{ error estimate of} \\ & \quad\;\; (\text{e.g., } \theta_K^2 = A_K(\varphi_K, \varphi_K) \text{ in the spirit of } (1.14)) \\[2mm] \theta \;\; &= \;\; \left(\sum_{K=1}^{N} \theta_K^2\right)^{\frac{1}{2}} = \text{global error estimate } (\approx \|e\|_{1,\Omega}) \\[2mm] \Lambda \;\; &= \;\; \left(\sum_{K=1}^{N} \Lambda_K^2\right)^{\frac{1}{2}} \end{aligned}$$

We define a target error index,

$$\eta_T \equiv \eta_{\text{TARGET}} \approx \frac{\|e\|_1}{\|u\|_1}$$

and an intermediate error level

$$\eta_I = \gamma \eta_T$$

γ is a parameter ranging from 5 to 10. For example let $\eta_T = .01$ (1 percent error) and $\eta_I = .07$.

Step 1. Introduce an initial mesh Ω_{h_0} of N_0 elements sufficiently fine to fall in the asymptotic part of the convergence curve for h-refinements. Solve the problem on the initial mesh and use an *a posteriori* error estimator to estimate the global and local errors θ_0 and θ_{0_K}.

$$\|e\|_1^2 \approx \theta_0^2 = \sum_{K=1}^{N_0} \theta_{0_K}^2$$

where N_0 is the number of elements in the initial mesh and θ_{0_K} is the element level error indicator. Now from the orthogonality of the error to $S_0^{hp}(\Omega) = \left(= S^{hp}(\Omega) \cap H_0^1(\Omega)\right)$

$$\|u\|_1^2 \approx \|U_0\|_1^2 = \left\|u^{h_0}\right\|_1^2 + \theta_0^2$$

$\|u_0\|_1$ can be used to compute an approximation of the initial error index, $\eta_0 \approx \frac{\theta_0}{\|U_0\|_1}$. Setting μ_K to μ_0, ν_K to ν_0 we compute

$$\theta_0^2 \approx \|e\|_1^2 = \frac{h^{2\mu_0}}{p_0^{2\nu_0}} \sum_{K=1}^{n_0} \Lambda_K^2 = \frac{h_0^{2\mu_0}}{p_0^{2\nu_0}} \Lambda^2$$

which yield Λ. For local values, take $\Lambda_K = \theta_{0K}\Lambda/\theta_0$.

Step 2. For an optimal h-adaptive mesh to achieve η_I we need to reduce the global error and equidistribute the element errors. This is achieved by computing for each element in the old mesh the number n_K of new elements required in the area of the old mesh covered by element Ω_K:

$$n_K = \left[\frac{\Lambda_K^2 N_I h_0^{2\mu}}{p_0^{2\nu} \theta_I^2}\right]^{\frac{1}{\beta\mu+1}}$$

where N_I is the number of elements in the intermediate mesh, β is a parameter depending on the dimension of the problem and $\theta_I = \eta_I\|U_0\|_1$ is the estimated global error for the intermediate mesh Ω_{h_I}. The parameter $\beta = 2$ for one dimension, one for two dimensions, and $2/3$ for three dimensions. Since $\sum_{K=1}^{N_0} n_K = N_I$, we may solve for n_K and N_I using a few Newton-Raphson iterations (great accuracy in this step is not important). Having obtained n_K, we redistribute inside the element using a weighting scheme based on errors in partitions of the initial element. If $n_K < 1$ we can consider unrefinements. Now solve the problem on the h-adapted mesh and compute *a posteriori* error estimate θ_I at the global level and θ_{if} at the element level.

Step 3. We now construct a third mesh by computing a distribution of polynomial degrees p_{T_f} to form a mesh Ω_{h_f} which will deliver the target error η_T. At this stage, an intermediate mesh Ω_{h_I} is available with nonuniform mesh sizes h_f and with N_I elements on which the intermediate error θ_I is achieved and equidistributed. The actual error distribution to compute Λ_f on the h-adapted mesh Ω_{h_I} is

$$\Lambda_f^2 = \frac{p_0^{2\nu}\theta_f^2\Lambda^2}{h_f^{2\mu}\theta_{If}^2}$$

The target error must satisfy

$$\theta_T^2 = \left(\eta_T\,\|U_0\|_1^2\right) = \sum_{f=1}^{N_I}\frac{h_{If}^{2\mu_0}\Lambda_f^2}{p_{T_f}^{2\nu_0}}$$

Now use p refinements to achieve the target error by computing p_{T_f} using Λ_f and setting

$$p_{T_f}^{2\nu_0} = \frac{h_f^{2\mu_0}\Lambda_f^2 N_I}{\theta_T^2}$$

We thus enrich p on each element of Ω_{h_I} to obtain Ω_{h_f}.

Finally, we solve the problem on Ω_{h_f} and compute an estimate of the final error index η_F. If $\eta_F \leq \eta_T$ the computation is terminated; otherwise steps 1–4 are repeated with higher γ.

1.5.3 DOMAIN DECOMPOSITION AND PARALLELIZATION

We next consider the issue of embedding in the 3–step scheme a strategy for domain decomposition and parallel computation. The scheme described here is based on an error–based decomposition strategy under development by Patra and Oden (1993). One simple version is embodied in the following algorithm:

0. Let M be the number of processors available, which is presumed here to be equal to the number of subdomains in a domain decomposition scheme. Let θ denote the estimated global error and D_I the set of elements in the Ith partition, with $H_K = \{= \Omega_J \in \mathcal{P} : \partial\Omega_J \cap \partial\Omega_K \neq \emptyset\}$. For $n = 2$, compute

1. $\varepsilon^2 = \theta^2/M$
2. $1 \to I$
3. For $K = 1$ to N_{elem} (number of elements) do

$$\theta_K = \min_{H_K}\theta_J$$

$$E_K^2 + \theta_K^2 \to E_K^2$$

4. If

$$E_K < \varepsilon,\ D_I = D_I \cup \{\Omega_J \in H_K\}$$

$$\text{find } H_{K+1},\ \text{set } K + 1 \to K$$

Else

$$I + 1 \to I,\ D_I = D_I \cup \{\Omega_K\}$$

$$K + 1 \to K$$

End For

This scheme may do a poor job in minimizing interface communication costs, but it does perform a reasonable strategy for load balancing. It is possible to also provide weights which bias the load when elements contain high p orders. Refinements of this strategy are under study.

Having partitioned the domain, an iterative solver is used to solve the algebraic problem. One scheme under study is the Layton-Rabier scheme (1993) based on Peaceman-Rachford iteration. For SPD systems, such as

$$\boldsymbol{Ax} = \boldsymbol{b}$$

we solve

$$
\begin{aligned}
(\mu\boldsymbol{I} + \boldsymbol{A}_1)\boldsymbol{x}^{n+\frac{1}{2}} &= (\mu\boldsymbol{I} - \boldsymbol{A}_2)\boldsymbol{x}^n + \boldsymbol{b} \\
(\mu\boldsymbol{I} + \boldsymbol{A}_2)\boldsymbol{x}^{n+1} &= (\mu\boldsymbol{I} - \boldsymbol{A}_1)\boldsymbol{x}^{n+\frac{1}{2}} + \boldsymbol{b}
\end{aligned}
\tag{1.20}
$$

for $\mu > 0$ and $\boldsymbol{A} = \boldsymbol{A}_1 + \boldsymbol{A}_2$. On a distributed system, the algorithm is implemented as follows:

1. For a domain with M partitions, define $\boldsymbol{A}_1 = \boldsymbol{A}_{\text{red}}$ and $\boldsymbol{A}_2 = \boldsymbol{A}_{\text{black}}$ in such a way that domains Ω_1 and Ω_2 have no interaction. Assemble (list) all element matrices $\boldsymbol{A}_K$ into $\boldsymbol{A}_2$, $\boldsymbol{A}_{\text{black}}$. matrices
2. Set $\boldsymbol{b} = \boldsymbol{b}_{\text{red}} + \boldsymbol{b}_{\text{black}}$
3. Pick $\mu > 0$ (which can be estimated *a priori* and choose initial iterates $\boldsymbol{x}_r^0$ and $\boldsymbol{x}_b^0$ ($r = $ red, $b = $ black).
4. Form $(\mu\boldsymbol{I} + \boldsymbol{A}_{\text{red}})$, $(\mu\boldsymbol{I} - \boldsymbol{A}_{\text{red}})$ $(\mu\boldsymbol{I} + \boldsymbol{A}\text{black}) = \boldsymbol{B}$, $(\mu\boldsymbol{I} - \boldsymbol{A}_{\text{black}})$.
5. Construct LU-factorization of $(\mu\boldsymbol{I} + \boldsymbol{A}_{\text{red}})$ and $(\mu\boldsymbol{I} + \boldsymbol{A}_{\text{black}})$.
6. Calculate in parallel

$$\boldsymbol{r}_B = \boldsymbol{b}_{\text{black}} - \boldsymbol{A}_{\text{black}}\boldsymbol{x}_b^k$$

7. Transfer data to connected components.
8. Solve in parallel

$$(\mu\boldsymbol{I} - \boldsymbol{A}_{\text{red}})\,\boldsymbol{x}_r^k = \boldsymbol{r}_b$$

9. Repeat with "black" for "red" and "red" for "black".
10. If $\|\boldsymbol{x}_b^{k+1} - \boldsymbol{x}_b^k\| + \|\boldsymbol{x}_r^{k+1} - \boldsymbol{x}_r^k\| \le TOL$, stop.

A study of this scheme is given in Patra and Oden (1993).

1.6   SAMPLE NUMERICAL RESULTS

In this final section, selected results of numerical experiments are given.

1.6.1   A MODEL NAVIER-STOKES PROBLEM

To test the error estimation theory, we consider a model problem of viscous flow in a square domain, $\Omega = (0, 1)^2$, with data

$$\boldsymbol{f} = (\boldsymbol{u} \cdot \boldsymbol{\nabla})\boldsymbol{u} - \boldsymbol{\nabla} \cdot 2\boldsymbol{D}(\boldsymbol{u}) + \boldsymbol{\nabla}P \quad \text{in} \quad \Omega$$

$$\boldsymbol{w} = \boldsymbol{u} \quad \text{on} \quad \partial\Omega$$

where

$$u_1 = (x_1 \sin(2x_1x_2) + x_2 \cos(2x_1x_2)) \, e^{x_1^2 - x_2^2}$$

$$u_2 = (x_1 \cos(2x_1x_2) - x_2 \sin(2x_1x_2)) \, e^{x_1^2 - -x_2^2}$$

$$P = \sin(x_1x_2) \, e^{x_1^2 - x_2^2}$$

The exact velocity and pressure fields are shown in Figs. 1.1a and b.

A sequence of hp-meshes were used in solving this problem, and global effectivity indices,

$$\xi = \frac{\theta}{\|(e, E)\|_*}$$

are tabulated in Table 1.1. Note that even for coarse meshes and high error, remarkable effectivity indices of 1.091 to 1.0064 are obtained.

Table 1.1: Effectivity Indices

Number of Elements	p	Global Effectivity Index
4	2	1.091
16	2	1.083
64	2	1.0237
4	3	1.0471
16	3	1.0064

Also noteworthy are the excellent local effectivity indices obtained through this error estimation technique. A typical result is shown in Fig. 1.2 which depicts a distribution of local effectivity indices for the case of 16 elements and uniform $p = 3$. The minimum value attained is 0.867 and the maximum is 1.021 with a standard deviation of 0.0458.

1.6.2 BACKSTEP CHANNEL FOR $RE = 300$

The classical backstep channel is solved with the Texas 3–Step scheme described in Section 1.5. Geometry and an initial mesh with uniform quadratic elements is indicated in Fig. 1.3 and various meshes and estimated errors for steps 1–3 in the three–step algorithm are illustrated in Figs. 1.4–1.8. Representative flow results are given in Fig. 1.9.

We observe that the scheme was able to detect and resolve the recirculation zones in the flow. The estimated global error of 45 percent in the original mesh was reduced to under 10 percent in the final mesh, this error being scaled as a percentage of the energy $\||(u_0, P_0)\||$ in the Stokes solution corresponding to the initial data. In each

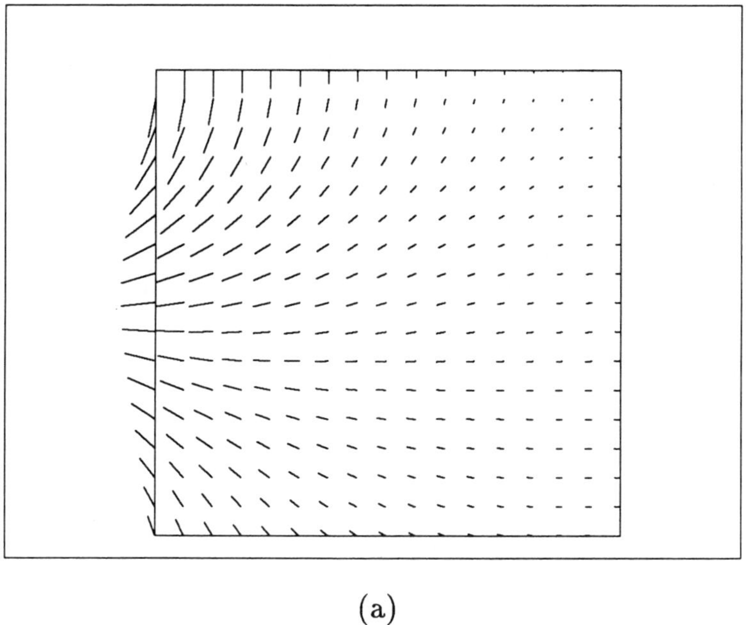

(a)

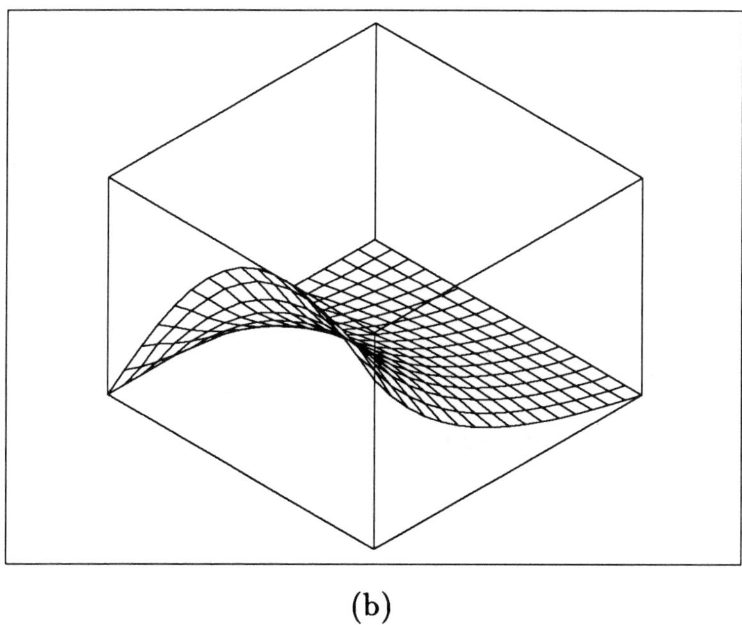

(b)

Figure 1.1   Velocity and pressure profiles for model problem.

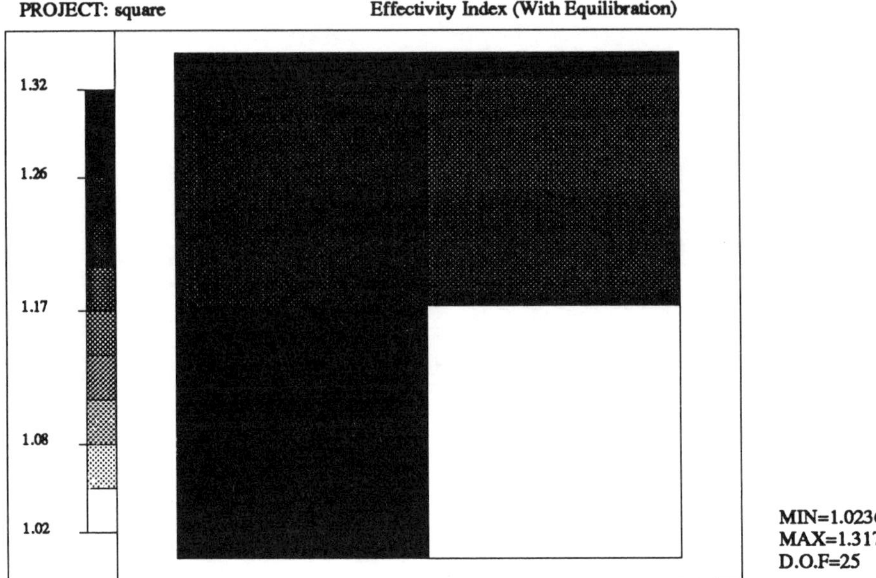

Figure 1.2   Distribution of effectivity indices for uniform $p = 3$ mesh.

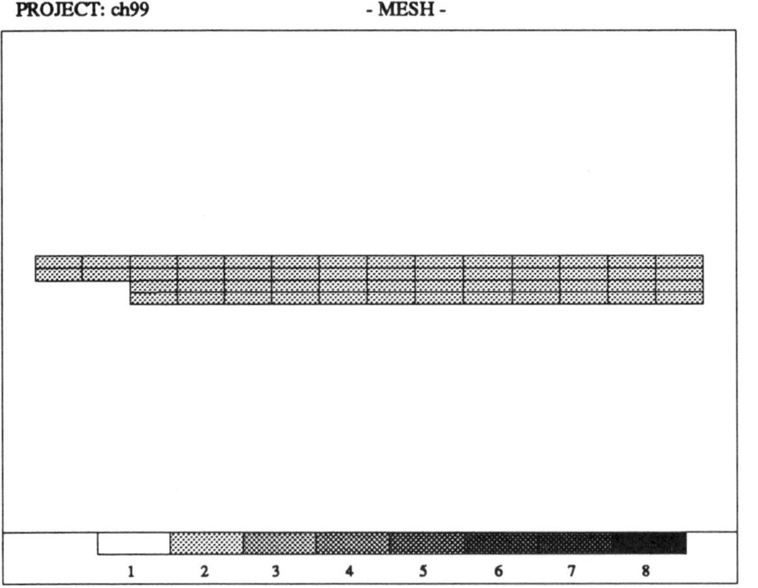

Figure 1.3   Geometry and initial mesh (uniform $p = 2$) for backstep channel problem. A parabolic inflow $(u, v) = (U_0 y(1 - y/b), 0))$ is prescribed at the left and balanced by a computed outflow at the right which is essentially parabolic. In this initial mesh, elements are rectangles $[a, b] = [1, 2.5]$, and U_0 and b are selected so that $Re = 300$. The shaded legend in this figure indicates the p-level (the order p_K) in each element and, in this case, all $p_K = 2$.

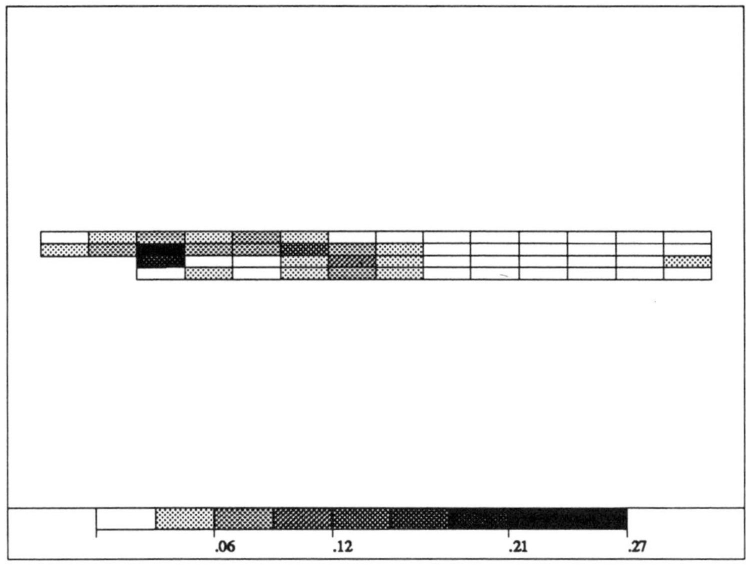

Figure 1.4   Estimated error distribution for initial mesh.

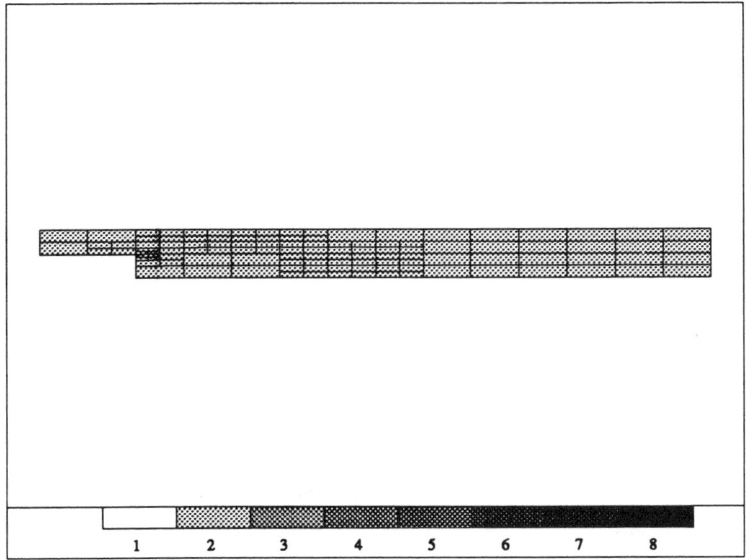

Figure 1.5   Intermediate mesh.

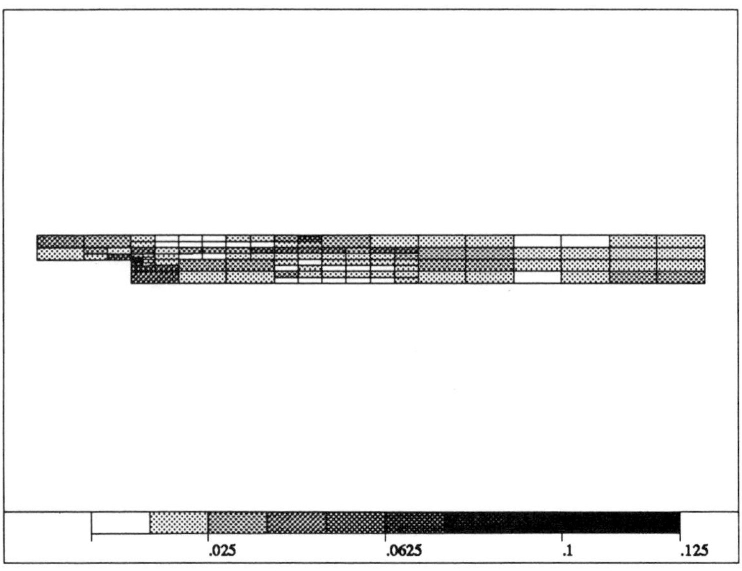

Figure 1.6   Estimated errors for intermediate mesh.

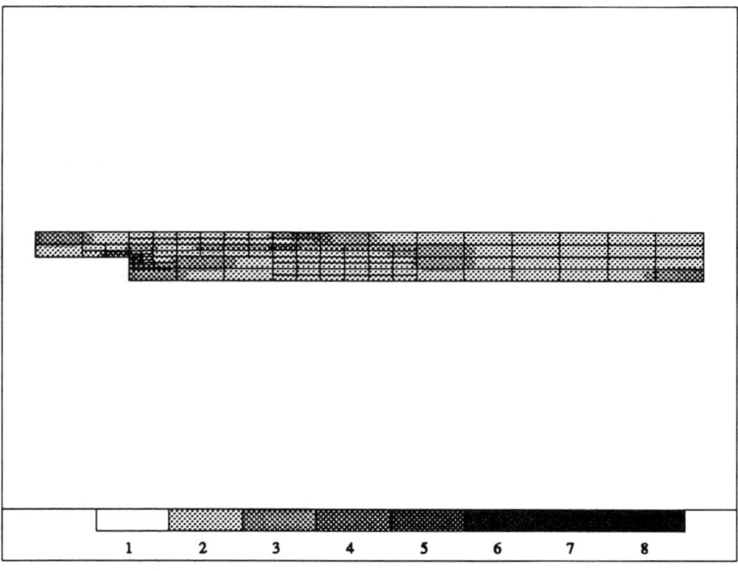

Figure 1.7   Final mesh in 3–step algorithm; note that the algorithm refines the mesh near the corner and at the primary and secondary recirculation zones, using elements with p up to $p_K = 4$.

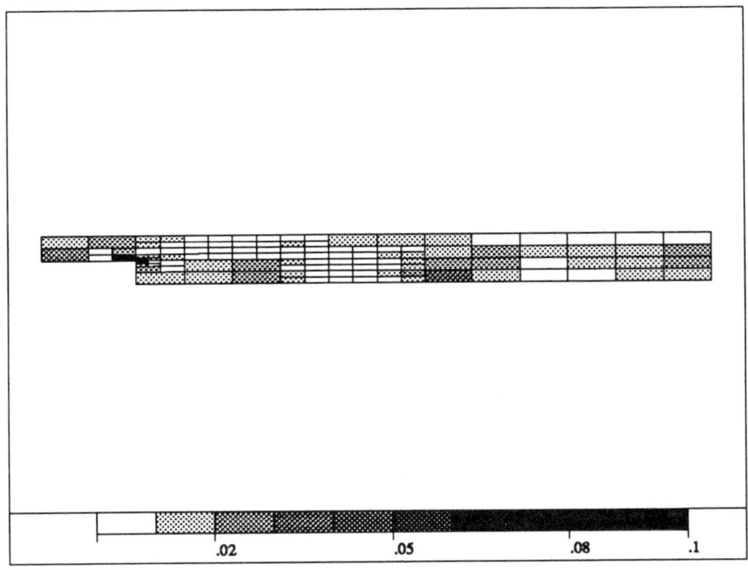

Figure 1.8 Final error distribution.

step, the solution was integrated using the pressure–based scheme of Section 1.4 to steady state. Error estimation required around 9 percent of the total computing time. Results are in excellent agreement with published data.

1.7 CONCLUDING COMMENTS

The ability to accurately and efficiently estimate the numerical error in computer simulations of viscous flows can represent a significant advance in computational fluid dynamics. With a good estimate of elementwise error, near–optimal meshes can be generated which can produce good solutions with few unknowns relative to standard methods. Moreover, the analyst can be provided with an independent measure of the quality of his solutions, or of the actual effects of mesh refinement.

In the present paper, a rigorous error estimate for a class of solutions of the Navier-Stokes equations is used as the basis for an *hp*-finite element scheme for solving those equations. The availability of both *a priori* and *a posteriori* estimates allows the development of an efficient *hp*-adaptive strategy which can be implemented in three steps. Also, error distribution can be used as a basis for load balancing for parallel computations, and one such parallel adaptive strategy is discussed in this paper. Preliminary numerical results show that significant speed-ups can be attained using these approaches.

Much additional work is needed to refine and generalize the ideas described here. Among issues deserving further attention are: extension of the error estimation technique to transient cases and to steady cases wherein multiple solutions exist;

                                                                J. T. Oden

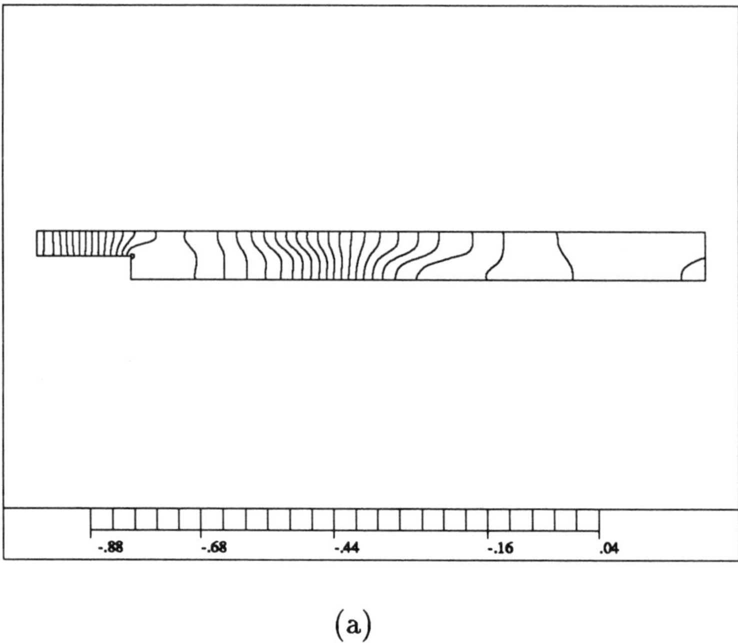

(a)

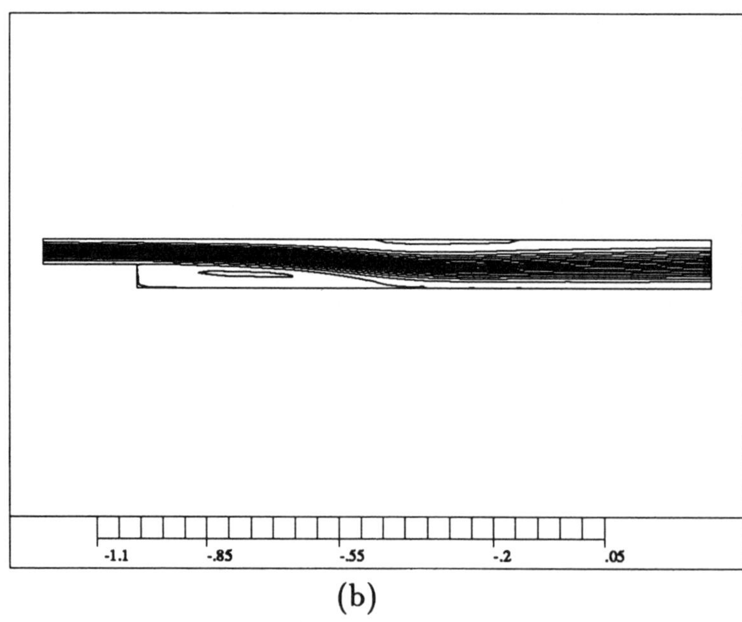

(b)

Figure 1.9   Final computed pressure (a) and stream function (b) for mesh 3.

development of "stopping" strategies which involve indexing the error in a way that allows the user to judge the actual error as a percentage of some current global parameter such as energy; additional improvements in the adaptive strategy and its parallelizations are needed and there is a need for measuring errors in other norms and in estimating errors in individual components in the solution. Reliable directionally-dependent error estimators are also needed to allow for mesh refinements that follow significant flow features.

In addition, much work remains to be done on the development of efficient flow solvers, particularly high–order solvers that function on hp meshes.

With some advances in each of these research areas, a breakthrough in computational fluid dynamics should be at hand, which could advance the field far beyond its present status and far faster than the rate the field has developed over the last two decades.

REFERENCES

Ainsworth, M., and Oden, J. T., (1992), A Procedure for *A Posteriori* Error Estimation for *h-p* Finite Element Methods, *Computer Methods in Applied Mechanics and Engineering*, **101**, pp. 73–96.

Ainsworth, M., and Oden, J. T., (1993(i)), A Unified Approach to *A Posteriori* Error Estimation Using Element Residual Methods, *Numerische Math.*, **54**.

Ainsworth, M., and Oden, J. T., (1993(ii)), *A Posteriori* Error Estimates for the Stokes Problems, **TICOM Report**, TR-93-1, Austin.

Ainsworth, M., and Oden, J. T., and Wu, W. (1993), *A Posteriori* Error Estimation for Applications in Elastostatics, in *Journal of Numerical Mathematics* (to appear).

Chorin, A., (1968), Numerical Solutions of the Navier-Stokes Equations, in *Mathematics of Computation*, **73**, pp. 341–353.

Demkowicz, L. F., Oden, J. T., Rachowicz, W., and Hardy, O., (1989), Toward a Unified Approach to *hp* Adaptive Finite Element Methods, Part 1: Constrained Approximation. In *Computer Methods in Applied Mechanics and Engineering*, **77**, No. 1, pp. 79–112.

Gresho, P. M., (1990), On the Theory of Semi–Implicit Projection Methods for Viscous Incompressible Flow and its Implementation via a Finite Element Methods that Also Introduces a Nearly Consistent Mars Matrix: Part 1. Theory. In *International Journal for Numerical Methods in Fluids*, **11**, pp. 587–620.

Layton, W., Rabier, P. J., Maubach, J., and Sunmonu, A., (1993), Parallel Finite Element Methods (preprint).

Oden, J. T., Demkowicz, L. F., Rachowicz, W., and Westermann, T., (1989), Toward a Unified Approach to *hp* Adaptive Finite Element Methods, Part 2. *A Posteriori* Error Estimates. In *Computer Methods in Applied Mechanics and Engineering*, **77**, No. 2, pp. 113–180.

Oden, J. T., Wu, W., and Ainsworth, M., *A Posterori* Error Estimates for the Navier-Stokes Equations. In TICOM Report 93-05, Austin and, to appear in *Computer Methods in Applied Mathematics and Engineering*.

Patra, A., and Oden, J. T., (1993), Toward Parallel *hp*-Adaptive Strategies, TICOM Report, TR 93-5, Austin (in press).

Temam, R., **Navier-Stokes Equations, Theory and Numerical Analysis**, (1985), second printing, 3rd rev. ed., North Holland, Amsterdam.

Wu, W., Oden, J. T., and Ainsworth, M., (1993), Error Estimation and Adaptive Methods for Incompressible Viscous Flow Simulations (in preparation).

Wu, W., (1993), HP-Adaptive Methods for Incompressible Viscous Flow Problems, Ph.D. Dissertation, The University of Texas, Austin.

Chapter 2

Reliability of Computational Mechanics

I. Babuška

Institute for Physical Science and Technology,
University of Maryland,
College Park, Maryland 20742, U.S.A.

2.1 INTRODUCTION

Computational analysis is directed toward the *quantitative* description of physical phenomena. It is, for example, aimed at avoiding failures, replacing costly physical experimentation and allowing optimal design to take place.

The computation solves numerically a *mathematically formulated problem* (model). The reliability of the computed data is dependent on both the reliability of the mathematical model and the accuracy of the numerical solution.

By *the reliability of numerical* treatment we mean that the computed data are close to the data of the *exact solution of the mathematical problem*. The control of this error involves *a priori* and especially *a posteriori* error estimation. This will be discussed in Section 2.4 of this paper.

The problem of *the reliability of the mathematical model* is much more complex and not easy to quantify in strict mathematical terms. Obviously this has to be intimately related to the aims of the analysis.

In practice the process of computational analysis has the following phases.

a) Identification of the available information and aims of the analysis.

b) General formulation of the problem and its heuristic analysis leading to various simplifications (e.g. by neglecting various factors).

c) Selection of a specific formulation or of a family of formulations which should be considered.

d) Quantitative specification of the data in the mathematical formulations used, for example, specification of various coefficients.

e) Further simplification leading to the specific problem to be solved numerically.

f) Numerical solution.

g) Interpretation of the results, very often in the frame of graphics.

In Section 2.2 we discuss a problem of plasticity addressing phases (a)-(d) mentioned above. In Section 2.3 we address the problem of dimensional reduction in conjunction with the types of problem arising in plate and shell theory which relates to phase (e). Finally in Section 2.4 we will discuss the problem of selecting a robust *a posteriori* estimator of the error of the finite element solution and this relates to the phase (f).

The Mathematics of Finite Elements and Applications
Edited by J. R. Whiteman © 1994 John Wiley & Sons Ltd

2.2 THE PROBLEM OF PLASTICITY {BABUŠKA *et al.*(1993(i)), LI (1992)}

In this section we shall address plasticity in its most simple setting - a one dimensional problem of plasticity. For more see Babuška *et al.* (1993(i)), Bathe (1982), Bonnetier (1988), Chaboche (1986), Drucker (1991), Johnson (1976), Li (1992), Mioshi (1985), Ono (1990), Wang *et al.* (1987), Zuckowski (1981), and references there.

Let $I = (0,1)$ and $W = (0,T)$. The one dimensional problem of (quasistatic) plasticity is to find $u(x,t), \sigma(x,t), \epsilon(x,t), (x,t) \in I \times W$ such that

$$\frac{\partial \sigma}{\partial x}(x,t) = f(x,t) \tag{2.1a}$$

$$\frac{\partial u}{\partial x}(x,t) = \epsilon(x,t) \tag{2.1b}$$

$$u(x,0) = \sigma(x,0) = 0 \tag{2.1c}$$

$$u(0,t) = 0 \tag{2.1d}$$

and

Problem A

$$u(1,t) = h(t) \tag{2.2a}$$

Problem B

$$\sigma(1,t) = g(t). \tag{2.2b}$$

In addition the constitutive law relating σ and ϵ is needed. We will consider the class of constitutive laws based on existence of an internal variable (vector $\alpha(x,t)$) which characterizes the state (memory) of the material. Denoting by $\dot{\sigma} = \frac{\partial \sigma}{\partial t}, \dot{\epsilon} = \frac{\partial \epsilon}{\partial t}$, the constitutive law is described in the form of ordinary differential equations

$$\dot{\sigma}(x,t) = F(\sigma, \dot{\sigma}, \epsilon, \alpha)\dot{\epsilon} \tag{2.3a}$$

$$\dot{\alpha}(x,t) = G(\sigma, \dot{\sigma}, \epsilon, \alpha)\dot{\epsilon} \tag{2.3b}$$

The analytic forms of F and G are derived by various principles, heuristics, physics, etc. Various parameters (parameter vector κ) are present in F and G.

Of the very many laws proposed in the literature, see e.g., Ono (1990), Zuckowski (1981) and references there, we shall consider the following four:

a) The bilinear kinematic hardening law

$$\begin{cases} \dot{\sigma} = E\dot{\epsilon} \\ \dot{\alpha} = 0 \end{cases} \quad \text{for } (\sigma, \alpha) \in \mathcal{E} \tag{2.4a}$$

$$\begin{cases} \dot{\sigma} = E_p\dot{\epsilon} \\ \dot{\alpha} = E_p\sqrt{\dfrac{E - E_p}{E E_p}}\,\dot{\epsilon} \end{cases} \quad \text{for } (\sigma, \alpha) \in \mathcal{P} \tag{2.4b}$$

where

$$\mathcal{E} = \left\{ (\sigma, \alpha) \,\middle|\, \begin{array}{l} \left|\sigma - \sqrt{\dfrac{E E_p}{E - E_p}}\,\alpha\right| < \sigma_y \\[1em] \text{or} \\[0.5em] \sigma - \sqrt{\dfrac{E E_p}{E - E_p}}\,\alpha = \sigma_y \quad \text{and } \dot{\sigma} \le 0 \\[1em] \text{or} \\[0.5em] -\sigma + \sqrt{\dfrac{E E_p}{E - E_p}}\,\alpha = \sigma_y \quad \text{and } \dot{\sigma} \ge 0 \end{array} \right\} \tag{2.5a}$$

$$\mathcal{P} = \left\{ (\sigma, \alpha) \,\middle|\, \begin{array}{l} \sigma - \sqrt{\dfrac{E E_p}{E - E_p}}\,\alpha = \sigma_y \quad \text{and } \dot{\sigma} > 0 \\[1em] -\sigma + \sqrt{\dfrac{E E_p}{E - E_p}}\,\alpha = \sigma_y \quad \text{and } \dot{\sigma} < 0 \end{array} \right\} \tag{2.5b}$$

This law has one internal variable and depends on a three dimensional parameter vector $\kappa = (E, E_p, \sigma_y)$. In addition there is the initial condition $\alpha(0) = \alpha_0$ of the internal variable. If the material is "virgin" then we can assume $\alpha_0 = 0$. In general the initial condition $\alpha(0) = \alpha_0$ depends on the (usually unknown) history.

The form (2.4),(2.5) is a typical form of (2.3) and distinguishes between the elastic region $\mathcal{E}$ and plastic region $\mathcal{P}$.

Many constitutive laws (forms) have been proposed. Nonetheless in order to provide the rigorous proof guaranteeing the well-posedness of the problem (2.1), (2.2) (and its analogue in 2 and 3 dimensions) based on these laws one needs to impose some conditions on the functions F, G. In the engineering literature many laws which do not satisfy such restrictions have been proposed. The kinematic law (2.4), (2.5) formulated above satisfies appropriate conditions. For more see Bonnetier (1988), Li (1992), Mioshi (1985). Further we will consider:

b) The isotropic hardening law

This law has again one internal variable α and a three dimensional parameter vector $\kappa = (E, E_p, \sigma_y)$. We remark that the values E, E_p, σ_y used in the kinematic and isotropic laws have physical meaning; namely E, E_p are respectively modulae of elasticity and plasticity and σ_y is the yield stress.

For the exact form of the isotropic law see e.g., Bathe (1982), Li (1992), Mioshi (1985). We will not present it here, but mention only that this law satisfies the conditions mentioned above which guarantee the well-posedness of the plasticity problem based on it.

c) The Chaboche law

This law has four internal variables and depends on a five dimensional parameter vector κ. For the virgin material the initial conditions for two internal variables are the same and are included in the parameter vector κ. The other two conditions are taken to be zero.

This law, which is one of the family proposed by Chaboche (1986), *does not* satisfy the mathematical conditions mentioned above.

d) The B-L law

This has two internal parameters and a six dimensional parameter vector κ. It satisfies the mathematical conditions mentioned above, see Li (1992).

The major problem for the use of these laws is the determination of the parameter vector κ and the initial condition α_0. In Babuška *et al.* (1993(i)) an extensive experimental study is made for the aluminium alloys 5454-H32 and 5086-H32 in sheet form ($0.2''$ thickness). Considering Problem A with $f(x,t) = 0$, we can experimentally determine $\sigma(t)$ for any $h(t)$. This was done using 84 samples with various $h(t)$ having 500 cycles (1000 reversals) and 40000 data reported for every sample. We considered both periodic and random functions $h(t)$. Given the parameter vector κ (and initial conditions for the internal variable) we define the error (depending on $h(t)$;

$$\theta(\kappa) = \frac{\max_t |\sigma_\kappa - \sigma_s)|}{\max_t \frac{1}{2}|\sigma_\kappa + \sigma_s)|}\%. \tag{2.6}$$

Here σ_κ is the computed stress utilizing the parameter vector κ and σ_s is the measured stress for the sample s. The maximum was taken over all data in the "window" of various cycles (0-10, 0-20, 0-50, 0-100, 0-200, 0-500, 10-20, 20-50, 50-100, 100-200, 200-500). The optimal κ_0^s for the given $h(t)$, the law and given sample was determined by the minimization of θ given in (2.6). Finally we computed κ_0 as the average of κ_0^s over various sets of samples.

Table 2.1 shows the accuracy θ for the optimal parameter κ_0 and for κ_H, where κ_H is the the parameter vector determined from data available in the material handbooks. We report here only the data for the window 0-500 and virgin material initial data.

Note that no handbook values are available for the B-L law. More on this is given in Babuška *et al.* (1993(i)) where we address also the reproducibility question analyzing the difference in the stresses $\sigma(t)$ induced by the same $h(t)$ on two samples of the same material. The mean was of order 5%.

So far we have considered Problem A when $h(t) = u(1,t)$ was given. We can also deal with Problem B when $\sigma(1,t)$ is taken from experiments and compare the experimental and predicted values of $u(1,t)$. In Table 2.2 we report the error θ for the window (0-10) when κ_0 and κ_H were used.

Table 2.1 Values of θ for κ_0 and κ_H for the window 0-500, virgin material based on Problem A.

Law	$\kappa = \kappa_0$			$\kappa = \kappa_H$		
	Mean	Min	Max	Mean	Min	Max
kinematic	24.28	17.92	29.47	33.24	29.43	39.91
isotropic	24.27	19.15	30.74	34.13	29.82	42.73
Chaboche	12.26	9.34	15.21	21.36	17.34	29.01
B-L	12.60	10.12	15.04			

Table 2.2 The error of the solution for Problem B.

Law	$\kappa = \kappa_0$			$\kappa = \kappa_H$		
	Mean	Min	Max	Mean	Min	Max
kinematic	84.13	55.9	101.8	89.9	51.13	112.3
isotropic	141.0	110.1	172.9	89.9	52.3	113.2
Chaboche	76.6	32.9	119.2	133.7	34.66	190.1

The tables (and more information in Babuška *et al.* (1993(i))) show that the uncertainty in the results due to the constitutive law is already very large in one dimension. The error can be expected to be much larger in higher dimensions, especially when nonproportional stresses occur, which of course in practice should be expected.

Let us mention that the material properties depend strongly on the material form (sheet, plate, bar, etc). For example the yield stress for aluminium 5454-H32 in sheet (0.2") and plate (0.4") form differs by 25% see e.g., Wang *et al.* (1987). We remark that no statistical data are reported and it is not clear on how many replicas the data in Wang *et al.* (1987) are based. This is also typical for all available data in the one dimensional setting. In fact we have not found any statistical data in the literature (see in this connection Aoki *et al.* (1990)).

From the computational point of view, three dimensional plasticity problems are numerically solvable on today's powerful computers. Nevertheless the reliability of these computations has to be questioned. In fact it seems that the usual deterministic computational analysis in many cases does not have much practical meaning because of above mentioned uncertainties. The only effective way is to determine brackets on the data of interest based on available information. This can be done using a Monte Carlo approach (if the variance matrix for the coefficients is available - which usually it is not) or computation of the worst brackets approach (see e.g., Bonnetier (1988)). In addition the initial values of the internal parameters which depend on the material's history have to be specified.

Let us illustrate the Monte Carlo approach for obtaining the brackets of interest. We compute from the data (see Babuška *et al.* (1993(i))) the variation matrix (based on a log Gauss assumption to ensure positivity of the data of interest) and simulate 10 random vectors κ. In the Figs. 2.1-2.4 we show the envelope of the peak tension and compression stress for a particular $h(t)$ in Problem A for the window 0-500 when the virgin assumption is used. In addition the Figs. show the error of a particular sample when second and fifth simulation was used (dotted line).

Analogously we computed the envelope for the peak of the displacement shown for

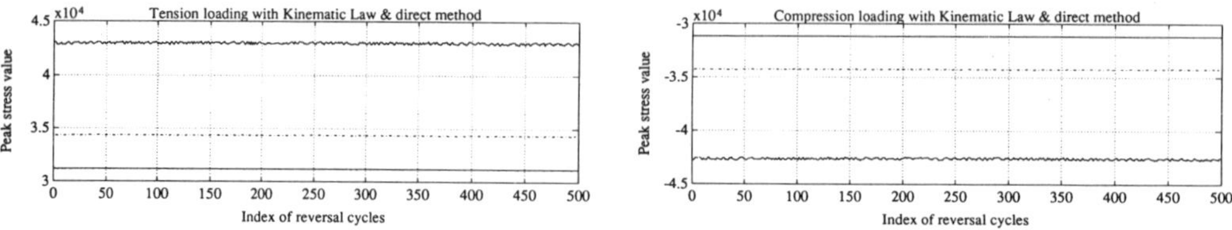

Fig. 2.1 The envelope of the peak stresses for the kinematic law.

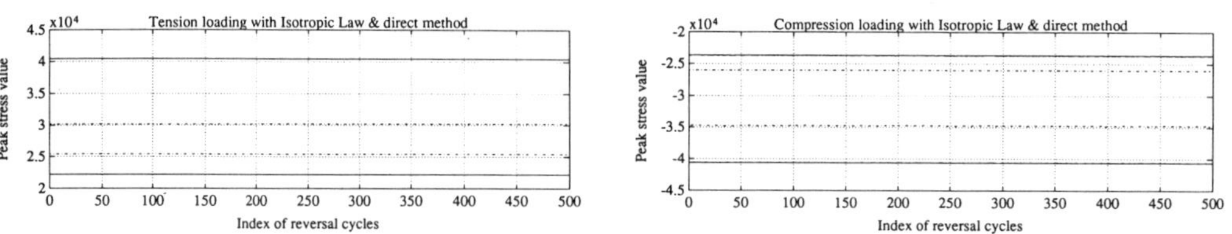

Fig. 2.2 The envelope of the peak stresses for the isotropic law.

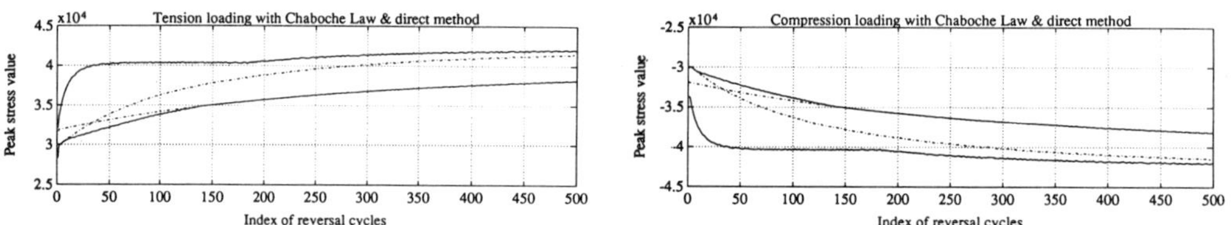

Fig. 2.3 The envelope of the peak stress for the Chaboche law.

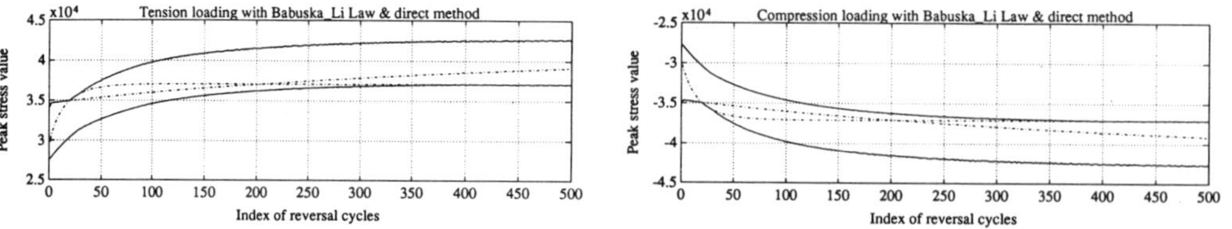

Fig. 2.4 The envelope of the peak stresses for the B-L law.

the window 0-10 (stream) when 10 simulated data were used for a particular $g(t)$ in Problem B.

From the Tables 2.1, 2.2 and Fig. 2.1–2.8 we see that in the most simple one dimensional case the uncertainty in the computed data is very high, as it is if many experimental data are available. Of course this uncertainty could effect different conclusions in different ways.

Our examples show that the quantitative reliability could also be very low if there is no numerical error present in the computations and many experimental data are available and that the usual standard deterministic approach could have very little practical value. (see also Babuška (1986)).

2.3 THE PROBLEM OF HIERARCHIC MODELLING {BABUŠKA *et al.*(1993(ii)), BABUŠKA AND SCHWAB (1993(iii))}

Typical problems of engineering analysis are those of plates and shells; in particular today problems of laminated plates and shells are of special interest. In the framework of Section 2.1, the general formulation under consideration is the 3 dimensional problem of linear elasticity. The simplified problem solved numerically is two dimensional and exploits the fact that the domain is thin.

Many two dimensional models have been proposed in the literature (see e.g., Ambartsumyan (1991), Babuška *et al.* (1991(i)), (1992(ii)), (1992(iii)) and (1993(ii)), Babuška and Schwab (1993(iii)) and (1993(iv)), Gilewski and Radwanska (1991), Naghdi (1972), Noor and Burton (1989), Panc (1975), Schwab (1989), Vekua (1982),Vogelius and Babuška (1981) and references there).

The computational results obtained from these can be both similar and different, depending on which data are of interest; see Babuška *et al.* (1992(ii)).

We will address here, for simplicity only, a model problem for obtaining the heat conduction in a plate.

Let $\omega \subset \mathbb{R}^2$ be a Lipschitz domain with piecewise smooth boundary γ. For $d > 0$ let $\Omega_d = \omega \times (-d/2, d/2), \Gamma = \gamma \times (-d/2, d/2)$ and $R_{\pm} = \{(x_1, x_2, x_3)|(x_1, x_2) \in \omega, x_3 = \pm d/2\}$. We will be interested in the problem

$$\begin{aligned} -\Delta u &= 0 & &\text{on } \Omega_d \\ u &= 0 & &\text{on } \Gamma \\ \frac{\partial u}{\partial n} &= f(x_1, x_2) & &\text{on } R_{\pm} \end{aligned}$$

where n is the outward unit normal. We will assume that $f \in L_2(\omega)$.

Let

$$H := \{u \in H^1(\Omega)| \ \text{trace} \ u = 0 \ on \ \Gamma\}$$

$$B(u, v) = \int_{\Omega} (\nabla u, \nabla v) dx_1 dx_2 dx_3$$

$$F(v) = \int_{\omega} f(x_1, x_2)(v(x_1, x_2, d/2) + v(x_1, x_2, -d/2)) dx_1 dx_2.$$

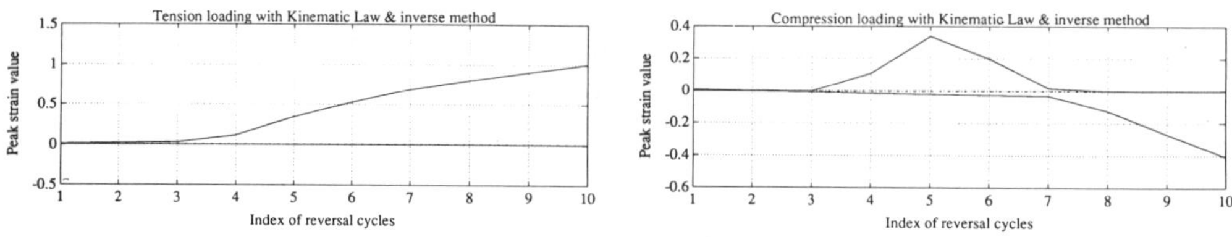

Fig. 2.5 The envelope of the peak stress for the kinematic law.

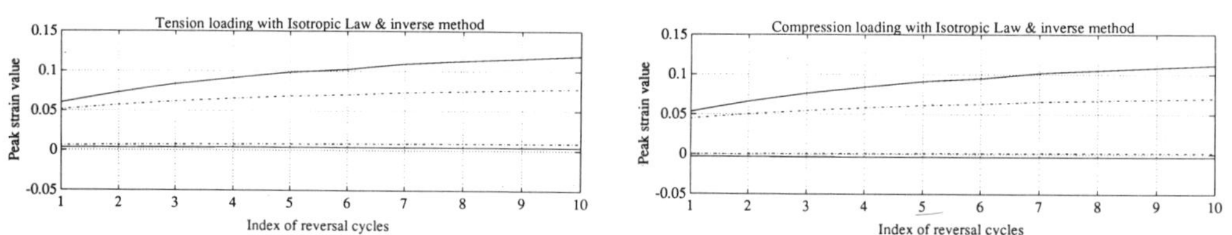

Fig. 2.6 The envelope of the peak stress for the isotropic law.

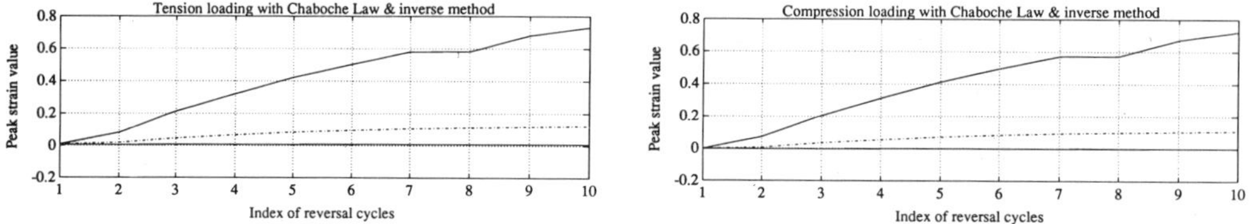

Fig. 2.7 The envelope of the peak strain for the Chaboche law.

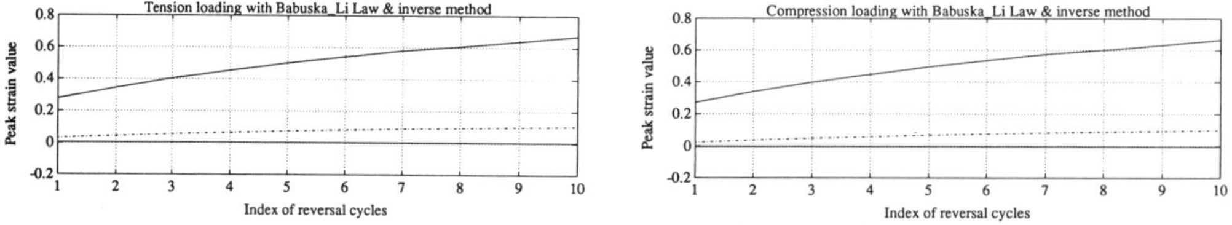

Fig. 2.8 The envelope of the peak strain for the B-L law.

Our problem is to find $u \in H$ such that

$$B(u, v) = F(v), \quad \forall v \in H. \tag{2.7}$$

We will now approximate the problem by a hierarchy of plate models. Denote by

$$P = \{\omega_i \subset \omega, 1 \le i \le n\}$$

with $\omega_i \cap \omega_j = \emptyset$ $\ for \ i \ne j$ and

$$\bar{\omega} = \bigcup_{i=1}^{n} \bar{\omega}_i$$

For a vector $q = (q_1, \cdots, q_n)$ of nonnegative integers and a dense sequence of symmetric linearly independent functions $\{\phi_j(z)\}_{j=0}^{\infty} \subset H^1(-1, 1)$ we define

$$S(P, q) := \left\{ u \in H \,|\, u|_{\omega_i} = \sum_{j=0}^{q_i} U_j^{(i)}(x_1, x_2)\phi_j\left(\frac{2x_3}{d}\right), \omega_i \in P \right\}.$$

Then $S(P, q) \subset H$ and the (P, q)-model is the following boundary value problem: Find $u(P, q) \in S(P, q)$ such that

$$B(u(P, q), v) = F(v), \quad \forall v \in S(P, q).$$

Asymptotic behavior (with respect to $d \to 0$) has been analyzed e.g., in Schwab (1989), Vogelius and Babuška (1981) where also the problem of optimal selection of Φ_j was discussed in a general setting.

In our particular case the optimal selection is $\Phi_j(z) = P_{2j}(z), j = 0, 1, \cdots$ when $P_j(z)$ is the Legendre polynomial. The optimal choice is also known for general laminated materials.

Our goal is now the following:

a) Find an *a posteriori* error indicator for the model solution, i.e. the norm of the difference of the model solution and the 3 dimensional solution.

b) The error indicator (estimator) should be a guaranteed one, and it should be asymptotically (with respect to $d \to 0$) and spectrally (with respect to $q \to \infty$) exact for a restricted set of input data.

c) It should allow adaptive selection of ω_i and q_i.

Remark 1

We address here only the error of the models, i.e., we assume that the solution of the model has been found exactly. In practice of course we would have to consider also the error of the approximate solution of the model problem.

Remark 2

We do not discuss here the task of the numerical solution of the system of differential equation stemming from the modelling. For the various interactive procedures we refer to Babuška and Schwab (1993(iv)).

Let $0 < \varphi(x_1, x_2) \in W^{1,\infty}(\omega)$ and let

$$Q = \max_{i=1,2} \left\| \frac{\partial \varphi^2}{\partial x_i} \Big/ \varphi^2 \right\|_{L^\infty(\omega)} < \infty \tag{2.8}$$

and

$$|||e|||_\varphi = \left(\int_\Omega |\nabla e|^2 \varphi^2 dx_1 dx_2 dx_3 \right)^{1/2}. \tag{2.9}$$

We will be interested in the estimation of the error $e_q = u(P, q) - u$ where u is the solution of the original three dimensional problem. We can use in (2.9) for example

$$\varphi(x_1, x_2) = exp\frac{\alpha}{d^\rho} \left(|x_1 - x_1^0| + |x_2 - x_2^0| \right), \tag{2.10}$$

$(x_1^0, x_2^0) \in \omega, 0 < \rho < 1$ and α properly chosen.

For simplicity of exposition let us assume that $P = \{\omega\}$ and instead of $u(P, q)$ we will write $u(q)$. The general case is discussed in Babuška and Schwab (1993(iii)).

Now we define the error estimator $\mathcal{E}_q(u(q))$

$$\mathcal{E}_q(u(q)) = \sqrt{\frac{d}{2q+3}} \left(\int_\omega r^2 \varphi^2 dx_1 dx_2 \right)^{1/2}$$

where

$$r(x_1, x_2) = f - \frac{\partial u(q)}{\partial n}(x_1, x_2, d/2)$$

and

$$\mathcal{E}_q(u(q)) \approx |||e_q|||_\varphi.$$

To assess the quality of this estimator we will define the effectivity index

$$\theta_q(u(q)) = \frac{\mathcal{E}_q(u(q))}{|||u - u(q)|||_\varphi}.$$

We will say that $\mathcal{E}_q$ is an upper (lower) estimator if $\theta > 1 (\theta < 1$, respectively). Further we will call the estimator (κ_1, κ_2) proper with respect to a class T of data f if

$$0 \leq \kappa_1 \leq \theta_q \leq \kappa_2 < \infty$$

holds for all data $f \in T$.

The estimator $\mathcal{E}_q$ is asymptotically spectrally exact on the set of data T_d with respect to T_q if $\theta_q \to 1$ as $d \to 0$ or equivalently if $\theta_q \to 1$ as $q \to \infty$.

Now we have the following theorem.

Theorem 2.1 {Babuška and Schwab (1993(iii))}
   a)*If $T = L_2(\omega)$ then*

$$\theta \geq \left(1 - dQ/\pi\sqrt{2}\right) \left(1 + \frac{\sqrt{2}}{\pi} dQ \left(1 + \frac{\sqrt{2}}{\pi} dQ\right)\right)^{1/2}$$

b) If $T_\beta := \{f|\; |||r|||_{1,\varphi} \,/\, |||r|||_{0,\varphi} \le \beta < \infty\}$ where $|||r|||_{k,\varphi}^2 = \int_\omega |\nabla_{x_1,x_2}^k r|^2 \varphi^2 dx_1 dx_2$ then

$$\theta \le \left(1 + \frac{3}{2}\frac{d^2}{(2q+e)^2 - 4}(\beta^2 + Q^2)\right)^{1/2}.$$

If moreover $\varphi \equiv 1$ then $Q = 0$ and the factor $\frac{3}{2}$ can be replaced by $\frac{1}{2}$.

Theorem 2.1 shows that the estimator is (κ_1, κ_2) proper with known κ_1 and κ_2. If we select the exponential weight φ given in (2.10) then the estimator is asymptotically exact on T_β. We see from the theorem that we get an upper estimator for a large class of data; namely we have only to assume that $f \in L_2(\omega)$. On the other hand to get $\theta \sim 1$ we need more smoothness of the solution. Selection of φ as in (2.10) leads essentially to an error of a local character on the area of diameter of order d.

So far we have addressed asymptotic exactness with respect to $d \to 0$. Let us formulate now the theorem when d is fixed and $q \to \infty$. We have

Theorem 2.2 {Babuška and Schwab (1993(iii))}

a) If $f \in T_\beta, \beta = \bar\beta(q/d)^{1-\epsilon}, \epsilon > 0, \bar\beta > 0$ independent of d, q then

$$\theta(q) \le \left(1 + C_1 \left(\frac{d}{q}\right)^{2\epsilon}\right)^{1/2}$$

where C_1 is independent of q and d.

b) Let $\lambda_k, \varphi_k(x_1, x_2)$ denote the k-th eigenpair of $-\Delta$ in ω with $\varphi_k|_\gamma = 0$. Let further $r(x_1, x_2) = \sum_k \rho_k \varphi_k(x_1, x_2)$ and assume f is such that $\rho_1 \ne 0$ and $\sum_{k\ge 2}\left(1 + \frac{d^2\lambda_k}{\pi^2}\right)^{-1}\phi_k\frac{|\rho_k|}{|\rho_1|} \le (\sqrt{2} - 1)(1 - q^{-1+\epsilon})$ for some $\epsilon > 0$. Then

$$\theta(q) \ge (1 + C_2 Q)(1 + 2C_2 Q(1 + 2C_2 Q))^{1/2}$$

where Q is as in (2.8) and

$$C_2 = \frac{d}{\sqrt{2}\{(2q+3)^2 - 4\}^\epsilon}$$

and Φ_k is the best constant in

$$\|\varphi_k/\varphi_1\|_{L^\infty(\omega)} \le \Phi_k$$

Theorems 2.1 and 2.2 are related to the energy norm. In Babuška and Schwab (1993(iii)) analogous theorems were proven for the $L_2(\Omega)$ norm and the general case of laminated plates.

Let us now show some examples for the case $\omega = (-1, 1)$ and $\Omega = \omega \times (-d/2, d/2)$ (see Babuška *et al.* (1993(ii))).

Example 2.1

Let $f = cos\frac{\pi}{2}x$. Then we can compute the effectivity index explicitly and get

$$\theta_q^2 = 1 + \frac{d^2\pi}{m_q} + 0(d^4) \tag{2.11}$$

Table 2.3 The coefficients m_q

q	0	1	2	3	4	5	6
m_q	240	360	936	1768	2856	4200	5800

Table 2.4 The effectivity index and the relative error
for the solution with a boundary layer

d	d = 0.2		d = 1.0		d = 2.0	
q	θ	$\|e\|\%$	θ	$\|e\|\%$	θ	$\|e\|\%$
0	1.011	9.88	1.063	42.53	1.140	87.65
1	1.46	0.505	1.463	5.14	1.454	12.291
2	1.778	0.197	1.778	2.01	1.778	4.736
3	1.933	0.112	1.933	1.14	1.933	2.681
4	1.911	0.079	1.908	0.803	1.912	1.883
5	1.760	0.064	1.743	0.655	1.760	1.529
6	1.554	0.056	1.511	0.587	1.554	1.358
7	1.348	0.053	1.292	0.557	1.348	1.269

where the coefficients m_q are listed in the Table 2.3

We clearly see the asymptotic exactness of the estimator.

Example 2.2

Let $f = 1$. Then the solution has a boundary layer and f belongs to the class T_β with large β. In the Table 2.4 we show the effectivity index for various d.

We see that the effectivity index grows for increasing q and achieves the maximum at $q \sim 3$ and then decreases. The behavior can be explained as the combination of the influence of large β and the spectral accuracy.

Let us now divide ω into subdomains. In our case when $\omega = (-1, 1)$ we will divide ω into subdomains $I_i = (x_{i-1}, x_i)$ and will assume $\varphi = \varphi_i, \varphi_i = 1$ on I_i and $\varphi_i = 0$ on $I_j, i \neq j$. This of course is not exactly in our theoretical framework. Nevertheless we use this selection of φ as an approximation. By this we understand the estimator on I_i as an indicator and use it for the adaptive selection of q_i. Theorems 2.1 and 2.2 essentially claim that the error (in our case indicators) does not essentially depend on the solution in the distance of larger than few thicknesses from the domain I_i.

The indicators are then used for the construction of the adaptive model Babuška *et al.* (1993(ii)).

2.4 THE PROBLEM OF *A POSTERIORI* ERROR ESTIMATION IN THE FINITE ELEMENT METHOD {BABUŠKA *et al.*(1993(v))}

Since the first papers Babuška *et al.* (1981), Babuška and Rheinboldt (1978(i)), and (1978(ii)) in *a posteriori* error estimation in finite element method many estimators, especially for the energy norm error, have been proposed.

The main part of the estimator is the (elemental) error indicator η_τ associated with the element τ of the mesh T_h. The indicator depends only on the computed finite element solution, in the particular element τ and its neighbours, and on the input data. The error estimator $\mathcal{E}(Z_0)$ in a subdomain $Z_0 \subset \Omega$ is

$$\mathcal{E}(Z_0) = \left(\sum_{\substack{\tau \in T_h \\ \tau \subset Z_0}} \eta_\tau^2 \right)^{1/2}, \qquad (2.12)$$

which is typical for the error measured in the energy norm and is directed to achieve

$$\||e_h\||_{Z_0} \approx \mathcal{E}(Z_0)$$

where $e_h = u_{Ex} - u_{FE}$ and $\|| \cdot \||_{Z_0}$ is the energy norm on Z_0.

As in Section 2.1 we introduce the effectivity index

$$\theta(Z_0) = \frac{\mathcal{E}(Z_0)}{\||e_h\||_{Z_0}} \qquad (2.13)$$

and call $\mathcal{E}(Z_0)C_L$ (with respect to C_U) proper if

$$C_L\mathcal{E}(Z_0) \leq \||e_h\||_{Z_0} \leq C_U\mathcal{E}(z_0) \qquad (2.14)$$

which obviously leads to

$$\frac{1}{C_U} \leq \theta \leq \frac{1}{C_L}.$$

In addition we can say that the estimator is asymptotically exact if $C_L, C_U \to 1$ as $\||e_h\||| \to 0$ for a particular set of input data. For more detail on numerical examples we refer to Babuška *et al.* (1993(v)).

The majority of the existing error estimators can be classified into the following categories:

a) *Residual estimators.* In this class are error estimators which are computed in terms of the solution of local problems which are local analogues of a higher order finite element approximation of the original problem. The data for the local problem are supplied by the local residuals. Residual types have been studied by many investigators, see e.g., Ainsworth and Craig (1992), Ainsworth and Oden (1993), Babuška *et al.* (1992(i)), Babuška *et al.* (1981), Babuška *et al.* (1991(ii)), Babuška and Rheinboldt (1978(i)), Babuška and Rheinboldt (1978(ii)), Babuška and Yu (1987), Bank and Weiser (1985), Duran and Rodriguez (1992), Ladeveze and Leguillon (1983), Oden *et al.* (1989), Shephard *et al.* (1989), Stroboulis and Hague (1992), Verfurth (1989).

Estimators which require the solution of a local problem are called implicit. Those which are constructed by estimating the magnitude directly from the residue are called explicit .

b) *Flux projections estimators.* Estimators of this type employ some local averaging technique to extract a post-processed flux from the finite element solution. The post-processed flux is then compared with the flux computed directly from the finite element solution in order to obtain an estimate of the error. Estimators based on this idea are discussed in Zienkiewicz and Zhu (1992), Zienkiewicz and Zhu (1987).

c) The estimators based on the applications of various *extrapolation approaches.*

The asymptotic exactness is in general almost impossible to obtain in practice because it needs smooth solutions and essentially uniform meshes of special type. In practice the robustness of the estimator is needed. This means that the effectivity index should be more or less independent of the mesh and solution and not be either too large or too small.

Although mathematical analysis of the asymptotic exactness and the properness is possible (in fact many papers address these questions), the robustness is not easy to address quantitatively. This is the reason that in the literature many ad-hoc numerical benchmarks are presented.

The performance of the error estimators depends on various factors.

a) Whether Z_0 is inside Ω i.e. it is not close to the boundary.

b) If $\bar{Z}_0 \cap \partial\Omega \neq 0$ i.e., Z_0 is in the boundary region.

c) Whether the solution is smooth in the neighbourhood of Z_0.

d) What is the mesh like. Whether or not it is more or less locally uniform and of what topological type.

In this section we will address the problem of comparing the robustness of the estimators for the case when $\bar{Z}_0 \subset \Omega$, the mesh is cell-piecewise invariant in $\bar{Z}_0$ and the solution is smooth. We will see that the results are not too sensitive as to whether or not the above assumptions are exactly satisfied in practical cases.

Let us consider the heat conduction problem

$$-\sum_{ij=1}^{2} \frac{\partial}{\partial x_i} k_{ij} \frac{\partial u}{\partial x_j} = f \qquad (2.15)$$

(in Babuška *et al.* (1993(v)) we analyzed also the analogous elasticity problem). In (2.15) $K = \{k_{ij}\}$ is the conductivity matrix. For simplicity we assume $f = 0$ and deal with linear elements. For the general case see Babuška *et al.* (1993(v)).

We will consider the following types of mesh.

Let $0 < H < H^0, x^0 = (x_1^0, x_2^0) \in \Omega$ and

$$S(x^0, H) = \{x = (x_1, x_2) | |x_i - x_i^0| < H, i = 1, 2\}$$

with $\bar{S}(x^0, H^0) \subset \Omega$.

Let now $S(x^0, H^0)$ be covered by the square cells $C(x^{i,j}, h) = S(x^{i,j}, h)$ of size h, and further on $S(x^0, H^0)$ let the mesh T_h be identical in every cell - i.e., it is cell - translation invariant. Let us also assume that the solution is smooth on $\bar{S}(x^0, H^0)$.

Then it can be shown that the performance of a particular error estimator is (asymptotically for $h \to 0$) related to is performance *on a single cell when the exact solution is a polynomial of one degree higher than the degree of the elements.* More precisely (see Babuška *et al.* (1993(v))) the following theorem holds. We formulate the assertion for linear elements.

Theorem 2.3.*Let $H_2 < H_1 < H < H^0$ and let H_1 be sufficiently small. Let further the following assumptions hold*

   a) $|D^\alpha u| \leq K < \infty, 0 \leq |\alpha| \leq 3, \alpha = (\alpha_1, \alpha_2)$.

b)   *If $a_\alpha := (D^\alpha u))(x^0), \alpha = (\alpha_1, \alpha_2), |\alpha| \le 2$*
then

$$\sum_{|\alpha|=2} a_\alpha^2 > 0.$$

c)   *On $S(x^0, H_2) = Z_0$*

$$|||e_h|||_{S(x^0,H_2)} \le C \ h^\beta H_2, \quad \beta = \frac{16}{9}$$

d) $CH_2^\alpha \le h \le C_1 H_2^\alpha, \alpha = \frac{9}{5}$
e) $Q = \sum_{|\alpha|=2} a_\alpha x_1^{\alpha_2} x_2^{\alpha_2}.$
Then for the effectivity index defined by (2.13) we get with $Z_0 = S(x^0, H_2)$

$$\theta(Z_0, u) = \frac{\mathcal{E}(z_0, u_h)}{|||e_h|||_{z_0}} =$$
$$= \frac{\mathcal{E}(C(x^0, h)\zeta_h)}{|||\Phi|||_{C(x^0,h)}}(1 + 0(h^{1/9})$$
$$= \theta(C(x^0, 1), \zeta_h)(1 + 0(h^{1/9})).$$

Here

$$\zeta_h = Q^{INT} + z$$

where Q^{INT} is the interpolant of Q on $C(x^0, 1), z \in \mathcal{S}^1_{PER}(C(x^0, 1))$, and $\mathcal{S}^1_{PER}(C(x^0, 1))$ is the space of all periodic piecewise linear functions on the unit cell $C(x^0, 1)$ and the master mesh such that

$$B_{C(x^0,1)}(z, v) = B_{C(x^0,1)}(Q - Q^{INT}, v) \ \ \forall v \in \mathcal{S}^1_{PER}(C(x^0, 1))$$

and

$$|||\Phi|||_{C(x^0,1)} = |||Q - Q^{INT} - z|||_{C(x^0,1)}$$

Some mild conditions have to be placed on the form of the estimator $\mathcal{E}$. These conditions are satisfied for all estimators used in the practice.

Theorem 2.3 allows the study of the bounds of the effectivity index of various estimators and their robustness related to the topology of the meshes, anisotropy of the material etc. (analogous theorems also hold for elasticity problems).

In Fig. 2.9 we show the four topologies of the mesh cells. We have analyzed the robustness of the estimators with respect to the aspect ratio, relation of the principal anisotropy axis and the mesh orientation.

In Babuška *et al.* (1993(v)) we analyzed 9 estimators suggested in the literature. Among other things we computed upper and lower bounds for an orthotropic material with the relation $k_{\max}/k_{\min} = 1000$ and the angle orientation with respect to the mesh. (We maximized and minimized the effectivity index over all polynomials of degree 2) In the Fig. 2.10 a and b we show as an example illustrative results for various estimators for two mesh patterns, the Chevron and Union jack.

These estimators are of residual type. In Fig. 2.11 we report the results based on the flux projection method.

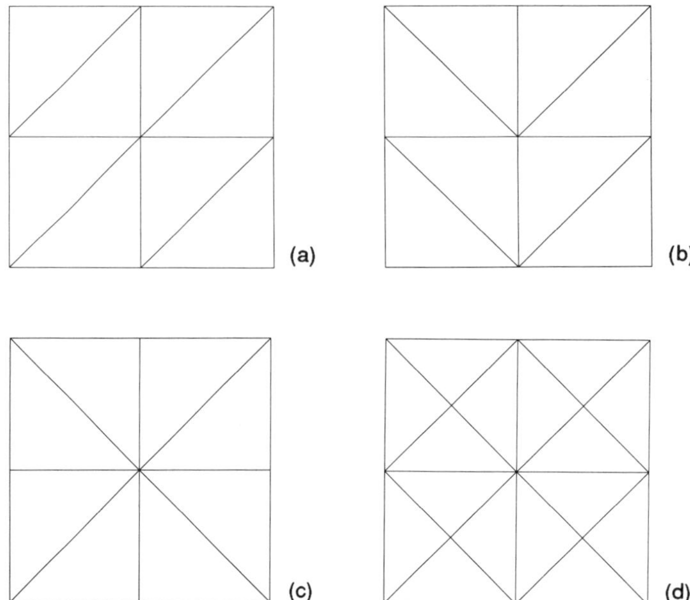

Fig. 2.9 The various pattern of the mesh. a) regular, b) chevron, c) union jack,
d) criss cross.

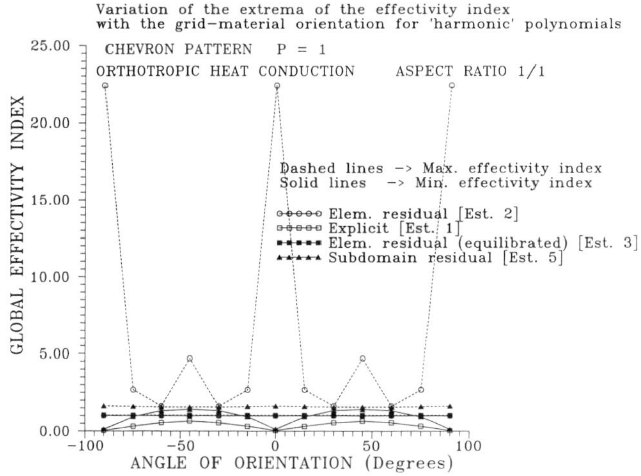

Fig 2.10a Bounds of θ for chevron pattern.

In the Figures the meaning of the headings is the following.

a) residual estimators (Fig. 2.10 ab).

Est1 = Explicit estimator Babuška *et al.* (1992(ii))

Est2 = Implicit estimator Bank and Weiser (1985), Oden *et al.* (1989)

Est3 = Implicit equilibrated estimator Ainsworth and Oden (1993)

Est5 = subdomain estimator Babuška *et al.* (1981)

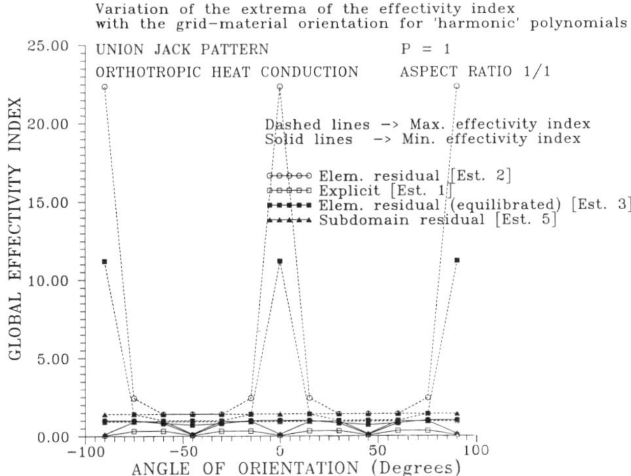

Fig 2.10b Bounds of θ for union jack pattern.

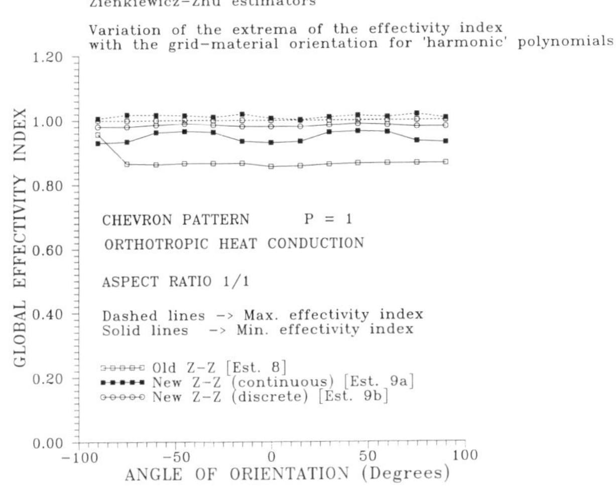

Fig 2.11 Bounds of θ for flux projection estimator.

b) flux estimators (2.14)

Est8 = Old ZZ estimator Zienkiewicz and Zhu (1987)

Est9a = New ZZ continuous estimator Zienkiewicz and Zhu (1992)

Est9b = New ZZ discrete estimator Zienkiewicz and Zhu (1992)

We have shown these figures only as examples. For more see Babuška *et al.* (1993(v)). The analysis suggests that the new ZZ estimator seems to the most robust.

Detailed computations show (Babuška *et al.* (1993(v))) that the bounds computed from the approach described here are also very robust for meshes which are not patchwise uniform. To show this we consider composite cells as shown in the Fig. 2.12a and 2.12b. Further we have incorporated these cell in larger cells with random elements. The effectivity index for the particular element τ is referred in the Table 2.5.

As was said there are many error estimators in the literature. It is necessary to

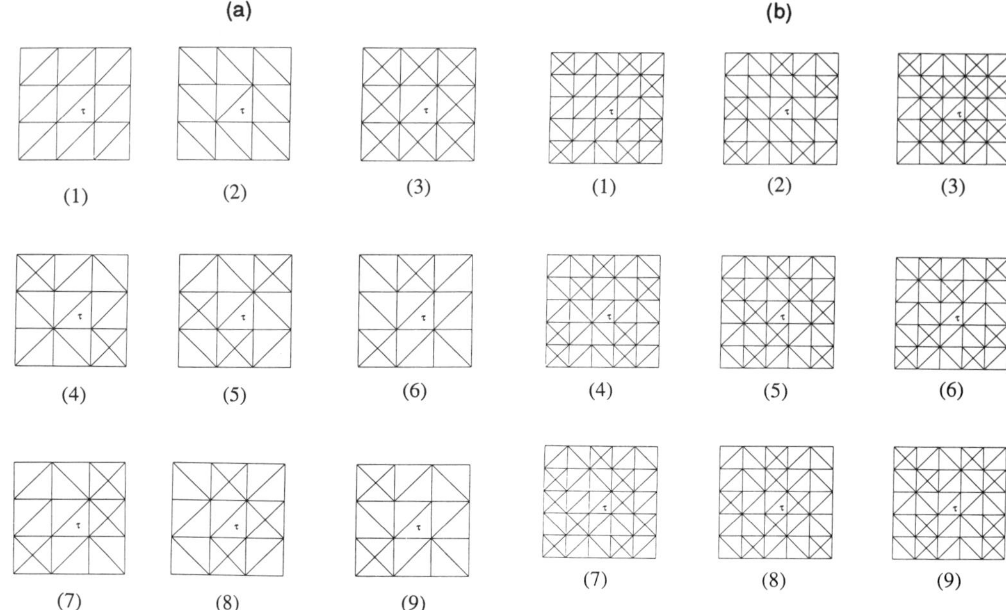

Fig. 2.12a and **2.12b** Sequence of the meshes.

Table 2.5 The effectivity index θ of the element τ for different meshes.

cell type	three cell		five cell		three cell+ bigger mesh		five cell+ bigger mesh	
	min	max	min	max	min	max	min	max
1	1.00	1.00	0.98	1.00	0.99	1.00	0.98	1.00
2	1.00	1.41	1.00	1.37	1.00	1.40	1.00	1.36
3	1.00	1.10	1.00	1.11	1.00	1.10	1.00	1.12
4	0.90	1.21	0.88	1.21	0.89	1.21	0.88	1.20
5	0.98	1.26	0.98	1.29	0.98	1.29	0.98	1.28
6	0.94	1.01	0.92	1.00	0.93	1.00	0.91	0.99
7	0.95	1.17	0.94	1.15	0.95	1.18	0.93	1.14
8	0.99	1.15	0.99	1.15	0.99	1.18	0.99	1.14
9	0.99	1.04	0.99	1.04	0.99	1.05	0.99	1.04

analyse these in relation to robustness, and in adaptive codes to adjust the mesh generator so that the meshes produce indicators of high reliability.

REFERENCES

Ambartsumyan S.A., 1991, Theory of anisotropic plates. *Hemisphere Publ. Corp. New York.*
Ainsworth M. and Craig A., 1992, *A posteriori* error estimators in the finite element method, *Numer.Math,* **60** 429-463
Ainsworth M. and Oden J. T., 1993, A procedure for a posteriori error estimation for *hp*

finite element method. *Comp.Math.Appl.Mech. Engg.* 10, 73-96.

Aoki H., Yamanouchi H., and Kato B. , 1990, Scatter on Mechanical Properties of Structural Steel recently studied in Japan. *NIST Spec Tech. Publ.* 796, 265-275

Babuška I., 1986, Uncertainties in Engineering Design Mathematical Theory and Numerical Experience in the *Optimum Shape,* ed. T.A. Bennett *M.E. Botkin, Plenum Press, New York,* 171-197

Babuška I., Duran R., and Rodriguez R. , 1992(i), Analysis of the efficiency of an *a posteriori* error estimator for linear triangular finite elements. *SIAM J. Numer. Anal.,* 20(4), 947-964.

Babuška I., Jerina K., Li Y. , and Smith P. , 1993(i), Quantitative assessment of the accuracy of constitutive laws for plasticity with an emphasis on cyclic deformation *Tech. Note BN - 1146, Institute for Physical Science and Technology, University of Maryland, To be presented at 1993 ASME Winter Annual Meetings, New Orleans, Nov. 28-December 3, 1993.*

Babuška I., Lee I., and Schwab C. , 1993(ii), On the *a posteriori* estimation of the modelling error for the heat conduction in a plate and its use for adaptive hierarchical modellings. *Tech. Note BN - 1145. Institute for Physical Science and Technology, University of Maryland, College Park, U.S.A.*

Babuška I. and Li L., 1991(i), Hierarchical modelling of plates *Comp. and Struct.,* 40, 419-430.

Babuška I. and Li L., 1992(ii), The problem of plate modeling: Theoretical and computational results. *Comp. Math. Appl. Mech. Engg.,* 100, 249-273.

Babuška I. and Miller A., 1981, *A posteriori* error estimates and adaptive techniques for the finite element method. *Tech. Note BN-968 Institute for Physical Science and Technology University of Maryland, College Park, U.S.A..*

Babuška I., Plank L., and Rodriguez R. , 1991(ii), Quality assessment of the *a posteriori* error estimation in finite elements. *Finite Elements in Analysis and Design,* 11, 285-305.

Babuška I. and Rheinboldt W.C., 1978(i), Error estimates for adoptive finite element computations. *SIAM J. Numer. Anal.,* 15(4), 736-754.

Babuška I. and Rheinboldt W.C., 1978(ii), *A posteriori* error estimates for the finite element method. *Int. J. Num. Meth. Engg.,* 12, 1597-1615

Babuška I. and Schwab C., 1993(iii), *A posteriori* error estimates for hierarchical models of elliptic boundary value problems on thin domains. *Tech Note BN-1148 Institute for Physical Science and Technology, University of Maryland, College Park.*

Babuška I. and Schwab C., 1993(iv), On Hierarchic Schwarz algorithm, to appear

Babuška I., Stroboulis T., and Upadhyay C.S. , 1993(v), A model study of the quality of *a posteriori* estimators for linear elliptic problems, Part Ia, error estimation in the interior of patchwise uniform grids of triangles. *Tech Note BN - 1147 Institute for Physical Science and Technology, University of Maryland, College Park, U.S.A.*

Babuška I., Szabo B., and Actis R. , 1992(iii), Hierarchical models for laminated composites. *Int. of Num. Meth. Engg.,* 33, 503-536.

Babuška I. and Yu D., 1987, Asymptotically exact *a posteriori* error estimator for biquadratic elements. *Finite Element in Analysis and Design,* 7, 341-354.

Bank R.E. and Weiser A., 1985, Some *a posteriori* error estimators for elliptic partial differential equations. *Math. Comp.,* 44, 283-31.

Bathe K. J., 1982, Finite element procedures in engineering analysis. *Prentice Hall,*   New Jersey.

Bonnetier E., 1988, Mathematical treatment of uncertainties appearing in the formulation of some models of plasticity. *Ph.D. Dissertation, University of Maryland, College Park, U.S.A.*

Chaboche J.L., 1986, Time independent constitutive theories for cyclic plasticity. *Int. J. of Plasticity ,* 2, 1049-188.

Ciarlet P.G., 1990, Plates and junctions in elastic multi-structures. An asymptotic analysis. *Masson, Paris, Springer, Berlin.*

Drucker D.C., 1991, Constitutive relations for solids. Retrospect and prospect. in *Constitutive Laws for Engineering Materials. Proceedings of the Third International Conference on Constitutive Laws for Engineering Materials. Theory and Applications.* eds. C.S. Desai, E. Krempl, G. Frantzskemis, H. Sadatamanesh, *ASME Press,*   *New York, 3-11.*

Duran R. and Rodriguez R., 1992, On the asymptotic exactness of Bank-Weiser estimator. *Num. Math.*, **62**, 297-303

Gilewski W. and Radwanska M., 1991, A survey of finite element models for the analysis of moderately thick shields. *Finite Elements in Analysis and Design*, **9**, 1-21.

Johnson C., 1976, Existence theorems for plasticity problems. *J.Math. Pure et Appl.*, **55**, 431-444.

Ladeveze P. and Leguillon D., 1983, Error estimate procedure in the finite element method and applications. *SIAM J. Numer. Anal.*, **20**, 485-501.

Li Y., 1992, The theoretical and numerical problems of constitutive laws of plasticity and their applications. *Ph. D. Thesis, University of Maryland, College Park, U.S.A.*.

Mioshi T., 1985, Foundations of Numerical Analysis of Plasticity. *Lecture Notes in Numerical and Applied Analysis, North Holland.*

Naghdi P.M., 1972, The theory of shells and plates. *Handbook of Physics.* eds. S. Flügge and C. Truesdell., **9**(2), 425-640.

Noor A.K. and Burton S.K., 1989, Assessment of shear deformation theories for multilayered composite plates. *Appl.Mech. Rev.*, **42**, 1-12.

Oden J.T., Demkowicz L., Rachowicz W. , and Westermann T.A. , 1989, Toward a universal *hp* adaptive finite element strategy: Part 2, *A posteriori* error estimates. *Comp.Meth. Appl. Mech. Engg*, **77**, 113-180.

Ono N., 1990, Recent topics in constitutive modeling of cyclic plasticity and viscoplasticity. *Appl. Mech. Rev.*, **43**, 283-295.

Panc K., 1975, Theory of elastic plates. *Academia Prague* .

Schwab C., 1989, Dimensional reduction method for elliptic boundary value problems. *Ph.D. Thesis, University of Maryland, College Park, Maryland, USA.*

Shephard M.S., Niu Q., and Bachannan P.L. , 1989, Some results using stress projectors for error indication in: *Adaptive Methods for Partial Differential Equations.* eds. J.E. Flaherty, P.J. Paslow, M.S. Shephard J.D. Vasilakis etc. *SIAM, Philadelphia.*

Stroboulis T. and Hague K.A., 1992, Recent experiences with error estimation and adaptivity Part I Review of error estimation for scalar elliptic problems *Comp. Meth. Appl. Mech. Engg.*, **97**, 399-436.

Vekua I.N., 1982, Some general methods for construction of various shell theories. *Nauka, Moscow.*

Verfurth R., 1989, A posteriori error estimators for the Stokes equation. *Numer. Math.*, **55**, 309-325.

Vogelius M. and Babuška I., 1981, On a dimensional reduction method I. The optimal selection of basis functions. *Math. Comp.*, **37**, 47-68.

Wang W.A., Bucci R.J., Stentz R.H. , and Conway J.B. , 1987. Tensile and strain controlled fatigue data for certain aluminium alloys in application in the transportation industry. *SAE Tech. Pep. Ser N 870094 Internat. Congress and Exposition, Detroit.*

Zienkiewicz O.C. and Zhu J.Z., 1992, The superconvergent patch recovery and *a posteriori* error estimates. Part 1. The recovery techniques. Part 2. Error estimates and adaptivity. *Int. J. Num. Meth. Engg.*, **33**, 1331-1364, 1365-1382.

Zienkiewicz O.C. and Zhu J.Z., 1987, A simple error estimator and the adaptive procedure for practical engineering analysis. *Int.J. Num.Meth. Engg.* 337-354.

Zuckowski M., 1981, Combined loading in the theory of plasticity. P.W.N. *Polish Scientific Publisher, Warsaw.*

Chapter 3

Numerical Techniques for Problems of Quasistatic and Dynamic Viscoelasticity

S.Shaw, M.K.Warby and J.R.Whiteman.

BICOM, Institute of Computational Mathematics,
Brunel University, Uxbridge, England

3.1 INTRODUCTION.

This paper is concerned with the effective numerical treatment of problems of quasistatic and dynamic viscoelastic stress analysis. We begin in this section with a brief introduction to the equations governing both dynamic and quasistatic response of solid viscoelastic media along with established results for linear quasistatic analysis. Section 3.2 illustrates the deficiencies of the linear viscoelastic model by comparing experimental uniaxial data for a real material with the prediction of the mathematical model and then moves on to build in a nonlinear formulation which is of the type described by Schapery (1969(ii)). By suitably choosing some of the unknown functions we show how a slight modification of a model due to Knauss and Emri (1981) is able to describe the available experimental data adequately. A key feature in this formulation is the concept of a reduced, internal or intrinsic time for a material. The paper proceeds with Section 3.3 which implements an algorithm suitable for simulating the dynamic response of viscoelastic structures, numerical results are given illustrating convergence rates and finally we close by suggesting an alternative approach.

In much of the theory of classical mechanics the physical media are classified as either solids or fluids; a fluid usually being defined as a substance which cannot sustain a shear force. It is taken as read that a substance which is not a fluid is a solid and *vice-versa*. However, in practice there is no precise distinction between these categories. Many structural engineering materials, notably concrete and many polymers, are classified *viscoelastic*, the term arising because they exhibit both solid and fluid like properties. Typically, when a load is suddenly applied to, for example, a polymer there will be an instantaneous deformation and internal load system, the *stress*, set up. If the load remains constant the stress does also and yet the deformation may continue indefinitely or asymptotically approach some constant state and the material is said to *creep*. The result of this is that over very short timescales the medium appears to be a solid whilst over longer timescales the creep, or flow, is more akin to fluidic behaviour, see the data in Cessna (1971) or Maranyl Nylon 66 (see references) (Maranyl is a registered trademark of I.C.I plc.) for example. Alternatively, if the medium is held in some constant state of deformation the stress will decay with time. This is known

The Mathematics of Finite Elements and Applications
Edited by J. R. Whiteman © 1994 John Wiley & Sons Ltd

as *stress relaxation*, for an illustration of this see the curves in Ferry (1980, appendix D). These phenomena mean that classical elasticity theory may not be successfully applied to such materials when the time dependent behaviour is of interest (although it will of course suffice for the instantaneous response). On the other hand the usual constitutive relation between the deformation, *strain*, and stress used to model fluid behaviour (notably Newton's law of viscosity, in which the shear stress is proportional to the rate of shear strain.) is found not to be appropriate either. In fact it is observed in experiments that the stress in a viscoelastic body, in general, depends not solely on the current strain but on the entire strain history. Mathematically interpreted this implies that the stress may be expressed as a functional of the strain (an inverse relationship also holds) and for this reason viscoelastic materials are said to possess *memory*. For more background to the phenomenological theory of viscoelasticity see Golden and Graham (1988), Ferry (1980), Christensen (1971), Lockett (1972) and, for a rigorous derivation of the theory of linear viscoelasticity, Gurtin and Sternberg (1962).

We shall be concerned with an isotropic, compressible, solid viscoelastic body, the interior of which when undeformed occupies the region $\Omega \subset \mathbb{R}^n$, where $n = 2, 3$, with boundary $\partial\Omega$. A point in this reference configuration is $\boldsymbol{x} = (x_i)_{i=1}^n$; we consider methods for the numerical determination of the displacement $\boldsymbol{u} = (u_i)_{i=1}^n$ of a point at $\boldsymbol{x}$ which results when the body is acted upon by external forces $\boldsymbol{f} = (f_i)_{i=1}^n$ and surface tractions $\boldsymbol{g} = (g_i)_{i=1}^n$ under isothermal conditions. We assume that $\nabla \boldsymbol{u}$ is small enough so that the resulting strain is completely characterized by the infinitesimal strain tensor $\underline{\underline{\epsilon}}(\boldsymbol{u}) = \big(\epsilon_{ij}(\boldsymbol{u})\big)_{i,j=1}^n$ which is derived from the displacement field via

$$\epsilon_{ij}(\boldsymbol{u}) := \frac{1}{2}\left(\frac{\partial u_i}{\partial x_j} + \frac{\partial u_j}{\partial x_i}\right). \tag{3.1}$$

Evidently $\underline{\underline{\epsilon}}(\boldsymbol{u})$ is symmetric, also the induced stress in the body is given by the symmetric Cauchy stress tensor $\underline{\underline{\sigma}} = (\sigma_{ij})_{i,j=1}^n$. Obviously for (3.1) to make sense $\boldsymbol{u}$ must possess certain regularity. However, for ease of exposition, we shall not pursue this point further but rather assume that $\boldsymbol{u}$ is smooth enough to justify all that follows.

The governing equations of the body are derived by applying Newton's second law and can be shown to yield the following classical system of partial differential equations (here, and subsequently, we employ the Einstein convention):

$$\begin{aligned}
\rho u_i''(\boldsymbol{x}, t) - \sigma_{ij,j}(\boldsymbol{x}, t) &= f_i(\boldsymbol{x}, t), & \boldsymbol{x} \in \Omega,\ t \in \mathcal{I} := (\hat{t}, T], \\
u_i(\boldsymbol{x}, t) &= 0, & \boldsymbol{x} \in \Gamma_D,\ t \in \mathcal{I}, \\
\sigma_{ij}(\boldsymbol{x}, t)n_j(\boldsymbol{x}) &= g_i(\boldsymbol{x}, t), & \boldsymbol{x} \in \Gamma_N,\ t \in \mathcal{I}, \\
u_i(\boldsymbol{x}, \hat{t}) &= u_i^0(\boldsymbol{x}), & \boldsymbol{x} \in \Omega, \\
u_i'(\boldsymbol{x}, \hat{t}) &= u_i^1(\boldsymbol{x}), & \boldsymbol{x} \in \Omega,
\end{aligned} \tag{3.2}$$

where: $i, j = 1, \ldots, n$, ρ is the mass density of the body, $\boldsymbol{n} = (n_i)_{i=1}^n$ is (almost everywhere) the unit outward directed normal to $\partial\Omega$, $\Gamma_N \cup \Gamma_D \equiv \partial\Omega$ but have empty intersection, the primes denote time differentiation and $\hat{t}$ is a reference time prior to which it is assumed that the material is quiescent, that is $\underline{\underline{\epsilon}}\big(\boldsymbol{u}(\boldsymbol{x}, s < \hat{t})\big) \equiv 0$. Also it is necessary to assume that the boundary and initial conditions are compatible on $\partial\Omega$ at $t = \hat{t}$.

Frequently in both theoretical studies and practical analysis of viscoelastic media the accelerations u_i'' are deemed to be negligible, particularly when the f_i and g_i are timewise constant, or only slowly varying, and the structure is not susceptible to vibration. In this case (3.2) is reduced to:

$$
\begin{aligned}
-\sigma_{ij,j}(\boldsymbol{x},t) &= f_i(\boldsymbol{x},t), \quad \boldsymbol{x}\in\Omega,\ t\in\mathcal{J}:=[\hat{t},T], \\
u_i(\boldsymbol{x},t) &= 0, \quad \boldsymbol{x}\in\Gamma_D,\ t\in\mathcal{J}, \\
\sigma_{ij}(\boldsymbol{x},t)n_j(\boldsymbol{x}) &= g_i(\boldsymbol{x},t), \quad \boldsymbol{x}\in\Gamma_N,\ t\in\mathcal{J}.
\end{aligned}
\tag{3.3}
$$

The equations (3.2) and (3.3) are fundamental statements of Newtonian mechanics, and as such are applicable to any physical continuum. Thus, as different materials respond in distinct ways when subjected to load, these equations by themselves only partly describe the problem. To complete the description the *constitutive relations* specific to the material must be given. These relations describe how the stress at a point in the body at a given time t depends upon the deformation which, in solid mechanics, is usually characterized by the displacement. In order to complete our formulations of the viscoelastic problem we must now specify the interdependence of stress and strain.

In the case of a linear viscoelastic material the functional relationship between $\underline{\underline{\sigma}}$ and $\underline{\underline{\epsilon}}$ is usually derived by assuming that the limit of a Boltzmann type of superposition of linear elastic responses at times $s < t$ may be employed. This leads to (see for example Shaw, Warby, Whiteman, Dawson and Wheeler (1993(i)), Golden and Graham (1988) or Gurtin and Sternberg (1962)) the following commonly encountered forms, the formal equivalence of which is readily shown by partial integration:

$$
\begin{aligned}
\boldsymbol{\sigma}(\boldsymbol{x},t) &= D(\boldsymbol{x},t,t)\epsilon\big(\boldsymbol{u}(\boldsymbol{x},t)\big) - \int_{\hat{t}}^{t} \frac{\partial D}{\partial s}(\boldsymbol{x},t,s)\epsilon\big(\boldsymbol{u}(\boldsymbol{x},s)\big)\,ds \\
&= D(\boldsymbol{x},t,\hat{t})\epsilon\big(\boldsymbol{u}(\boldsymbol{x},\hat{t})\big) + \int_{\hat{t}}^{t} D(\boldsymbol{x},t,s)\frac{\partial\epsilon}{\partial s}\big(\boldsymbol{u}(\boldsymbol{x},s)\big)\,ds,
\end{aligned}
\tag{3.4}
$$

where we have collected the stress and strain tensor components into

$$
\begin{aligned}
\boldsymbol{\sigma} &:= \big(\sigma_{11},\sigma_{22},\sigma_{33},\sigma_{12},\sigma_{13},\sigma_{23}\big)^T, \\
\boldsymbol{\epsilon} &:= \big(\epsilon_{11},\epsilon_{22},\epsilon_{33},2\epsilon_{12},2\epsilon_{13},2\epsilon_{23}\big)^T,
\end{aligned}
\tag{3.5}
$$

for the case $n = 3$ (the reduction to $n = 2$ being obvious). Here D is the constitutive or stress relaxation matrix the entries of which comprise the viscoelastic analogues of the Lamé coefficients that appear in linear elasticity theory. For example, in the two dimensional case ($n = 2$) we have

$$
\begin{aligned}
D(\boldsymbol{x},t,s) &:= D_\lambda(\boldsymbol{x},t,s) + D_\mu(\boldsymbol{x},t,s), \\
D_\lambda(\boldsymbol{x},t,s) &:= \lambda(\boldsymbol{x},t,s)\begin{pmatrix} 1 & 1 & 0 \\ 1 & 1 & 0 \\ 0 & 0 & 0 \end{pmatrix}, \\
D_\mu(\boldsymbol{x},t,s) &:= \mu(\boldsymbol{x},t,s)\begin{pmatrix} 1 & 0 & 0 \\ 0 & 1 & 0 \\ 0 & 0 & 1/2 \end{pmatrix}.
\end{aligned}
\tag{3.6}
$$

When the material exhibits a temporally constant Poisson ratio (ν) these viscoelastic functions may be factored thus:

$$\frac{\lambda(x,t,s)}{\lambda_0(x)} = \frac{\mu(x,t,s)}{\mu_0(x)} = \phi(x,t,s), \tag{3.7}$$

giving a single *stress relaxation function* ϕ, and the material is termed *synchronous*; otherwise it is termed *asynchronous* and the timewise independence of λ and μ implies that the Poisson ratio is time dependent.

Commonly in studies of viscoelasticity the material is assumed to be *non-ageing*. In this case λ and μ are no longer dependent upon t and s individually but rather on the elapsed time $t - s$, so that we now have

$$\begin{aligned}
\lambda(x,t,s) &\equiv \lambda(x,t-s), \\
\mu(x,t,s) &\equiv \mu(x,t-s), \\
\phi(x,t,s) &\equiv \phi(x,t-s),
\end{aligned} \tag{3.8}$$

with the latter applicable only to synchronous materials. When the rate of strain is zero, i.e. the strain is constant in time, the second equation of (3.4) implies that $\sigma(\cdot,t)$ is essentially proportional to $D(\cdot,t,\cdot)$. Now from physical observations it is known that the stress must "relax", or decay, and asymptotically approach some timewise constant value. It is easily deduced therefore that λ and μ must be monotonically increasing, non-negative, functions of s for any fixed t. This is known as the *fading memory hypothesis*. (In passing we remark that there apparently exist some pathological materials which exhibit a negative Poisson ratio, in which case λ will be non-positive but $|\lambda|$ will behave in the way just described.) Henceforth we shall consider only non-ageing models.

When either of the constitutive laws (3.4) are coupled to (3.2) we have the dynamic formulation of linear viscoelasticity which we shall term "viscodynamics" (by analogy with the term "elastodynamics" used to describe the analogous linear elasticity problem) and this we consider later in Section 3.3. On the other hand if either of the constitutive laws (3.4) are coupled to (3.3) the result is termed the quasistatic formulation which we now briefly review.

Taking (3.3) and combining with the first of (3.4), dropping the x dependence and assuming $\hat{t} = 0$ the quasistatic non-ageing problem may be given the following weak formulation: for

$$V := \{v \in \left(\mathcal{H}^1(\Omega)\right)^n : \quad v = 0 \quad \text{on } \Gamma_D\}, \tag{3.9}$$

find $u(x,t) \in \mathcal{L}_2(\mathcal{J};V)$ such that

$$a\big(u(x,t),v(x)\big) + \int_0^t b\big(t-s;u(x,s),v(x)\big)\, ds = \big(f(t),v\big) + \langle g(t),v\rangle, \quad \forall v \in V, \tag{3.10}$$

where, for $v, w \in V$, $s \in J$ we have set

$$(w, v) := \int_\Omega w_i v_i \, d\Omega, \quad \forall w, v \in \big(\mathcal{L}_2(\Omega)\big)^n,$$

$$\langle w, v \rangle := \int_{\Gamma_N} w_i v_i \, d\Gamma_N, \quad \forall w, v \in \big(\mathcal{L}_2(\partial\Omega)\big)^n,$$

$$a(w, v) := \int_\Omega \lambda(0)\nabla\cdot w \nabla\cdot v + \mu(0)\epsilon_{ij}(w)\epsilon_{ij}(v) \, d\Omega, \quad \forall w, v \in V, \qquad (3.11)$$

$$b(t - s; w, v) := -\int_\Omega \lambda_s(t - s)\nabla\cdot w \nabla\cdot v$$
$$+ \mu_s(t - s)\epsilon_{ij}(w)\epsilon_{ij}(v) \, d\Omega, \quad \forall w, v \in V.$$

Using now the standard Galerkin methodology to effect spatial discretization we partition Ω into N_E finite elements, such that $\Omega \equiv \cup_{i=1}^{N_E}\Omega_i$ (i.e. for simplicity we assume a polyhedral boundary), with a total of N_N nodes and replace V with a finite dimensional subspace $V^h := \otimes_{i=1}^n \mathrm{span}\{\varphi_j(x)\}_{j=1}^{N_N}$, comprising piecewise polynomials of degree r. Thus we obtain, in the usual way, a semidiscrete approximating problem. Now discretizing the time interval into

$$J^k := \{0 = t_0 < \cdots < t_i < \cdots < t_J = T\} \subset J \qquad (3.12)$$

and employing a trapezoidal rule numerical integration to discretize in time we are led to the fully discrete problem: find $\big(u^{hk}(x)\big)_j \in V^h$ for each $t_j \in J^k$ such that

$$a\big(\big(u^{hk}(x)\big)_j, v(x)\big) + \sum_{i=0}^j \varpi_{ij} b\big(t_j - t_i; \big(u^{hk}(x)\big)_i, v(x)\big) = L_j \quad \forall v \in V^h, \qquad (3.13)$$

where the ϖ_{ij} are the quadrature weights for the trapezoidal rule with $\varpi_{00} \equiv 0$, and L_j contains the loading and traction terms. Full details of this scheme are given in Shaw et.al. (1993(i)) together with an error analysis which reveals that, under appropriate assumptions,

$$\max_{t_j \in J^k} \big\| u(x, t_j) - \big(u^{hk}(x)\big)_j \big\|_1 \leq C(h^r + k^2), \qquad (3.14)$$

where $\|\cdot\|_p$ is the norm on $\big(\mathcal{H}^p(\Omega)\big)^n$ and the discretization parameters h and k denote the spatial meshwidth and timestep respectively.

In order to implement the numerical scheme just described it is clearly necessary to know the forms of $\lambda(t - s)$ and $\mu(t - s)$ for the material in question. These viscoelastic functions are typically modelled by a finite Dirichlet series, that is, dropping the spatial dependence, with N_L, N_M and N_F suitably chosen positive integers,

$$\lambda(t - s) = \sum_{i=0}^{N_L - 1} L_i \exp\big(-l_i(t - s)\big),$$

$$\mu(t - s) = \sum_{i=0}^{N_M - 1} M_i \exp\big(-m_i(t - s)\big), \qquad (3.15)$$

$$\phi(t - s) = \sum_{i=0}^{N_F - 1} C_i \exp\big(-\alpha_i(t - s)\big),$$

for non-negative constants $L_i, l_i, M_i, m_i, C_i, \alpha_i$. Alternatively, expressions of the form $C_0(t - s + C_1)^{-\alpha}$ (the "power law") are often used in theoretical studies, see Golden and Graham (1988) for example. From a practical point of view it is not the expression which is important but it is how well the expression leads to a model which is sufficiently "close" to the experimental data. Thus we are free to choose the form to take for λ and μ (and therefore ϕ when appropriate), and from a numerical point of view (3.15) has an important advantage over the power law in terms of being able to "carry" the entire strain history during a computation using only $N_L + N_M$ arrays of the same dimension as the solution vector. This is based on a recursion and is illustrated in, for example, Shaw *et al.* (1993(i)) or Zienkiewicz and Watson (1966), but this expedient is limited to use only in the linear non-ageing case.

The relations just given describe the response of the material by giving the stress in terms of the strain. In the context of a displacement formulation of a stress problem this is the required form, but in some applications it may be more convenient to express the strain in terms of the stress. For example, in a creep test the stress is the known, or controlled, quantity and it is the strain which is measured. It may be shown, see Golden and Graham (1988) or Gurtin and Sternberg (1962), that the integral operators in (3.4) may be inverted to give two dual expressions of strain in terms of stress history. In the case of a synchronous non-ageing material we could write such a pair as:

$$
\begin{aligned}
\sigma(t) &= D_0\left(\phi(t)\epsilon(0) + \int_{\hat{t}}^t \phi(t - s)\frac{\partial\epsilon}{\partial s}(s)\,ds\right) \\
\epsilon(t) &= D_0^{-1}\left(\psi(t)\sigma(0) + \int_{\hat{t}}^t \psi(t - s)\frac{\partial\sigma}{\partial s}(s)\,ds\right),
\end{aligned}
\tag{3.16}
$$

where D_0 is constructed in an obvious way from λ_0 and μ_0, as in (3.6), and ψ is the *creep function* for the material. Choosing $\hat{t} = 0$ and taking the Laplace transform, $L : w(t \in \mathbb{R}^+) \mapsto \hat{w}(p \in \mathbb{C})$, of these it follows that ϕ and ψ are related by

$$
\begin{aligned}
\phi(0)\psi(0) &= 1 \\
p^2\hat{\phi}(p)\hat{\psi}(p) &= 1.
\end{aligned}
\tag{3.17}
$$

We shall make use of these relations in Section 3.2.

The numerical scheme described above has been implemented using viscoelastic functions as in (3.15) for data arising from creep experiments on concrete. In Troxell, Davis and Kelly (1968, Chap. 12, §22, Fig. 12.10) some creep curves for a particular concrete mix are given and the effect of ageing is clearly demonstrated. Some of the concrete test specimens (in this case cylinders) were loaded axially and compressively at age 28 days to 300, 600 and 900 psi, whilst others were loaded at age 3 months to 600, 900 and 1200 psi. The resulting creep curves show that the older material is stronger, in that it does not creep as far as the younger. According to Troxell et. al. this ageing effect is "switched on" by a change in the loading:

Previous loading history influences the creep at a later age caused by an increase in sustained load. The later-age creep rates appear to be dependent on the relative load level at early ages; the lower the initially applied load, the greater the creep rate at later ages due to increase of the load.

This notwithstanding, and purely as a numerical experiment, we have attempted to model the quasistatic creep response of the 3 month concrete with a synchronous, (3.7), linear non-ageing, (3.15), model assuming plane stress and, for concrete, typical values of Poisson ratio, $\nu = 0.17$, and mass density, $\rho = 2650\text{kg/m}^3$, (ρ is, of course, not needed for quasistatic analysis, but we shall require it in Section 3.3), and the relaxation function

$$\phi(t) = 677\,823 + 877\,289 \exp(-0.01t) + 151\,989 \exp(-0.001t), \qquad (3.18)$$

(where the coefficients are in psi, and the units of time are days) which has been derived by interpolation from the diagram given by Troxell et.al. Figure 3.1 shows the numerical results, using both linear and logarithmic time scales, and in these the lines marked with a circle correspond to experimental data and the solid lines to the computed results (the discrete computed data has been interpolated to yield the continuous curves). In this case we have combined the first part of (3.4) with (3.3) to produce the algorithm (3.13). The results demonstrate that the linear model predicts the deformation within this loading regime indicating that there is, in this case, linear viscoelastic behaviour.

Whilst the above results are encouraging, the linear model is of course not suited to many viscoelastic materials and loading regimes as the data in the forthcoming section illustrate, and therefore nonlinear constitutive relations are required. Models of this type are now discussed.

3.2 NONLINEAR VISCOELASTICITY

3.2.1 PREAMBLE

The derivation of the linear viscoelastic models just described in Section 3.1 was based on the assumptions that the strain is small and that the Boltzmann superposition is valid. Clearly, if a material is deformed in such a way that either of these assumptions fails, then nonlinear models should be considered instead. If the strain is too large then geometric nonlinearity will occur. On the other hand if the relation between the stress measure and the strain measure used is nonlinear, then we have constitutive nonlinearity. The geometric nonlinearity and the constitutive nonlinearity are usually independent effects. Important questions thus concern the range of validity of the linear models and what the correct form of nonlinear models should be, i.e. how should the linear models be generalized to a more widely applicable nonlinear form? As we will indicate in this section, this is not a question which can be properly or satisfactorily answered. To understand why this is the case, the derivation of constitutive relations needs to be considered, see e.g. Schapery (1969(i)). In texts on continuum mechanics the relations are derived using the laws of thermodynamics and, although theoretically well founded, are usually far too complicated to be incorporated into a computational model or to be associated with an experimental program where material parameters are determined. Thus for practical reasons alone, simplifications must be made and to this date there is no generally accepted simplification which is adequate for all viscoelastic materials in all situations. In this paper we consider only strains which are small and hence we only require a simplification which is valid in this case. A

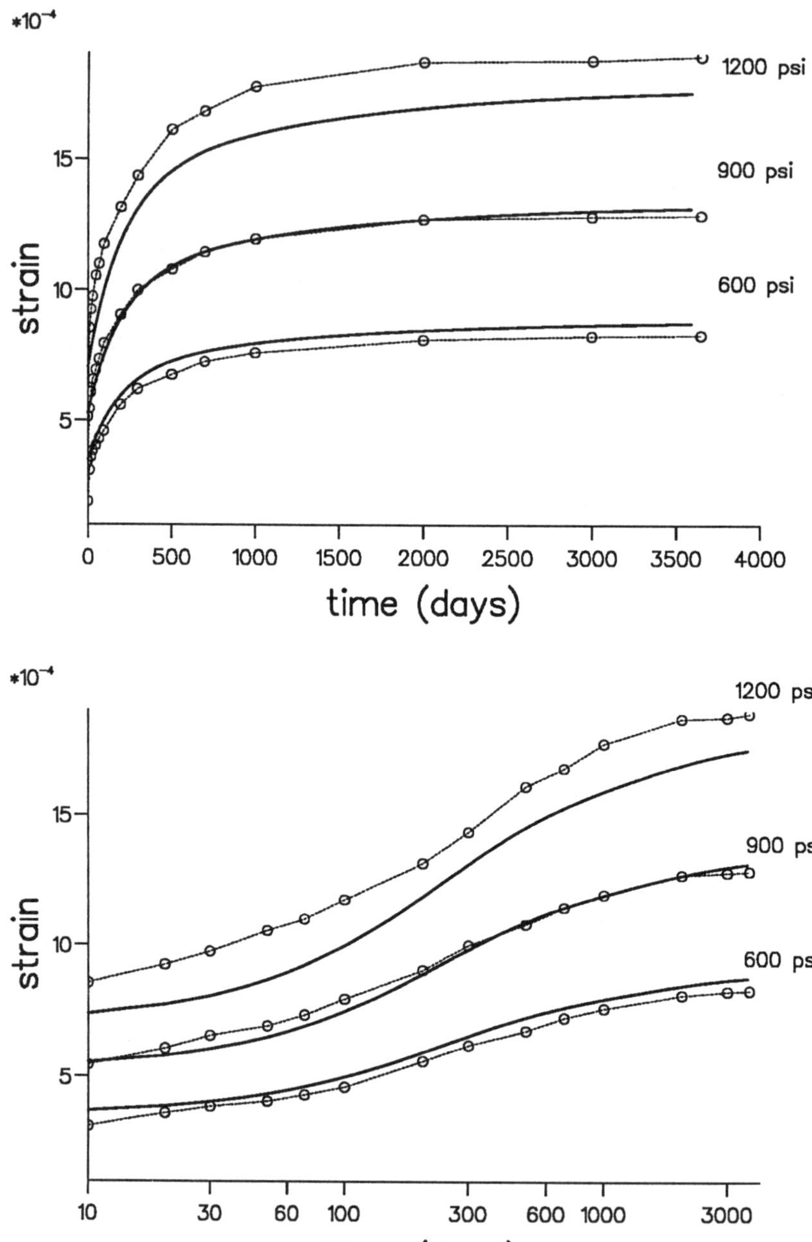

Figure 3.1 Comparison of computed creep results with experimental data given by Troxell et.al. (1968). Computations performed with a single 8 noded quadrilateral element, minimally constrained to prevent rigid body rotation, and a time step of 10 days. The strain illustrated here is along the centreline of the element in the direction of the indicated traction.

practical question then concerns what can be regarded as small. If we consider the extent of smallness as covering the range over which the infinitesimal strain can be used to represent the actual strain adequately, then from typical engineering applications strains as large as 10% or even 20% come into this catagory. Only when strains are larger than this is the problem regarded as geometrically nonlinear. However the implication of our previous statement concerning simplifying the general theory in the small strain case is that we anticipate constitutive nonlinearity to be a feature before geometric nonlinearity is encountered. This was the situation observed in Shaw *et al.* (1993(i)) in treating Nylon 66. Those results indicated that constitutive nonlinearity occurs when the strain is as low as 2% or 3% and therefore much below the onset of geometric nonlinearity. The purpose of this section is to present a specific form that the nonlinear constitutive model may take for this magnitude of strain and to demonstrate how to fit the model to the given experimental data for Nylon 66.

3.2.2 NONLINEAR VISCOELASTIC MODELS

The particular model that we use is summarized in Schapery (1969(ii)) with the details of the derivation given in the longer reports Schapery (1969(i)) and (1966). In these models the simplification of the general theory arises from expanding certain quantities about their equilibrium state and then only retaining the first few terms. More specifically the Gibb's free energy is introduced and then expanded with respect to those independent state variables which may be identified with the strain components and temperature. The theory covers the general three dimensional case but as in Schapery (1969(i)) we concentrate here mainly on the uniaxial situation.

Let $E(t) = E_0\phi(t)$ denote Young's modulus of relaxation in the linear case, i.e. for a uniaxial tensile test with $\sigma_{ij} = 0$ for all i and j except $i = j = 1$, $\epsilon_{ij} = 0$ for $i \neq j$ and $\epsilon_{22} = \epsilon_{33} = -\nu\epsilon_{11}$ we have from the second part of (3.4), with $\hat{t} = 0$,

$$\sigma_{11}(t) = E_0\phi(t)\epsilon_{11}(0) + E_0 \int_0^t \phi(t-\tau)\frac{\partial}{\partial\tau}(\epsilon_{11})d\tau, \quad t \in \mathcal{J}. \tag{3.19}$$

The simplification summarized in Schapery (1969(ii)) gives the nonlinear form of the relation as

$$\sigma_{11}(t) = h_1 h_2 E_0\big(\phi(\zeta(t)) - \phi(\infty)\big)\epsilon_{11}(0) + h_0 E_0\phi(\infty)\epsilon_{11}(t)$$
$$+ h_1 E_0 \int_0^t \big(\phi(\zeta(t) - \zeta(\tau)) - \phi(\infty)\big)\frac{\partial}{\partial\tau}(h_2\epsilon_{11})d\tau, \tag{3.20}$$

where h_0, h_1 and h_2 are functions of ϵ_{11} and

$$\zeta(t) = \int_0^t \frac{dt'}{A(\epsilon_{11}(t'), \sigma_{11}(t'))}, \quad A(\epsilon_{11}(t'), \sigma_{11}(t')) > 0 \tag{3.21}$$

is the so called reduced, internal or intrinsic time for the material. When $h_0 = h_1 = h_2 = A = 1$ the nonlinear model simplifies to the linear model (3.19) which is thus a special case of (3.20) and (3.21). In addition as A depends on both stress and strain, the relation only implicitly defines the stress components. The nonlinear generalization thus uses the same relaxation function ϕ as in the linear model but since the arguments are now given by the reduced time $\zeta(t)$ instead of the actual time t we say that the material's time scale has been changed as a result of the deformation that the material has undergone.

As has already been mentioned the results of Shaw et al. (1993(i)) when modelling Nylon 66, demonstrate the limitations of the linear model and the need for other (nonlinear) models in order to fit all of the available experimental creep data. For this particular material the basic practical question is whether or not we can determine functions h_0, h_1, h_2 and $\mathcal{A}$ which lead to a significantly better fit of the data than that obtained with the linear model. Since the linear model is found to be satisfactory for the creep data resulting from the smallest applied loads, it would seem reasonable to consider functions h_0, h_1, h_2 and $\mathcal{A}$ which are "close" to unity. Hence in seeking a suitable nonlinear model we choose $h_0 = h_1 = h_2 = 1$ but allow a more general form for the function $\mathcal{A}$ in the reduced time relation. Thus the constitutive relation (3.20) becomes

$$\sigma_{11}(t) = E_0\phi(\zeta(t))\epsilon_{11}(0) + E_0 \int_0^t \phi(\zeta(t) - \zeta(\tau))\frac{\partial}{\partial \tau}(\epsilon_{11})d\tau \qquad (3.22)$$

and we retain (3.21). Knauss and Emri (1981,1987) use a model of this type with the function $\mathcal{A}$ depending only on the dilatation (volume change) θ with $\theta = (1 - 2\nu)\epsilon_{11}$ for a uniaxial tensile test on a material with constant Poisson's ratio ν. They further assume that the nonlinear effect in a purely mechanical deformation arises solely as a consequence of the change in "free volume" (the space between the molecules) as the material is deformed. Thus in the models of Knauss the function $\mathcal{A}$ in the reduced time expression depends only on θ. The specific equation used to specify this relation is known as Dolittle's equation, see Ferry (1980), which is

$$\log(\mathcal{A}) = \frac{-b\theta}{f_0(f_0 + \theta)}, \qquad (3.23)$$

where b and f_0 are material constants. When $f_0 \gg \theta$ then (3.23) gives

$$\log(\mathcal{A}) \approx \left(\frac{-b}{f_0^2}\right)\theta \qquad (3.24)$$

and thus there is only one constant to consider. In the Knauss papers it is further shown that a constitutive model as given by (3.21)-(3.23) can fit certain experimental data for material such as polyvinylacetate in various loading and unloading regimes.

However, for the experimental data that we have for Nylon 66, we find that in order to produce an adequate constitutive relation it is necessary for $\mathcal{A}$ also to depend on the current stress component σ_{11}. A way of achieving this relation is described in Subsection 3.2.3.

3.2.3 FITTING A CONSTITUTIVE MODEL TO ACTUAL CREEP DATA

The experimental creep response of the material Nylon 66, to various uniaxially applied stresses is reproduced here from the data for Maranyl (see references) in Fig. 3.2. In Shaw et al. (1993(i)) it was found that the linear model could approximate the creep curves adequately in the lower range of applied stress, see Fig. 3.3, but as the stress increases the accuracy of the linear model deteriorates. This can be easily observed without doing any rigorous analysis since it would appear that the lowest 5 curves in Fig. 3.2 are roughly a linear rescaling of each other whereas at the higher applied stresses the material appears to creep significantly more rapidly than at the lowest applied stress. To be able to visualise this observation more clearly we show in Fig. 3.3 the experimental curves and the linear model predictions for the creep data corresponding to 500, 2500, 4500, 6000 and 7500 psi. In Fig. 3.3 the linear model predictions were determined by first fitting a function of the form

$$\epsilon_{11}(t) = \sigma_0 \psi(t), \quad \psi(t) = B_0 + \sum_{1}^{M} B_i e^{-t/\tau_i} \tag{3.25}$$

to the experimental creep curve corresponding to $\sigma_0 = 1000$ psi (which is in the linear range of response) and then plotting the curves of $\sigma_1 \psi(t)$ for $\sigma_1 = 500, 2500, 4500, 6000$ and 7500. From the figure it is evident, as already indicated, that the linear model gives a good fit at the lower applied stresses but becomes increasingly inaccurate as the stress level increases.

We consider now the nonlinear model described in Subsection 3.2.2 and begin with the determination of the relaxation function ϕ from the creep data given in Fig. 3.2. This is achieved by using (3.17) which relates the transforms of the creep function ψ and the relaxation function ϕ. For the particular values of the coefficients $B_0, \ldots B_4$, $\tau_1, \ldots, \tau_4$ of (3.25) listed with Fig. 3.3, this leads to the function

$$E_0 \phi(t) = C_0 + \sum_{1}^{M} C_i e^{-t/\alpha_i} \tag{3.26}$$

with $C_0, C_1, \ldots, C_4 = 640.8, 1342, 918.7, 1195, 250.9$ (psi), $\alpha_0, \alpha_1, \ldots, \alpha_4 = 70.32$, 718.9, 4344, 33557 (hours) and where we normalise the function ϕ so that $\phi(0) = 1$. To complete the description of the nonlinear model (3.21)–(3.23) we also need to determine the reduced time function $\mathcal{A}$. However for the moment we will suppose that this is known and we consider the computation required to determine the creep response of the model.

For the nonlinear model described by (3.21)–(3.23) the creep response, i.e. the strain $\epsilon_{11}(t)$, for a given constant stress σ_0 satisfies

$$\frac{\sigma_0}{E_0} = \phi(\zeta(t))\epsilon_{11}(0) + \int_0^t \phi(\zeta(t) - \zeta(\tau))\frac{\partial}{\partial \tau}(\epsilon_{11}(\tau))d\tau, \tag{3.27}$$

where

$$\zeta(t) = \int_0^t \frac{dt'}{\mathcal{A}(\theta(t'), \sigma_0)}, \quad \theta(t') = (1 - 2\nu)\epsilon_{11}(t'). \tag{3.28}$$

From (3.27) and (3.28) it will not in general be possible to determine ϵ_{11} analytically as a function of time and thus even for this simplest deformation numerical methods are required to compute the response of the material. We now describe an appropriate scheme.

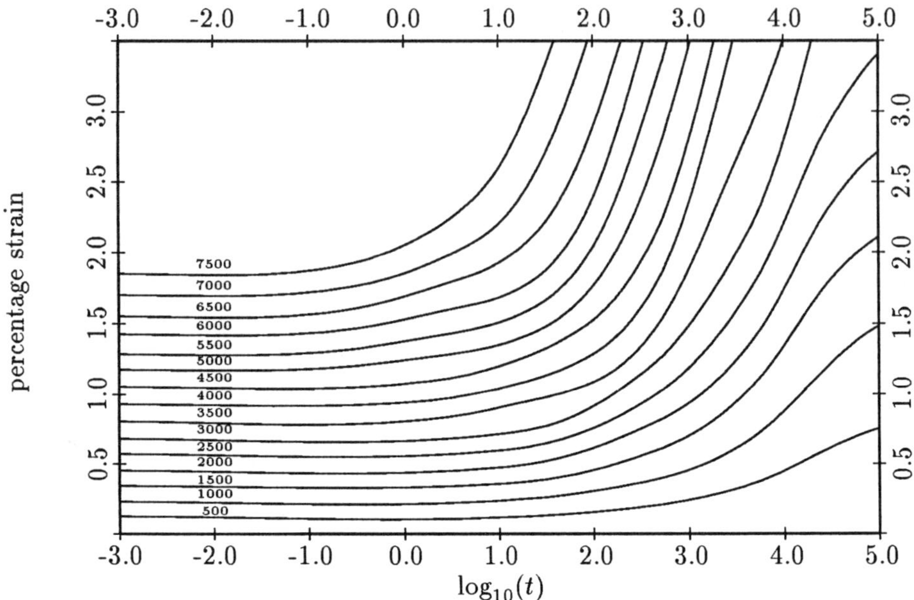

Figure 3.2 Nylon 66 creep curves produced from experimental data

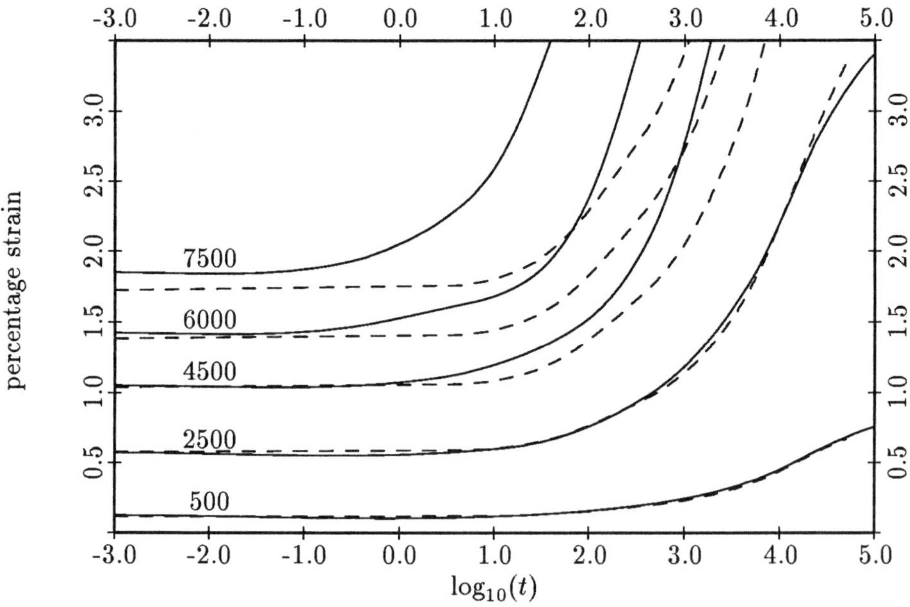

Creep curve coefficients: $M = 4$, $\tau_1, \tau_2, \tau_3, \tau_4 = 10^2, 10^3, 10^4, 5 \times 10^4$ and $B_0, B_1, B_2, B_3, B_4 = 0.1561, -0.9149 \times 10^{-2}, -0.1118 \times 10^{-1}, -0.5376 \times 10^{-1}, -0.5897 \times 10^{-1}$

Figure 3.3 Comparison of experimental results (solid lines) and linear model (dashed lines)

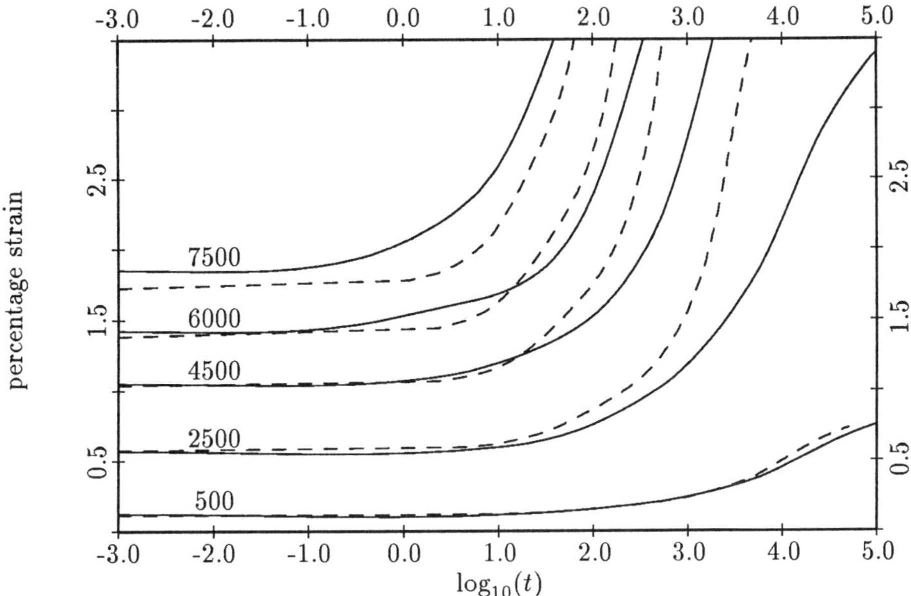

Figure 3.4 Comparison of experimental results (solid lines) with Knauss type model (dashed lines) with $A = 500$

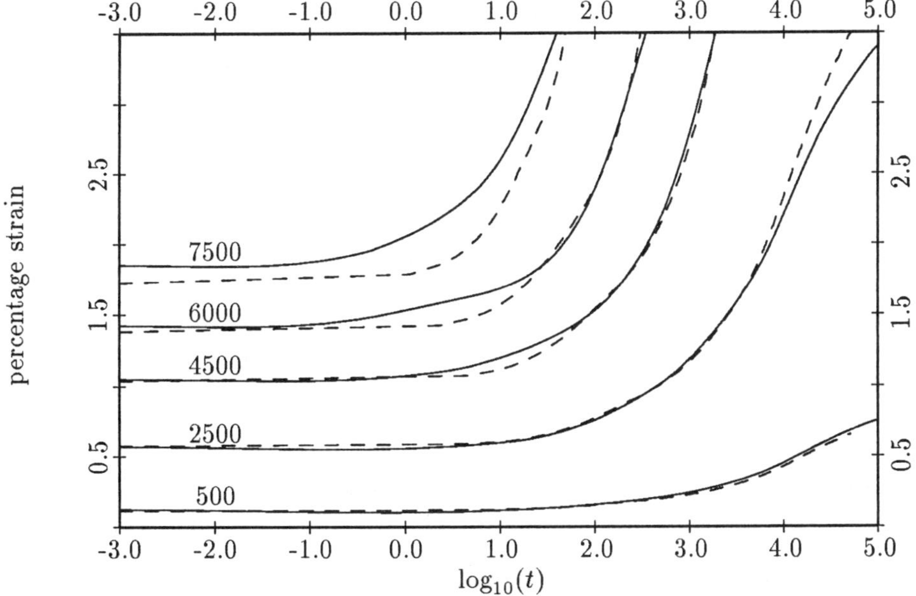

Figure 3.5 Comparison of experimental results (solid lines) with Schapery type model (dashed lines) with $A' = 0.1$ and $\sigma_{\text{ref}} = 2000$

Corresponding to the instantaneously imposed stress σ_0 we obtain the initial strain of $\epsilon_{11}(0) = \sigma_0/E_0$ given that we have the normalisation $\phi(0) = 1$. To obtain the subsequent approximate strains we subdivide the time interval $[0,t]$ by $0 = t_0 < t_1 < \ldots < t_N = t$ and we use the central difference/trapezoidal approximation

$$\int_{t_{i-1}}^{t_i} \phi(\zeta(t) - \zeta(\tau)) \frac{\partial}{\partial \tau}(\epsilon_{11}(\tau))d\tau$$
$$\approx \frac{1}{2}\left(\phi(\zeta(t) - \zeta(t_i)) + \phi(\zeta(t) - \zeta(t_{i-1}))\right)\left(\epsilon_{11}(t_i) - \epsilon_{11}(t_{i-1})\right). \tag{3.29}$$

Similarly we use the trapezoidal rule to approximate the reduced time relation giving

$$\zeta(t_N) = \zeta(t_{N-1}) + \frac{1}{2}\left(\frac{1}{\mathcal{A}(\theta(t_N), \sigma_0)} + \frac{1}{\mathcal{A}(\theta(t_{N-1}), \sigma_0)}\right)(t_N - t_{N-1}),$$
$$\theta(t_i) = (1 - 2\nu)\epsilon_{11}(t_i). \tag{3.30}$$

Thus (3.29) is nonlinear in $\epsilon_{11}(t_N)$ due to the dependence of $\zeta(t_N)$ on $\epsilon_{11}(t_N)$ as given in (3.30). The solution $\epsilon_{11}(t_N) > \epsilon_{11}(t_{N-1})$ to (3.27) assuming the discretization (3.29) and (3.30) is then computed by a standard root finding procedure.

The successive computation of the strains $\epsilon_{11}(t_0)$, $\epsilon_{11}(t_1)$, ..., using the procedure outlined above involves progressively more and more computational effort because of the presence of the history term. This is of course a feature of viscoelastic problems. To control the total amount of computation required to obtain the approximate strain over the complete time range, we adaptively choose the time steps $t_i - t_{i-1}$. This is done by starting the procedure with small time steps, which are then doubled or halved at each stage as necessary in order that the changes in strain per step are approximately the same. Thus in this way the time step is small when we predict $\partial \epsilon_{11}/\partial t$ to be large and conversely the time step is large when $\partial \epsilon_{11}/\partial t$ is predicted to be small. With the normalisation $\phi(0) = 1$ and with $\epsilon_{11}(0)$ being the initial strain, the final strain is given by $\epsilon_{11}(0)/\phi(\infty)$. Thus as a guide to how this adaption is implemented the computation given corresponds to trying to achieve a change in strain of $0.05\epsilon_{11}(0)/\phi(\infty)$ at each step.

The strain response that is obtained depends of course on the dependence of the function $\mathcal{A}$ in (3.28) on ϵ_{11} and σ_0. Various numerical tests have been performed in order to investigate whether or not a simple form of this function can reasonably describe the experimental data for Nylon 66. Initially the tests were performed with the Knauss model form for $\mathcal{A}$ given in (3.23) or (3.24). In the case $f_0 \gg \theta$, $\log(\mathcal{A})$ is essentially a linear function of θ, i.e.

$$\log(\mathcal{A}) = -A\theta, \quad A \geq 0. \tag{3.31}$$

The linear response corresponds to $A = 0$. In the papers of Knauss typical values for b and f_0 are 0.6 and 0.025 respectively which give $A = b/f_0{}^2 = 960$. Thus in Fig. 3.4 we give calculated creep curves corresponding to $A = 500$ and note that the Knauss model is qualitatively better than the linear model but again fails to fit all the curves given. To achieve a better fit we find it is necessary for $\mathcal{A}$ to depend additionally on the applied stress level σ and for these experiments we choose the form to be

$$\log(\mathcal{A}) = -A'\theta(\sigma - \sigma_{\text{ref}}), \tag{3.32}$$

where A' and σ_{ref} are constants. The results of numerical experiments with the values $A' = 0.1$ and $\sigma_{\text{ref}} = 2000$ (with stress measured in psi) are given in Fig. 3.5.

As a result of the work of this section we have produced a nonlinear constitutive relation appropriate to Nylon 66 for the stress and strain levels indicated.

3.3 VISCODYNAMICS.

We begin in this section with a qualitative overview of the nature of the response of viscoelastic materials under dynamic loading conditions and then move on to numerical algorithms. We close, in the final subsection, by suggesting an alternative discretization.

3.3.1 OVERVIEW.

When there is reason to suspect that a viscoelastic structure is susceptible to vibration, or when the body forces or surface tractions vary rapidly or oscillate in time, the quasistatic formulation of the continuum problem (3.3) is likely to fail, or only be of limited use, due to the neglect of the inertia term u'' in (3.2). The inclusion of this term can lead to phenomena that have no parallel in elastodynamic structural response. We cannot hope to do justice here to the body of theory developed concerning viscodynamic response but refer the reader instead to, for example, Christensen (1971, Chap. 4), Hunter (1983, Chaps. 14 and 15), Mainardi (1982) and Kolsky (1956). We shall be content here simply to allude to the nature of the dynamic response of a viscoelastic material by using a simple, but illustrative, analytic result.

In elastodynamics the system of governing equations is a generalization of the wave equation, which in one space dimension takes its simplest form,

$$\begin{aligned}
u''(x,t) &= c^2 u_{xx}(x,t) \quad x \in \mathbb{R},\ t > 0, \\
u(x,0) &= u^0(x), \\
u'(x,0) &= u^1(x),
\end{aligned} \tag{3.33}$$

where u^0 and u^1 are given functions. It is well known that the solution to this equation can be no smoother than the initial data (in contrast to elliptic and parabolic equations), indeed the D'Alembert solution to (3.33) in the case $u^1 \equiv 0$ is $u(x,t) = \frac{1}{2}\big(u^0(x - ct) + u^0(x + ct)\big)$, showing that, for any u^0, the initial profile is split equally with each "half" travelling in the $\pm x$ direction, respectively, with speed c. When the spatial domain is bounded, e.g. $x \in (-L, L)$ and Dirichlet boundary data is specified the initial profile is still a travelling wave but appears as only a standing wave. This is due to the repeated interaction of the characteristic curves, $x \pm ct$, with the boundary. We shall numerically illustrate viscoelastic standing waves in a beam later in Subsection 3.3.2, and contrast them with elastic waves, but first we shall explore, in a qualitative way, the nature of the dynamic response of a viscoelastic material.

In the absence of body forces the elastodynamic equations (i.e. (3.2) combined with Hooke's law $\sigma = D\epsilon$, see, for example, Hunter (1983), or any introductory text on elasticity) may be re-cast as:

$$\nabla\Big((\lambda + \mu)\nabla^2\eta - \rho\eta''\Big) + \nabla\times\Big(\frac{\mu}{2}\nabla^2\gamma - \rho\gamma''\Big) = 0, \tag{3.34}$$

where the displacement is written in terms of its irrotational and solenoidal components:

$$u(x,t) = \nabla\eta(x,t) + \nabla\times\gamma(x,t), \tag{3.35}$$

so that η and γ are, respectively, a scalar and a vector potential. The elastodynamic equation is satisfied provided that η and γ solve the two indicated wave equations. Thus there are two possible speeds, c_η and c_γ, at which disturbances may propagate through the material, these are given by:

$$c_\eta = \sqrt{\frac{\lambda + \mu}{\rho}}, \qquad c_\gamma = \sqrt{\frac{\mu}{2\rho}}. \tag{3.36}$$

A displacement field given by $\nabla\eta$ is known as dilatational, since, in general, $\nabla^2\eta$ is not zero; therefore the body exhibits time dependent volume changes. On the other hand $\nabla\cdot u$ is identically zero for a displacement field given by $\nabla\times\gamma$; this type of displacement field is known as equivoluminal and as a particular case may describe shear waves, where a disturbance propagates orthogonal to the displacement, see for example Hunter (1983, Chap. 14).

When applied to the quasistatic formulation of (non-ageing) viscoelasticity the correspondence principle, see Golden and Graham (1988), provides a means of reducing the viscoelastic problem to an elastic problem parameterized by the transform variable. Thus, in a sense, quasistatic viscoelasticity does not present any fundamentally new difficulties. Viscodynamic formulations, however, do not admit such an expedient because the integral transform of the problem does not yield an elastic problem. In this case the temporal evolution of the response of a viscoelastic structure is, although of a similar nature in terms of the transport of initial disturbances, rather different to that of an otherwise identical elastic structure. The addition of the history integral represents an internal damping mechanism and although initial profiles are still propagated, they become time dependent. As an example we consider the following simple, but illuminating, one dimensional problem.

Let $x := x_1$, $\sigma := \sigma_{11}$, $\epsilon := \epsilon_{11}$, $\epsilon_{22} = \epsilon_{33} = -\nu\epsilon_{11}$ and $\sigma_{ij} = \epsilon_{ij} \equiv 0$ otherwise. We seek solutions of the form

$$u(x,t) = \tilde{u}(x)\exp(\imath\omega t), \tag{3.37}$$

where $\imath^2 = -1$, to the one dimensional form of the governing equation (3.2) combined with the first of (3.4) with $\hat{t} = -\infty$:

$$\rho u''(x,t) - \frac{\partial\sigma}{\partial x}(x,t) = 0 \quad (x,t) \in \mathbb{R} \times \mathbb{R},$$

$$\sigma(x,t) = E(0)\epsilon(x,t) - \int_{-\infty}^{t} \frac{\partial E}{\partial s}(t-s)\epsilon(x,s)\,ds, \tag{3.38}$$

$$\epsilon(x,t) = \frac{\partial u}{\partial x}(x,t),$$

$$E(t) = E_0 + E_1\exp(-\alpha t) \quad \text{(Young's modulus of relaxation)}.$$

We find (using, for example, integral transform techniques, or following Hunter (1983, Chap. 15) and working with complex moduli) that the displacement has the form

$$u(x,t) = A^{\pm}\exp\left(\imath\omega(t \pm x/\xi(\omega)) \pm \omega x\zeta(\omega)\right), \tag{3.39}$$

where A^+ and A^- are arbitrary constants (the use of ζ should *not* be confused with the use as a reduced time in Section 3.2). Now defining

$$
\begin{aligned}
a &:= (E_0 + E_1)/\rho \\
b &:= -\alpha E_1/\rho \\
\gamma &:= \alpha + b/a \\
r &:= \sqrt{(\gamma\alpha + \omega^2)^2 + \omega^2(\gamma - \alpha)^2} \\
\beta &:= \operatorname{ArcTan}\left(\frac{\omega(\gamma - \alpha)}{\gamma\alpha + \omega^2}\right)
\end{aligned}
\tag{3.40}
$$

we may write the phase velocity and attenuation factor (using the terminology of Kolsky (1956)) as

$$
\begin{aligned}
\xi(\omega) &= \frac{\sqrt{a(\gamma^2 + \omega^2)}}{\sqrt{r}\cos\beta/2}, \\
\zeta(\omega) &= \frac{\sqrt{r}|\sin\beta/2|}{\sqrt{a(\gamma^2 + \omega^2)}},
\end{aligned}
\tag{3.41}
$$

respectively. The interpretation of (3.39) is that of a wave travelling to the left (or right, depending upon the choice of sign) with an amplitude decaying in the direction of travel. Taking the linear elastic limit: $\alpha \to 0 \Rightarrow b = \gamma = 0$, $a = E/\rho$, where $E := E_0 + E_1$, we see that (3.39) reduces to

$$
u_e(x, t) = A^\pm \exp\left(\imath\omega\left(t \pm x\sqrt{E/\rho}\right)\right),
\tag{3.42}
$$

with the subscript e denoting "elasticity". Comparison of (3.42) with (3.39) is revealing; in the linear elastic case the waves move with speed $\sqrt{E/\rho}$ (this is comparable with c_η, the irrotational displacement field wave speed given by (3.36)) and have constant amplitude, also this speed depends only upon the material parameters of inertia and stiffness. Thus any initial profile will propagate along the x-axis at a constant speed and retain its shape. In the viscoelastic case the wave speed clearly depends upon the frequency, but also if we observe that the quantity $\omega/\xi(\omega)$ represents the spatial frequency we can see that the rate of attenuation is different for different harmonics. From this we conclude that, for a viscoelastic material, the initial profile cannot, in general, retain its shape, since each of its Fourier components travel at a different speed and become attenuated at a different rate.

3.3.2 NUMERICAL ALGORITHM.

Returning now to (3.2) and combining with the first part of (3.4), taking $\hat{t} = 0$, we use the notation introduced in Section 3.1 and give the resulting problem the weak formulation: find $u \in C^2(\mathcal{I}; \mathcal{V})$ such that

$$
(\rho u''(t), v) + a(u(t), v) + \int_0^t b(t - s; u(s), v)\, ds = (f(t), v) + \langle g(t), v\rangle \quad \forall v \in \mathcal{V}, \tag{3.43}
$$

with the Cauchy data:

$$
u(0) = u^0 \qquad u'(0) = u^1. \tag{3.44}
$$

For a study of topics such as existence and uniqueness see for example Miller (1978) or Dafermos (1970). We discretize in space in the same way as described for the quasistatic problem in Section 3.1 and obtain the semidiscrete approximating problem: find $u^h(x, t) \in C^2(\mathcal{I}; \mathcal{V}^h)$ such that

$$MU''(t) + AU(t) + \int_0^t B(t - s)U(s)\, ds = F(t), \qquad (3.45)$$

with appropriate Cauchy data u^{0h} and u^{1h}. Here $U(t)$ is a time dependent array of coordinates in $\mathcal{V}^h$ and $F(t)$ is derived from the body forces and surface tractions. The *mass matrix*, M, is derived by assembly of the element mass matrices, M_i, which are determined from

$$M_i = \rho \int_{\Omega_i} \left(\Phi_i^T \Phi_i\right)(x)\, d\Omega_i, \qquad 1 \le i \le N_E, \qquad (3.46)$$

where, in two space dimensions, for element ℓ possessing N_ℓ nodes,

$$\Phi_\ell(x) := \begin{pmatrix} \bar{\varphi}_{\ell_1} & 0 & \cdots & \bar{\varphi}_{\ell_{N_\ell}} & 0 \\ 0 & \bar{\varphi}_{\ell_1} & \cdots & 0 & \bar{\varphi}_{\ell_{N_\ell}} \end{pmatrix}(x), \qquad (3.47)$$

with the $\{\ell_i\}_{i=1}^{N_\ell}$ indicating the nodes on element Ω_ℓ for $1 \le \ell \le N_E$ and the $\bar{\varphi}$ indicating the restriction of the basis functions to Ω_ℓ – in practice such an integral would be carried out on a reference element using the standard polynomial basis functions. The matrices A and $B(t - s)$ are essentially identical with the stiffness matrix of linear elasticity except that the Lamé coefficients λ and μ are replaced by the respective viscoelastic functions $\lambda(0)$, $\mu(0)$ in A and $-\lambda_s(t - s)$, $-\mu_s(t - s)$ in $B(t - s)$.

Discretization into the time domain is accomplished by first defining the discrete time interval $\mathcal{I}_j^k$ as

$$\mathcal{I}_j^k := \{t_i = ik \in \mathcal{I} : j \le i \le N - 1\}, \qquad 0 \le j \le N - 1, \qquad (3.48)$$

for a constant time step k and an integer $T/k = N \ge 2$, and then setting up the following notation for a function w defined on $\mathcal{I}_0^k \cup \{T\}$:

$$\begin{aligned} w_i &:= w(t_i), & 0 \le i \le N, \\ \partial_t w_i &:= \frac{w_i - w_{i-1}}{k}, & 1 \le i \le N, \\ \partial_t^2 w_{i+1} &:= \frac{w_{i+1} - 2w_i + w_{i-1}}{k^2}, & 1 \le i \le N - 1, \\ \Delta w_i &:= \theta_b w_{i-1} + (1 - \theta_b - \theta_f)w_i + \theta_f w_{i+1}, & 1 \le i \le N - 1 \\ \Delta w_0 &:= w_0, \end{aligned} \qquad (3.49)$$

for parameters $\theta_b, \theta_f \in [0, 1/2]$. Finally we replace the integral in (3.45) with the trapezoidal rule. The fully discrete approximation to the solution of (3.43) is defined to be a sequence $\left\{\left(u^{hk}(x)\right)_i\right\}_{i=0}^N$, $(u^{hk})_i \in \mathcal{V}^h$ for each i, with $(u^{hk})_0$ and $(u^{hk})_1$ obtained from the Cauchy data, such that for $t_i \in \mathcal{I}_1^k$

$$M\partial_t^2 U^{i+1} + A\Delta U^i + \sum_{j=0}^i \varpi_{ij} B(t_i - t_j)\Delta U^j = \Delta F_i, \qquad (3.50)$$

where $U^i \approx U(t_i)$ and the ϖ_{ij} are the weights associated with the trapezoidal rule.

In Shaw, Warby and Whiteman (1993(ii)) we have analyzed this discrete scheme, in the case $\theta_b = \theta_f = 1/4$, for a prototype of (3.43) and found that, providing the initial conditions are suitably approximated and the exact solution is smooth enough, we have the following error bound

$$\max_{1 \leq m \leq N-1} \left\{ \left\| u'_{m+1/2} - \partial_t(u^{hk})_{m+1} \right\|_0 + h \left\| u_{m+1/2} - \frac{(u^{hk})_{m+1} + (u^{hk})_m}{2} \right\|_1 \right\}$$
$$\leq C(h^{r+1} + k^2),$$
$$(3.51)$$

where r is the degree of the piecewise polynomials comprising $\mathcal{V}^h$. In terms of approximating the initial conditions the analysis suggests that it is not sufficient that $(u^{hk})_0$ and $(u^{hk})_1$ be "close" to $u(0)$ and $u(k)$, respectively, but rather they need to be close to the Ritz-Volterra projections of $u(0)$ and $u(k)$ into $\mathcal{V}^h$, see Cannon and Lin (1990) or Lin et. al. (1991); this is a natural generalization of the Ritz projection, Wheeler (1973), to problems with memory. Now, working with the coordinates in $\mathcal{V}^h$ for simplicity, the implication is that to obtain U^0 we must solve an elliptic problem and to obtain U^1 we must solve an elliptic-Volterra problem. From a computational point of view this is a rather cumbersome task. A much simpler approach is to set $U(0) = u^0$ and $U'(0) = u^1$ at the nodes of Ω and then obtain starting values for the discrete scheme via:

$$U^0 = U(0),$$
$$U^1 = U(0) + kU'(0) + \frac{k^2}{2}U''(0), \qquad (3.52)$$

where the second derivative is obtained from the differential equation as

$$U''(0) = M^{-1}\big(F(0) - AU(0)\big). \qquad (3.53)$$

This is the approach taken in the numerical experiment detailed below and the results clearly indicate that the expected rate of convergence is achieved.

To examine the numerical performance of (3.50) we could use the 1-D solution given earlier. However a much more general test of the algorithm is provided by using the following, physically meaningless, exact solution to (3.2). We set the material parameters to be:

$$\begin{aligned}
\lambda(t) &:= 3.7 + 2.34\exp(-0.34t) + 1.4\exp(-2.0t),\\
\mu(t) &:= 1.1 + 1.9\exp(-3.0t), \qquad\qquad\qquad (3.54)\\
\rho &:= 11,
\end{aligned}$$

and specify the body forces and surface tractions to be such that the displacement components are given by

$$
\begin{aligned}
u_1(\boldsymbol{x}, t) &= 3\sin\left(x_1 + \frac{1}{2}\left(x_2 - \frac{1}{2}\right)\right)\sin\left(\frac{\pi}{4}t\right), \\
u_2(\boldsymbol{x}, t) &= 5\sin\left(2x_1\left(x_2 - \frac{1}{2}\right)\right)\sin\left(\frac{\pi}{2}t\right).
\end{aligned}
\tag{3.55}
$$

We define $\Omega := (0,1)^2$ and take the discrete time interval as

$$
\mathcal{I}_2^k := \{2k, 3k, \ldots, mk : \quad m \text{ is such that } (m-1)k \le 20 \le mk\},
\tag{3.56}
$$

that is, although we shall subsequently use a variety of different time steps we have chosen a basic value of T in (3.2) as 20. For the finite element discretization we choose 8 noded quadrilateral elements which have shape functions (on the reference element) of the form $\sum_{ij}\alpha_{ij}x_1^i x_2^j$ with $\alpha_{22} \equiv 0$. Traction boundary conditions are specified over the entire boundary with the displacements $u_1\big((0,1/2), t\big)$, $u_2\big((0,1/2), t\big)$ and $u_2\big((1,1/2), t\big)$ held at each time step to prevent rigid body rotation. The errors resulting from a series of computations for various values of k using 1, 2×2, 4×4 and 8×8 elements (corresponding to $h = 1$, $1/2$, $1/4$ and $1/8$) are shown for both displacement and stress in Tables 3.1 and 3.2. The errors are measured in the discrete norms:

$$
\begin{aligned}
\|\boldsymbol{u} - \boldsymbol{u}^{hk}\| &:= \max_{t \in \mathcal{I}_2^k} \sqrt{\sum_{i=1}^{2} \|u_i - u_i^{hk}\|_0^2}, \\
\|\boldsymbol{\sigma} - \boldsymbol{\sigma}^{hk}\| &:= \max_{t \in \mathcal{I}_2^k} \sqrt{\|\sigma_{12} - \sigma_{12}^{hk}\|_0^2 + \sum_{i=1}^{2} \|\sigma_{ii} - \sigma_{ii}^{hk}\|_0^2},
\end{aligned}
\tag{3.57}
$$

that is, we do not explicitly examine the initial conditions. The norm $\|\cdot\|_0$ is calculated with an order $O(h^4)$ accurate quadrature rule.

Table 3.1 Displacement errors for $\theta_b = \theta_f = 1/4$, to 6 d.p.

mesh	time step, k					
size, h	0.8	0.4	0.2	0.1	0.05	0.025
1	1.824351	.540676	.344133	.359084	.365779	.367625
1/2	1.893735	.553656	.127908	.029442	.021661	.023687
1/4	1.905864	.556537	.138251	.033750	.007907	.002020
1/8	1.979547	.555757	.139076	.034630	.008600	.002109

The final columns of the tables illustrate, to varying degrees, the order $O(h^3)$ convergence of the displacements and order $O(h^2)$ convergence of the stresses whilst the bottom rows demonstrate the order $O(k^2)$ convergence for both quantities. Also, the diagonals of Table 3.2 show the order $O(h^2 + k^2)$ convergence of the stress. Thus,

Table 3.2 Stress errors for $\theta_b = \theta_f = 1/4$, to 5 d.p.

mesh	time step, k					
size, h	0.8	0.4	0.2	0.1	0.05	0.025
1	20.07950	6.63703	4.93868	4.98360	5.02207	5.03131
1/2	26.47485	4.30625	1.24920	1.08880	1.10123	1.10595
1/4	37.03759	5.42741	1.16148	.30793	.26503	.26465
1/8	49.60644	7.16702	1.17208	.27612	.08216	.07250

even for our very straightforward approximation of the initial conditions the theoretical convergence rates are achieved.

Finally we close this subsection with a simple numerical demonstration of the damping effect that the history integral has on the governing equations, we shall, qualitatively, contrast the response of a material possessing memory with that of a linear elastic material. For this we consider a 2-D beam, that is we let

$$\Omega := \{0 < x_1 < 1\} \times \{0 < x_2 < 0.4\}, \tag{3.58}$$

be the region of the beam which is fixed at its ends such that $\forall x_2 \in [0, 0.4]$

$$u_1\big((0, x_2), t\big) = u_1\big((1, x_2), t\big) = u_2\big((0, x_2), t\big) = u_2\big((1, x_2), t\big) = 0, \tag{3.59}$$

with zero tractions prescribed elsewhere and we assume the initial conditions

$$\begin{aligned} \boldsymbol{u}^0 &= (0, 0.01 \sin \pi x)^T \\ \boldsymbol{u}^1 &= (0, 0)^T. \end{aligned} \tag{3.60}$$

Now, for demonstrative purposes we base our material parameters on those for concrete given in Section 3.1, but in order to present clear results it is necessary to scale the material density by a factor 10^4, that is, $\rho = 2650 \times 10^4 \text{kg/m}^3$. Again we assume plane stress and set, initially,

$$\phi(t) = 677\,823 + 877\,289 \exp(-0.01t) + 151\,989 \exp(-0.001t) \tag{3.61}$$

so as to emulate a viscoelastic material, and record the evolution of the numerical approximation to the displacement $u_2\big((0.5, 0.4), t\big)$. Now, degenerating (3.61) to the linear elastic case, that is,

$$\phi(t) = 677\,823 + 877\,289 + 151\,989, \tag{3.62}$$

we again record the vertical displacement and plot these two responses in Figures 3.6 and 3.7. We can see immediately that in the elastic case the response takes the form of a standing wave which oscillates "forever" whilst in the viscoelastic case the standing wave is initially identical with the elastic wave but after some time the memory begins to damp the response - eventually to zero. If we had not scaled the density of the concrete the period of oscillation would be much smaller as compared with the time taken for the viscoelastic response to decay, and the curve in Figure 3.6 would not have shown this oscillation clearly. This fact that the viscoelastic response operates

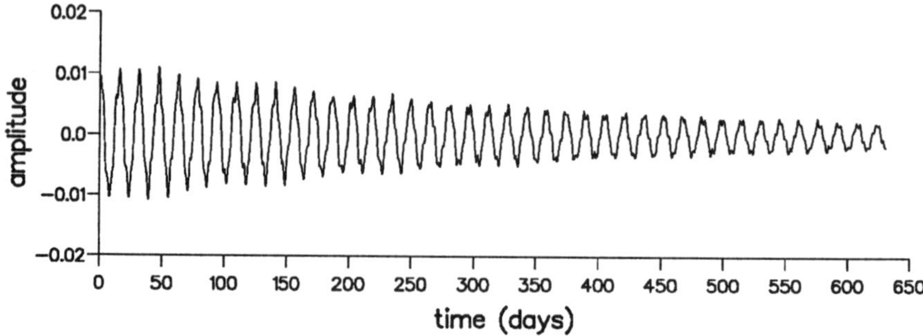

Figure 3.6 Qualitative viscoelastic response, time step $= 1/2$ day, with weights $\theta_b = \theta_f = 1/4$.

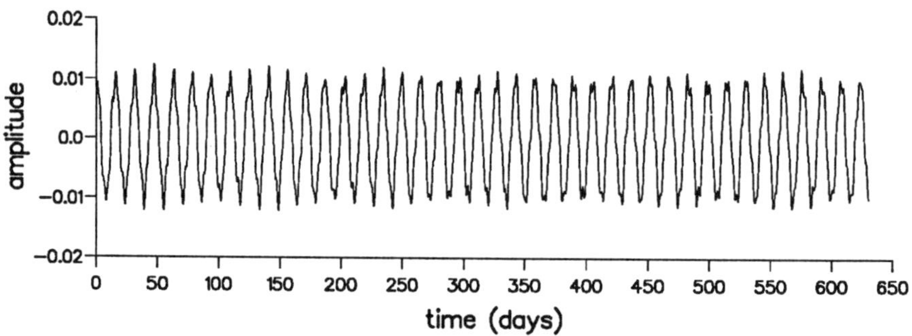

Figure 3.7 Qualitative elastic response, time step $= 1/2$ day, with weights $\theta_b = \theta_f = 1/4$.

over very long time scales could mean that in reality the viscoelastic decay would be "swamped" by other effects such as atmospheric drag, and since no material is perfectly elastic (some degree of internal friction must always exist) it may be that, in practice, the viscoelastic response of such an oscillating beam would be indistinguishable from the corresponding response of a similar elastic beam.

The formulation of the viscoelastic problem just presented utilises the first equation of (3.4); in closing we remark that choosing the second of (3.4) instead allows a very different interpretation of the governing differential equation. In particular, it is not necessary to consider the displacements at all, it is the *velocities* which become the primary "unknowns". Specifically we set $w(x,t) := u'(x,t)$ in the governing equation (3.2) and the second of (3.4), construct a weak formulation and carry out the finite element approximation in the usual manner, this yields a semidiscrete problem of the form

$$MW'(t) + \int_0^t \tilde{B}(t-s)W(s)\,ds = F(t) - \tilde{A}(t)U(0) \tag{3.63}$$

with $U(0)$ and $W(0)$ obtained from u^0 and u^1. A reasonable method of time discretization would now seem to be a finite difference replacement for the derivative and a trapezoidal replacement for the of the integral coupled with a Crank-Nicolson type of averaging at two consecutive time levels. Unfortunately we have, at the present time, no analysis or numerical results to indicate the validity of this approach but the potential value is clear: the presence of discrete first derivatives only would allow for variable length time steps and since the Crank-Nicolson averaging is symmetric we might reasonably expect $O(k^2)$ convergence for the time discretization. In such a case this *parabolic* formulation would be superior to the formulation (3.50) which requires a constant time step in order to preserve accuracy.

3.4 CONCLUSIONS

Finite element methods in space together with discrete time approximations have been applied to problems of linear viscoelasticity using various constitutive relations. In the linear case for both quasistatic and dynamic problems, theoretical error estimates have been derived which are valid under suitable regularity assumptions. In the quasistatic case numerical results obtained indicate that the constitutive relation is appropriate to concrete whilst it is only suitable for a polymeric material such as Nylon 66 for either limited ranges of load or short times. This last observation has demonstrated the need for nonlinear constitutive relations which are suited to implementation in numerical algorithms. To this end the numerical experiments of Section 3.2 indicate the possible applicability of a simplified Schapery model in this context.

ACKNOWLEDGEMENT

The work of Whiteman was supported, in part, by the United States Army Research Standardization Group, Europe, under contract No. DAJA45–89–C–003, and that of Shaw by the Science and Engineering Research Council. This support is gratefully acknowledged. Also the authors would like to express their gratitude to the Brunel University computer center, the facilities of which were used to reproduce substantial parts of this paper and to R.A. Schapery, J.R. Walton and M.F. Wheeler for many stimulating and helpful conversations over a number of years.

REFERENCES

Cannon, J.R. and Lin Y., (1990), A Priori $\mathcal{L}_2$ Error Estimates for Finite-Element Methods for Nonlinear Diffusion Equations with Memory. *SIAM J. Numer. Anal.*, **27**: 595—607.

Cessna (jr), L.C., (1971), Stress-Time Superposition of Creep Data for Polypropylene and Coupled Glass-Reinforced Polypropylene. *Polymer Engineering and Science*, **11**: 211—219.

Christensen, R.M., (1971), Theory of Viscoelasticity, An Introduction. *Academic Press, London.*

Dafermos, C.M., (1970), An Abstract Volterra Equation with Applications to Linear Viscoelasticity. *J. Diff. Eqns.*, **7**: 554—569.

Ferry, J.D., (1980), Viscoelastic Properties of Polymers. *John Wiley. New York.*

Golden, J.M. and Graham G.A.C., (1988), Boundary Value Problems in Linear Viscoelasticity. *Springer-Verlag, Berlin.*

Gurtin, M.E. and Sternberg E., (1962), On the Linear Theory of Viscoelasticity. *Arch. Rat. Mech. Anal.*, **11**: 291—356.

Hunter, S.C., (1983), Mechanics of Continuous Media, Ellis Horwood series, Mathematics and its Applications, second edition, *Ellis Horwood Ltd., Chichester.*

Knauss, W.G. and Emri I.J., (1981), Nonlinear Viscoelasticity Based on Free Volume Considerations. *Computers and Structures,* **13**: 123—128.

Knauss, W.G. and Emri I.J., (1987), Volume Change and the Nonlinearly Thermoviscoelastic Constitution of Polymers. *Polymer Engineering and Science,* **27**: 86—100.

Kolsky, H., (1956), The Propagation of Stress Pulses in Viscoelastic solids. *Phil. Mag. eighth series,* **1**: 693—710.

Lin, Y., Thomée V. and Wahlbin L.B., (1991), Ritz-Volterra Projections to Finite Element Spaces and Applications to Integrodifferential and Related Equations. *SIAM J. Numer. Anal.,* **28**: 1047—1070.

Lockett, F.J., (1972), Nonlinear Viscoelastic Solids. *Academic Press, London.*

Mainardi, F. (editor), (1982), Wave Propagation in Viscoelastic Media, Research notes in mathematics, Pitman advanced publishing program, *Pitman, London.*

Maranyl, Nylon 66, ——. Technical service note N109, Moulding Powders Group, I.C.I. Plastics Division, Welwyn Garden City, Herts, U.K.

Miller, R.K., (1978), An Integrodifferential Equation for Rigid Heat Conductors With Memory. *J. Math. Anal. Appl.,* **66**: 313—332.

Schapery, R.A., (1966), A Theory of Nonlinear Thermoviscoelasticity Based on Irreversible Thermodynamics, Proceedings 5*th* U.S. Nat. Cong. Appl. Mech., ASME, **111**.

Schapery, R.A., (1969(i)), Further Development of a Thermodynamic Constitutive Theory: Stress Formulation. *Purdue University, report No. AA&ES 69-2.*

Schapery, R.A., (1969(ii)), On the Characterization of Nonlinear Viscoelastic Materials. *Polymer Engineering and Science.* **9**: 295—310.

Schapery, R.A., (1992), On Nonlinear Viscoelastic Constitutive Equations for Composite Materials. *Proceedings of the VII International Congress on Experimental Mechanics, Las Vegas.*

Shaw, S., Warby M.K. and Whiteman J.R., (1992), Finite Element Methods with Recovery for Problems of Viscoelasticity. *Numerical Methods in Engineering and the Applied Sciences.* H. Alder, J.C. Heinrich, S. Lavanchy, E. Onate and B. Suarez (editors), CIMNE, Barcelona.

Shaw, S., Warby M.K., Whiteman J.R., Dawson C. and Wheeler M.F., (1993(i)), Numerical Techniques for the Treatment of Quasistatic Solid Viscoelastic Stress Problems. *BICOM tech. rep. 93/2, Brunel University, England.* (Also submitted for publication to *Computational Methods in Applied Mechanics and Engineering.*)

Shaw, S., Warby M.K. and Whiteman J.R., (1993(ii)), An Error Bound via the Ritz-Volterra Projection for a Fully Discrete Approximation to a Hyperbolic Integral Equation. *BICOM tech. rep. 93/4, Brunel University, England.*

Troxell, G.E., Davis H.E. and Kelly J.W., (1968), Composition and Properties of Concrete, McGraw Hill Civil Engineering series, second edition. *McGraw Hill, New York.*

Wheeler, M.F., (1973), A Priori $\mathcal{L}_2$ Error Estimates for Galerkin Approximations to Parabolic Partial Differential Equations. *SIAM J. Numer. Anal.,* **4**: 723—759.

Zienkiewicz, O.C. and Watson M., (1966), Some Creep Effects in Stress Analysis with Particular Reference to Concrete Pressure Vessels. *Nuclear Engineering and Design,* **4**: 406—412.

Chapter 4

Numerical Simulation of Non-Newtonian Flow through an Axisymmetric Contraction

A. Baloch, P. Townsend and M.F. Webster

University of Wales Institute of non-Newtonian Fluid Mechanics,
Department of Computer Science,
University College of Swansea, SA2 8PP, U.K.

4.1 INTRODUCTION

This article investigates flow phenomena attributed to non-Newtonian fluids on encountering complex changes in the flow domain. Consideration is given to the simulation of axisymmetric flows through abrupt and rounded-corner contraction geometries at a contraction ratio of four to one. Of particular interest are the effects on the flow of change in rheological material properties and impact of relatively small changes in re-entrant corner characteristics. In the simulations, use is made of a special class of fluids which have a constant shear viscosity but exhibit strain thickening, that is, the viscosity of such a fluid is an increasing function of extension rate. It is found that the flow patterns observed for an abrupt corner can be quite different from those in which the corner has been rounded.

Contraction flows of viscoelastic materials have attracted much interest in the literature. Such flows occur in many industrial processes, and it is important to try to predict the precise form such flows might take, and in particular, the nature of any recirculation that might be present. Much of the literature is concerned with computer simulations which make use of constitutive models designed to represent both viscous and elastic, and sometimes, shear-thinning behaviour. In such flows one of the dominant features which occurs is the rapid extension of the fluid as it enters the contraction. In the present paper, we investigate the use of a model which attempts to represent the strain thickening which would occur under such a rapid extension. In pure shear the viscosity is constant and normal stresses are identically zero. Such a model has already been used in a similar context by Debbaut and Crochet (1988).

The particular behaviour of interest in this flow is the apparent appearance in experimental studies of a lip vortex near the entrance of the contraction, under certain flow conditions and fluid parameters. It is of interest to investigate whether such an effect is an elastic effect or a consequence of strain thickening. The importance of the severe flow conditions near the re-entrant corner of the contraction and their effect on the dynamic structure of the flow are investigated by changing the sharp corner to one of smooth curvature, and observing the consequential changes in the flow behaviour.

The Mathematics of Finite Elements and Applications
Edited by J. R. Whiteman © 1994 John Wiley & Sons Ltd

4.2   GOVERNING EQUATIONS

The incompressible flow of non-Newtonian inelastic materials is governed by equations for conservation of mass and momentum transport, which may be stated as

$$\nabla \cdot \mathbf{v} = 0 \tag{4.1}$$

$$\rho \frac{\partial \mathbf{v}}{\partial t} = \nabla \cdot \{2\mu(\dot\varepsilon)\mathbf{D}\} - \rho \mathbf{v} \cdot \nabla \mathbf{v} - \nabla p \tag{4.2}$$

where $\mathbf{v}$ is the velocity vector, p is the isotropic pressure, ρ is the fluid density, $\mathbf{D}$ is the rate of deformation tensor. Following Debbaut and Crochet (1988), we take the fluid viscosity μ to be a function of extension rate $\dot\varepsilon$, defined as follows:

$$\mu(\dot\varepsilon) = \mu_0 \cosh(m\lambda\dot\varepsilon)$$

where μ_0 is a constant viscosity, m is a material constant, and λ is a time constant. Here we may write $\dot\varepsilon = 3I_3/I_2$, where $I_2 = \frac{1}{2}\mathrm{trace}(\mathbf{D}^2)$ and $I_3 = \det(\mathbf{D})$.

It should be noted that this model is useful to try to identify extensional effects in axisymmetric geometries, but does not exhibit any special effects in plane flows, where I_3 is identically zero. Such a model is not therefore suitable for use in general flow situations.

4.3   NUMERICAL ALGORITHM

To simulate these flow problems, analytical methods are out of the question and we turn instead to numerical methods, and in particular, a time-marching primitive variable Taylor-Galerkin/pressure-correction scheme. A full description of this numerical scheme is given in Townsend and Webster (1987) and Hawken, Tamaddon-Jahromi, Townsend and Webster (1990). The derivation follows firstly a temporal discretisation through a Taylor series expansion, and subsequently, a Galerkin spatial discretisation (see Donea (1984)). An operator splitting technique of pressure-correction type generates multiple stages to the solution process and satisfies the incompressibility constraint, for more details see Peyret and Taylor (1983). This leads to a second order semi-implicit time-marching scheme similar to a two-step Lax and Wendroff (1960) method, but applied in a finite element context. Here a predicator-corrector pair of equations are employed where the diffusion terms are treated implicitly in a Crank-Nicolson manner, whilst convection terms are treated explicitly. Such a scheme has the advantage of being computationally tractable for large scale problems in that it is predominantly vector-based, and does not require the storage of large matrices, as is the case with many fully implicit coupled schemes. Initially at a first fractional stage a nondivergence-free velocity field is computed using, for second order time accuracy, two half time-steps. At a first half time step $(n + 1/2)$ we compute velocity from data at time level (n), while in a second half time step we advance to calculate an intermediate non-divergence-free velocity field, using the solutions at (n) and $(n+1/2)$ time steps. A Poisson equation for the pressure difference emerges at a second stage, the solution of which determines a divergence-free velocity field at a third and final stage to complete a time step cycle. For the finite element

approximation we employ a mixed Galerkin approximation in the weak formulation
of the problem, adopting piecewise quadratic shape functions for velocity components
and piecewise linear shape functions for pressure over a triangular subdivision of the
spatial domain. The equations at stages one and three are governed by augmented
mass matrices which are solved iteratively, while at stage two, a pressure stiffness
matrix equation is solved through a direct Choleski method.

To characterise the flow simulations we introduce non-dimensional variables and
scales as $\mathbf{x}(r,z) = \mathbf{x}^*(r,z)R$; $\mathbf{v} = V\mathbf{v}^*$; $p = \rho V^2 p^*$; $t = \frac{R}{V}t^*$, with non-dimensional
numbers Re and We defined as

$$Re = \frac{\rho RV}{\mu_0}, \quad We = \lambda \frac{V}{R} \tag{4.3}$$

where V and R are, respectively, characteristic velocity and length scales, taken to be
the maximum, fully developed velocity in the smaller tube, and the radius of that tube.
We follow the notation used by Debbaut and Crochet (1988), although, as pointed out
in that paper, the implication that We is a Weissenberg number is inappropriate here,
in the absence of memory effects. It is also useful to define an elasticity number E
given by

$$E = \frac{We}{Re} = \frac{\lambda \mu_0}{\rho R^2}. \tag{4.4}$$

E is a particularly appropriate parameter for comparison with experimental results,
where, usually, observations are recorded for a sequence of increasing values of the
flowrate. In such circumstances, E is constant and Re is directly proportional to the
flowrate.

Employing the derivation above the non-dimensional system of equations in fully
discrete form becomes:

Stage 1a:

$$[\frac{2}{\Delta t}\mathbf{M} + \frac{1}{2}\mathbf{S}](\mathbf{U}^{n+\frac{1}{2}} - \mathbf{U}^n) = [-\frac{1}{Re}\mathbf{S}\mathbf{U} - \mathbf{N}(\mathbf{U})\mathbf{U} + \mathbf{L}^\dagger\mathbf{P}]^n + \frac{1}{2}(\mathbf{F}^{n+\frac{1}{2}} + \mathbf{F}^n)$$

Stage 1b:

$$[\frac{1}{\Delta t}\mathbf{M} + \frac{1}{2}\mathbf{S}](\mathbf{U}^* - \mathbf{U}^n) = [-\frac{1}{Re}\mathbf{S}\mathbf{U} + \mathbf{L}^\dagger\mathbf{P}]^n - [\mathbf{N}(\mathbf{U})\mathbf{U}]^{n+\frac{1}{2}} + \frac{1}{2}(\mathbf{F}^{n+1} + \mathbf{F}^n)$$

Stage 2:

$$\theta\mathbf{K}\mathbf{Q}^{n+1} = -\frac{1}{\Delta t}\mathbf{L}\mathbf{U}^*$$

Stage 3:

$$\frac{1}{\Delta t}\mathbf{M}(\mathbf{U}^{n+1} - \mathbf{U}^*) = \theta\mathbf{L}^\dagger\mathbf{Q}^{n+1}$$

where $\mathbf{U}^n$ and $\mathbf{P}^n$ are nodal velocity and pressure vectors at time t^n respectively,
$\mathbf{Q}^{n+1} = \mathbf{P}^{n+1} - \mathbf{P}^n$, $\mathbf{U}^*$ is an intermediate non-divergence-free velocity vector, $\mathbf{M}$ is
a mass matrix, $\mathbf{N}(\mathbf{U})$ is a convection matrix, $\mathbf{K}$ is a pressure stiffness matrix, $\mathbf{L}$ is
a divergence/pressure gradient matrix and $\mathbf{S}$ is a momentum diffusion matrix. The

system matrices are defined as

$$\mathbf{M}_{ij} = \int_{\Omega} \phi_i \phi_j r d\Omega$$

$$\mathbf{N(U)}_{ij} = \int_{\Omega} \phi_i (U_l^k \phi_l \frac{\partial \phi_j}{\partial x_k}) r d\Omega$$

$$\mathbf{K}_{kl} = \int_{\Omega} \nabla \psi_k \nabla \psi_l r d\Omega$$

$$(\mathbf{L}_k)_{li} = \int_{\Omega} \psi_l \nabla^k \phi_i r d\Omega$$

$$\mathbf{S}_{ij} = \left[\begin{array}{cc} \mathbf{S}_{rr} & \mathbf{S}_{rz} \\ \mathbf{S}_{rz}^{\dagger} & \mathbf{S}_{zz} \end{array} \right]_{ij}$$

where

$$(\mathbf{S}_{rr})_{ij} = \int_{\Omega} \cosh(mWe\dot{\varepsilon})[2\frac{\partial \phi_i}{\partial r}\frac{\partial \phi_j}{\partial r} + \frac{\partial \phi_i}{\partial z}\frac{\partial \phi_j}{\partial z} + 2\frac{\phi_i \phi_j}{r^2}]r d\Omega$$

$$(\mathbf{S}_{rz})_{ij} = \int_{\Omega} \cosh(mWe\dot{\varepsilon})[\frac{\partial \phi_i}{\partial r}\frac{\partial \phi_j}{\partial z}]r d\Omega$$

$$(\mathbf{S}_{zz})_{ij} = \int_{\Omega} \cosh(mWe\dot{\varepsilon})[\frac{\partial \phi_i}{\partial r}\frac{\partial \phi_j}{\partial r} + 2\frac{\partial \phi_i}{\partial z}\frac{\partial \phi_j}{\partial z}]r d\Omega$$

Here $\phi_i(x)$, $\psi_k(x)$ are piecewise quadratic and linear shape functions respectively, and r is the radial dimension. Repeated indices imply summation taken over i, j for all velocity nodal points, and k, l for all vertex pressure nodal points on the triangular meshes. The superscript $\dagger$ indicates the transpose of the matrix and $\mathbf{F}^n$ is a forcing function vector due to boundary conditions at time t^n, which vanishes here. To give the precise second order form of the pressure-correction algorithm the Crank-Nicolson coefficient θ is taken as one half.

4.4   PROBLEM SPECIFICATION

Experimental evidence (cf. Boger and Binnington (1990)) indicates that symmetrical flows may be considered and so we compute solutions on only the upper half geometry. The simulation starts from quiescent initial conditions for any particular elasticity value and flowrate. Also a continuation process through prior steady-state flowrate solutions may be employed as artificial start-up conditions for higher flowrates to accelerate the convergence to a new steady-state. A fully developed parabolic velocity profile is imposed at inlet, at outlet a traction free normal condition is enforced consistent with a fixed pressure datum, and vanishing tangential traction and radial velocity are imposed on the line of symmetry. No-slip is assumed on the bounding channel walls. Appropriate boundary conditions for the radial and axial velocity components, v_r and v_z, are shown in the schematic diagram, presented in Figure 4.1a.

The results presented here are for simulations of flow through two 4:1 contraction geometries, as shown in Figure 4.1. The first has an abrupt re-entrant corner, whereas

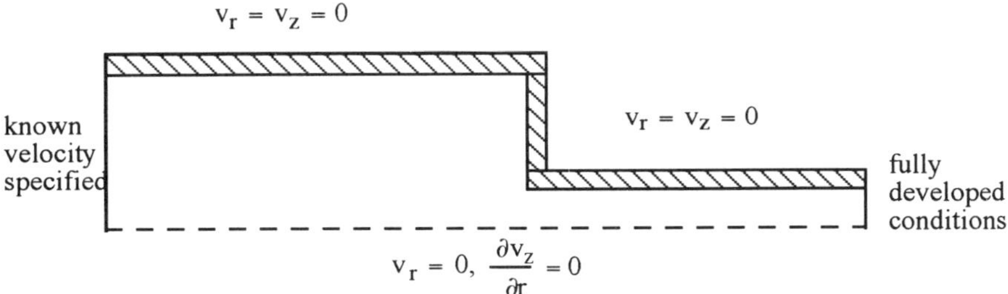

Figure 4.1 (a): Schematic diagram 4:1 circular contraction and boundary conditions.

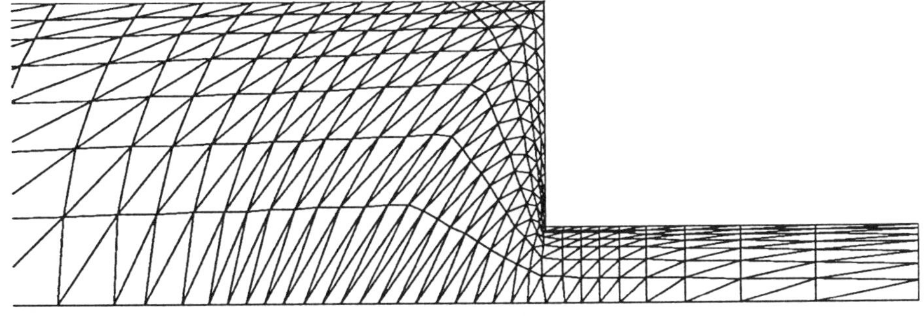

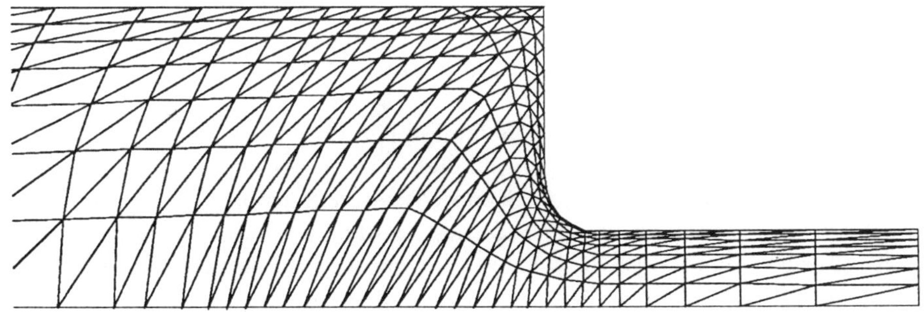

Figure 4.1 (b): Finite element meshes for abrupt(above) and rounded-corner(below) circular contractions.

in the second, the corner has been rounded to form a smooth entrance to the contraction by fitting a circle quadrant with radius 0.72 of the narrow tube radius. One expects the fluid to experience a much more rapid extension near the corner in the first case compared with that of the second. In order to avoid the influence of inlet and outlet boundary conditions a sufficiently large upstream length of $30R$ and downstream length of $50R$ have been chosen. In the present investigation for both abrupt and rounded-corner meshes the total number of elements, nodes and degrees of freedoms are 713, 1542 and 3499 respectively. Illustrations of typical meshes employed are displayed in Figure 4.1b.

4.5 RESULTS

The simulations carried out show that, for low elasticity of the order of E=0.1, as the flowrate is increased over the range $Re = 0.01$ to 0.15, an initial salient corner vortex grows and strengthens. This growth is almost negligible in the case of the rounded corner, but very significant for the sharp corner case. In the latter, although the size of the vortex in the axial direction remains essentially the same, the vortex now spreads across the entire salient corner, right up to the lip of the contraction. At higher values of the elasticity, of order 0.2, at a Reynolds number of about unity, the vortex splits into two separate vortices, namely the original corner vortex and a so-called lip vortex near the upstream portion of the re-entrant corner. As the flowrate is increased, the salient corner vortex decays and disappears, being dominated by the lip vortex which grows to fill the whole of the salient corner. The latter stages of the process are illustrated in Figure 4.2 (on the left).

Of interest is the equivalent development for the rounded corner, also shown in Figure 4.2 (on the right). Initially there appears to be very little activity in this case, and for the range of parameters given above, the salient corner vortex is small and almost static. If, however, one increases the value of E, equivalent in the experiments to using a more strain thickening material, or alternatively, if one increases the Reynolds number, that is, the flowrate, one eventually reaches a level where the flow pattern development described for the sharp corner is repeated for the rounded case. The transition stage is apparent in Figure 4.2, illustrating the emergence of separate lip and salient corner vortices.

It is clear that the rounding of the contraction corner, an apparently minor change in the flow geometry, leads to significant changes in flow patterns. The fluid entering the abrupt contraction experiences much more severe extension near the corner than in the rounded case, and this leads to the formation of a lip vortex. For higher flowrates or more strain thickening materials, the damping effect of the smoothness of the corner is eventually insufficient to prevent the formation of the lip vortex observed earlier for the sharp corner.

4.6 CONCLUSIONS

The effects of strain thickening have been investigated on the flow patterns observed in axisymmetric contractions. High extension near an abrupt corner leads to the

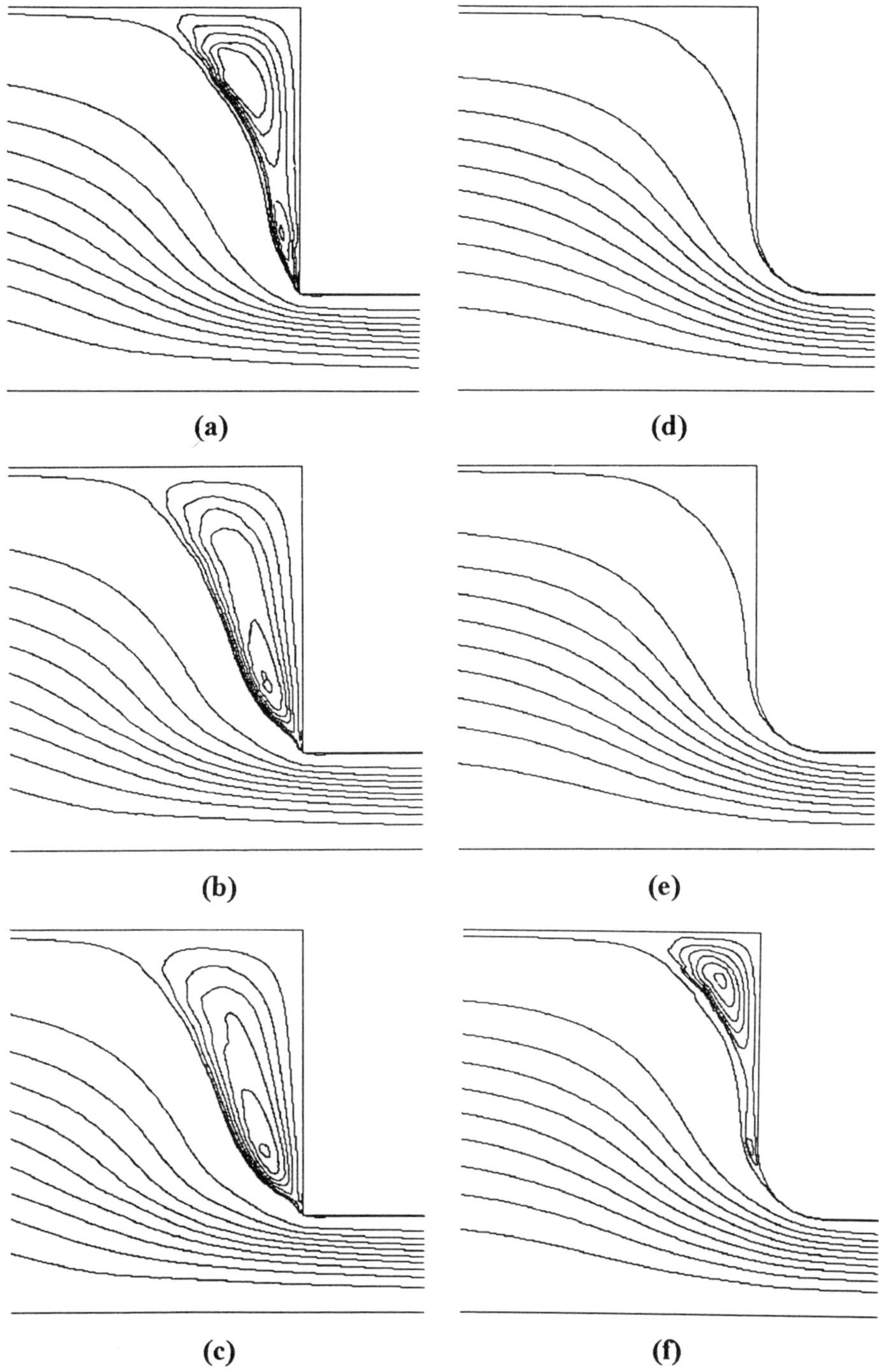

Figure 4.2 Streamline projections for $E = 0.2$ with $Re = 1.8$, (a) and (d); $Re = 2.4$, (b) and (e); and $Re = 3$, (c) and (f).

formation of a lip vortex which grows and dominates the salient corner vortex. The latter decays and disappears. This mechanism is initially suppressed by rounding the contraction corner, but only until the product of the elasticity and Reynolds number reaches sufficiently high values, when the same flow development is then observed.

REFERENCES

Boger D.V. and Binnington R.J. (1990), Circular entry flows of fluid M1. *J. Non-Newtonian Fluid Mech.*, **35**: 339–360.

Debbaut B. and Crochet M.J. (1988), Extensional effects in complex flows. *J. Non-Newtonian Fluid Mech.*, **30**: 169–184.

Donea J. (1984), A Taylor-Galerkin method for convective transport problems. *Int. J. Numer. Meth. Eng.*, **20**: 101–119.

Hawken D.M., Tamaddon-Jahromi H.R. ,Townsend P. and Webster M.F. (1990), A Taylor-Galerkin-based algorithm for viscous incompressible flow. *Int. Journal Num. Meth. Fluids*, **10**: 327–351.

Lax P.D. and Wendroff B. (1960), System of conservation laws. *Comm. Pure Appl. Math.*, **13**: 217-235.

Peyret R. and Taylor T.D. (1983), *Computational Methods for Fluid Flow.* Springer-Verlag, New York.

Townsend P. and Webster M.F. (1987), An algorithm for the three-dimensional transient simulation of non-Newtonian fluid flows. In G.N. Pande and J. Middleton, editors, *Proc. Int. Conf. Num. Meth. Eng.: Theory and Applications*, II: T12/1–11, Swansea. NUMETA, Nijhoff, Dordrecht.

Chapter 5

Approximate Approximations

V. Maz'ya

Department of Mathematics, Linköping University,
S-581 83 Linköping, Sweden

5.1 INTRODUCTION

The starting idea is to represent an arbitrary function as a linear combination of basis functions, which, in contrast to splines, form an approximate partition of unity. The approximations obtained do not converge as the mesh size tends to zero. The lack of the convergence is compensated for, first of all, by the flexibility in the choice of the basis functions and by the simplicity of the generalization to the multidimensional case. Another, and probably the most important, advantage is the possibility to obtain explicit formulae for values of various integral and pseudodifferential operators of mathematical physics applied to the basis functions.

This property, when used for solving elliptic boundary value problems, leads to a new approach to the discretization of boundary integral equations, which is not based upon the decomposition of the boundary into elements. In this approach the coefficients of the resulting algebraic system depend only on the coordinates of a finite number of boundary points; hence the name Boundary Point Method (BPM) seems quite natural.

The first step of the BPM consists of the approximation of the boundary data and of the potential densities by a linear combination of basis functions, each of them being either concentrated near a particular boundary point or decreasing rapidly with the distance from this point. In the second step the calculation of the potentials, whose densities are the basis functions, is carried out. I discuss only one concrete variant of the BPM which leads to an algebraic system with coefficients that are explicitly expressed - up to a simple quadrature.

The approximation method can be employed also to solve various non-stationary problems. Its effectiveness is demonstrated by the numerical solution of Cauchy problems for quasilinear parabolic equations. An important feature of the algorithms is that they are both explicit and stable under much milder restrictions to the time step, depending on the size of the space grid, in comparison with usual explicit difference schemes.

5.1.1 AN EXAMPLE

Recently I proposed an approximation method based upon the use of approximate

The Mathematics of Finite Elements and Applications
Edited by J. R. Whiteman © 1994 John Wiley & Sons Ltd

partitions of unity (Maz'ya (1991)). As an illustrative example consider the following approximation formula for a function u on $\mathbb{R}^1$

$$\mathcal{M}_h^{(0)}u(x) = \frac{1}{\sqrt{\pi\mathcal{D}}} \sum_{m=-\infty}^{\infty} u(mh) \exp\left(-\frac{(x-mh)^2}{\mathcal{D}h^2}\right) \tag{5.1}$$

with two positive parameters, "small" h and "large" $\mathcal{D}$.

If $u \in C^2(\mathbb{R}^1)$, then the Taylor expansion

$$u(mh) = u(x) + u'(x)(mh - x) + u''(x_m)\frac{(mh - x)^2}{2}$$

for some x_m between x and mh leads to

$$\mathcal{M}_h^{(0)}u(x) - u(x) = u(x)\left(\theta\left(\frac{x}{h},\mathcal{D}\right) - 1\right) + u'(x)\frac{\mathcal{D}h}{2}\theta'\left(\frac{x}{h},\mathcal{D}\right) +$$

$$\frac{1}{2\sqrt{\pi\mathcal{D}}} \sum_{m=-\infty}^{\infty} u''(x_m)(mh - x)^2 \exp\left(-\frac{(x-mh)^2}{\mathcal{D}h^2}\right),$$

where

$$\theta(\xi,\mathcal{D}) = \frac{1}{\sqrt{\pi\mathcal{D}}} \sum_{m=-\infty}^{\infty} \exp\left(-\frac{(\xi-m)^2}{\mathcal{D}}\right) = 1 + 2\sum_{\nu=1}^{\infty} e^{-\pi^2\mathcal{D}\nu^2} \cos 2\pi\nu\xi$$

The last relation is a special case of the Poisson summation formula

$$\frac{1}{\mathcal{D}^{n/2}} \sum_{m\in\mathbb{Z}^n} \eta\left(\frac{x-m}{\sqrt{\mathcal{D}}}\right) = \sum_{\nu\in\mathbb{Z}^n} \mathcal{F}\eta\left(\sqrt{\mathcal{D}}\nu\right) e^{2\pi i(x,\nu)}, \quad x\in\mathbb{R}^n,$$

with the Fourier transform

$$\mathcal{F}\eta(\lambda) = \int_{\mathbb{R}^n} \eta(x)e^{-2\pi i(x,\lambda)}\,dx,$$

which is valid if η is continuous and

$$|\eta(x)| \le A(1 + |x|)^{-n-\delta}, \quad |\mathcal{F}\eta(\lambda)| \le A(1 + |\lambda|)^{-n-\delta}$$

for some $\delta > 0$ (Stein and Weiss (1971)).

Furthermore,

$$\theta'\left(\frac{x}{h},\mathcal{D}\right) = -4\pi \sum_{\nu=1}^{\infty} \nu e^{-\pi^2\mathcal{D}\nu^2} \sin 2\pi\nu\frac{x}{h}$$

and

$$\frac{1}{2\sqrt{\pi\mathcal{D}}} \left| \sum_{m=-\infty}^{\infty} u''(x_m)(mh - x)^2 \exp\left(-\frac{(x-mh)^2}{\mathcal{D}h^2}\right) \right| \le$$

$$\|u''\|_{C(\mathbb{R}^1)}\frac{\mathcal{D}h^2}{4}\left(1 + 2\sum_{\nu=1}^{\infty}(1 - 2\pi^2\mathcal{D}\nu^2)e^{-\pi^2\mathcal{D}\nu^2}\cos 2\pi\nu\frac{x}{h}\right).$$

(We write this and other similar inequalities with the maximum norms of derivatives over $\mathbb{R}^1$ for the sake of simplicity. Actually, only the values of derivatives in a small neighbourhood of the point x are relevant). Hence, we obtain the asymptotic form

$$\mathcal{M}_h^{(0)}u(x) - u(x) \doteq 2u(x)e^{-\pi^2\mathcal{D}}\cos 2\pi\frac{x}{h} + R_h^{(0)}(x) \tag{5.2}$$

with

$$\left|R_h^{(0)}(x)\right| \le \frac{\mathcal{D}h^2}{4}\|u''\|_{C(\mathbb{R}^1)}$$

for small h and large $\mathcal{D}$. We remark that if, for example, $\mathcal{D} = 2$ then $\exp(-\pi^2\mathcal{D}) = 0.0000000026753$. Hence the right-hand side of (5.2) is also a good approximation to $\mathcal{M}_h^{(0)}u(x) - u(x)$. In what follows we denote such relations as approximate identities and use the sign $\doteq$.

On the other hand, from (5.2) we see that for fixed $\mathcal{D}$ the sum $\mathcal{M}_h^{(0)}u(x)$ approximates $u(x)$ with the order $O(h^2)$ only if $h \ge h_0 > 0$, where the constant h_0 depends on $\mathcal{D}$, due to the saturation term $2u(x)\exp(-\pi^2\mathcal{D})\cos 2\pi x/h$.

We shall see that a quite general class of approximation formulas of the type (5.1) shows such a behaviour. For that reason we call such processes *approximate approximations*.

5.1.2 FURTHER EXAMPLES

A more accurate approximation of $u(x)$ is given by

$$\mathcal{M}_h^{(1)}u(x) = \frac{1}{2\sqrt{\pi\mathcal{D}}}\sum_{m=-\infty}^{+\infty}u(mh)\left(3 - \frac{2}{\mathcal{D}h^2}(x - mh)^2\right)\exp\left(-\frac{(x - mh)^2}{\mathcal{D}h^2}\right) \tag{5.3}$$

which is equivalent to

$$\mathcal{M}_h^{(1)}u(x) = \mathcal{M}_h^{(0)}u(x) - \frac{\mathcal{D}h^2}{4}\frac{d^2}{dx^2}\mathcal{M}_h^{(0)}u(x). \tag{5.3'}$$

Here

$$\mathcal{M}_h^{(1)}u(x) - u(x) \doteq 2u(x)e^{-\pi^2\mathcal{D}}(1 + \pi^2\mathcal{D})\cos 2\pi\frac{x}{h} + R_h^{(1)}(x),$$

where the remainder term satisfies the estimate

$$\left|R_h^{(1)}(x)\right| \le \frac{\mathcal{D}^2h^4}{4}\|u^{(4)}\|_{C(\mathbb{R}^1)}. \tag{5.4}$$

Another example of a "fourth order" approximation is provided by the formula

$$\mathcal{L}_h u(x) = \left(\frac{e}{\pi \mathcal{D}}\right)^{1/2} \sum_{m=-\infty}^{\infty} u(mh)\cos(\sqrt{2}\tau_m)\exp(-\tau_m^2). \tag{5.5}$$

Here $\tau_m = (x - mh)/\sqrt{\mathcal{D}}h$ and

$$\mathcal{L}_h u(x) - u(x) \doteq 2u(x)e^{-\pi^2 \mathcal{D}}\cosh\left(\sqrt{2\mathcal{D}}\pi\right)\cos 2\pi\frac{x}{h} + R_h(x)$$

with

$$|R_h(x)| \leq 0.03\mathcal{D}^2 h^4 \|u^{(4)}\|_{C(\mathbb{R}^1)}.$$

Approximations of the type (5.1), (5.3), (5.5) are also effective for non-smooth functions, but, naturally, the error estimate becomes worse. For instance, functions in the Hölder class C^α admit an $O(h^\alpha)$-approximation (modulo $O(\exp(-\pi^2\mathcal{D}))$, of course).

The above approximations are easily extended to the multidimensional case. Let, for example, u be a function on $\mathbb{R}^n$. The analogue of (5.1) is the formula

$$\mathcal{M}_h^{(0)}u(x) = \frac{1}{(\pi\mathcal{D})^{n/2}}\sum_{m\in\mathbb{Z}^n} u(mh)\,\exp\left(-\frac{|x-mh|^2}{\mathcal{D}h^2}\right). \tag{5.6}$$

Similarly to (5.1) this is an approximation of order $O(h^2)$ modulo $2n\exp(-\pi^2\mathcal{D})$. A better approximation is given by

$$\mathcal{M}_h^{(1)}u(x) = \frac{1}{(\pi\mathcal{D})^{n/2}}\sum_{m\in\mathbb{Z}^n} u(mh)\,\left(\frac{n+2}{n} - \frac{|x-mh|^2}{\mathcal{D}h^2}\right)\exp\left(-\frac{|x-mh|^2}{\mathcal{D}h^2}\right), \tag{5.7}$$

whereas a "sixth order" approximation is obtained with

$$\mathcal{M}_h^{(2)}u(x) = \frac{1}{2(\pi\mathcal{D})^{n/2}}\sum_{m\in\mathbb{Z}^n} u(mh)\left(\frac{(n+4)(n+2)}{2} - \right.$$

$$\left. (n+4)\frac{|x-mh|^2}{\mathcal{D}h^2} + \frac{|x-mh|^4}{\mathcal{D}^2 h^4}\right)\exp\left(-\frac{|x-mh|^2}{\mathcal{D}h^2}\right). \tag{5.8}$$

In general, an approximation of order $2N+2$ is given by

$$\mathcal{M}_h^{(N)}u(x) = \frac{1}{(\pi\mathcal{D})^{n/2}}\sum_{m\in\mathbb{Z}^n} u(mh)L_N^{\left(\frac{n}{2}\right)}\left(\frac{|x-mh|^2}{\mathcal{D}h^2}\right)\exp\left(-\frac{|x-mh|^2}{\mathcal{D}h^2}\right), \tag{5.9}$$

where

$$L_N^{(\alpha)}(\xi) = e^\xi \xi^{-\alpha}\left(\frac{d}{d\xi}\right)^N \left(e^{-\xi}\xi^{N+\alpha}\right), \quad \alpha > -1,$$

are the generalized Laguerre polynomials.

In particular, if $n = 1$ then (5.9) takes the form

$$\mathcal{M}_h^{(N)} u(x) = \frac{(-1)^N}{\sqrt{\pi \mathcal{D}} N! 2^{2N+1}} \sum_{m=-\infty}^{+\infty} u(mh) \tau_m^{-1} H_{2N+1}(\tau_m) \exp\left(-\tau_m^2\right), \qquad (5.10)$$

where τ_m is as before and the H_{2N+1} are the Hermite polynomials

$$H_{2N+1}(\xi) = -e^{\xi^2} \left(\frac{d}{d\xi}\right)^{2N+1} e^{-\xi^2}.$$

An interesting feature of the formula (5.9) is its additive structure

$$\mathcal{M}_h^{(N)} u(x) = \mathcal{M}_h^{(N-1)} u(x) +$$

$$(\pi \mathcal{D})^{-n/2} \sum_{m \in \mathbb{Z}^n} u(mh) L_N^{\left(\frac{n-2}{2}\right)} \left(\frac{|x - mh|^2}{\mathcal{D}h^2}\right) \exp\left(-\frac{|x - mh|^2}{\mathcal{D}h^2}\right) =$$

$$\mathcal{M}_h^{(N-1)} u(x) + \frac{(-1)^N}{N!} \left(\frac{\mathcal{D}h^2}{4}\right)^N \Delta^N \mathcal{M}_h^{(0)} u(x), \qquad (5.11)$$

where Δ is the Laplace operator. In Section 5.2 we shall see that the last relation is especially useful for obtaining high order approximations for various multidimensional operators of mathematical physics.

5.1.3 APPROXIMATION ESTIMATES

Consider a more general approximation process

$$\mathcal{L}_h^{(N)} u(x) = \mathcal{D}^{-n/2} \sum_{m \in \mathbb{Z}^n} u(mh) \eta\left(\frac{x - mh}{\sqrt{\mathcal{D}}h}\right), \qquad (5.12)$$

where η is a real-valued continuous function, rapidly decreasing at infinity such that for some $k_0 > 0$ and $\delta > 0$

$$|\eta(x)| \le A(1 + |x|)^{-k_0 - n - \delta},$$

and for $N \le k_0$

$$\int_{\mathbb{R}^n} \eta(x) dx = 1, \qquad \int_{\mathbb{R}^n} x^\alpha \eta(x) dx = 0, \quad 1 \le |\alpha| \le N - 1.$$

Furthermore we assume that $\mathcal{D}$ can be chosen such that the series

$$\sum_{\nu \in \mathbb{Z}^n} |\mathcal{F}(t^\alpha \eta(t))(\sqrt{\mathcal{D}}\nu)|, \quad 0 \le |\alpha| \le N - 1,$$

converge.

Here and henceforth we use the notations $x^\alpha = x_1^{\alpha_1} \cdots x_n^{\alpha_n}$, $|\alpha| = \alpha_1 + \cdots + \alpha_n$, $\alpha! = \alpha_1! \cdots \alpha_n!$, $\partial^\alpha = \partial^{|\alpha|}/\partial x_1^{\alpha_1} \cdots \partial x_n^{\alpha_n}$, $D^\alpha = (i^{-1}\partial)^\alpha$.

Theorem 5.1 (Maz'ya and Schmidt (1993)). *Let* $u \in C^k(\mathbb{R}^n)$, $1 \le k \le k_0$. *Then there exists a constant* c_η *such that*

$$\left| \mathcal{L}_h^{(N)} u(x) - u(x) \right| \le c_\eta (\sqrt{\mathcal{D}} h)^k \sum_{|\alpha|=k} \frac{\|\partial^\alpha u\|_{C(\mathbb{R}^n)}}{\alpha!} +$$

$$\sum_{|\alpha|=0}^{\min(k,N)-1} (\sqrt{\mathcal{D}} h)^{|\alpha|} \frac{|\partial^\alpha u(x)|}{\alpha!} \sum_{\nu \in \mathbb{Z}^n} |\mathcal{F}(t^\alpha \eta(t))(\sqrt{\mathcal{D}}\nu)| +$$

$$\sum_{|\alpha|=\min(k,N)}^{k-1} (\sqrt{\mathcal{D}} h)^{|\alpha|} \frac{|\partial^\alpha u(x)|}{\alpha!} \left| \max_{|y| \le h} \sum_{m \in \mathbb{Z}^n} \left(\frac{y - mh}{\sqrt{\mathcal{D}} h} \right)^\alpha \eta\left(\frac{y - mh}{\sqrt{\mathcal{D}} h} \right) \right|.$$

The last sum appears only in the case $k > N$.

Similar inequalities can be proved for the approximation of the derivatives $\partial^\beta u(x)$ by $\partial^\beta \mathcal{L}_h^{(N)} u(x)$, for functions u in the Hölder space $C^{k,\lambda}(\mathbb{R}^n)$ and for the approximation in Sobolev norms if $u \in H^s(\mathbb{R}^n)$, $s > n/2$.

Theorem 5.1 shows that the difference $\mathcal{L}_h^{(N)} u(x) - u(x)$ can be estimated by terms of the order $O(h^{\min(k,N)})$ and higher and by the sum

$$\sum_{|\alpha|=0}^{\min(k,N)-1} (\sqrt{\mathcal{D}} h)^{|\alpha|} \sum_{\nu \in \mathbb{Z}^n \setminus \{0\}} |\mathcal{F}(t^\alpha \eta(t))(\sqrt{\mathcal{D}}\nu)|, \tag{5.13}$$

which in general does not converge to zero as $h \to 0$. One of the basic ideas of applying approximate approximations in numerical mathematics consists of choosing η and $\mathcal{D}$ such that (5.13) is small compared with the $O(h^{\min(k,N)})$-term for h relevant to practical applications.

The higher order approximation (5.11) can be extended to the case of general η as

$$\mathcal{D}^{-n/2} \sum_{m \in \mathbb{Z}^n} u(mh) \eta_N \left(\frac{x - mh}{\sqrt{\mathcal{D}} h} \right),$$

where

$$\eta_N = \sum_{|\alpha| \le N-1} \frac{1}{\alpha!} \left(\frac{1}{2\pi} \partial \right)^\alpha \left. \frac{1}{\mathcal{F}\eta(\lambda)} \right|_{\lambda=0} D^\alpha \eta, \quad N = 1, 2, \ldots,$$

Here the error estimate is $O(h^N)$ modulo a negligible term.

5.1.4 SOME COMMENTS

In all examples considered to date, (5.13) is mainly determined by the value of $\exp(-\pi^2 \mathcal{D})$, which makes the choice $\mathcal{D} = 2$ reasonable in numerical calculations. Since in general $\mathcal{D}$ determines the number of summands in (5.12), necessary to evaluate the approximate value of $u(x)$ with a given precision, it is of great practical interest, especially for the multi-dimensional case, to reduce $\mathcal{D}$. We show two simple possibilities for the first example (5.1).

At the points $x_k = \left(k + \frac{1}{4}\right)h$, $k \in \mathbb{Z}$, we have

$$\mathcal{M}_h^{(0)}u(x_k) - u(x_k) \doteq -2u(x_k)e^{-4\pi^2\mathcal{D}} - 2\mathcal{D}\pi h u'(x_k)e^{-\pi^2\mathcal{D}} + R_h^{(0)}(x_k),$$

so that by computing $\mathcal{M}_h^{(0)}u$ at the knots x_k we can take a smaller value of $\mathcal{D}$.

The other possibility is the modification of the kernel function $\exp(-x^2/\mathcal{D})$ in order to decrease the Fourier transform at $\lambda = \sqrt{\mathcal{D}}$. This leads to the summation

$$\mathcal{L}_h u(x) = \frac{\left(1 - 2e^{-2\mathcal{D}\pi^2}\right)^{-1}}{\sqrt{\pi\mathcal{D}}} \sum_{m=-\infty}^{+\infty} u(mh)\left(1 - 2e^{-\mathcal{D}\pi^2}\cos\left(2\pi\frac{x - mh}{h}\right)\right)e^{-\tau_m^2}$$

and we obtain the approximate identity

$$\mathcal{L}_h u(x) - u(x) \doteq \frac{2}{2 - e^{2\mathcal{D}\pi^2}}\cos\left(4\pi\frac{x}{h}\right)u(x) + R_h(x)$$

with

$$|R_n(x)| \leq \frac{\mathcal{D}h^2}{4}\|u''\|_{C(\mathbb{R}^1)}.$$

Another application of Theorem 5.1 leads to estimates with better constants for smooth functions. For example, (5.4) can be improved for $u \in C^6(\mathbb{R}^1)$ to the estimate

$$|R_n^{(1)}(x)| \leq \frac{\mathcal{D}^2 h^4}{32}|u^{(4)}(x)| + 0.006\mathcal{D}^3 h^6\|u^{(6)}\|_{C(\mathbb{R}^1)}.$$

The generalization of the approximation process described above to meshes different from the cubic one is very attractive. Whereas the case of a constant affine grid is an almost obvious consequence of the results mentioned to date, the approximation on variable meshes still holds some open problems. We cite here one partial result.

Theorem 5.2 (Maz'ya and Schmidt (1993)). *Let $y : \mathbb{R}^1 \to \mathbb{R}^1$ be a sufficiently smooth parametrization of $\mathbb{R}^1$, $y'(t) \geq c > 0$, and $\Delta_h = \{y(mh) = y_m\}_{m=-\infty}^{+\infty}$. Suppose that η satisfies the conditions of Theorem 5.1. Then for any $u \in C^k(\mathbb{R}^1), k \leq N$, the approximations*

$$\mathcal{L}_{\Delta_h}^{(N)}u(x) = \frac{1}{\sqrt{\mathcal{D}}} \sum_{m=-\infty}^{+\infty} u(y_m)\eta\left(\frac{x - y_m}{h\sqrt{\mathcal{D}}y'(mh)}\right)$$

are of order $O(h^k)$ modulo small saturation terms.

Finally it should be remarked that the approximate approximation is in fact of local nature. Since the functions $\eta((x - mh)/\sqrt{\mathcal{D}}h)$ are concentrated in a neighbourhood of x only a finite number of summands $u(mh)\eta((x - mh)/\sqrt{\mathcal{D}}h)$ contributes to the value of $\mathcal{L}_h^{(N)}u(x)$ at this point. Moreover, this number does not depend on h (but decreases together with $\mathcal{D}$), thus $\mathcal{L}_h^{(N)} u(x)$ depends only on the behaviour of $u(y)$ for $|y - x| \le ah$ with fixed a for all interesting h. Therefore we can derive local estimates of the form

$$\left| \mathcal{L}_h^{(N)} u(x) - u(x) \right| \le ch^k \max_{|y-x| \le ah} \sum_{|\alpha| = k} |\partial^\alpha u(y)| + \epsilon_k \|u\|_{C^{k-1}(\mathbb{R}^n)},$$

where ϵ_k can be estimated by very small constants for appropriately chosen η and $\mathcal{D}$.

The lack of convergence, which is not perceptible in practical applications, is compensated for, first of all, by the flexibility in the choice of the basis functions and by the simplicity of the generalization to the multidimensional case in contrast to spline approximations and other finite element procedures. Another, and probably the most important, advantage is the possibility for obtaining explicit formulas for values of various integral and pseudodifferential operators of mathematical physics applied to the basis functions.

5.2 THE ACTION OF INTEGRAL OPERATORS ON BASIS FUNCTIONS

It is well known that the numerical treatment of potential and other integral operators with singularities is an essential and, especially in the multidimensional case, a very time-consuming part of different methods for the solution of many practical problems. In this section we demonstrate by examples that the value of such operators acting on basis functions, introduced in Section 5.1, can be either explicitly obtained via effectively computable expressions or computed by simple one-dimensional quadrature of well-behaved functions. By using high order approximations of the potential density $u(x)$ it is then possible to develop new classes of high order approximate formulas for potentials which proved their effectiveness in a number of numerical experiments.

5.2.1 THE HILBERT TRANSFORM

For $\mu(x) = (\pi\mathcal{D})^{-1/2} \exp(-x^2/\mathcal{D}h^2)$ we get

$$\mathcal{H}\mu(x) = \frac{1}{\pi} \int_{\mathbb{R}^1} \frac{\mu(t)}{t - x} dt = -\frac{1}{\sqrt{\pi\mathcal{D}}} e^{-\tau^2} \mathrm{erfi}(\tau),$$

where $\tau = x/\sqrt{\mathcal{D}}h$ and

$$\mathrm{erfi}(\tau) = \frac{2}{\sqrt{\pi}} \int_0^\tau e^{t^2} dt.$$

Hence

$$\mathcal{H}u(x) \doteq -\frac{1}{\sqrt{\pi\mathcal{D}}} \sum_{m \in \mathbb{Z}} u(mh) \left\{ e^{-\tau_m^2} \mathrm{erfi}(\tau_m) + \right.$$

$$\left[\left(\frac{1}{2} - \tau_m^2\right) e^{-\tau_m^2} \mathrm{erfi}(\tau_m) - \frac{\tau_m}{2}\right] + \cdots\right\}$$

with $\tau_m = (x - mh)/\sqrt{\mathcal{D}}h$. The first term in the curly brackets gives the accuracy $O(h^2)$ which is improved to $O(h^4)$ by the term in the square brackets.

5.2.2 THE LOGARITHMIC POTENTIAL

For $\mu(x) = (\pi\mathcal{D})^{-1} \exp(-|x|^2/\mathcal{D}h^2)$, $x \in \mathbb{R}^2$, we derive

$$\mathcal{L}\mu(x) = \frac{1}{2\pi} \int_{\mathbb{R}^2} \log\frac{1}{x-y}\mu(y)dy = \frac{h^2}{4\pi}\left(\gamma - \log\mathcal{D}h^2 - \int_0^{r^2}\frac{1-e^{-t}}{t}dt\right),$$

where $r = |x|/\sqrt{\mathcal{D}}h$ and $\gamma = 0.577215\ldots$ is the Euler constant.

We approximate u by $\mathcal{M}_h^{(N)}u(x)$ to obtain

$$\mathcal{L}u(x) \doteq \frac{1}{\pi\mathcal{D}}\sum_{m\in\mathbb{Z}^2} u(mh)\mathcal{L}\left(L_N^{(1)}(r_m^2)e^{-r_m^2}\right)$$

with $r_m^2 = |x - mh|^2/\mathcal{D}h^2$. ¿From (5.11) we see that

$$\mathcal{L}\left(L_N^{(1)}(r_m^2)e^{-r_m^2}\right) = \mathcal{L}\left(L_{N-1}^{(1)}(r_m^2)e^{-r_m^2}\right) - \frac{(-1)^N}{N!}\left(\frac{\mathcal{D}h^2}{4}\right)^N \Delta^{N-1}e^{-r_m^2} =$$

$$\mathcal{L}\left(L_{N-1}^{(1)}(r_m^2)e^{-r_m^2}\right) + \frac{1}{N}\frac{\mathcal{D}h^2}{4}L_{N-1}^{(0)}(r_m^2)e^{-r_m^2}.$$

Hence

$$\mathcal{L}u(x) \doteq \frac{h^2}{4\pi}\sum_{m\in\mathbb{Z}^2} u(mh)\{\gamma - \log\mathcal{D}h^2 - \int_0^{r_m^2}\frac{1-e^{-t}}{t}dt$$

$$+e^{-r_m^2}\left(1 + \frac{1}{2}(1-r_m^2) + \frac{1}{3}(1 - 2r_m^2 + r_m^4) + \cdots + \frac{1}{N}L_{N-1}^{(0)}(r_m^2)\right)\}$$

gives a cubature formula of order $O(h^{2N+2})$. Furthermore we note that the singular integral

$$\frac{1}{2\pi}\frac{\partial^2}{\partial x_1 \partial x_2}\int_{\mathbb{R}^2}\log\frac{1}{|x-y|}u(y)dy$$

is approximated by

$$\frac{1}{\pi\mathcal{D}}\sum_{m\in\mathbb{Z}^2} u(mh)\frac{(x_1 - m_1h)(x_2 - m_2h)}{\mathcal{D}h^2}\left(r_m^{-4}(1 - (1+r_m^2)e^{-r_m^2}) + e^{-r_m^2} +\right.$$

$$\frac{1}{2}\left(\frac{d}{d\tau}\right)^2\left((1-\tau)e^{-\tau}\right)\bigg|_{\tau=r_m^2} + \cdots + \frac{1}{N}\left(\frac{d}{d\tau}\right)^2\left(L_{N-1}^{(0)}(\tau)e^{-\tau}\right)\bigg|_{\tau=r_m^2}\right) \tag{5.14}$$

with $x = (x_1, x_2)$, $m = (m_1, m_2)$.

5.2.3 NEWTONIAN POTENTIAL

For $\mu(x) = (\pi\mathcal{D})^{-3/2}\exp(-|x|^2/\mathcal{D}h^2)$, $x \in \mathbb{R}^3$, we obtain

$$\mathcal{N}\mu(x) = \frac{1}{4\pi}\int\limits_{\mathbb{R}^3}\frac{\mu(y)}{|x-y|}dy = \frac{h^2}{2\mathcal{D}^{1/2}\pi^{3/2}}\frac{1}{r}\int\limits_0^r e^{-t^2}dt$$

with $r = |x|/\sqrt{\mathcal{D}}h$.

Acting as in 5.2.2 we see that

$$\mathcal{N}u(x) \doteq \frac{h^2}{2\mathcal{D}^{1/2}\pi^{3/2}}\sum_{m\in\mathbb{Z}^3}u(mh)\left\{\frac{1}{r_m}\int\limits_0^{r_m}e^{-t^2}dt + \right.$$

$$e^{-r_m^2}\left.\left(L_0^{(\frac{1}{2})}(r_m^2) + \frac{1}{2}L_1^{(\frac{1}{2})}(r_m^2) + \cdots + \frac{1}{N}L_{N-1}^{(\frac{1}{2})}(r_m^2)\right)\right\} =$$

$$= \frac{h^2}{2\mathcal{D}^{1/2}\pi^{3/2}}\sum_{m\in\mathbb{Z}^3}u(mh)\left\{\frac{1}{r_m}\int\limits_0^{r_m}e^{-t^2}dt + e^{-r_m^2}\sum_{j=1}^{N}\frac{(-1)^j}{2^{2j}j!}\frac{H_{2j-1}(r_m)}{r_m}\right\}$$

with $r_m = |x - mh|/\sqrt{\mathcal{D}}h$ gives an approximation of order $O(h^{2N+2})$.

Table 5.1 The approximation of the Newtonian potential with the density $u(x) = -\Delta\exp(-1/(1-|x|^2)_+)$ at $x = 0$ by taking only the first term in the sum

Computed value	h	Relative error
0.3289	2^{-3}	10.57 %
0.3644	2^{-4}	0.928 %
0.3667	2^{-5}	0.294 %
0.3676	2^{-6}	0.07329%
0.36781	2^{-7}	0.01831%

Table 5.2 By taking two terms in the sum we obtain better results. One can compare them with the exact value 0.367879441...

Computed value	h	Relative error
0.3468	2^{-3}	5.7 %
0.3688	2^{-4}	0.25 %
0.36787	2^{-5}	0.001166 %
0.3678793	2^{-6}	0.00001179%
0.36787943	2^{-7}	0.00000208%

5.2.4 THE DIFFRACTION POTENTIAL

Consider the potential

$$\mathcal{S}u(x) = \frac{1}{4\pi}\int\limits_{\mathbb{R}^3}\frac{e^{ik|x-y|}}{|x-y|}u(y)dy, \quad \Im k = 0.$$

For the same μ as in 5.2.3 we find

$$\mathcal{S}\mu(x) = \frac{h^2}{4\pi\mathcal{D}^{1/2}} e^{-\mathcal{D}h^2 k^2/4} q_h\left(\frac{|x|}{\mathcal{D}^{1/2}h}\right),$$

with

$$q_h(r) = r^{-1}\Re\left(e^{i\mathcal{D}^{1/2}hkr}\,\mathrm{erf}\left(r + \frac{i}{2}\mathcal{D}^{1/2}hk\right)\right) + ir^{-1}\sin(\mathcal{D}^{1/2}hkr),$$

where $\mathrm{erf}(z) = \frac{2}{\sqrt{\pi}}\int_0^z e^{-t^2}\,dt$. Hence the approximations

$$\mathcal{S}^{(0)}u(x) = \frac{h^2}{4\pi\mathcal{D}^{1/2}} e^{-\mathcal{D}h^2 k^2/4} \sum_{m\in\mathbb{Z}^3} u(mh)q_h(r_m)$$

and

$$\mathcal{S}^{(1)}u(x) = \frac{h^2}{4\pi\mathcal{D}^{1/2}} \sum_{m\in\mathbb{Z}^3} u(mh)\left(\frac{2}{\pi^{1/2}}e^{-r_m^2} + \left(1 + \frac{\mathcal{D}h^2 k^2}{4}\right) e^{-\mathcal{D}h^2 k^2/4} q_h(r_m)\right)$$

evaluate $\mathcal{S}u(x)$ with the orders $O(h^2)$ and $O(h^4)$, respectively.

5.2.5 3D-ELASTIC POTENTIAL

The potential $\mathcal{E}\mathbf{u} = ((\mathcal{E}\mathbf{u})_1, (\mathcal{E}\mathbf{u})_2, (\mathcal{E}\mathbf{u})_3)$ is defined by

$$(\mathcal{E}\mathbf{u})_i(x) = \frac{1}{16\pi\mu(1-\sigma)} \int_{\mathbb{R}^3} \left(\frac{3 - 4\sigma}{|x-y|}\delta_i^j + \frac{(x_i - y_i)(x_j - y_j)}{|x-y|^2}\right) u_j(y)\,dy,$$

where σ denotes Poisson's ratio and μ is the shear modulus.

The sums

$$(\mathcal{E}\mathbf{u})_i^{(1)}(x) = \frac{h^2}{\alpha} \sum_{m\in\mathbb{Z}^3} u_j(mh)\left\{\delta_i^j\left[\frac{1 + 2(3 - 4\sigma)r_m^2}{r_m^3}\int_0^{r_m} e^{-t^2}\,dt - \frac{e^{-r_m^2}}{r_m^2}\right] + \right.$$

$$\left. \omega_i\omega_j\left[\frac{2r_m^2 - 3}{r_m^3}\int_0^{r_m} e^{-t^2}\,dt + \frac{3e^{-r_m^2}}{r_m^2}\right]\right\}$$

and

$$(\mathcal{E}\mathbf{u})_i^{(2)}(x) = (\mathcal{E}\mathbf{u})_i^{(1)}(x) + \frac{h^2}{\alpha} \sum_{m\in\mathbb{Z}^3} u_j(mh)\left\{\delta_i^j\left[-\frac{1}{r_m^3}\int_0^{r_m} e^{-t^2}\,dt + \right.\right.$$

$$\left. \frac{4(1-\sigma)r_m^2+1}{r_m^2}e^{-r_m^2}\right] + \omega_i\omega_j\left[\frac{3}{r_m^3}\int_0^{r_m}e^{-t^2}dt - \frac{2r_m^2+3}{r_m^2}e^{-r_m^2}\right]\right\}$$

are the approximations of order $O(h^2)$ and $O(h^4)$, respectively. Here

$$\omega_i = \frac{x_i - m_i h}{|x - mh|}, \quad \alpha = 16\pi^{3/2}\mu(1-\sigma)\mathcal{D}^{1/2}.$$

5.2.6 SQUARE ROOT OF THE LAPLACIAN

In fracture mechanics, electrostatics and hydrodynamics one encounters the first order pseudodifferential operator

$$(-\Delta)^{1/2}u(x) = \frac{-\Delta_x}{2\pi}\int_{\mathbb{R}^2}\frac{u(y)dy}{|x-y|},$$

where Δ is the two-dimensional Laplacian. By using (5.6) this operator is approximated by the formula

$$(-\Delta)^{1/2}u(x) \doteq \frac{(\pi\mathcal{D})^{-1/2}}{\mathcal{D}h}\sum_{m\in\mathbb{Z}^2}u(mh)e^{-r_m^2/2}\left((1-r_m^2)\,I_0\left(\frac{r_m^2}{2}\right) + r_m^2 I_1\left(\frac{r_m^2}{2}\right)\right),$$

where I_0 and I_1 are Bessel functions with imaginary argument and $r_m = |x-mh|/\sqrt{\mathcal{D}h}$ as before.

5.2.7 THE CAUCHY PROBLEM FOR THE WAVE AND THE HEAT EQUATIONS

The method can also be employed to solve various nonstationary problems. For example, an immediate application of (5.6) to the Cauchy problem

$$u_{tt} - \Delta_x u = f(x,t), t > 0;$$

$$u(x,0) = \varphi(x), \quad u_t(x,0) = \psi(x), \quad x \in \mathbb{R}^3,$$

is the approximation formula

$$u(x,t) \doteq (\pi\mathcal{D})^{-3/2}\sum_{m\in\mathbb{Z}^3}d_m^{-1}\Big((\mathcal{A}(t+d_m) - \mathcal{A}(t-d_m))\varphi(mh)+$$

$$(\mathcal{B}(t+d_m) - \mathcal{B}(t-d_m))\psi(mh) + \int_0^t (\mathcal{B}(\tau+d_m) - \mathcal{B}(\tau-d_m))f(mh,\tau)d\tau\Big),$$

where $d_m = |x - mh|$ and

$$\mathcal{A}(\lambda) = \frac{\lambda}{2} e^{-\lambda^2/\mathcal{D}h^2}, \quad \mathcal{B}(\lambda) = \frac{\mathcal{D}h^2}{4} e^{-\lambda^2/\mathcal{D}h^2}.$$

Quite similarly the solution of the Cauchy problem

$$u_t - \Delta_x u = f(x,t), \quad t > 0;$$

$$u(x,0) = \varphi(x), \quad x \in \mathbb{R}^3$$

admits the approximate representation

$$u(x,t) \doteq \left(\frac{h}{\sqrt{\pi(4t + \mathcal{D}h^2)}} \right)^3 \sum_{m \in \mathbb{Z}^3} e^{-d_m^2/(4t + \mathcal{D}h^2)} \varphi(mh) +$$

$$\sum_{m \in \mathbb{Z}^3} \int_0^t \left(\frac{h}{\sqrt{\pi(4(t-\tau) + \mathcal{D}h^2)}} \right)^3 e^{-d_m^2/(4(t-\tau) + \mathcal{D}h^2)} f(mh,\tau) d\tau \tag{5.15}$$

In Sections 5.3 and 5.4 the effectiveness of the above approximations for the numerical solution of the Cauchy problem for nonlinear parabolic equations and for the Navier-Stokes equations is shown.

5.3 TIME-MARCHING ALGORITHMS FOR NON-LINEAR PARABOLIC EQUATIONS

The material of this section is borrowed from the paper (Maz'ya and Karlin (1993)), where some semianalytic time-marching algorithms for numerical solution of quasilinear parabolic equations based upon approximate approximations were developed. An important feature of the algorithms is that they are both explicit and stable under much milder restrictions to the time step, depending on the size of the grid, in comparison with usual explicit difference schemes (Richtmyer and Morton (1967)). The algorithms give the time step approximation $u_h(x, n\tau)$ in the form of an analytic function, which enables one to differentiate the approximate solution explicitly with respect to x. Crucial issues which arise under the computer implementation of the method as well as numerical experiments are discussed in (Maz'ya and Karlin (1993)).

5.3.1 ONE-STEP TIME MARCHING ALGORITHMS

We start with the equation

$$u_t - \nu u_{xx} = \frac{\partial}{\partial x} \left(f(x,t,u) \right) \quad \text{for} \quad x \in \mathbb{R}^1, \quad t > 0. \tag{5.16}$$

which can be rewritten in the equivalent form

$$u(x,T) = u(x,t) + \int_{\mathbb{R}^1} \mathcal{P}(x - \xi, T)(u(\xi,t) - u(x,t))d\xi +$$

$$\frac{\partial}{\partial x} \int_t^T \int_{\mathbb{R}^1} \mathcal{P}(x - \xi, T - \lambda) f(\xi, \lambda, u(\xi, \lambda)) d\xi d\lambda, \qquad (5.17)$$

where $T > t \geq 0$ and $\mathcal{P}$ is the Poisson kernel.

We split the time interval into the subintervals $[(n-1)\tau, n\tau]$, $n = 1, 2, \ldots$, and obtain

$$u(x, n\tau) = u(x, (n-1)\tau) + \int_{\mathbb{R}^1} \mathcal{P}(x - \xi, n\tau)(u(\xi, (n-1)\tau) - u(x, (n-1)\tau))d\xi +$$

$$\frac{\partial}{\partial x} \int_{(n-1)\tau}^{n\tau} \int_{\mathbb{R}^1} \mathcal{P}(x - \xi, n\tau - \lambda) f(\xi, \lambda, u(\xi, \lambda)) d\xi d\lambda. \qquad (5.18)$$

By (5.1) and (5.15) the first integral on the right can be approximated by

$$\frac{h}{\sqrt{\pi}\,(4\nu\tau + Dh^2)^{1/2}} \sum_{m=-\infty}^{\infty} u_h(mh, (n-1)\tau) \exp\left(-\frac{(x - mh)^2}{4\nu\tau + Dh^2}\right) -$$

$$\frac{1}{(\pi D)^{1/2}} \sum_{m=-\infty}^{\infty} u_h(mh, (n-1)\tau) \exp\left(-\frac{(x - mh)^2}{Dh^2}\right) \qquad (5.19)$$

with the error $O(\tau h^2)$.

Again using (5.1) we evaluate the third term in the right-hand side of (5.18) as

$$\frac{1}{(\pi D)^{1/2}} \frac{\partial}{\partial x} \int_{(n-1)\tau}^{n\tau} \int_{\mathbb{R}^1} \mathcal{P}(x - \xi, n\tau - \lambda) \times$$

$$\sum_{m=-\infty}^{\infty} f(mh, \lambda, u(mh, \lambda)) \exp\left(-\frac{(\xi - mh)^2}{Dh^2}\right) d\xi d\lambda \qquad (5.20)$$

by making an error $O(\tau h^2)$. Here and elsewhere in mentioning error estimates we assume that the data and the solutions are sufficiently smooth.

In developing the simplest one-step explicit time marching procedure we replace (5.20) by

$$\frac{1}{(\pi D)^{1/2}} \sum_{m=-\infty}^{\infty} f(mh, (n-1)\tau, u(mh, (n-1)\tau)) \times$$

$$\frac{\partial}{\partial x} \int\limits_{(n-1)\tau}^{n\tau} \int\limits_{\mathbb{R}^1} \mathcal{P}(x - \xi, n\tau - \lambda) \exp\left(-\frac{(\xi - mh)^2}{Dh^2}\right) d\xi \, d\lambda. \tag{5.21}$$

Obviously the inner integral over $\mathbb{R}^1$ is equal to

$$\frac{D^{1/2}h}{(4\nu(n\tau - \lambda) + Dh^2)^{1/2}} \exp\left(-\frac{(x - mh)^2}{4\nu(n\tau - \lambda) + Dh^2}\right),$$

which along with the identity

$$\int\limits_{(n-1)\tau}^{n\tau} \exp\left(-\frac{(x - mh)^2}{4\nu(n\tau - \lambda) + Dh^2}\right) \frac{(x - mh)d\lambda}{(4\nu(n\tau - \lambda) + Dh^2)^{3/2}} =$$

$$\frac{\sqrt{\pi}}{4\nu}\left\{\text{erf}\left(\frac{x - mh}{D^{1/2}h}\right) - \text{erf}\left(\frac{x - mh}{(4\nu\tau + Dh^2)^{1/2}}\right)\right\}$$

leads to the explicit form of (5.21)

$$\frac{h}{2\nu} \sum_{m=-\infty}^{\infty} f(mh, (n-1)\tau, u_h(mh, (n-1)\tau)) \times$$

$$\left\{\text{erf}\left(\frac{x - mh}{(4\nu\tau + Dh^2)^{1/2}}\right) - \text{erf}\left(\frac{x - mh}{D^{1/2}h}\right)\right\}. \tag{5.22}$$

Thus we have arrived at the step-by-step formula

$$u_h(x, n\tau) = u_h(x, (n-1)\tau) +$$

$$\frac{1}{\sqrt{\pi}} \sum_{m=-\infty}^{\infty} u_h(mh, (n-1)\tau) \left.\frac{e^{-\Xi_m^2(\lambda)}}{(\kappa(\lambda) + D)^{1/2}}\right|_{\lambda=0}^{\lambda=\tau} +$$

$$\frac{h}{2\nu} \sum_{m=-\infty}^{\infty} f(mh, (n-1)\tau, u_h(mh, (n-1)\tau)) \text{erf}\left(\Xi_m(\lambda)\right)\Big|_{\lambda=0}^{\lambda=\tau}. \tag{5.23}$$

Here and in the sequel

$$\kappa(\lambda) = \frac{4\nu\lambda}{h^2}, \quad \Xi_m(\lambda) = \frac{\frac{x}{h} - m}{(\kappa(\lambda) + D)^{1/2}}.$$

By summarising what was said on error estimates we conclude that the algorithm just constructed gives the error $O(\tau^2 + \tau h^2)$ at each time step.

To obtain a similar more precise procedure we use the approximation formula (5.3). By duplicating previous arguments we arrive at the time marching procedure

$$u_h(x, n\tau) = u_h(x, (n-1)\tau) +$$

$$\frac{1}{\sqrt{\pi}} \sum_{m=-\infty}^{\infty} u_h(mh, (n-1)\tau) \left(\kappa(\lambda) + \frac{3}{2}D - D\Xi_m^2(\lambda)\right) \left.\frac{e^{-\Xi_m^2(\lambda)}}{(\kappa(\lambda) + D)^{3/2}}\right|_{\lambda=0}^{\lambda=\tau} +$$

$$\frac{h}{2\nu} \sum_{m=-\infty}^{\infty} f(mh, (n-1)\tau, u_h(mh, (n-1)\tau)) \times$$

$$\left.\left\{\frac{D}{\sqrt{\pi}}\Xi_m(\lambda) \frac{e^{-\Xi_m^2(\lambda)}}{\kappa(\lambda) + D} + \mathrm{erf}\,(\Xi_m(\lambda))\right\}\right|_{\lambda=0}^{\lambda=\tau} \tag{5.24}$$

with accuracy $O(\tau^2 + \tau h^4)$.

The above algorithms can be readily modified for the more general equation

$$u_t - \nu u_{xx} = \frac{\partial}{\partial x}\left(f(x, t, u)\right) + F(x, t, u) \quad \text{for} \quad x \in \mathbb{R}^1, \quad t > 0, \tag{5.25}$$

The only difference in comparison with (5.23) is the appearance of the term

$$-\frac{h^2}{2\sqrt{\pi}\nu} \sum_{m=-\infty}^{\infty} F(mh, (n-1)\tau, u_h(mh, (n-1)\tau)) \times$$

$$(\kappa(\lambda) + D)^{1/2} \left.\left\{e^{-\Xi_m^2(\lambda)} + \sqrt{\pi}\Xi_m(\lambda)\mathrm{erf}\,(\Xi_m(\lambda))\right\}\right|_{\lambda=0}^{\lambda=\tau}.$$

By (5.3') the analogous extra term for process (5.24) takes the form

$$-\frac{h^2}{4\sqrt{\pi}\nu} \sum_{m=-\infty}^{\infty} F(mh, (n-1)\tau, u_h(mh, (n-1)\tau)) \times$$

$$(\kappa(\lambda) + D)^{1/2} \left.\left\{\frac{2\kappa(\lambda) + D}{\kappa(\lambda) + D}e^{-\Xi_m^2(\lambda)} + 2\sqrt{\pi}\Xi_m(\lambda)\mathrm{erf}\,(\Xi_m(\lambda))\right\}\right|_{\lambda=0}^{\lambda=\tau}.$$

5.3.2 PREDICTOR-CORRECTOR PROCEDURES

One of the possibilities to improve the time resolution is to apply the predictor-corrector technique combined with approximations (5.1) and (5.3).

To avoid cumbersome formulae we restrict ourselves to equation (5.16).

For the predictor approximation we take the function $\overline{u}_h(x, n\tau)$ defined by the right-hand side of (5.23). As an approximation for $f(mh, \lambda, u(mh, \lambda))$ we choose the linear function

$$[(n-1)\tau, n\tau] \ni \lambda \rightarrow f(mh, (n-1)\tau, u_h(mh, (n-1)\tau)) +$$

$$(\lambda - (n-1)\tau)\frac{f(mh, n\tau, \overline{u}_h(mh, n\tau)) - f(mh, (n-1)\tau, u_h(mh, (n-1)\tau))}{\tau}.$$

Then function (5.20) is approximated by the sum of (5.22) and

$$\frac{1}{(\pi D)^{1/2}\tau} \sum_{m=-\infty}^{\infty} (f(mh, n\tau, \overline{u}_h(mh, n\tau)) -$$

$$f(mh, (n-1)\tau, u_h(mh, (n-1)\tau))) \times$$

$$\frac{\partial}{\partial x} \int_{(n-1)\tau}^{n\tau} \int_{\mathbb{R}^1} (\lambda - (n-1)\tau)\mathcal{P}(x - \xi, n\tau - \lambda) \exp\left(-\frac{(x-mh)^2}{Dh^2}\right) d\xi d\lambda.$$

The derivative with respect to x of the last double integral equals

$$\frac{\partial}{\partial x} \int_{(n-1)\tau}^{n\tau} \exp\left(-\frac{(x-mh)^2}{4\nu(n\tau - \lambda) + Dh^2}\right) \frac{D^{1/2}h(\lambda - (n-1)\tau)}{(4\nu(n\tau - \lambda) + Dh^2)^{1/2}} d\lambda =$$

$$\frac{h^3(\pi D)^{1/2}}{4\nu^2} \left\{ (\kappa(\tau) + D)\left(\Xi_m^2(\tau) + \frac{1}{2}\right) \mathrm{erf}\,(\Xi_m(\lambda)) + \right.$$

$$\left. \frac{1}{\sqrt{\pi}}(\kappa(\lambda) + D)\Xi_m(\lambda)e^{-\Xi_m^2(\lambda)} \right\}\Bigg|_{\lambda=0}^{\lambda=\tau}.$$

The corrected values $u_h(x, n\tau)$ of the solution are found from the explicit formula

$$u_h(x, n\tau) = \overline{u}_h(x, n\tau) + \frac{h^3}{4\nu^2\tau} \sum_{m=-\infty}^{\infty} (f(mh, n\tau, \overline{u}_h(mh, n\tau)) -$$

$$f(mh, (n-1)\tau, u_h(mh, (n-1)\tau))) \left\{ (\kappa(\tau) + D)\left(\Xi_m^2(\tau) + \frac{1}{2}\right) \mathrm{erf}\,(\Xi_m(\lambda)) + \right.$$

$$\left. \frac{1}{\sqrt{\pi}}(\kappa(\lambda) + D)\Xi_m(\lambda)e^{-\Xi_m^2(\lambda)} \right\}\Bigg|_{\lambda=0}^{\lambda=\tau}. \tag{5.26}$$

The intermediate predicted values are calculated by

$$\overline{u}_h(x, n\tau) = u_h(x, (n-1)\tau) +$$

$$\frac{1}{\sqrt{\pi}} \sum_{m=-\infty}^{\infty} u_h(mh, (n-1)\tau) \frac{e^{-\Xi_m^2(\lambda)}}{(\kappa(\lambda) + D)^{1/2}}\Bigg|_{\lambda=0}^{\lambda=\tau} +$$

$$\frac{h}{2\nu} \sum_{m=-\infty}^{\infty} f(mh, (n-1)\tau, u_h(mh, (n-1)\tau))\mathrm{erf}\,(\Xi_m(\lambda))\Bigg|_{\lambda=0}^{\lambda=\tau}. \tag{5.27}$$

We put $u_h(x, 0) = \varphi(x)$. The resulting procedure (5.26), (5.27) has the accuracy $O(\tau^3 + \tau h^2)$.

Similarly to the proof of (5.24) we apply (5.3′) to (5.26) and (5.27) and arrive at the time marching algorithm

$$u_h(x, n\tau) = \overline{u}_h(x, n\tau) + \frac{h}{4\nu^2\tau} \sum_{m=-\infty}^{\infty} (f(mh, n\tau, \overline{u}_h(mh, n\tau)) -$$

$$f(mh, (n-1)\tau, u_h(mh, (n-1)\tau))) \times$$

$$\left\{ \left(\frac{1}{2}\kappa(\tau) + (\kappa(\tau) + D)\Xi_m^2(\tau) \right) \operatorname{erf}(\Xi_m(\lambda)) + \right.$$

$$\left. \frac{1}{2\sqrt{\pi}}\Xi_m(\lambda)(2\kappa(\lambda) + D + 4D\Xi_m^2(\lambda))e^{-\Xi_m^2(\lambda)} \right\} \Bigg|_{\lambda=0}^{\lambda=\tau},$$

where

$$\overline{u}_h(x, n\tau) = u_h(x, (n-1)\tau) +$$

$$\frac{1}{\sqrt{\pi}} \sum_{m=-\infty}^{\infty} u_h(mh, (n-1)\tau) \left(\kappa(\lambda) + \frac{3}{2}D - D\Xi_m^2(\lambda) \right) \frac{e^{-\Xi_m^2(\lambda)}}{(\kappa(\lambda) + D)^{3/2}} \Bigg|_{\lambda=0}^{\lambda=\tau} +$$

$$\frac{h}{2\nu} \sum_{m=-\infty}^{\infty} f(mh, (n-1)\tau, u_h(mh, (n-1)\tau)) \times$$

$$\left\{ \frac{D}{\sqrt{\pi}}\Xi_m(\lambda) \frac{e^{-\Xi_m^2(\lambda)}}{\kappa(\lambda) + D} + \operatorname{erf}(\Xi_m(\lambda)) \right\} \Bigg|_{\lambda=0}^{\lambda=\tau}$$

The accuracy is of the order $O(\tau^3 + \tau h^4)$.

5.3.3 THE CASE OF TWO SPACE DIMENSIONS

All the algorithms introduced above have direct analogs for the multidimensional case. We restrict ourselves to the equation with two space variables

$$u_t - \nu\Delta u = \operatorname{div} \mathbf{f}(x, t, u) \quad \text{for} \quad x \in \mathbb{R}^2, \quad t > 0 \tag{5.28}$$

and to the simplest algorithms similar to (5.23) and (5.24).

We rewrite (5.28) as

$$u(x, T) = u(x, t) + \int_{\mathbb{R}^2} \mathcal{P}(x - \xi, T)(u(\xi, t) - u(x, t))d\xi +$$

$$\operatorname{div} \int_t^T \int_{\mathbb{R}^2} \mathcal{P}(x - \xi, T - \lambda)\mathbf{f}(\xi, \lambda, u(\xi, \lambda))d\xi d\lambda.$$

(compare with (5.17)) and employ approximation (5.6) to evaluate the integrals. By the same arguments as in Section 5.3.1 we arrive at the computational formula

$$u_h(x, n\tau) = u_h(x, (n-1)\tau) + \sum_{m \in \mathbb{Z}^2} u_h(mh, (n-1)\tau) \left. \frac{e^{-z(\lambda)}}{\pi(\kappa(\lambda) + D)} \right|_{\lambda=0}^{\lambda=\tau} +$$

$$\frac{h^2}{2\pi\nu} \sum_{m \in \mathbb{Z}^2} \left\langle \mathbf{f}(mh, (n-1)\tau, u_h(mh, (n-1)\tau)) \,,\, \frac{(x - mh)}{|x - mh|^2} \right\rangle e^{-z(\lambda)} \Bigg|_{\lambda=0}^{\lambda=\tau} , \qquad (5.29)$$

where

$$z(\lambda) = \left| \frac{x}{h} - m \right|^2 (\kappa(\lambda) + D)^{-1}.$$

The accuracy is $O(\tau^2 + \tau h^2)$ at each time step.

The more accurate approximation

$$\mathcal{M}_h^{(1)} u(x) = \mathcal{M}_h^{(0)} u(x) - \frac{\mathcal{D} h^2}{4} \Delta \mathcal{M}_h^{(0)} u(x)$$

transforms (5.29) to an algorithm with accuracy $O(\tau^2 + \tau h^4)$:

$$u_h(x, n\tau) = u_h(x, (n-1)\tau) +$$

$$\sum_{m \in \mathbb{Z}^2} u_h(mh, (n-1)\tau) \{\kappa(\lambda) - 2D - Dz(\lambda)\} \left. \frac{e^{-z(\lambda)}}{\pi(\kappa(\lambda) + D)^2} \right|_{\lambda=0}^{\lambda=\tau} +$$

$$\sum_{m \in \mathbb{Z}^2} \left\langle \mathbf{f}(mh, (n-1)\tau, u_h(mh, (n-1)\tau)) \,,\, \frac{(x - mh)}{|x - mh|^2} \right\rangle \times$$

$$\{\kappa(\lambda) + D - Dz(\lambda)\} \left. \frac{h^2 e^{-z(\lambda)}}{2\pi\nu(\kappa(\lambda) + D)} \right|_{\lambda=0}^{\lambda=\tau} . \qquad (5.30)$$

By using

$$\mathcal{M}_h^{(1)} u(x) = \mathcal{M}_h^{(0)} u(x) - \frac{\mathcal{D} h^2}{4} \Delta \mathcal{M}_h^{(0)} u(x) + \frac{\mathcal{D}^2 h^4}{32} \Delta^2 \mathcal{M}_h^{(0)} u(x)$$

we arrive in the same manner at the next algorithm with accuracy $O(\tau^2 + \tau h^6)$:

$$u_h(x, n\tau) = u_h(x, (n-1)\tau) +$$

$$\sum_{m \in \mathbb{Z}^2} u_h(mh, (n-1)\tau) \frac{\kappa^2(\lambda) - 3D^2 + (\kappa(\lambda) + 3D - Dz(\lambda))^2}{2\pi(\kappa(\lambda) + D)^3} e^{-z(\lambda)} \Bigg|_{\lambda=0}^{\lambda=\tau} +$$

$$\sum_{m \in \mathbb{Z}^2} \left\langle \mathbf{f}(mh, (n-1)\tau, u_h(mh, (n-1)\tau)) \,, \, \frac{(x - mh)}{|x - mh|^2} \right\rangle \times$$

$$\frac{\kappa^2(\lambda) - 2D^2 + (\kappa(\lambda) + 2D - Dz(\lambda))^2}{4\pi\nu(\kappa(\lambda) + D)^2} h^2 e^{-z(\lambda)} \bigg|_{\lambda=0}^{\lambda=\tau} . \tag{5.31}$$

Maybe it is of interest to note that in contrast to the one-dimensional case, computational formulae (5.29), (5.30) and (5.31) do not include any special functions. Moreover, since the exponentials are factorizable with respect to variables x_1 and x_2, their values may be precomputed before the march in time. The factors only occupy two one-dimensional arrays of the length l_0.

To test the algorithms, we have computed some model problems. For instance, we have compared the 2nd and the 6th order algorithms (5.29) and (5.31) by solving equation (5.28) for

$$f_1 = \frac{1}{2}(v^2 - u^2 + \frac{\nu_1}{\nu}x_1 v), \quad f_2 = \frac{1}{2}(v^2 - u^2 + \frac{\nu_1}{\nu}x_2 v),$$

where $v(x) = \exp\left(-|x|^2/4\nu_1\right)$. The function v coincides with the stationary solution of equation (5.28). We solved the Cauchy problem for (5.28) with the initial values

$$u(x, 0) = \exp\left(-|x - x^{(0)}|^2/4\nu_1\right).$$

The functions $u_h - v$ computed by algorithms (5.29) and (5.31) are depicted in Fig. 5.1 and Fig. 5.2, respectively.

We have used $\nu_1 = \nu = 0.5, t = 10, \tau = 0.05, x_1^{(0)} = x_2^{(0)} = -0.07$. For the algorithm (5.29) the space grid step was $h = 0.2$ while for (5.31) it was $h = 0.5556$. The accuracy was practically the same, but the computer time required was ten times less in the last case.

5.4 APPLICATION OF APPROXIMATE APPROXIMATIONS FOR SOLVING THE NON-STATIONARY NAVIER-STOKES EQUATIONS

Consider the Cauchy problem for the Navier-Stokes equations

$$\mathbf{u}_t - \nu\Delta\mathbf{u} - \nabla p + \frac{\partial}{\partial x_k}(u_k\mathbf{u}) = 0 \tag{5.32}$$

$$\mathrm{div}\,\mathbf{u} = 0, \quad \mathbf{u}(x, 0) = \varphi(x), \quad x \in \mathbb{R}^2.$$

Using the continuity equation we find that the pressure is expressed by

$$p = \frac{\partial^2}{\partial x_i \partial x_k}\Delta^{-1}(u_i u_k) \tag{5.33}$$

Rewrite (5.32) as

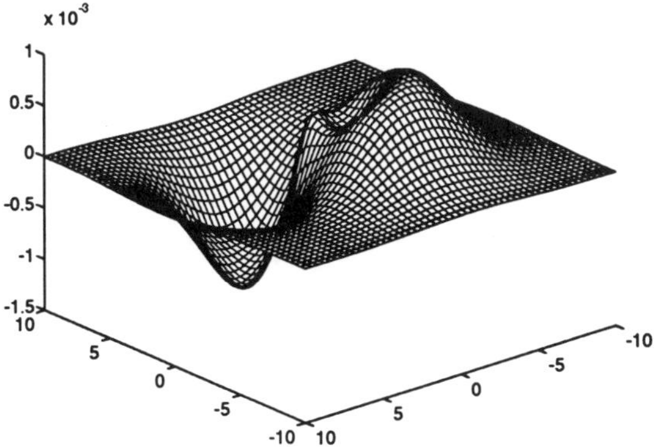

Figure 5.1

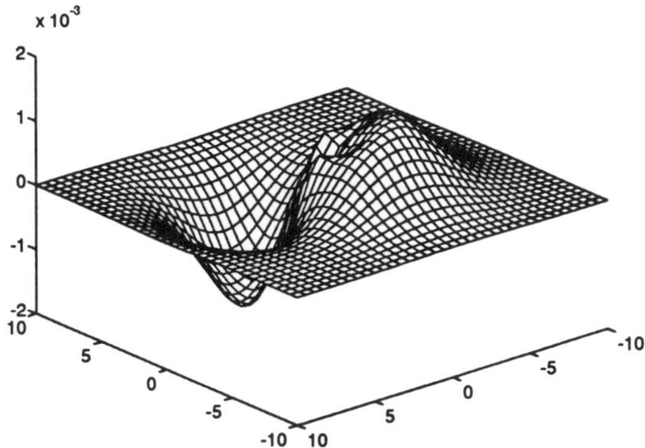

Figure 5.2

$$\mathbf{u}(x,T) = \mathbf{u}(x,t) + \int\limits_{\mathbb{R}^2} \mathcal{P}(x-\xi,T)(\mathbf{u}(\xi,t) - \mathbf{u}(x,t))d\xi +$$

$$\nabla_x \int\limits_t^T \int\limits_{\mathbb{R}^2} \mathcal{P}(x-\xi,T-\lambda)p(\xi,\lambda)d\xi d\lambda -$$

$$\frac{\partial}{\partial x_k} \int\limits_t^T \int\limits_{\mathbb{R}^2} \mathcal{P}(x-\xi,T-\lambda)(u_k\mathbf{u})(\xi,\lambda)d\xi d\lambda. \tag{5.34}$$

By (5.33) we transform this equation to the form

$$u_l(x,T) = u_l(x,t) +$$

$$\int_{\mathbb{R}^2} \mathcal{P}(x-\xi,T)(u_l(\xi,t)-u_l(x,t))d\xi - \left(\delta_{il} - \Delta_x^{-1}\frac{\partial^2}{\partial x_i \partial x_l}\right) F_i,$$

where

$$F_i = \mathrm{div}\int_t^T \int_{\mathbb{R}^2} \mathcal{P}(x-\xi,T-\lambda)(u_i\mathbf{u})(\xi,\lambda)d\xi d\lambda, \quad i=1,2.$$

By (5.29) the functions F_i can be approximated as follows

$$F_i = \frac{h^2}{2\pi\nu}\sum_{m\in\mathbb{Z}^2}\left\langle (u_i\mathbf{u})(mh,(n-1)\tau,u_h(mh,(n-1)\tau)),\frac{(x-mh)}{|x-mh|^2}\right\rangle e^{-a(\lambda)|x-mh|^2}\Bigg|_{\lambda=0}^{\lambda=\tau},$$

where $a(\lambda) = (4\nu\lambda + \mathcal{D}h^2)^{-1}$.

It remains to calculate

$$W_{ikl} = \Delta_x^{-1}\frac{\partial^2}{\partial x_i \partial x_l}\left(\frac{x_k - m_k h}{|x-mh|^2}e^{-a(\lambda)|x-mh|^2}\right).$$

By setting $y = x - mh$ we have

$$W_{ikl} = -\frac{1}{2}\Delta_y^{-1}\frac{\partial^3}{\partial y_i \partial y_k \partial y_l}\left(\int_{a(\lambda)|y|^2}^{\infty} e^{-t}\frac{dt}{t}\right) = -\frac{1}{2a(\lambda)}\frac{\partial^3}{\partial y_i \partial y_k \partial y_l}\left(v(a(\lambda)|y|^2)\right),$$

where v satisfies the equation

$$\Delta(v(|Y|^2)) = \int_{|Y|^2}^{\infty} e^{-t}\frac{dt}{t}.$$

Hence

$$4v'(\xi) = \int_\xi^\infty e^{-t}\frac{dt}{t} + \frac{1-e^{-\xi}}{\xi}. \tag{5.35}$$

Furthermore

$$W_{ikl} = -\frac{\partial^2}{\partial y_i \partial y_k}v'(a(\lambda)|y|^2)y_l =$$

$$2a(\lambda)v''(a(\lambda)|y|^2)\left(\delta_{kl}y_i + \delta_{il}y_k + \delta_{ik}y_l\right) - 4a^2(\lambda)v'''(a(\lambda)|y|^2)y_i y_k y_l$$

which along with (5.35) implies

$$W_{ikl} = \frac{1 - e^{-a(\lambda)|y|^2}}{2a(\lambda)|y|^4}\left(\delta_{kl}y_i + \delta_{il}y_k + \delta_{ik}y_l - \frac{4y_i y_k y_l}{|y|^2}\right) + y_i y_k y_l \frac{e^{-a(\lambda)|y|^2}}{|y|^4}.$$

Hence

$$\left(\delta_{il} - \Delta_x^{-1}\frac{\partial^2}{\partial x_i \partial x_l}\right)F_i \doteq \frac{h^2}{2\pi\nu}\sum_{m\in\mathbb{Z}^2}(u_i u_k)(mh, (n-1)\tau)\times$$

$$\left\{\left(\delta_{il} - \frac{(x_i - m_i h)(x_l - m_l h)}{|x - mh|^2}\right)\frac{x_k - m_k h}{|x - mh|^2}e^{-z(\lambda)}\bigg|_{\lambda=0}^{\tau} - \right.$$

$$\left(\frac{x_i - m_i h}{|x - mh|}\delta_{kl} + \frac{x_k - m_k h}{|x - mh|}\delta_{il} + \frac{x_l - m_l h}{|x - mh|}\delta_{ik} - \right.$$

$$\left.4\frac{(x_i - m_i h)(x_k - m_k h)(x_l - m_l h)}{|x - mh|^3}\right)\frac{(1 - e^{-z(\lambda)})(\kappa(\lambda) + D)}{2|x - mh|^3}\bigg|_{\lambda=0}^{\tau}\bigg\} \tag{5.36}$$

with the notation introduced in Section 5.3.4.

It remains to notice that by (5.29) the first integral in the right-hand side of (5.34) can be approximated as

$$\sum_{m\in\mathbb{Z}^2}\mathbf{u}_h(mh, (n-1)\tau)\frac{e^{-z(\lambda)}}{\pi(\kappa(\lambda) + \mathcal{D})}\bigg|_{\lambda=0}^{\tau}.$$

By combining this with (5.36) we arrive at the time marching algorithm for the velocity components

$$u_{h,l}(x, n\tau) = u_{h,l}(x, (n-1)\tau) + \sum_{m\in\mathbb{Z}^2}u_{h,l}(mh, (n-1)\tau)\frac{e^{-z(\lambda)}}{\pi(\kappa(\lambda) + \mathcal{D})}\bigg|_{\lambda=0}^{\tau} -$$

$$\frac{h^2}{2\pi\nu}\sum_{m\in\mathbb{Z}^2}(u_{h,i}u_{h,k})(mh, (n-1)\tau)\times$$

$$\left\{\left(\delta_{il} - \frac{(x_i - m_i h)(x_l - m_l h)}{|x - mh|^2}\right)\frac{x_k - m_k h}{|x - mh|^2}e^{-z(\lambda)}\bigg|_{\lambda=0}^{\tau} - \right.$$

$$\left(\frac{x_i - m_i h}{|x - mh|}\delta_{kl} + \frac{x_k - m_k h}{|x - mh|}\delta_{il} + \frac{x_l - m_l h}{|x - mh|}\delta_{ik} - \right.$$

$$\left.4\frac{(x_i - m_i h)(x_k - m_k h)(x_l - m_l h)}{|x - mh|^3}\right)\frac{(1 - e^{-z(\lambda)})(\kappa(\lambda) + D)}{2|x - mh|^3}\bigg|_{\lambda=0}^{\tau}\bigg\} \tag{5.37}$$

The accuracy of this algorithm is $O(\tau^2 + \tau h^2)$ at each time step.

When the approximate velocity is found, the pressure is approximated by

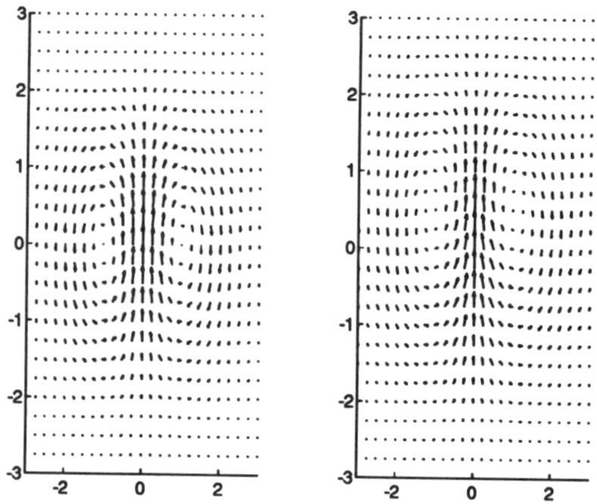

Figure 5.3 Velocity vectors at $t = 0$ (left) and $t = 2$ (right).

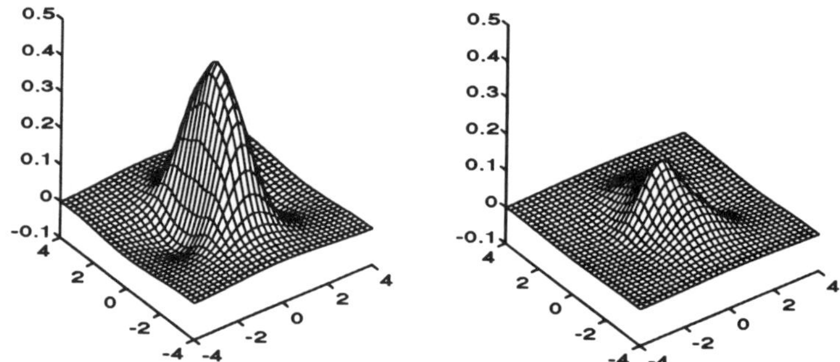

Figure 5.4 Pressure distribution at $t = 0$ (left) and $t = 2$ (right).

$$p_h(x, n\tau) = -\frac{h^2}{2\pi} \sum_{m \in \mathbb{Z}^2} (u_{h,i} u_{h,k})(mh, n\tau) \left\{ \frac{(1 - e^{-z(0)})}{z(0)} \delta_{ik} + \right.$$

$$\left. \frac{2}{\mathcal{D}h^2} \frac{(1 + z(0))e^{-z(0)}}{z^2(0)} (x_i - m_i h)(x_k - m_k h) \right\} \tag{5.38}$$

with the accuracy $O(h^2)$ due to (5.33) and (5.14).

Some results of the numerical solution of the Cauchy problem (5.32) with $\varphi_1 = x_1 x_2 \exp(-|x|^2/2)$, $\varphi_2 = (1 - x_1^2) \exp(-|x|^2/2)$ are depicted in figures 5.3 and 5.4 ($\nu = 0.1$, $\tau = 0.01$, $h = 0.05$).

By using the higher order approximation (5.11) we easily modify the above algorithm to obtain the accuracy $O(\tau^2 + \tau h^{2N})$, $N > 1$. This error estimate can be improved with respect to t by using, for example, the predictor-corrector technique in the spirit of Section 5.3.2.

5.5 BOUNDARY POINT METHOD

The possibility to obtain explicit formulae for values of various integral and pseudodifferential operators of mathematical physics applied to basis functions leads to a new approach for the discretization of boundary integral equations, which is not based upon the decomposition of the boundary into elements. In this approach the coefficients of the resulting algebraic system depend only on the coordinates of a finite number of boundary points and the direction of the normal at these points; hence the name Boundary Point Method (BPM) seems quite natural (Maz'ya (1991)).

The first step of the BPM consists of the approximation of the potential densities by a linear combination of basis functions each being either concentrated near a particular boundary point or decreasing rapidly with the distance from this point. The calculation of the potentials, whose densities are the basis functions, is carried out in the second step. Of course both steps can be performed in many different ways.

In what follows I discuss only one particular variant of the BPM which leads to an algebraic system with coefficients which are expressed explicitly — up to a simple quadrature. The role of a basis function is played by the density of the normal distribution shifted to each of the selected boundary points. When calculating the double layer potential of such a basis function one replaces the surface integration by integration over the tangent plane at an appropriate boundary point. In the two-dimensional case the coefficients of the resulting algebraic system are represented by the function erfc, whereas in the three-dimensional case a certain one dimensional integral containing a Bessel function has to be calculated. Although I treat only the Dirichlet problem for the Laplacian, it is clear that the method can be applied to a large class of problems.

Here I restrict myself to a heuristic exposition, making no attempt at a mathematical justification of the method and estimating errors.

Let u be the solution of the Dirichlet problem

$$\Delta u = 0 \text{ in } \Omega, \quad u = b \text{ on } \Gamma. \tag{5.39}$$

Here $\Omega \subset \mathbb{R}^m$, $(m > 1)$ is a domain with smooth boundary Γ, Δ is the Laplacian in $\mathbb{R}^m$, and b is a smooth function given on Γ.

The solution u of problem (5.39) can be represented in the form of a double-layer potential

$$(W\varphi)(x) = \int_\Gamma T(x, y)\varphi(y)d_y\Gamma. \tag{5.40}$$

We use the notation $T(x, y) = \frac{\partial}{\partial n(y)}E(x, y)$, where $n(y)$ is the outward unit normal at the point $y \in \Gamma$. Let $E(x, y)$ denote the fundamental solution of the Laplace equation specified by

$$E(x, y) = \begin{cases} (2\pi)^{-1} \log|x - y| & \text{for } m = 2, \\ -(m - 2)^{-1}\omega_m^{-1}|x - y|^{2-m} & \text{for } m \geq 3, \end{cases}$$

where $\omega_m = 2\pi^{m/2}/\Gamma(m/2)$ is the area of the unit sphere in $\mathbb{R}^m$.

As usual, from (5.40) we obtain the boundary integral equation

$$\frac{1}{2}\varphi(x) + (W\varphi)(x) = b(x), \quad x \in \Gamma. \tag{5.41}$$

Let $\{x_k\}_{k=1}^{N}$ be a collection of boundary points and let $\{h_k\}_{k=1}^{N}$ be a collection of positive constants having the dimension of length. For example, h_n may be chosen as the average distance from x_n to the neighbouring points of $\{x_k\}_{k=1}^{N}$.

Let φ be a function on Γ. For the approximation of φ we use

$$\sigma(x) = \sum_{k=1}^{N} c_k g_k(x),$$

where the basis functions are given by

$$g_k(x) = \exp(-\frac{(x - x_k)^2}{Dh_k^2}), \quad x \in \mathbb{R}^m,$$

and D is a dimensionless positive constant.

Then

$$u(x) \approx \sum_{k=1}^{N} c_k (Wg_k)(x), \quad x \in \Omega, \tag{5.42}$$

and by (5.41)

$$\sum_{k=1}^{N} c_k(\frac{1}{2}g_k(x) + (Wg_k)(x)) \approx b(x), \quad x \in \Gamma$$

In particular

$$\sum_{k=1}^{N} c_k(\frac{1}{2}g_k(x_j) + (Wg_k)(x_j)) \approx b(x_j), \quad j = 1, \ldots, N \tag{5.43}$$

We shall give approximate analytic expressions for the coefficients of this algebraic system using only the points $x_1, \ldots, x_N$ and the unit normals $n_k = n(x_k)$, $k = 1, \ldots, N$.

Let Γ_k be the tangent plane to Γ at x_k and let

$$(\Psi_k g_k)(x) = \int_{\Gamma_k} T(x, y) g_k(y) d_y \Gamma_k,$$

with $n(y) = n(x_k)$ in the definition of $T(x, y)$.

We use the approximate identity:

$$(Wg_k)(x_j) \approx (\Psi_k g_k)(x_j) + \frac{1}{2}g_k(x_j)\mathrm{sign}((x_j - x_k) \cdot n_k). \tag{5.44}$$

which can be obtained by the following plausible argument.

Let $(Wg_k)^+(x_j)$ and $(Wg_k)^-(x_j)$ be the interior and exterior limit values of $(Wg_k)(x_j)$ as $x \to x_j$. By x_j^+ and x_j^- we denote two points inside and outside Ω placed on the normal to $\partial\Omega$ at the point x_j and such that $h_j \gg |x_j^\pm - x_j| \gg h_j^2$. Then in the case $(x_j - x_k) \cdot n_k < 0$, we have

$$(Wg_k)^+(x_j) \approx (Wg_k)(x_j^+) \approx (\Psi_k g_k)(x_j)$$

which along with the identity

$$(Wg_k)(x_j) = (Wg_k)^+(x_j) - \frac{1}{2}g_k(x_j)$$

leads to (5.44). If $(x_j - x_k) \cdot n_k > 0$, then

$$(Wg_k)^-(x_j) \approx (Wg_k)(x_j^-) \approx (\Psi_k g_k)(x_j)$$

and by the relation

$$(Wg_k)(x_j) = (Wg_k)^-(x_j) + \frac{1}{2}g_k(x_j)$$

we arrive at (5.44). In the third case when $(x_j - x_k) \cdot n_k = 0$, we have

$$(Wg_k)^+(x_j) \approx (\Psi_k g_k)^+(x_j) = \frac{1}{2}g_k(x_j).$$

Since $Wg_k = (Wg_k)^+ - \frac{1}{2}g_k$ on $\partial\Omega$ we obtain $(Wg_k)(x_j) = 0$, which together with the identity $(\Psi_k g_k)(x_j) = 0$ implies (5.44).

The same argument applies to the following approximation for Wg_k in Ω

$$(Wg_k)(x) \approx \frac{1}{2}g_k(x)[1 + \text{sign}((x - x_k) \cdot n_k)] + (\Psi_k g_k)(x).$$

By (5.43) and (5.44) a discrete form of the boundary integral equation (5.41) is given by the linear algebraic system

$$\sum_{k=1}^{N} a_{jk}c_k = b(x_j), \quad 1, ..., N, \tag{5.45}$$

where

$$a_{jk} = \frac{1}{2}g_k(x_j)[1 + \text{sign}((x_j - x_k) \cdot n_k)] + (\Psi_k g_k)(x_j) \tag{5.46}$$

One can easily verify that

$$(\Psi_k g_k)(x_j) = H\left(\frac{z_{jk}}{\sqrt{D}h_k}, \frac{\rho_{jk}}{\sqrt{D}h_k}\right), \tag{5.47}$$

where

$$z_{jk} = (x_j - x_k) \cdot n_k, \quad \rho_{jk} = |x_j - x_k - z_{jk} \cdot n_k|.$$

In the two-dimensional case we find

$$H(z,\rho) = -\frac{1}{2}\mathrm{sign}(z)\Re[\exp(-(\rho + i|z|)^2)\mathrm{erfc}(|z| - i\rho)] \qquad (5.48)$$

where erfc is the complementary error function $1 - \mathrm{erf}$.

In the three-dimensional case we obtain

$$H(z,\rho) = -\mathrm{sign}(z) \int_0^\infty \exp(-t^2 - 2|z|t)J_0(2\rho t)dt, \qquad (5.49)$$

where J_0 is the Bessel function of the first kind and order zero.

The approximation for the solution of the boundary value problem takes the form:

$$u(x) \approx \sum_{k=1}^N c_k \{\frac{1}{2}g_k(x)[1 + \mathrm{sign}((x - x_k) \cdot n_k)] +$$

$$+ H(\frac{(x - x_k) \cdot n_k}{\sqrt{D}h_k}, \frac{|x - x_k - (x - x_k) \cdot n_k|}{\sqrt{D}h_k})\}, \quad x \in \Omega,$$

where $c_1, \ldots, c_N$ satisfy algebraic system (5.45).

Some numerical experiments with the BPM are described in (Maz'ya (1991)).

ACKNOWLEDGEMENTS

I would like to thank Dr. Vladimir Karlin for the time and effort devoted to computer implementation of the time marching algorithms in Section 5.4.

I wish to express my gratitude to Dr. Gunther Schmidt for invaluable help and useful discussions.

I am grateful to Dr. Naum Khutoryansky for interesting comments on numerical aspects and practical perspectives of the BPM. The name itself, Boundary Point Method, appeared in one of our conversations.

REFERENCES

Maz'ya, V.G., (1991), A new approximation method and its applications to the calculation of volume potentials. Boundary point method. *3 DFG – Kolloquium des DFG – Forschungsschwerpunktes "Randelementmethoden"*, 30 Sept. – 5 Oct. 1991, Schloß Reisensburg, S. 18.

Stein, E.M. and Weiss G., (1971), Introduction to Fourier analysis on Euclidean spaces. Princeton University Press.

Maz'ya, V. and Schmidt G., (1993), Error estimates for approximate approximations. (to appear).

Maz'ya, V. and Karlin V., (1993), Semianalytic Time-Marching Algorithms for Non-Linear Parabolic Equations. *Department of Mathematics, University of Linköping*, Preprint LiTH–MAT–R–93–06.

Richtmyer, R.D. and Morton K.W., (1967), Difference methods for initial-value problems. John Wiley and Sons.

Maz'ya, V., (1991), Boundary Point Method. *Department of Mathematics, University of Linköping*, Preprint LiTH–MAT–R–91–44.

Chapter 6

A New Paradigm for Adaptive Finite Element Methods

Claes Johnson

Mathematics Department, Chalmers University of Technology,
412 96 Göteborg, Sweden

6.1 INTRODUCTION

In this paper we give an overview of our recent work together with K. Eriksson, P. Hansbo and A. Szepessy on adaptive finite element methods. The paper continues the previous survey Johnson and Hansbo (1992), where numerical examples and references into the literature on adaptive finite elements can be found. A detailed account of our work will be presented in the monograph Erikkson, Hansbo and Johnson (1993(iv)). Here we seek to exhibit the basic ideas underlying our work, which we believe to give a new "paradigm" for adaptivity in the finite element method based on the cornerstones:

$$\text{strong stability,} \tag{6.1a}$$

$$\text{Galerkin orthogonality.} \tag{6.1b}$$

Strong stability is a "new" concept, which will be explained below.

An adaptive finite element method is a finite element method together with an algorithm, referred to as an *adaptive algorithm*, for automatic quantitative control of the discretization error in the finite element method in a norm and on a tolerance level given by the user. Ideally the adaptive algorithm should be *reliable* in the sense that the error control should be guaranteed, and also *efficient* in the sense that the computational work should be minimal. The basic problem may be said to be to find an algorithm for determining the "right" computational mesh for the given problem, error norm and tolerance level. Adaptivity opens new possibilities to solve problems with complex solutions with features of different scales, such as boundary layers, shocks and reaction zones. Further, quantitative error control is clearly of great importance in applications: For instance, the user should be able to demand and obtain error control in the maximum norm for the stresses on the tolerance level 1% in a stress analysis computation. Successful solutions to the basic problem of adaptive error control will have a profound influence on the finite element software of tomorrow. Today, very few codes have adaptive features and quantitative error control is available for only a very limited class of problems. Tomorrow, a code which is not adaptive will not be competitive. An intense activity in creating new adaptive codes, or supplying old

The Mathematics of Finite Elements and Applications
Edited by J. R. Whiteman © 1994 John Wiley & Sons Ltd

codes with adaptive features is to be expected in the next few years. It appears that the basic mathematical tools required to realize quantitative error control for finite element methods for large classes of problems are now available, and a period of rapid development in this area is to be expected.

6.2   THE FINITE ELEMENT METHOD

The finite element method is a general method for the numerical solution of the partial differential equations and integral equations which are the basic mathematical models in science and engineering. The possibility of solving partial differential equations numerically using computers has brought a revolutionary new technique to many areas of science and engineering and has led to the development of new academic and industrial fields, such as computational solid mechanics, computational fluid mechanics, computational physics, computational chemistry, computational biology etc. . The finite element method is the dominating numerical method in many of these areas with rapid expansion into others. The finite element method may be described as the combination of

$$\text{Galerkin's method,} \tag{6.2a}$$

$$\text{piecewise polynomial approximation.} \tag{6.2b}$$

We recall that Galerkin's method essentially corresponds to seeking an approximate solution $u_{h,p}$ in a finite dimensional 'trial space' $V_{h,p}$ determined through a system of equations obtained by 'multiplying' the given equation by functions varying over a finite dimensional 'test space' $W_{h,p}$, with in the model case $V_{h,p} = W_{h,p}$. The finite dimensional space $V_{h,p}$ is constructed using piecewise polynomial functions of degree $p = p(x,t)$ on a 'mesh' or 'triangulation' of space or space-time of mesh size $h = h(x,t)$, subdividing the given domain into elements K (triangles, quadrilaterals, tetrahedrons etc. .) of diameter h_K, where for $(x,t) \in K$, $h(x,t) \sim h_K$ and $p(x,t)$ is the degree of the polynomials on K.

The combination (6.2a,b) has had an astounding success : Tens of thousands of scientific papers and thousands of computer codes have been produced with this basis. The range of applications of the finite element method has been steadily expanding since its introduction in the late 1950's with focus on solid and structural mechanics. Today the finite element method is applied to a wide range of partial differential equations, including elliptic, parabolic and hyperbolic type equations, linear and nonlinear, arising in many areas of applications, such as solid mechanics, fluid mechanics, aerodynamics, electromagnetics, heat conduction, reaction-diffusion, scattering, wave propagation etc.

6.3   GALERKIN'S METHOD: THE "UNIVERSAL" METHOD

Broadly speaking the partial differential equations arising in science and engineering may be classified as being of elliptic, parabolic or hyperbolic type, or combinations thereof. As an important example we recall that stationary diffusion-dominated convection-diffusion problems are elliptic, nonstationary such problems are parabolic and convection-dominated such problems are of hyperbolic type. Further, the wave equation is hyperbolic and the equations of fluid and aerodynamics (the Navier Stokes equations) are parabolic-hyperbolic depending on the balance between viscous and convective effects. The finite element method in the basic form (6.2a,b) has found extensive applications to elliptic and parabolic problems. Modified forms of the Galerkin method including weighted least squares and artificial viscosity were introduced during the 1980's extending the range of application of the finite element method to hyperbolic problems. Today the finite element method is applicable to the whole range of partial differential equations of elliptic, parabolic or hyperbolic type. This brings a new unification into the area of numerical methods for differential equations, where essentially the same method, that is Galerkin's method in various forms, is used in both solid and fluid mechanics. In fact, evidence has gathered in recent years indicating that Galerkin methods are more "universal" than believed before. For instance, discretization in not only space but also time, or more generally in "space-time", is today realized through Galerkin procedures with important advantages. In particular, the idea of Galerkin in space-time makes it possible to unite the traditionally separated worlds of Eulerian and Lagrangean methods and basically "all" methods of computational fluid dynamics (finite difference, finite volume, particle) may be viewed as special cases of Galerkin in space-time. This is a central theme underlying our work.

To sum up, it appears that Galerkin's method with piecewise polynomials in space-time, is a truely "universal" method and certainly a fundamental cornerstone of Numerical Analysis.

6.4   THE ROLE OF MATHEMATICS IN THE DEVELOPMENT OF THE FINITE ELEMENT METHOD

Mathematics has certainly played an important role in the development of finite element methods, but a large number of finite element codes, also for complex problems, have been created without going too much into the mathematical intricacies of the method related to stability, error estimates etc. . The basic scheme (6.1a,b), together with some insight into the physics of the problem treated, has often been enough to serve as the basis to produce a code. However, in adaptivity with the aim of quantitative error control, it appears to be necessary to dig deeper into the mathematical structure built by (6.1a,b), and thus adaptivity seems to pose a challenge in the development of the finite element method that can only be met with some deeper insight into the mathematical structure of the finite element discretization.

6.5 THE TWO LEGS OF NUMERICAL ANALYSIS: *A PRIORI* AND *A POSTERIORI* ERROR ESTIMATES

The "fundamental theorem of Numerical Analysis" is usually presented as follows

$$\text{consistency} + \text{stability} \rightarrow \text{convergence.} \qquad (6.3)$$

Here "convergence" usually results from an *a priori* error estimate, where the error is estimated in terms of the exact solution and a discretization parameter representing the mesh size. The *a priori* error estimate is obtained combining "stability" of the discrete problem with "consistency" in the sense that the truncation error tends to zero as the mesh size tends to zero. The "truncation error" is obtained inserting the exact solution into the discrete equations. The "fundamental theorem of Numerical Analysis" may thus be expressed alternatively as follows:

$$\text{small truncation error} + \text{stability of discrete problem} \rightarrow \text{error small.} \qquad (6.4)$$

This may be described as the *a priori* "leg", and is traditionally considered to give "the" theoretical foundation of Numerical analysis. However, there is another leg equally important, where the error is estimated in *a posteriori* form in terms of the approximate solution. Symbolically this may be expressed as:

$$\text{small residual} + \text{stability of continuous problem} \rightarrow \text{error small,} \qquad (6.5)$$

where the residual is obtained inserting the approximate solution into the exact equations. Thus, Numerical Analysis has two legs, one *a priori* leg and one *a posteriori* leg. Both are fundamental and serve different purposes. Surprisingly, in the traditional presentation, the *a posteriori* leg is largely missing, at least in the case of numerical methods for differential equations. The reason for this unsatisfactory state of affairs is not completely clear, but may be related to the domination of finite difference methods in the early development of numerical methods for differential equations: For difference methods it seems "impossible" to insert the discrete solution into the exact equation and compute the residual. On the other hand, inserting the exact solution into the discrete equations to compute the "truncation error" is the standard starting point for the finite difference convergence analysis. For finite element methods there is no technical difficulty in computing the residual and the *a posteriori* leg is as natural and fundamental as the *a priori* leg. The adaptive methods are built on the *a posteriori* leg and they are analyzed using the *a priori* leg. Our work is based on this principle.

6.6 A NEW PARADIGM FOR ADAPTIVITY: GALERKIN + STRONG STABILITY

In the series of papers Erikkson (1993)-Johnson and Szepessy (1993), we present a new approach based on (6.4) to the design and analysis of reliable and efficient adaptive algorithms for finite element methods for a large class of problems, including elliptic, parabolic and hyperbolic problems, linear as well as nonlinear problems. The applications include

- linear scalar convection-diffusion problems in the whole range from diffusion-dominated elliptic-parabolic problems to convection-dominated hyperbolic problems, both stationary and time-dependent,
- linear elliptic problems: elasticity, viscous flow,
- second order linear wave equations,
- nonlinear reaction-diffusion,
- incompressible and compressible flow,
- nonlinear monotone elliptic problems modelling unilateral contact and elasto-plasticity,
- systems of ordinary differential equations.

We emphasize the generality of the setting presented. The success of the finite element method itself is largely connected with the generality of the method making it possible to attack virtually all partial differential equations arising in applications with the same methodology summarized in Section (6.1) above.

Our approach to adaptivity has the same degree of generality: The same basic methodology is used for "all" problems, whether of elliptic, parabolic or hyperbolic type covering a wide range of applications. Of course, there are differences related to the fundamental structural differences between the different types of partial differential equations, but the basic ingredients are the same. As far as we know this is the first time such a unified approach to adaptivity for finite element methods is realized. The adaptive algorithms are based on *a posteriori* error estimates, where the error is estimated in terms of the computed solution. The adaptive algorithms are analyzed through *a priori* error estimates involving the exact solution.

The fundamental ingredients to prove the *a posteriori* and *a priori* error estimates are (6.1a,b). The concept of strong stability is used to connect the residual of the finite element solution or the truncation error to the discretization error using the special feature of finite element methods: Galerkin orthogonality. The analogous connection for discretization methods not based on Galerkin orthogonality, is established using "weak stability". This is the standard approach taken, which in many cases leads to non-optimal results. Strong stability measures certain norms of certain derivatives of the solution of a perturbation equation in terms of norms of given data. The term "strong" refers to the fact that derivatives of the solution of the perturbation equation are involved, as compared to standard or weak stability measuring the norm of the perturbation itself. Strong stability is designed to measure the effect on the error of the special perturbations arising in Galerkin methods, while weak stability is designed to measure the influence of general perturbations. The advantage of strong stability coupled with Galerkin orthogonality is in many cases an essential improvement in the quality and sharpness of the *a posteriori* and *a priori* error estimates with a corresponding improvement of the efficiency in the performance and analysis of the adaptive method.

This is achieved by taking advantage of the special nature of Galerkin perturbations reflecting the Galerkin orthogonalities. As a familiar manifestation of this phenomenon, let us mention the mass conservation property of finite element schemes.

In the case of elliptic problems and energy norms, strong stability is the same as stability in the energy norm, and is thus a very familiar concept, while for other norms and problems strong stability is indeed a "new" concept. It is directly connected to the norm used on the residual of the finite element solution or the truncation error of the exact solution, allowing a weaker norm on these quantities than using standard weak stability and leading to improved results.

There is another important feature of our work, namely the distinction between the following sources of error in the finite element method:

- Galerkin discretization error,
- quadrature error.

The Galerkin discretization error is related to the polynomial approximation in the finite element method, assuming that the integrals arising from the variational formulation are evaluated exactly. The quadrature error comes from evaluating these integrals using numerical quadrature. The advantage of the separation of these two errors is that the Galerkin errors accumulate in a different (better) way because of the Galerkin orthogonalities, and thus improved results may be obtained this way. Further, in an adaptive setting the two errors may be controlled independently, which may increase efficiency. The standard error analysis for finite element methods for elliptic problems uses the splitting of Galerkin and quadrature errors, but the idea is generally applicable, and in fact seems to give an important new "principle" in Numerical Analysis. The available error analysis for finite difference methods for partial or ordinary differential equations does not contain this idea: In this traditional setting, no distinction is made between errors resulting from "discretizing derivatives" and quadrature errors. As a result the usual error estimates for finite difference methods are not sharp.

The mathematical content of our work may be summarized as a study of *a posteriori* and *a priori* error estimates resulting from

- the interplay between strong stability and Galerkin orthogonality,
- quantitative estimates of strong stability for various classes of problems.

The remaining part of our work is concerned with algorithmic questions concerning the realization of the adaptive algorithms built on the *a posteriori* error estimates.

Adaptivity for finite element methods with mathematical basis was initiated in the mid 1970's by Babuška and coworkers. The main emphasis in this work, which has been continued into present time with contributions by e.g Bank, Verfurth, Oden, and others, is on linear elliptic problems with error control in the energy norm. In our work we go outside this 'classical' framework and consider adaptivity also for parabolic and hyperbolic problems and error control in other norms than energy norms. The novelty of our work is the consistent use for "all" problems of the principle: strong stability plus Galerkin orthogonality, and the separation of Galerkin and quadrature errors.

6.7 THE BASIC PROBLEM IN ADAPTIVITY. RELIABILITY AND EFFICIENCY.

Let us now in more precise terms formulate the basic problem concerning adaptive finite element methods. Let then $\mathcal{P}$ be a given stationary boundary value problem, written symbolically as follows

$$Au = f, \tag{6.6}$$

where A is a partial differential operator on a domain Ω in space including (homogeneous) boundary conditions, f is given data and u is the exact solution. A finite element method $\mathcal{P}_{h,p}$ for $\mathcal{P}$ basically takes the form: Find $u_{h,p} \in V_{h,p}$ such that

$$\int_\Omega Au_{h,p}v \ dx = \int_\Omega fv \ dx \quad \forall v \in V_{h,p}, , \tag{6.7}$$

where $V_{h,p}$ is a certain finite element space of dimension n defined on a "mesh" $T_{h,p}$ where the "mesh function" $(h, p) = (h(x), p(x))$ gives the local mesh size $h = h(x)$ and the local degree $p = p(x)$ of the piecewise polynomial approximation. The relation (6.7) gives an $n \times n$ system of equations for the determination of $u_{h,p}$. We may also add to the mesh function information of the local orientation-stretching of the mesh. We may then think of the mesh $T_{h,p}$ essentially being defined by its mesh function (h, p). In fact, the input to the mesh generator is the geometry of the domain to be meshed and the mesh function.

The basic problem in adaptivity for finite element methods may be formulated as follows: Suppose $\mathcal{P}$ is a given (initial)-boundary value problem with given data f and corresponding exact solution u. Let $\mathcal{P}_{h,p}$ be a finite element method for $\mathcal{P}$ based on piecewise polynomial approximation on a family $\{T_{h,p}\}$ of meshes with mesh functions $\{(h, p)\}$ and let $\{u_{h,p}\}$ be the corresponding family of finite element solutions. Let further $\| \cdot \|$ be a given norm and TOL > 0 a given tolerance. Then construct an algorithm $\mathcal{A}$ for finding a mesh $T_{\bar{h},\bar{p}}$, referred to as an *optimal* mesh, such that for the corresponding finite element solution $u_{\bar{h},\bar{p}}$, we have

$$\|u - u_{\bar{h},\bar{p}}\| \leq \text{TOL}, \tag{6.8}$$

and the work to compute $u_{\bar{h},\bar{p}}$ is nearly minimal (for definiteness we may assume that the computational work is proportional to the total number of degrees of freedom associated with $T_{h,p}$). This is a complex optimization problem which is not directly solvable, since the exact solution u is not known and the dependence of $\|u - u_{h,p}\|$ on (h, p) is very implicit. In general, both the local mesh size h and the local degree p of the polynomial approximation are variable in space and time and should be determined by the adaptive algorithm. We refer to adaptive algorithms in this generality as (h, p)-methods. Adaptive algorithms with p fixed or h fixed are referred to as h-methods and p-methods, respectively. Concerning h or (h, p)-adaptivity, note that in addition to the local mesh size also the local mesh orientation and stretching of the mesh (in space or space-time) may be adaptively controlled. For simplicity of notation, we assume that the mesh parameter h in general represents both the mesh size in space and time, as well as the mesh orientation/stretching.

Note that we seek an adaptive algorithm $\mathcal{A}$ which is both *reliable* and *efficient*, in the sense that the desired error control should be guaranteed, and the computational work should be nearly minimal. The basic problem in adaptivity is to construct adaptive algorithms that are both reliable and efficient.

Typically adaptive algorithms are of the form:

step 1   Set $i = 0$. Construct an initial mesh T_{h^0,p^0}.

step 2   Given a mesh T_{h^i,p^i}, compute the corresponding solution u_{h^i,p^i}

step 3   Check if the stopping criterion is satisfied. If satisfied stop. If not continue.

step 4   Determine a new mesh function (h^{i+1}, p^{i+1}).

step 5   Construct a new mesh $T_{h^{i+1},p^{i+1}}$ with the mesh function (h^{i+1}, p^{i+1}).

step 6   Set $i = i + 1$. Go to 2.

The critical ingredients of an adaptive algorithm of this type are:

1.   The stopping criterion of step 3.
2.   The construction of the new mesh function in step 4.
3.   The construction of the new mesh in step 5.

We emphasize the distinction between the stopping criterion and the design of the new mesh function. The stopping criterion is "simply" that the estimated error is less than the tolerance, while the construction of the new mesh function in case the stopping criterion is not satisfied, is more delicate and requires more information on the structure than "just" an estimate of the (local) error, cf. the discussion in Johnson (1988).

6.8   *A PRIORI* AND *A POSTERIORI* ERROR ESTIMATES.

Rational adaptive algorithms are based on *a posteriori* estimates of the form:

$$\|u - u_{h,p}\| \leq E_1(u_{h,p}, h, p, f), \tag{6.9}$$

where $E_1(u_{h,p}, h, p, f)$ is a quantity depending on the mesh function (h, p), the computed solution $u_{h,p}$ and the data f. Variants occur for nonlinear problems with a certain dependence on both u and $u_{h,p}$. *A priori* error estimates have the form

$$\|u - u_{h,p}\| \leq E_2(u, h, p), \tag{6.10}$$

where $E_2(u, h, p)$ is a quantity depending on the mesh function (h, p) and (derivatives of) the exact solution u. Let us now state the typical form of the *a priori* and *a posteriori* error estimates for (linear) model problems of elliptic, parabolic and hyperbolic type.

6.8.1   *ELLIPTIC PROBLEMS.*

The typical optimal *a priori* error estimate for a finite element method for a linear second order elliptic problem over a domain Ω takes the form (Eriksson (1993)):

$$\|u - u_{h,p}\|_{L_q(\Omega)} \leq C^i C_{h,p}^s \|h^{p+1} D^{p+1} u\|_{L_q(\Omega)},$$

or

$$\|\nabla(u - u_{h,p})\|_{L_q(\Omega)} \leq C^i C^s_{h,p} \|h^p D^{p+1} u\|_{L_q(\Omega)},$$

where $1 \leq q \leq \infty$,

$$D^{p+1} u = \max_{|\alpha|=p+1} |D^\alpha u|,$$

where C^i is an interpolation constant and $C^s_{h,p}$ a stability constant. The interpolation constant C^i depends on the polynomial degree p and the stretching of the elements in the underlying mesh $T_{h,p}$, and the stability constant $C^s_{h,p}$ measures certain (strong) stability properties of a linearized version of the given discrete problem. Typical *a posteriori* error estimates for finite element methods for second order elliptic problems take the form (Eriksson (1993),Erikkson and Johnson (1991)):

$$\|u - u_{h,p}\|_{L_q(\Omega)} \leq C^i C^s \|h^2 R(u_{h,p})\|_{L_q(\Omega)},$$

or

$$\|\nabla(u - u_{h,p})\|_{L_q(\Omega)} \leq C^i C^s \|h R(u_{h,p})\|_{L_q(\Omega)},$$

where $R(u_{h,p})$ is the *residual* which is basically the deviation from equality (properly evaluated) obtained inserting the finite element solution $u_{h,p}$ into the given continuous equation. The residual R in general has a contribution from the interior of each element and also a contribution from the boundary of each element. Further C^s is the strong stability constant for an associated continuous dual problem, to be commented on further below.

6.8.2 PARABOLIC PROBLEMS

The typical *a priori* error estimate for a parabolic problem formulated over a space-time domain $\Omega \times (0, T)$ and discretized with the discontinuous Galerkin method, takes the form (Erikkson and Johnson (1991)-(1992(iii))))

$$\|u - u_{h,p}\| \leq C^i C^s_{h,p} \left(\|h^{p+1} D^{p+1} u\| + C\|k u_t\| \right)$$

where $\| \cdot \| = \| \cdot \|_{L_\infty(0,T;L_2(\Omega))}$, $h = h(x,t)$ is the mesh size in space and $k = k(t)$ is the time step. Typical corresponding *a posteriori* error estimates for parabolic problems take the form (see again Eriksson et Johnson (1991)-Erikkson and Johnson (1992(iii))):

$$\|u - u_{h,p}\| \leq C^i C^s \left(\|h^2 R(u_{h,p})\| + C\|[u_{h,p}]\| \right),$$

where $[u_{h,p}]$ is the jump in time across a discrete time level. Here, C^s and $C^s_{h,p}$ are stability constants typically independent (up to inessential logarithms) of (h, p) and T.

6.8.3 HYPERBOLIC PROBLEMS

The typical *a priori error* estimate for the streamline diffusion finite element method for a stationary first order hyperbolic problem over a domain Ω takes the form (Erikkson and Johnson (1993(ii)) and (1993(iii))):

$$\|u - u_h\|_{L_2(\Omega)} \leq C^i C^s_{h,p} \|h^{p+\frac{1}{2}} D^{p+1} u\|_{L_2(\Omega)}.$$

The corresponding *a posteriori* error estimate takes the form

$$\|u - u_{h,p}\|_{L_2(\Omega)} \leq C^i C^s (\|\hat{\epsilon}^{-1} h^2 R(u_{h,p})\|_{L_2(\Omega)} + \|\hat{\epsilon}^\alpha \nabla u_{h,p}\|_{L_2}). \tag{6.11}$$

where $\hat{\epsilon}$ is an artificial viscosity (depending on h and $R(u_{h,p})$) and $0.5 \leq \alpha \leq 1$. For instance, in the simplest case with $\hat{\epsilon} = Ch$, corresponding to a first order monotone scheme, (6.11) takes the form

$$\|u - u_{h,p}\|_{L_2} \leq C^i C^s (\|h R(u_{h,p})\|_{L_2} + \|\hat{\epsilon}^\alpha \nabla u_{h,p}\|_{L_2}).$$

6.9 ADAPTIVE ALGORITHMS.

Our adaptive algorithms for quantitative control of the error $\|u - u_{h,p}\|$ are based on *a posteriori* error estimates of the form (6.9). The adaptive algorithm producing a final mesh $\mathcal{T}_{\hat{h},\hat{p}}$ seeks to (approximately) minimize the total number of degrees of freedom under the condition

$$E_1(h, p, u_{h,p}, f) \leq \text{TOL}. \tag{6.12}$$

This is again a nonlinear minimization problem, but this problem is more directly solvable since knowledge of u is not required and the dependence of $E_1(h, p, u_{h,p}, f)$ on (h, p) is more explicit. Usually this nonlinear minimization problem may be solved iteratively by seeking to equidistribute the element contributions in the quantity $E_1(h, p, u_{h,p}, f)$ with $u_{h,p}$ corresponding to a previous mesh. A very important feature of $E_1(h, p, u_{h,p}, f)$ is that this quantity contains information on the structure of the error as a function of (h, p), which usually makes it possible to solve the minimization problem, i.e., construct the adaptive algorithm, rather easily. Without sufficient information on the structure of the error, it appears difficult to design efficient adaptive algorithms. This means that algorithms based exclusively on estimating the error locally, without paying attention to the structure of the error, may be difficult to get to work for large classes of problems, cf the discussion in Johnson (1988).

Clearly, adaptive algorithms using (6.12) as stopping criterion will be reliable in the sense that, by the *a posteriori* error estimate (6.9), the error control (6.8) will be guaranteed. The adaptive algorithm will be efficient if the adaptively constructed mesh $\mathcal{T}_{\hat{h},\hat{p}}$ is close to the optimal mesh $\mathcal{T}_{\bar{h},\bar{p}}$, which is the same as requiring that $\mathcal{T}_{\hat{h},\hat{p}}$ is nowhere overly refined as compared to $\mathcal{T}_{\bar{h},\bar{p}}$. In the case without orientation/stretching, this is a question related to the quality of the *a posteriori* bound (6.9). To prove sharpness of (6.9), which is clearly required to give an efficient algorithm, we typically prove that

$$E_1(h, p, u_{h,p}, f) \leq C E_2(h, p, u), \tag{6.13}$$

where C is a constant and $E_2(h, p, u)$ is a sharp *a priori* error bound depending on h, p and the exact solution u, satisfying (6.10). If (6.10) is (reasonably) sharp and (6.13) holds, then it follows that $\mathcal{T}_{\hat{h},\hat{p}}$ is not globally overly refined as compared to $\mathcal{T}_{\bar{h},\bar{p}}$, up to the constant C in (6.13), which indicates efficiency. To prove that $\mathcal{T}_{\hat{h},\hat{p}}$ is not locally overly refined, which is the real test of efficiency in the case without orientation/stretching and which may be done for model equations of elliptic or parabolic type, localized forms of (6.13) is used, see Eriksson (1993), Eriksson et Johnson (1991).

6.10 THE CHOICE OF NORM FOR THE ERROR CONTROL

Concerning the choice of norm $\| \cdot \|$ in the error control, note that in our approach a variety of different norms are possible, depending on the nature of the problem; we are not in general restricted to the use of energy norms as is the case in other approaches to adaptivity. We recall that energy norms may be used for elliptic problems, while for parabolic and hyperbolic problems energy norms do not play the same role. For elliptic problems we may in addition to energy norms use L_q-norms, $1 \leq q \leq \infty$, for the solution itself or for derivatives of the solution, or weighted such norms. We may for instance use maximum norms for the displacements or the stresses in elasticity problems, which would give more precise error control of clear significance in applications than standard energy norm control, where the stresses are only controlled in a mean square sense. For time dependent problems we may use $L_\infty(L_q)$-norms, i.e., maximum-norm in time and L_q-norm in space, for the solution or derivatives thereof, or $L_2(L_2)$-norms in space-time. We emphasize the generality of our approach, giving the possibility of considering more general norms than energy norms and general problems, not only of elliptic type, but also parabolic and hyperbolic problems.

6.11 FUNDAMENTAL STRUCTURES: GALERKIN ORTHOGONALITY AND STRONG STABILITY

We now present in some detail the structure of the proofs of *a priori* or *a posteriori* error estimates for the finite element method based on the combination of strong stability and Galerkin orthogonality. The structure in the two cases is similar, but not identical. The structure of the proofs of the *a posteriori* error estimates underlying the adaptive algorithms is as follows:

1. Representation of the error in terms of the residual of the finite element solution and the solution of a continuous (linearized) dual problem.
2. Use of Galerkin orthogonality.
3. Local interpolation estimates for the dual solution.
4. Strong stability estimates for the continuous dual problem.

Let us give the steps 1-4 with some more detail in an abstract situation for a symbolic linear problem of the form

$$Au = f, \tag{6.14}$$

with corresponding discrete symbolic Galerkin problem: Find $u_h \in V_h$ such that

$$(Au_h, v) = (f, v) \quad \forall v \in V_h, \tag{6.15}$$

where $A : V \to V$ is a linear operator, V is a Hilbert space with scalar product $(\cdot, \cdot)$ and corresponding norm $\| \cdot \|$, and $V_h \subset V$ is a finite element space. Typically $V = L_2(\Omega)$, where Ω is a domain in R^n.

1. Error representation via a dual problem:

$$\|e\|^2 = (e, e) = (e, A^*\varphi) = (Ae, \varphi)$$

$$= (f - Au_h, \varphi) = -(R(u_h), \varphi),$$

where φ solves

$$A^*\varphi = e \equiv u - u_h,$$

with A^* the transpose of A.

2. Galerkin orthogonality: Using that $(Ae, v) = 0 \quad \forall v \in V_h$, gives

$$\|e\|^2 = -(R(u_h), \varphi - \pi_h\varphi),$$

where $\pi_h\varphi \in V_h$ is the interpolant of φ.

3. Interpolation estimate:

$$\|h^{-\alpha}(\varphi - \pi_h\varphi)\| \leq C^i\|\|\varphi\|\|,$$

where C^i is an interpolation constant and $\|\|\cdot\|\|$ is a strong norm, typically involving derivatives of order α.

4. Strong stability of the dual problem:

$$\|\|\varphi\|\| \leq C^s\|e\|,$$

where C^s is a stability constant.

Altogether, 1-4 gives the following *a posteriori* error estimate:

$$\|e\|^2 = (R(u_h), \pi_h\varphi - \varphi) \leq C^s C^i\|h^\alpha R(u_h)\| \ \|\|\varphi\|\|,$$

i.e.,

$$\|e\| \leq C^s C^i\|h^\alpha R(u_h)\|.$$

Clearly, the *a posteriori* error estimates obtained in this way are residual-based. Critical issues concern the choice of norm in the strong stability estimate and the related norm used to estimate the residual, see below. The data of the dual problem is the error itself in the case of L_2-norm control. In the case of energy norm control for elliptic problems, the solution of the dual problem coincides with the error itself, and the introduction of the dual problem may be avoided. Notice that the error representation usually gives valuable information concerning the structure of the error, which is used in the design of the adaptive algorithm.

The proofs of the *a priori* error estimates have a similar structure:

1. Representation of the error in terms of the exact solution and a discrete linearized dual problem.
2. Use of Galerkin orthogonality to introduce the truncation error in the error representation.
3. Local interpolation estimates for the truncation error.
4. Strong stability estimates for the discrete dual problem.

Let us again give the steps 1-4 with some more details with the set up (6.14) and (6.15) seeking an estimate in the L_2-norm $\|\cdot\|$.

1. Error representation via a discrete dual problem:

$$\|e_h\|^2 = (e_h, e_h) = (e_h, A^*\varphi_h) = (Ae_h, \varphi_h),$$

where $e_h = \pi_h u - u_h$ and

$$(v, A^*\varphi_h) = (v, e_h) \qquad \forall v \in V_h,$$

where $\pi_h u \in V_h$ is an interpolant of u.

2. Galerkin orthogonality: Using that $(Ae, v) = 0 \quad \forall v \in V_h$, gives

$$\|e_h\|^2 = (A(\pi_h u - u_h), \varphi_h) = (\pi_h u - u, A^*\varphi_h).$$

3. Interpolation estimate:

$$\|h^{-\beta}(u - \pi_h u)\| \leq C^i \|D^\beta u\|.$$

4. Strong stability of the discrete dual problem:

$$\|A^*\varphi_h\| \leq C_h^s \|e_h\|,$$

where C_h^s is a (discrete) stability constant.

Combining now 1-4, we get the an *a priori* error estimate of the form

$$\|u - u_h\| \leq (C_h^s + 1)C^i \|h^\beta D^\beta u\|. \tag{6.16}$$

Again, the interplay between the "strong" norm used in the strong stability involving A^*, and a corresponding "weak" norm used to estimate the truncation error, is crucial. Here the truncation error $\text{Trunc}_{(w)}(u, h)$, the "weak" truncation error, is defined by

$$\text{Trunc}_{(w)}(u, h) \equiv \sup_{v \in V_h} \frac{|(A(\pi_h u - u), v)|}{\|A^*v\|} = \|u - \pi_h u\|. \tag{6.17}$$

This definition involving the operator A^* should be compared with the typical definition of the "strong" truncation error $\text{Trunc}_{(st)}(u, h)$ used in the classical theory of finite difference methods, which is as follows:

$$\text{Trunc}_{(st)}(u, h) \equiv \sup_{v \in V_h} \frac{|(A(\pi_h u - u), v)|}{\|v\|}. \tag{6.18}$$

Estimates using this definition of the truncation typically are non-optimal with either powers of h lacking or requiring too much regularity of the exact solution u.

In both the case of *a priori* and *a posteriori* error estimates the (strong) stability of the dual problem is the critical issue. Note that in the case of an *a posteriori* error estimate, the stability of a continuous dual problem enters, while in the case of an *a priori* error estimate, we are concerned with the stability of a discrete dual problem. In both cases the stability of the dual problem reflects the error propagation properties of the discretization procedure: In the case of the *a posteriori* error estimates, the error is connected to the residual through the error representation formula, involving the continuous dual solution. In the case of *a priori* estimates, the error is connected to the truncation error (which is the interpolation error for the exact solution) through the discrete dual solution. Thus, the solution of the (discrete or continuous) dual problem plays a fundamental role in any attempt to control the discretization error. More precisely, it is the stability properties of the dual solution that matter. A particular feature of our methodology is the use of (new) strong stability estimates for the dual problem, which makes it possible to derive sharp error estimates which result in efficient adaptive algorithms. These new strong stability estimates involve control of certain derivatives of the solution of the dual problem (typically the leading derivatives in the dual problem) in terms of the data of the dual problem (whereas standard stability estimates, except in the case of energy norms, involve weaker such control).

6.12 THE STABILITY AND INTERPOLATION CONSTANTS C^s AND C^i

In the *a posteriori* error estimates underlying the adaptive algorithms, two types of constants enter as already indicated above, one set of constants C^s related to the stability estimates for the continuous dual problem, and one set of constants C^i related to local polynomial interpolation. These constants have to be determined approximatively in order to define the adaptive algorithm and control the error on the given tolerance level. The interpolation constants C^i depend on the shape of the elements and the local order of the polynomial approximation, but not (or only trivially so) on the particular problem $\mathcal{P}$ to be solved, and may thus be determined analytically or computationally once and for all for different classes of problems. The stability constants C^s, on the other hand, in general depend on the particular problem $\mathcal{P}$ (but not on the discretization, i.e., not on p and h) and for nonlinear problems, also on the exact solution, and it is less obvious how to estimate the C^s with one important exception: In energy norms the problem is trivial, since by definition we have $C^s = 1$ (in which case the interpolation constants C^i in fact contain a dependence on the energy norm, which is easy to take into account). In other cases it may be possible to obtain analytical estimates of the stability constants fairly easily. In general, however, we will have to rely on auxiliary computations to determine the actual values of the stability constants C^s. In Erikkson and Johnson (1992(ii)) we discuss different possibilities of computational evaluation of the stability constants. We note that this situation cannot be avoided without seriously limiting the framework: Either we have to restrict ourselves to elliptic problems and use energy norms, in which case $C^s = 1$ by definition, or we restrict ourselves to a certain class of (simple) problems for which analytical estimates are possible to obtain, or we expand the framework to include more general problems and pay the prize (which usually is not large) required to estimate the stability constants computationally. We recall that the stability properties of the dual continuous problem gives us the bridge between the computable residual of the discrete solution and the error itself, and thus it is necessary to estimate the stability of the dual problem, one way or the other, for instance by estimating the stability constants C^s. There is no way we can get around this difficulty if we want to design algorithms for automatic error control based on rational arguments; the question is only how much computational work will be required for this purpose. It appears that in many cases this work is small compared to the total work.

6.13 FROM QUALITATIVE TO QUANTITATIVE ERROR CONTROL

The basic goal in adaptivity is to construct algorithms for efficient and reliable automatic quantitative error control. To reach this goal it is necessary to take the stability properties of the underlying continuous problem into account. This appears to be possible to do using analytical or computational techniques for many linear problems, but for complex nonlinear problems it appears to be difficult to give a full theoretical justification guaranteeing reliability. This is because such a full justification would for instance for flow problems require a complete solution of the problem of hydrodynamic stability, a problem of extreme difficulty for which very little progress has been made. However, if we put the goal somewhat lower concerning the reliability the possibilities of verification of reliability increase dramatically: It appears possible to guarantee reliability (and reasonable efficiency) through the adaptive algorithm if the approximate solution is reasonably close to the exact solution, that is , the adaptive algorithm would make it possible (also for complex nonlinear problems) to take the step from qualitative to quantitative error control if the error tolerance is "small enough". Thus, for example, adaptive error control for compressible flow appears to be within reach, but the reliability may only be guaranteed if the error is small enough. In particular the reliability could not be fully guaranteed in the initial stages of an adaptive refinement process when the mesh used may be fairly coarse corresponding to a large tolerance.

6.14 A HIGHLIGHT: FROM exp(Re) TO Re IN CFD

Let us here briefly indicate one example where our results give considerable improvement of existing results. This concerns error estimates for the incompressible Navier-Stokes equations, the theoretical foundation of computational fluid mechanics CFD. In the classical *a priori* approach, the error estimates for discretization in time only, with the backward Euler method for simplicity, typically have the form

$$\max_{0 \leq t \leq T} \|(u - u_h)(\cdot, t)\| \leq \exp(CKT) \max_{0 \leq t \leq T} \|hu_{tt}(\cdot, t)\| \tag{6.19}$$

where $K = \|\nabla u\|_{L_\infty}$ is the maximum norm of the gradient of the velocity u, $(0, T)$ is the given time interval, C is a constant of moderate size, $h = h(t)$ is a measure of the local time step and $\| \cdot \|$ is the L_2-norm. In most cases of interest we have $KT >> 1$, in which case (6.19) has no actual content, because from all practical points of view $\exp(KT) = \infty$. The classical estimate (6.19) thus only covers the case of smooth flow-short time corresponding to $KT \sim 1$. In Johnson, Rannacher and Bowan (1993) we show that in certain cases (6.19) may be improved to

$$\max_{0 \leq t \leq T} \|(u - u_h)(\cdot, t)\| \leq CKT \max_{0 \leq t \leq T} \|hu_t(\cdot, t)\| \tag{6.20}$$

where the exponential has "disappeared" as well as one time-derivative of u. As far as we know, this is the first theoretical support for CFD in a case outside the realm of smooth flow-short time covered by classical theory. There is also an *a posteriori* analogue of (6.20) of essentially the form

$$\max_{0 \le t \le T} \|(u - u_h)(\cdot, t)\| \le CKT \max_{0 \le t \le T} \|[u_h](\cdot, t)\| \qquad (6.21)$$

The estimates (6.20) and (6.21) result from an analysis combining strong stability with Galerkin orthogonality. The *a posteriori* error estimate (6.21) directly gives an adaptive algorithm, which is the first adaptive algorithm with mathematical basis in CFD applicable in case outside the realm of smooth flow-short time. Note that in the case of smooth flow with $K = 1$, the relevant time scale is $T = Re$, where Re is the Reynolds number. Thus, the estimates (6.20) and (6.21) represent an improvement from $\exp(Re)$ to Re.

REFERENCES

Eriksson K. (1993), An adaptive finite element method with efficient maximum norm error control for elliptic problems, Preprint 1993-20, Department of Mathematics, Chalmers University of Technology.

Eriksson K. and Johnson C. (1991), Adaptive finite element methods for parabolic problems I: A linear model problem, *SIAM J. Numer. Anal.*, **28**, 43-77.

Eriksson K. and Johnson C. (1992(i)), Adaptive finite element methods for parabolic problems II: Optimal error estimates in $L_\infty L_2$ and $L_\infty L_\infty$, Preprint 1992-09, Department of Mathematics, Chalmers University of Technology.

Eriksson K. and Johnson C. (1993(i)), Adaptive finite element methods for parabolic problems III: Time steps variable in space, to appear.

Eriksson K. and Johnson C. (1992(ii)), Adaptive finite element methods for parabolic problems IV: A nonlinear problem, Preprint 1992-44, Department of Mathematics, Chalmers University of Technology.

Eriksson K. and Johnson C. (1992(iii)), Adaptive finite element methods for parabolic problems V: Long-time integration, Preprint 1992-46, Department of Mathematics, Chalmers University of Technology.

Eriksson K. and Johnson C. (1993(ii)), Adaptive streamline diffusion finite element methods for stationary convection-diffusion problems, *Math. Comp.*, **60**, 167-188.

Eriksson K. and Johnson C. (1993(iii)), Adaptive streamline diffusion finite element methods for time dependent convection-diffusion problems, Preprint 1993-23, Department of Mathematics, Chalmers University of Technology.

Eriksson K., Hansbo P. and Johnson C. (1993(iv)), Adaptive Finite Element Methods, North-Holland, to appear.

Johnson C. and Hansbo P. (1992), Adaptive finite element methods in computational mechanics, *Comput. Methods Appl. Meth. Engrg.*, **101**, 143-181.

Johnson C. (1988), Error estimates and adaptive time step control for a class of one step methods for stiff ordinary differential equations, *SIAM J. Numer. Anal.*, **25**, 908-926.

Johnson C. (1990), Adaptive finite element methods for diffusion and convection problems, *Comput. Methods Appl. Meth. Engrg.*, **82**, 301-322.

Johnson C. (1992), Adaptive finite element methods for the obstacle problem, M3AS 4, 483-487.

Johnson C., Rannacher R. and Boman M. (1993), Numerics and hydrodynamic stability: Towards error control in CFD, Preprint 1993-13, Department of Mathematics, Chalmers University of Technology, 1993.

Johnson C. and Szepessy A. (1993), Adaptive finite element methods for conservation laws Preprint 1991-31, Mathematics Department, Chalmers University of Technology, 1992.

Chapter 7

Mesh optimality criteria for adaptive finite element computations

E. Oñate and G. Bugeda

International Center for Numerical Methods in Engineering
Universidad Politécnica de Cataluña
08034 Barcelona, Spain

7.1 INTRODUCTION

In this paper a methodology for deriving adaptive mesh refinement (AMR) procedures is proposed. The basis of the approach is the decoupling of the concepts of error measure and mesh optimality criteria. This leads to the definitions of global and local error parameters from which the element refinement strategy can be simply obtained. In particular, the convergence rates of the global and local error norms influence the expressions of the element refinement parameter. It is shown that an inaccurate evaluation of this important parameter can lead to oscillations in the refinement process.

The methodology proposed is particularized for two mesh optimality criteria using the Zienkiewicz-Zhu error estimator (Zienkiewicz and Zhu (1987-1990)), and Zienkiewicz, Zhu, and Gong (1989)). First the standard criterion of equal distribution of the global error over all the elements is studied. A careful interpretation of the concepts of global error and optimal mesh leads in this case to an enhanced expression of the element refinement parameter slightly different from that typically used in the literature.

The second mesh optimality criterion studied is based in the equal distribution of the error per unit area or volume (i.e. the specific error). This strategy allows the concentration of more and smaller elements in zones where the gradients of the problem unknowns (i.e. stresses in structural problems) are higher, as should be expected from an engineering point of view.

Examples of application of the AMR methodology proposed to structural, potential flow and optimum shape design problems are presented.

7.2 ERROR ESTIMATION AND ACCEPTABLE SOLUTION

When dealing with adaptive mesh refinement finite element analysis the following two concepts should be clearly defined:

The Mathematics of Finite Elements and Applications
Edited by J. R. Whiteman © 1994 John Wiley & Sons Ltd

(a) Error estimator. Since the "exact" solution is not known, a method to approximately evaluate the error of the finite element solution should be defined.

(b) Acceptable solution. A finite element solution is "acceptable" if the estimated error satisfies some prescribed *global* and/or *local* conditions.

Both concepts are further extended in next sections.

7.2.1 ERROR ESTIMATOR

One of the most popular error estimators for elliptic problems is based on the error energy norm expressed as

$$\|e\| = \left[\int_\Omega [\sigma - \hat{\sigma}]^T D^{-1} [\sigma - \hat{\sigma}] d\Omega \right]^{\frac{1}{2}} \tag{7.1}$$

where σ are the exact "fluxes" (i.e. heat fluxes in a heat conduction problem or stresses in an elasticity problem, etc.), $\hat{\sigma}$ are the flux values obtained from the finite element solution and D is the constitutive matrix relating the fluxes with the solution gradient (i.e. temperature gradient, strains, etc.). For simplicity the values of σ will be identified with the *stresses* in a standard structural problem.

Since the exact stresses are usually not known they are approximated by

$$\sigma \simeq \sigma^* = N_\sigma \bar{\sigma}^* \tag{7.2}$$

where N_σ are stress interpolating functions and $\bar{\sigma}^*$ are nodal stress values obtained by either simple nodal averaging of the finite element values, local or global least square smoothing, or other adequate nodal stress recovery techniques Zienkiewicz and Zhu (1987), Oñate and Castro (1991), Zienkiewicz and Taylor (1991) and Zienkiewicz and Zhu (1992).

The "strain energy" U of the exact solution is estimated as

$$U = \|u\|^2 \simeq \int_\Omega \sigma^{*T} D^{-1} \sigma^* d\Omega + \int_\Omega [\sigma^* - \hat{\sigma}]^T D^{-1} [\sigma^* - \hat{\sigma}] d\Omega \tag{7.3}$$

Both norms $\|e\|$ and $\|u\|$ can be evaluated as the sum of their respective element contributions. Thus if n is the total number of elements in the mesh

$$\|e\|^2 = \sum_{i=1}^n \|e\|_i^2 \qquad \|u\|^2 = \sum_{i=1}^n \|u\|_i^2 \tag{7.4}$$

7.2.2 DEFINITION OF ACCEPTABLE SOLUTION

It is usually agreed that a solution is "acceptable" if the two following conditions are satisfied:

(a) The global error in energy norm is not greater than a specified value of the total strain energy:

$$\|e\| \leq \eta \|u\| \tag{7.5}$$

Where η is the user's specified value of the permissible relative global error.

Equation (7.5) allows a *global error parameter* ξ_g to be defined as

$$\xi_g = \frac{\|e\|}{\eta \|u\|} \tag{7.6}$$

Clearly the values $\xi_g \leq 1$ denote satisfaction of the global error criterion, whereas $\xi_g > 1$ indicates that further refinement is necessary.

A mesh refinement strategy based on ξ_g only will lead to an uniform refinement ($\xi_g > 1$) or derefinement ($\xi_g < 1$) of all element sizes (with the usual corrections to preserve the dimensions of the analysis domain). A local error indicator is therefore needed and this is defined next.

(b) The distribution of elements in the mesh satisfies a local condition, which can be expressed as

$$\|e\|_i = \|e\|_{r_i} \tag{7.7}$$

where $\|e\|_i$ is the actual error norm in each element i and $\|e\|_{r_i}$ is the "required" error norm in the element. The *local error indicator* $\bar{\xi}_i$ is defined as

$$\bar{\xi}_i = \frac{\|e\|_i}{\|e\|_{r_i}} \tag{7.8}$$

Note that a value of $\bar{\xi}_i = 1$ defines an "optimal" element size (in the sense of satisfaction of (7.7)), whereas $\bar{\xi}_i > 1$ and $\bar{\xi}_i < 1$ indicate that the size of element i needs further refinement or derefinement, respectively. The definition of the required error in each element $\|e\|_{r_i}$ is a key isue and it strongly affects the distribution of element sizes in the mesh. This definition can be based on different mesh optimality criteria and some of these are presented in a later section.

7.2.3 ELEMENT REFINEMENT PARAMETER

Generally in practice we will aim to satisfy both local and global conditions (a) and (b) defined in the previous section. This allows an *element refinement parameter* to be defined using (7.6) and (7.8) as

$$\xi_i = \bar{\xi}_i \xi_g = \frac{\|e\| \, \|e\|_i}{\eta \|u\| \, \|e\|_{r_i}} \tag{7.9}$$

The expressions of ξ_i given in (7.9) can be interpreted as the result of trying to satisfy the global and local error conditions in a successive manner. Equation (7.9) provides all the terms involved in this combined process and they could play individually a very different role as explained in next section.

7.3 MESH OPTIMALITY CRITERIA AND AMR PROCEDURES

7.3.1 MESH OPTIMALITY CRITERION BASED ON THE EQUAL DISTRIBUTION OF THE GLOBAL ERROR

A very popular criterion for elliptic problems is that a mesh is "optimal" if the distribution of the energy norm is equal between all elements Zienkiewicz and Zhu (1987-1990), Zienkiewicz, Zhu, and Gong (1989), and Zienkiewicz and Taylor (1991). On the basis of this assumption the required error for each element can be defined as the ratio between the global error and the total number of elements in the mesh. Thus noting that only the square of the error norm is additive we have

$$\|e\|_{r_i} = \frac{\|e\|}{\sqrt{n}} \tag{7.10}$$

Combining (7.8) and (7.10) yields the expression of the local error parameter as

$$\bar{\xi}_i = \frac{\|e\|_i}{\|e\|n^{-1/2}} \tag{7.11}$$

The element refinement parameter is obtained viz. equation (7.9) as

$$\xi_i = \bar{\xi}_i\xi_g = \frac{\|e\|_i}{\eta\|u\|n^{-1/2}} \tag{7.12}$$

The parameter ξ_i can now be readily interpreted as the ratio between the element error and the distributed value of the permissible error over the mesh. Clearly $\xi_i \geq 1$ will indicate that the element should be further refined, whereas $\xi_i < 1$ implies that both the local and global error conditions are satisfied.

Expression (7.12) is in fact identical to that used in Zienkiewicz and Zhu (1987). However, the multiplicative form (7.11) allows the correct AMR strategy to be derived. For that purpose the convergence rates of the element and global error norms will be considered next.

Let us assume that m is the degree of the shape function polynomials used in the interpolation of the basic unknowns (i.e. the displacements) m=1 for linear elements, m=2 for quadratic elements, etc.) and that only first derivatives are involved in the strain operator. It is then easy to find for the global error norm (7.1), Oñate, Castro and Kreiner (1991) and Oñate and Castro (1991),

$$\|e\| = \left[\int_\Omega O(h^m)D^{-1}O(h^m)d\Omega\right]^{1/2} \simeq O(h^m), \tag{7.13}$$

where h is the average element size of all the elements in the mesh. For the element error norm one obtains

$$\|e\|_i = \left[\int_{\Omega_i} O(h_i^m)D^{-1}O(h_i^m)d\Omega\right]^{1/2} \simeq O(h_i^m)\Omega_i^{1/2} \simeq O(h_i^{m+\frac{d}{2}}) \tag{7.14}$$

where h_i and d are, respectively, the existing element size and the number of dimensions of the problem ($d=1,2,3$ for 1D,2D and 3D problems, respectively).

Equations (7.13) and (7.14) are essential to define the new element size $\bar{h}_i$ in terms of the existing size using the expression

$$\bar{h}_i = \frac{h_i}{\beta_i} \qquad \text{with} \qquad \beta_i = \xi_g^{\frac{1}{m}} \bar{\xi}_i^{-\frac{2}{2m+d}} \tag{7.15}$$

Note that (7.15) can be interpreted as equivalent to applying a two step refinement (or derefinement) process in which the element sizes are *first* changed to fulfil the local mesh optimality criterion and *then* changed again to satisfy the global error criterion.

Many authors (Zienkiewicz and Zhu (1987,1989,1990), Zienkiewicz, Zhu, Liu, Morgan, and Peraire (1988), Zhu and Zienkiewicz (1988), Zienkiewicz, Zhu, and Gong (1989), Atamaz-Sibai and Hinton (1990), Atamaz-Sibai, Hinton, and Selman (1990), Selman, Hinton and Atamaz-Sibai (1990) and Hinton, Özakça and Rao (1990)) use a simpler expression for β_i based directly on the element refinement parameter ξ_i as

$$\beta_i = (C\xi_i)^{1/m'} \tag{7.16}$$

where C is a relaxation factor (generally $C=1$ is taken Zienkiewicz, Zhu, and Gong (1989)) and $m' = m$ except for elements adjacent to singularities where $m = \lambda$ is used (λ being the singularity strength).

The authors have found that the computation of β_i as given by (7.16) with $C = 1$ and $m' = m$ leads to a non consistent oscillatory mesh refinement. This is due to the violation of the convergence rate of the global and local error norms which enables the simultaneous satisfaction of both norms using a single element size parameter.

This problem disappears if the correct expression (7.15) for the element size parameter β_i is used (see examples and Oñate, Castro and Kreiner (1991), Oñate and Castro (1991), Bugeda and Oñate (1992(i-ii),1993(i))).

7.3.2 MESH OPTIMALITY CRITERION BASED ON THE EQUAL DISTRIBUTION OF THE SPECIFIC ERROR

An alternative criterion is to assume that a mesh is optimal if the square of the error per unit area (or volume) is the same over the whole mesh. It is clear then that in the optimal mesh

$$\frac{\|e\|_i^2}{\Omega_i} = \frac{\|e\|^2}{\Omega} \tag{7.17}$$

Comparing (7.17) and (7.7) gives the expression of the required error norm for each element as

$$\|e\|_{r_i} = \|e\| \left(\frac{\Omega_i}{\Omega}\right)^{1/2} \tag{7.18}$$

The local error parameter $\bar{\xi}_i$ is obtained now using (7.8) and (7.18) as

$$\bar{\xi}_i = \frac{\|e\|_i}{\Omega_i^{1/2}} \left[\frac{\|e\|}{\Omega^{1/2}}\right]^{-1} = \frac{\|e\|_i}{\|e\|} \left(\frac{\Omega}{\Omega_i}\right)^{1/2} \tag{7.19}$$

The element refinement parameter ξ_i is obtained from (7.6), (7.9) and (7.19) as

$$\xi_i = \bar{\xi}_i \xi_g = \frac{\|e\|_i}{\eta\|u\|}\left(\frac{\Omega}{\Omega_i}\right)^{1/2} \tag{7.20}$$

Note that equations (7.12) and (7.20) coincide if $\frac{\Omega}{\Omega_i} = n$, i.e. all elements are equal in size. This is however not the case for unstructured meshes which results in quite different mesh distributions for the two optimality criteria considered, as shown in the examples.

The way the element error is now defined eliminates its dependence with the element area, and the convergence rate of the element error can be deduced from (7.14) to be

$$\frac{\|e\|_i}{\Omega_i^{1/2}} \simeq O\left(h_i^m\right) \tag{7.21}$$

The new element size is obtained from (7.15) with the element size parameter β_i given now by

$$\beta_i = (\bar{\xi}_i \xi_g)^{1/m} = (\xi_i)^{1/m} \tag{7.22}$$

The expression for the element refinement parameter coincides now with that given in (7.16) for $C = 1$ and $m' = m$. Note, however, that (7.16) was introduced in the context of the mesh optimality criterion based on the equal distribution of the global error. The different forms presented here clarify the correct expressions of β_i to be used for each mesh optimality criterion chosen.

7.4 EXAMPLES

Three examples have been chosen to compare the different AMR strategies presented:
1) Thick cylinder under internal pressure.
2) Incompressible potential flow around an airfoil.
3) Optimization of a hook.

In all the examples the following notation is used for the different AMR strategies.

Strategy A. This is based on the criterion of equal distribution of the global error between all elements and the value of the element size parameter β_i given by equation (7.16) with $C = 1$ and $m' = m$.

Strategy B. This uses the same mesh optimality criterion as strategy A with the expression for the element size parameter as given by equation (7.15)

Strategy C. This is based on the criterion of equal distribution of the specific error with the element size parameter as given by equation (7.22)

7.4.1 ANALYSIS OF A THICK CIRCULAR CYLINDER UNDER INTERNAL PRESSURE

The first example is the analysis of the thick circular cylinder under internal pressure shown in Figure 7.1. Due to the symmetry of the problem only a quadrant has been studied under plane strain conditions. Linear elastic behaviour has been assumed with

$E = 1.0 \times 10^5$ and $\nu = 0.3$. A value of the permissible global error of $\eta = 5\%$ has been chosen. Standard 3 noded linear triangular elements are used in the analysis. This example is typical of elliptic problems and it has many analogies in heat flow, ground water flow, etc.

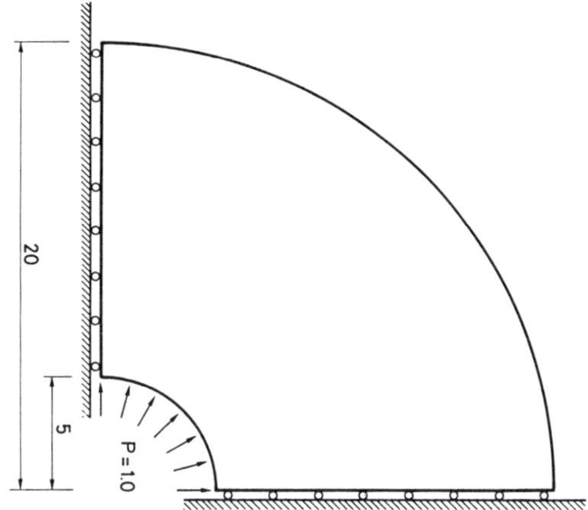

Figure 7.1 Thick circular cylinder under internal pressure. Geometry and loads

Figure 7.2 shows the sequence of refined meshes obtained with the three AMR strategies previously mentioned. The first column of Figure 7.2 shows the results obtained with strategy A. Note the oscillations in the AMR process clearly shown in the alternative re and derefinement of some mesh zones.

The second column of Figure 7.2 shows the results obtained with strategy B. Note that the refinement oscillations disappear and the AMR process converges in a consistent manner.

The third column shows the results for strategy C. It can be clearly seen that: (a) the AMR process converges without oscillations, and (b) this AMR strategy concentrates more and smaller elements in the vicinity of the internal edge (where the error is greater due to the higher stress gradients), whereas in the rest of the mesh bigger elements than in the previous case are allowed. The price to be paid is the increase in the total number of elements with respect to strategies A and B for the same global accuracy as shown in Table 7.1.

Table 7.1 also shows some characteristic results for each solution like the number of elements, the global error parameter ξ_g, the average value of the local error parameter $\left(\bar{\xi}_i^2\right)_a$ and its mean deviation $\left(\bar{\xi}_i^2\right)_\sigma$ over each mesh for the three AMR strategies used. From the numbers shown in this table we deduce:

- All AMR strategies converge to the global permissible error chosen.
- The mean deviation of strategy A oscillates and does not converge to zero.
- Strategies B and C converge to an "optimal mesh" characterized by the appropriate values $\left(\bar{\xi}_i^2\right)_a = 1.0$ and $\left(\bar{\xi}_i^2\right)_\sigma = 0.0$. However, the number of elements and their distribution in each mesh is very different for these two AMR strategies.

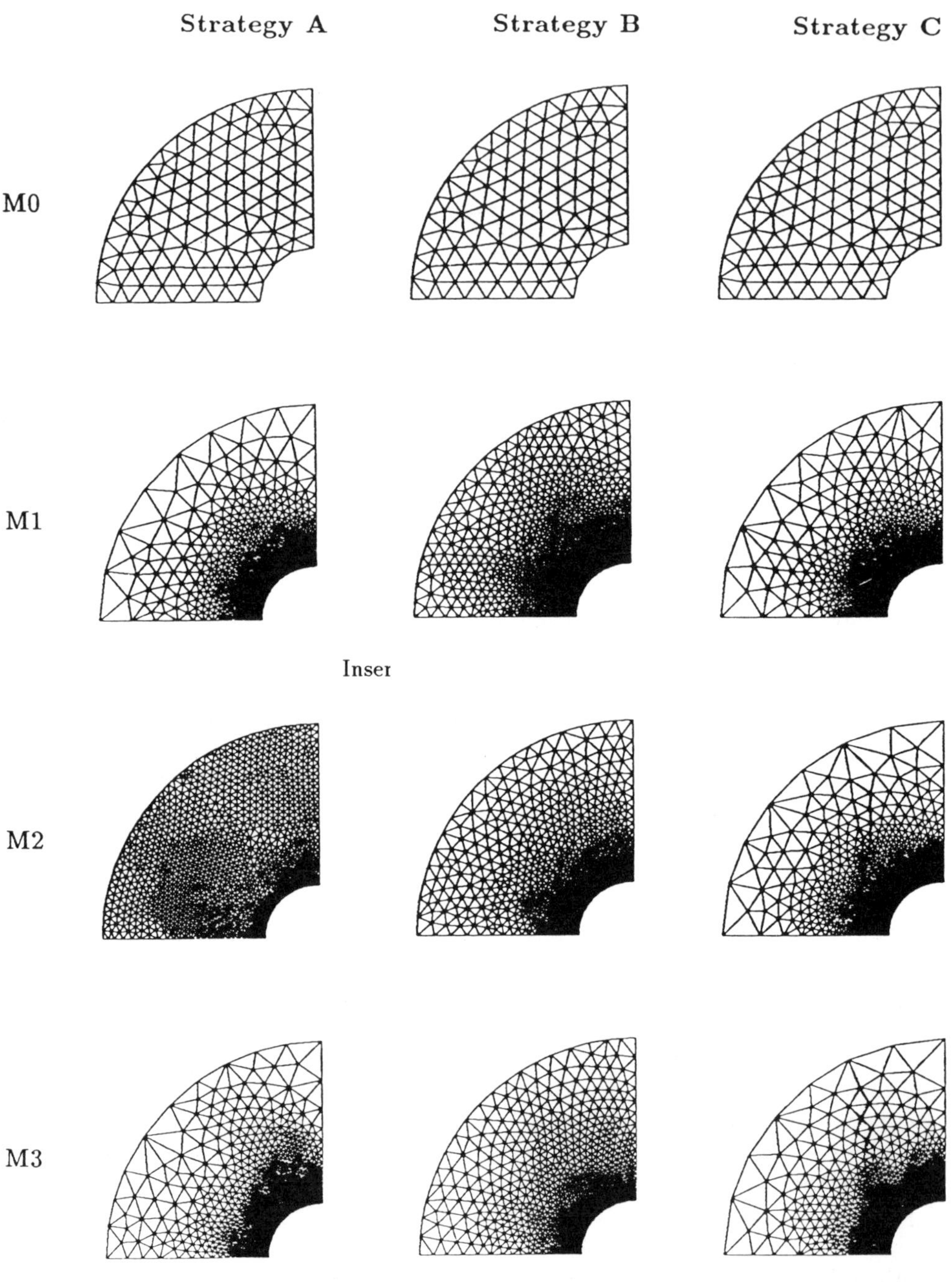

Figure 7.2 Thick circular cylinder under internal pressure. Sequence of meshes obtained with AMR strategies A, B, and C

Table 7.1 Thick circular cylinder under internal pressure. Some statistical results of the AMR strategies based on the equal distribution of the global error (Strategies A and B) and the specific error (Strategy C).

	Strategy A				Strategy B				Strategy C			
	NE	ξ_g	$(\bar{\xi}_i^2)_a$	$(\bar{\xi}_i^2)_\sigma$	NE	ξ_g	$(\bar{\xi}_i^2)_a$	$(\bar{\xi}_i^2)_\sigma$	NE	ξ_g	$(\bar{\xi}_i^2)_a$	$(\bar{\xi}_i^2)_\sigma$
M0	200	4.133	1.000	8.401	200	4.133	1.000	8.401	200	4.133	1.190	15.37
M1	2820	0.919	1.000	8.469	2180	0.839	1.000	0.604	3028	0.884	2.224	2.071
M2	2711	0.907	1.000	2.197	1838	0.909	1.000	0.189	6359	0.796	1.204	0.163
M3	2758	0.839	1.000	4.484	1797	0.925	1.000	0.193	6026	0.835	1.206	0.151

7.4.2 POTENTIAL FLOW AROUND AN AIRFOIL

The applicability of the AMR presented to fluid flow problems are shown in the analysis of the 2D flow around a classical Korn airfoil. An incompressible potential flow model with lifting has been used for this analysis. Details of this model can be found in Bugeda and Oñate (1992(ii)). The angle of attack has been taken equal to zero and the analysis domain has been discretized using standard 6 noded triangular elements. The exterior boundary of the domain has been defined at a distance of 10 chords from the profile.

The error energy norm and the energy of the exact solutions have been estimated as

$$\|e\|^2 \simeq \int_\Omega [v^* - \hat{v}]^T [v^* - \hat{v}] d\Omega \tag{7.23}$$

$$\|u\|^2 \simeq \int_\Omega v^{*^T} v^* d\Omega + \int_\Omega [v^* - \hat{v}]^T [v^* - \hat{v}] d\Omega \tag{7.24}$$

In (7.23) v is the velocity vector obtained from the final continuous potential solution as $v = \nabla\phi$. Also, as usual, $\hat{v}$ and v^* denote velocities obtained directly from the finite element solutions and after a nodal smoothing technique, respectively.

This case has been used to compare two different meshes with a similar number of degrees of freedom, but each one obtained using a different optimality criteria.

Figure 7.3a shows the mesh obtained after two remeshings using Strategy B and using a permissible global error of $\eta = 0.02\%$. This mesh has 2162 degrees of freedom and 1038 quadratic elements. In fact, the results from the first remeshing have a total amount of error smaller than the permissible one but a new remeshing has been performed in order to get a better error distribution over the domain.

Figure 7.3b shows the mesh obtained after two remeshings using Strategy C and using a permissible global error of $\eta = 0.2\%$. This mesh has 2133 degrees of freedom and 1019 quadratic elements. Again, the results from the first remeshing have a total amount of error smaller than the permissible one.

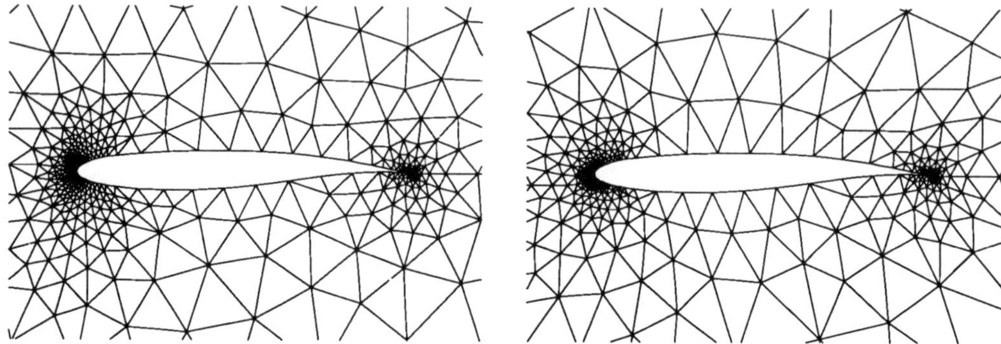

Figure 7.3 Analysis of a Korn airfoil. (a) Mesh obtained with AMR strategy $B(\eta = 0.02\%)$, 2162 quadratic triangles and 1038 DOF. (b) Mesh obtained with AMR strategy $C(\eta = 0.2\%)$, 2133 quadratic triangles and 1019 DOF

The computational cost of solving the problem is very similar for both meshes because the number of degrees of freedom is practically identical in both cases. Looking at Figures 7.3a and 7.3b it can be observed that Strategy C concentrates more and smaller elements in the vicinity of the leading edge (where the gradients of the velocities are higher), whereas in the rest of the domain the elements are bigger.

Figure 7.4 shows a plot of the C_p distribution obtained with both meshes around the leading edge. The mesh obtained with Strategy B has a smaller global error norm (as prescribed), but this error is concentrated in the zone where the gradients of the velocities are higher, as it can be observed from the oscillations in Figure 7.4. It is remarkable that although the mesh obtained with Strategy C has a higher global error norm, the C_p distribution around the leading edge is more accurate and no oscillations are observed.

From the last comments we conclude that for the same computational cost, Strategy B leads to a smaller global error norm, but Strategy C gives better results in the most interesting zones of the analysis domain.

7.4.3 OPTIMIZATION OF A HOOK

The authors have extended the concept of adaptive mesh refinement to optimum shape design problems. The basic idea is to project the computed sentivities of the finite element mesh coordinates and that of the error estimator from one design shape to the next one giving *a priori* knowledge of the error distribution on each enhanced design. This allows a finite element mesh to be built up for each design with a specified and controlled level of error. Details of this methodology can be found in Bugeda (1990), Bugeda and Oliver (1991,1993), Bugeda and Oñate (1992(i),1992(ii),1993(ii)).

The final example shows an application of these concepts to the optimization of the shape of the hook in order to minimize its weight. The initial shape, the applied load and the geometry definition points are shown in Figure 7.5. Twelve points are allowed to move for improvement of the shape, i.e nine of them can move horizontally, one can move vertically and the rest have been enforced to move along a straight line inclined 45°.

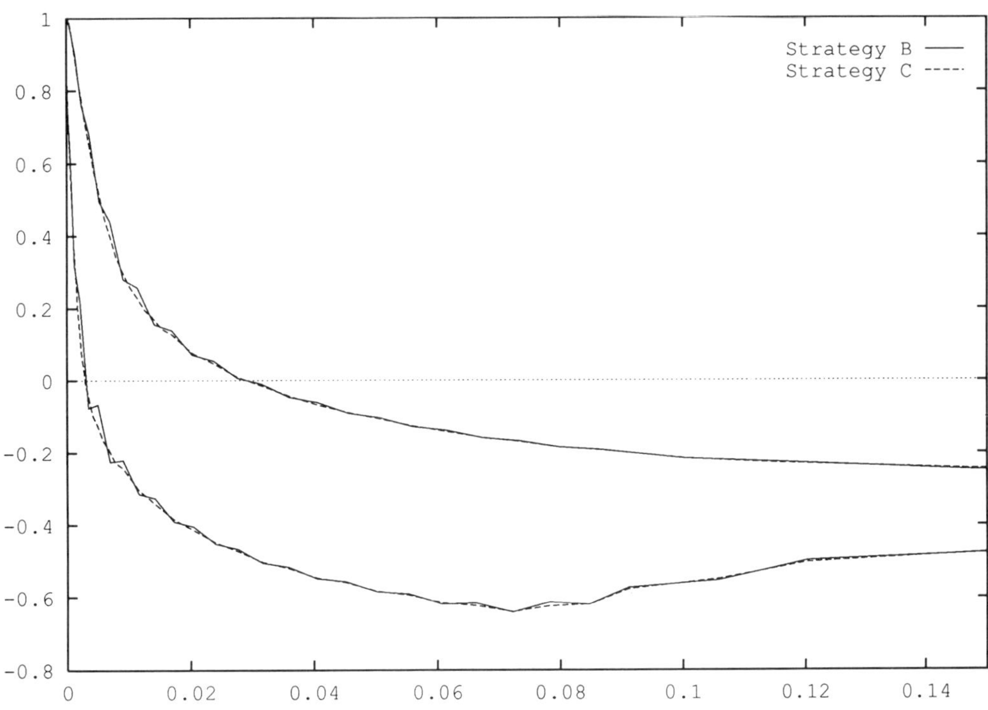

Figure 7.4 Analysis of a Korn airfoil. C_p distribution around the leading edge obtained using strategies B and C

A parabolic vertical load has been applied over the inner part of the hook with a resulting load of 630 kg. The material properties are: Young modulus E=2100000 kg/cm^2 and Poisson ratio $\nu = 0.3$. A plane stress model with 6 node quadratic triangular elements has been used. The global error level has been limited to 1%. The AMR criteria C based on the specific error distribution has been used.

The objective function is the weight of the hook. The maximum values of the Von Mises stresses have been constrained to 2000 kg/cm^2. This constraint has been applied to all the nodes placed along the boundary. The minimum thickness of the spike of the hook has been limited to 0.50 cm.

The algorithm converges after 130 iterations. Figure 7.6 shows the successive meshes and the designs corresponding to iterations 0, 10, 20, 30, 40, 50, 60, 75, 90, 105, 120 and 130 as well as the global error obtained in each solution. It can be observed how the optimization process displaces the vertical part of the hook until it coincides with the resultant of the load forces. This is due to the absence of bending moments over this part and thus its width can be considerably reduced. The curved part of the hook is thicker because of bending action producing high stresses on the boundaries. Futher details on this example can be found in Bugeda and Oñate (1993(ii)).

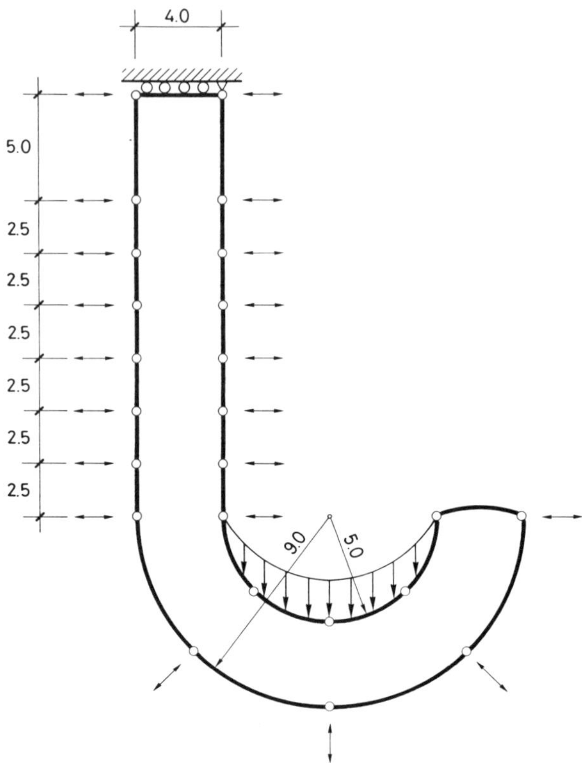

Figure 7.5 Hook optimization problem. Initial shape and parametrization

7.5 CONCLUDING REMARKS

The definition of the mesh optimality criterion has proved to be an essential point in the design of an effective AMR process.

The use of the criterion of equal distribution of error over the elements leads to a "smooth" distribution of the element sizes. This can be useful to prevent the use of meshes with large numbers of elements. However, this criterion is unable to concentrate elements in zones of high gradients only. It has also been shown that oscillations in the refinement process can appear in this case unless the element size parameter is consistently computed.

The criterion of equal distribution of specific error concentrates more and smaller elements in zones where important changes in the solution occur. This provides a more accurate solution in these regions which, in fact, are those of interest for the engineer. Therefore, this criterion seems to be the one to be recommended for practical purposes. The only drawback of this criterion is the large number of elements involved (this number tends to infinity if singularity points exist!). This can be overcome by appropriately prescribing the minimum element size or by introducing some limitations on the local error. These alternatives are currently being explored by the authors.

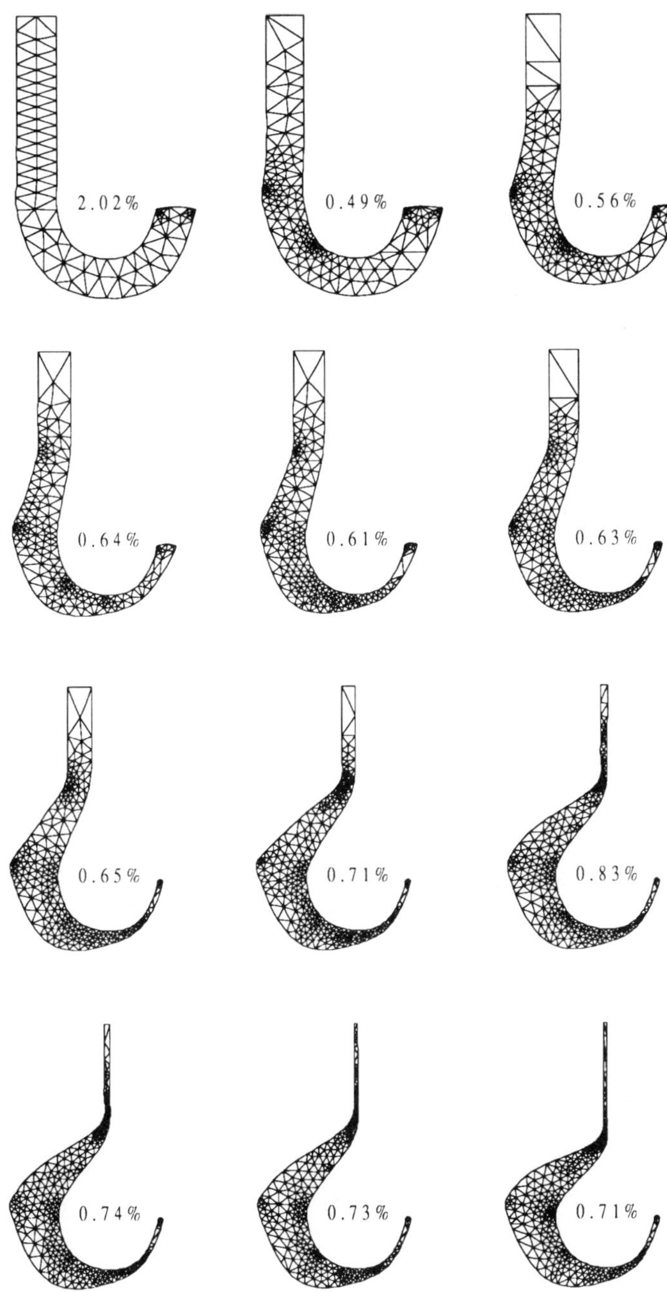

Figure 7.6 Hook optimization problem. Successive meshes and designs for iterations 0, 10, 20, 30, 40, 50, 60, 75, 90, 105, 120 and 130. The numbers denote the relative global error for the solution obtained for each design

ACKNOWLEDGEMENTS

The authors thank J. Castro and Prof. O.C. Zienkiewicz for many useful discussions and some of the ideas presented in this work.

REFERENCES

Zienkiewicz, O.C. and Zhu, J.Z. (1987), "A simple error estimator and adaptive procedure for practical engineering analysis", *Int. Num. Meth. Engrg.*, **24**, 337–357.

Zienkiewicz, O.C., Zhu, J.Z., Liu, I.C., Morgan, K. and Peraire, J. (1988), "Error estimates and adaptivity from elasticity to high speed compressible flow", in J.R. Whiteman, ed., MAFELAP 87, 483–512, Academic Press, New York.

Zhu, J.Z. and Zienkiewicz, O.C. (1988), "Adaptive techniques in the finite element method", *Comm. Appl. Numer. Methods*, **4**, 197–204.

Zienkiewicz, O.C., Zhu, J.Z. and Gong, N.G. (1989), "Effective and practical $h - p$ version adaptive analysis procedures for the finite element method", *Internat. J. Numer. Methods Engrg.*, **28**, 879–891.

Zienkiewicz, O.C., Zhu, J.Z. (1990), "The three R's of engineering analysis and error estimation and adaptivity", *Comp. Meth. in Appl. Mech. and Engng*, **82**, 95-113.

Zienkiewicz, O.C., Zhu, J.Z. (1989), "Error estimates and adaptive refinement for plate bending problems", *Internat. J. Numer. Methods Engng.*, **28**, 2839–53.

Atamaz-Sibai, W. and Hinton, E. (1990), "Adaptive mesh refinement with the Morley plate element", Proc. of NUMETA 90 conference held at Swansea 1990, **2**, 1044–57, Elserver App. Sc., London.

Atamaz-Sibai, W., Hinton, E. and Selman, A. (1990), "Adaptive mesh refinement with Mindlin-Reissner elements", *Proc. 2nd Int. Conf. Computer Aided Analysis and Design of Concrete Structures*, Apr. 1990, Zell-am-see, Austria, **1**, 303–15, Pineridge Press, Swansea, U.K.

Selman, A., Hinton, E. and Atamaz-Sibai, W. (1990), "Edge effects in Mindlin-Reissner plates using adaptive mesh refinement", *Eng. Comp.*, **7**. 3. 217–27.

Hinton, E. Özakça, M. and Rao, N.V.R. (1990), "Adaptive analysis of thin shells using facet elements", Int. Report CR/950/90, Univ. College of Swansea, U.K. April 1990.

Oñate, E., Castro, J. and Kreiner, R. (1991), "Error estimations and mesh adaptivity techniques for plate and shell problems", presented at the *3rd. International Conference on Quality Assurance and Standards in Finite Element Methods*, Stratford-upon-Avon, England, 10–12 September, 1991.

Oñate, E. and Castro, J. (1991), "Adaptive mesh refinement techniques for structural problems", published in *"The Finite Element Method in the 90's. A book dedicated to O. C. Zienkiewicz"*, E. Oñate, J. Periaux and J. Samuelsson (Eds.), Springer-Verlag and CIMNE, Barcelona.

Bugeda, G. (1990), "Utilización de técnicas de estimación de error y generación automática de mallas en procesos de optimización estructural", *Ph D. Thesis*, Universitat Polytècnica de Catalunya.

Bugeda, G. and Oliver, J. (1991), "Automatic adaptive remeshing for structural shape optimization", European Conf. on New Advances in *Computat. Struct. Mech.*, Giens, France, April, 1991.

Bugeda, G. and Oliver, J. (1993), "A general methodology for structural shape optimization problems using automatic adaptive remeshing" To appear in *Internat. J. Numer. Methods Engng.*

Bugeda, G. and Oñate, E. (1992(i)), "New adaptive techniques for structural problems", First European Conference on *Numerical Methods in Engineering*, Brussels, Belgium, September, 1992.

Bugeda, G. and Oñate, E. (1992(ii)), "Adaptive mesh refinement tecniques for aerodynamic problems", in *Numerical Methods in Engineering and Applied Sciences.* H. Alder, J. C. Heinrich, S. Lavanchy, E. Oñate and B. Suárez (Eds.), CIMNE-Barcelona 1992.

Wu, J., Zhu, J.Z., Smelter, J. and Zienkiewicz, O.C. (1990), "Error estimation and adaptivity in Navier-Stokes incompressible flow", *Computational Mechanics*, 6, 4, 259–71, 1990.

Zienkiewicz, O.C. and Taylor, R.L. (1989/1991), "The finite element method", Mc Graw Hill, **I**, 1989; **II**, 1991.

Peraire, J. (1986), "A finite element method for convection dominated flows", *Ph. D. Thesis*, Civil Eng. Dept. University College of Swansea, U.K.

Zienkiewicz, O.C. and Zhu, J.Z. (1992), "Superconvergence derivative recovery techniques and *a posteriori* error estimation in the finite element method. Part I and part II". *Internat. J. Numer. Methods Engng.*, **33**, 1331–82.

Bugeda, G. and Oñate, E. (1993(i)), "A study of mesh optimality criteria in adaptive finite element analysis." To be published in *Internat. J. Numer. Methods Engng.*

Bugeda, G. and Oñate, E. (1993(ii)), "A Methodology for Adaptive Mesh Refinement in Optimum Shape Design Problems", Research Report N° 31, CIMNE, Barcelona, March 1993.

Chapter 8

New Developments in Applications of *hp*-Adaptive BE/FE Methods to Elastic Scattering

L. Demkowicz* and J.T. Oden**

* Section of Applied Mathematics,
Cracow University of Technology, 31-155 Kraków, Poland
** The Texas Institute for Computational Mechanics,
The University of Texas at Austin,
Austin, TX 78712-1055, U.S.A.

8.1   INTRODUCTION

This paper is a presentation of the project reported in Demkowicz *et al.* (1991(i)), ... Demkowicz (1992). We begin with the formulation of the problem and a summary of principal design assumptions and follow with a sample, typical numerical example. A short summary of two currently investigated issues concludes the paper.

8.2   FORMULATION OF THE PROBLEM

Given a bounded domain $\Omega \subset \mathbb{R}^3$ with boundary Γ, we wish to find complex-valued

- total pressure
$$p(x) = p^{inc}(x) + p^s(x) \,, \ x \in \mathbb{R}^3 - \Omega \tag{8.1}$$

  with $p^{inc}(x), x \in \mathbb{R}^3$, a given *incident pressure*, prescribed in the whole $\mathbb{R}^3$ and $p^s(x)$, an unknown *scattered pressure* defined in the exterior of Ω;
- displacement vector
$$u(x) \,, \ x \in \Omega \tag{8.2}$$

satisfying the following equations

- Helmholtz equation in $\mathbb{R}^3 - \bar\Omega$

$$- \Delta p - k^2 p = 0 \tag{8.3}$$

- (Helmholtz) elasticity equations in Ω

$$\mathbf{div}(E : \epsilon(u)) + \rho_s \omega^2 u = 0 \tag{8.4}$$

The Mathematics of Finite Elements and Applications
Edited by J. R. Whiteman © 1994 John Wiley & Sons Ltd

- compatibility conditions on boundary Γ

$$\rho_f \omega^2 u_n = \frac{\partial p}{\partial n} \tag{8.5}$$

$$\sigma_n(\boldsymbol{u}) = -p$$

$$\boldsymbol{\sigma}_\tau(\boldsymbol{u}) = \boldsymbol{0}$$

- Sommerfeld condition at ∞

$$\mid \frac{\partial p^s}{\partial r} - ikp^s \mid = O(\frac{1}{r^2}) \tag{8.6}$$

where the following notation has been used : ω - time frequency, $k = \omega/c$ - wave number, with c - sound speed in fluid, $\epsilon = \epsilon(\boldsymbol{u})$ - infinitesimal strain tensor, $\boldsymbol{E}$ - tensor of elasticities, ρ_s - solid density, ρ_f - fluid density, $\boldsymbol{n}$ - external, normal unit vector to boundary Γ, $\sigma_n, \boldsymbol{\sigma}_\tau$ - normal and tangential stress components, r - distance from the origin, i - imaginary unit.

8.3 SOLUTION METHODOLOGY

The principal steps in the proposed methodology are as follows

1. The Helmholtz equation (8.3) and the Sommerfeld condition (8.6) are replaced with the equivalent Burton-Miller boundary integral equation on Γ. The choice of the Burton-Miller approach is essential in establishing full equivalence with the original formulation and avoiding the so called ficticious frequencies (compare Demkowicz *et al.* (1991(i)) ... Demkowicz *et al.* (1993(i)).
2. The resulting coupled system of the boundary integral equation and the original elasticity equations is restated in the following variational form

$$\begin{cases} \text{Find } (p, \boldsymbol{u}) \in H^{\frac{1}{2}}(\Gamma) \times \boldsymbol{H}^1(\Omega) & \text{such that} \\[2mm] a(p, q) + b(\boldsymbol{u}, q) = l(q) & \forall q \in H^{\frac{1}{2}}(\Gamma) \\[2mm] c(p, \boldsymbol{v}) + d(\boldsymbol{u}, \boldsymbol{v}) = 0 & \forall \boldsymbol{v} \in \boldsymbol{H}^1(\Omega) \end{cases} \tag{8.7}$$

with

$$\begin{aligned} a(p, q) &= \alpha\{\frac{1}{2} \int_\Gamma p(\boldsymbol{x})q(\boldsymbol{x})\, dS_{\boldsymbol{x}} - \int_\Gamma \int_\Gamma \frac{\partial \Phi}{\partial n_y}(\boldsymbol{x}, \boldsymbol{y})p(\boldsymbol{y})q(\boldsymbol{x})\, dS_y dS_{\boldsymbol{x}}\} \\[2mm] &+ \frac{(1-\alpha)i}{k}\{\int_\Gamma \int_\Gamma \Phi(\boldsymbol{x}, \boldsymbol{y})\, \mathrm{rot}_y p(\boldsymbol{y})\, \mathrm{rot}_{\boldsymbol{x}} q(\boldsymbol{x})\, dS_y dS_{\boldsymbol{x}} \\[2mm] &- k^2 \int_\Gamma \int_\Gamma \Phi(\boldsymbol{x}, \boldsymbol{y})n_{\boldsymbol{x}} \circ n_y p(\boldsymbol{y})q(\boldsymbol{x})\, dS_y dS_{\boldsymbol{x}}\} \end{aligned} \tag{8.8}$$

$$b(\boldsymbol{u}, q) = \alpha\rho_f \omega^2 \{\int_\Gamma \int_\Gamma \Phi(\boldsymbol{x}, \boldsymbol{y})n_I(\boldsymbol{x})u_I(\boldsymbol{y})q(\boldsymbol{x})\, dS_y dS_{\boldsymbol{x}}\} +$$

$$+ \; \frac{(1-\alpha)i\rho_f\omega^2}{k}\{\frac{1}{2}\int_\Gamma n_I(\boldsymbol{x})u_I(\boldsymbol{x})q(\boldsymbol{x})\,dS_{\boldsymbol{x}}$$

$$+ \; \int_\Gamma\int_\Gamma \frac{\partial\Phi}{\partial n_{\boldsymbol{x}}}(\boldsymbol{x},\boldsymbol{y})n_I(\boldsymbol{y})u_I(\boldsymbol{y})q(\boldsymbol{x})\,dS_{\boldsymbol{y}}\,dS_{\boldsymbol{x}}\} \tag{8.9}$$

$$l(q) \;=\; \alpha\int_\Gamma p^{inc}(\boldsymbol{x})q(\boldsymbol{x})\,dS_{\boldsymbol{x}} + \frac{(1-\alpha)i}{k}\int_\Gamma \frac{\partial p^{inc}}{\partial n_{\boldsymbol{x}}}(\boldsymbol{x})q(\boldsymbol{x})\,dS_{\boldsymbol{x}} \tag{8.10}$$

$$c(p,\boldsymbol{v}) \;=\; \int_\Gamma p(\boldsymbol{x})v_I(\boldsymbol{x})n_I(\boldsymbol{x})\,dS_{\boldsymbol{x}} \tag{8.11}$$

$$d(\boldsymbol{u},\boldsymbol{v}) \;=\; \int_\Omega E_{ijkl}\frac{\partial u_i}{\partial x_j}(\boldsymbol{x})\frac{\partial v_k}{\partial x_l}(\boldsymbol{x})\,d\boldsymbol{x} \tag{8.12}$$

$$\tag{8.13}$$

where

- $\Phi(\boldsymbol{x},\boldsymbol{y}) = \Phi(r) = \frac{e^{ikr}}{r}$ is the fundamental solution to the Helmoltz equation (8.3), $r = \mid \boldsymbol{x} - \boldsymbol{y} \mid$;
- $\frac{\partial\Phi}{\partial n_{\boldsymbol{x}}} = \Phi'(r)\frac{\partial r}{\partial n_{\boldsymbol{x}}}$, $\frac{\partial\Phi}{\partial n_{\boldsymbol{y}}} = \Phi'(r)\frac{\partial r}{\partial n_{\boldsymbol{y}}}$;
- $\mathbf{rot}_{\boldsymbol{x}}p(\boldsymbol{x}) = \nabla p(\boldsymbol{x}) \times \boldsymbol{n}_{\boldsymbol{x}}$;
- $\alpha \in (0,1)$ is an arbitrary parameter, in practice $\alpha = 1/2$;
- $H^{1/2}(\Gamma), \boldsymbol{H}^1(\Omega)$ are the standard Sobolev spaces.

3. Based on the variational formulation (8.7), the usual Galerkin approximation is applied. Given a set of basis functions on the surface Γ

$$e_i(\boldsymbol{x}),\; i = 1,\ldots,N,\; \boldsymbol{x} \in \Gamma \tag{8.14}$$

and a set of basis functions in the domain Ω

$$\phi_j(\boldsymbol{x}),\; j = 1,\ldots,M,\; \boldsymbol{x} \in \Omega \tag{8.15}$$

we introduce approximations

$$p(\boldsymbol{x}) = \sum_{i=1}^{N} p_i e_i(\boldsymbol{x})\,,\; q(\boldsymbol{x}) = \sum_{k=1}^{N} q_k e_k(\boldsymbol{x}) \tag{8.16}$$

$$p_i, q_k \in \mathbb{C}\,,\; i,k = 1,\ldots,N$$

$$u_I(\boldsymbol{x}) = \sum_{j=1}^{M} u_j^I \phi_j(\boldsymbol{x})\,,\; v_J(\boldsymbol{x}) = \sum_{l=1}^{M} v_l^J \phi_l(\boldsymbol{x})$$

$$u_j^I, v_l^J \in \mathbb{C}\,,\; I,J = 1,2,3\,,\; j,l = 1,\ldots,M$$

Upon replacing p and q, $\boldsymbol{u}$ and $\boldsymbol{v}$ in the variational formulation (8.7) with their approximations (8.15) we end up with a linear system of equations of the form

$$\begin{cases} \sum_{i=1}^{N} a_{ik}p_i + \sum_{I=1}^{3}\sum_{j=1}^{M} b_{jk}^I u_j^I = l_k & ,k = 1,\ldots,N \\[2mm] \sum_{i=1}^{N} c_{il}^J p_i + \sum_{I=1}^{3}\sum_{j=1}^{M} d_{jl}^{IJ} u_{jl}^I = 0 & ,J = 1,2,3,\; l = 1,\ldots,M \end{cases} \tag{8.17}$$

where the stiffness matrices $a_{ik}, \ldots, d_{jl}^{IJ}$ and the load vector l_k, are obtained by replacing in formulas (8.8)-(8.12) solution and test functions with the respective basis functions.

4. The Galerkin basis functions are constructed using adaptive hp-approximations based on the notion of constrained approximation (see Demkowicz *et al.* (1989)). An arbitrary, unstructured triangular grid may be used on the surface. The grid propagates into the body in the form of layers of prismatic solid elements. In this way full compatibility of the surface and body meshes is enforced.

   Elements may have an arbitrary, locally variable order of approximation p (in practice $p = 1, 2, \ldots, 9$) and may be h-refined. Prismatic body elements may be refined *vertically* into two ($h2$-refinement) or four ($h4$-refinement) elements or may be refined *horizontally* into two elements ("sandwich like", $h3$-refinement). Any vertical $h2$- or $h4$-refinement as well as p-refinement is accompanied by a corresponding refinement of surface elements. For details concerning the approximation and related data structure issues, we refer to Demkowicz *et al.* (1991(i)) ... Demkowicz *et al.* (1993(i)) and Demkowicz and Banaś (1993).

   In practice, initial meshes are generated using a multiblock, algebraic mesh generator based on a geometric modeling package (see Demkowicz *et al.* (1992)). Recent developments concerning the initial mesh generation and geometry representation are discussed in Demkowicz and Oden (1993(i))

5. The fluid-variable methodology is used to solve discrete system of equations (8.16). First the structural part of system (8.16) is solved for surface degrees of freedom of the displacement vector, using a frontal solver with a multiple right-hand side technique. The resulting structural impendance matrices are next used to form the final BE system of equations for the surface pressure. The accumulation for the final stiffness matrix is done in an element by element fashion, described in Demkowicz *et al.* (1993(i)), resulting in very essential memory savings. In practice only the matrix a_{ij} and the structural front are kept in the computer memory.

6. An hp-adaptive strategy, based on an a-posteriori residual error estimate is used to produce an optimal hp-mesh and the corresponding solution with a prescribed accuracy at a minimum computational cost (see Oden *et al.* (1993)). Once the solution on the scatterer is known, the classical Helmholtz representation formula is used to evaluate far-field values at desired points.

8.4   EXAMPLE OF A SOLUTION

As a typical example we present a solution of the scattering of a plane wave on an elastic spherical shell problem. The sphere problem was solved for wave number $k = 1.13$ close to the first resonant frequency (comp. Junger and Feit (1986)) using a simplified h4-adaptive strategy based on equilibriation of surface residuals, only. For precise specification of numerical data we refer to Demkowicz *et al.* (1993(i)). Only one layer of prismatic elements to model the shell was used. The optimal mesh is displayed in Figure 8.1, while a comparison with the exact solution along the vertical cross-section is shown in Figure 8.2.

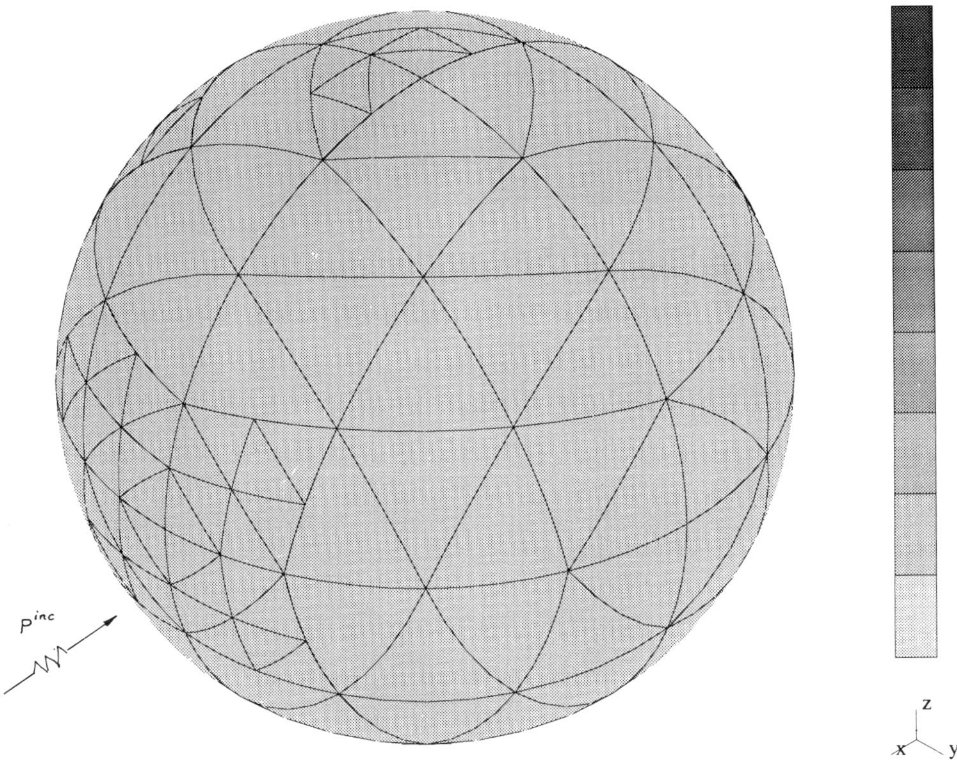

Figure 8.1 Scattering of a plane wave on a spherical shell problem for the first resonant frequency. Optimal mesh

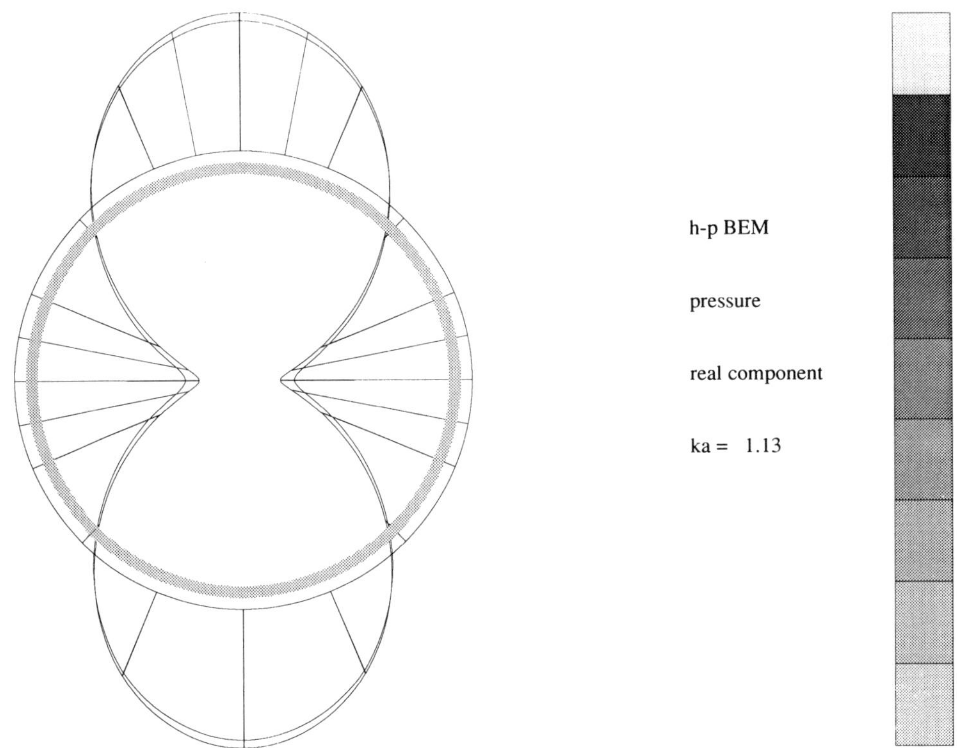

Figure 8.2 Scattering of a plane wave on a spherical shell problem for the first resonant frequency. Comparison of the final mesh approximate solution with the exact one

8.5 NEW DEVELOPMENTS

8.5.1 ONE-IRREGULAR MESHES STRATEGY FOR ANISOTROPIC h-REFINEMENTS

For simplicity, we shall restrict ourselves to the case of 2D rectangular meshes. Recall that an element vertex node is called *regular* if it constitutes a vertex for each of the neighbouring elements. If not, then for one of the neighbouring elements, the node is located in the interior of one of the element sides, and it is called *irregular*. If the maximum number of irregular nodes located on a (big) element side is equal m, we say that the mesh is m-irregular. In practice one admits only 1-irregular meshes, i.e. the irregular node is always located exactly in the middle of a big element side. In order to enforce the continuity of approximation along the side, the value of a degree of freedom (d.o.f.) at the irregular node is set to the average of the values of the corresponding big neighbour nodes, hereafter called the *parent nodes*. For that reason the irregular nodes are also called the *constrained* nodes.

In the case of isotropic h-refinements (elements refined into four *sons*), the 1-irregularity of the mesh is enforced using the well known *two-to-one algorithm*. If an element is to be refined then one checks whether all its neighbours are of equal or smaller size. If a big neighbour is encountered , the element is put on a waiting list and the whole procedure repeats with the big neighbour replacing the original element. If the currently considered element has no big neighbours then it is refined and the procedure continues with the last element from the waiting list. The algorithm is completed once the waiting list is empty, i.e. in particular the original element is refined.

In the case of h-isotropic refinements 1-irregularity of the mesh guarantees that parent nodes of an irregular (constrained) node are themselves always regular. We say that the node is *(single) 1-constrained.* Recursively, an irregular node is said to be *(multiple) n-constrained* if its parent nodes are *(multiple) (n-1)-constrained.* When anisotropic (vertical or horizontal) h-refinements are introduced, the one-to-two rule no longer guarantees that all irregular nodes are single constrained only. A typical situation is depicted in Figure 8.3, where each of A_n nodes is n-constrained. In fact, a closer look at the mesh reveals that, excluding the boundary nodes, the proposed refinement does not introduce a single new degree of freedom !

It becomes evident that in order to avoid the multiple constrained nodes, a new strategy for anisotropic refinements has to be worked out. One possible choice would be to allow for refinement of an element only if all its nodes are regular. The corresponding algorithm, however, produces *refinement- path dependent* meshes. A typical situation is shown in Figure 8.4. In the case of each mesh presented the middle element is to be refined into four sons through consecutive vertical (horizontal) refinement and then following horizontal (vertical) refinements of its sons. The two cases show that the final mesh will depend upon which of the anisotropic refinements is executed first.

The simple algorithm, we propose, does not prohibit the existence of double-constrained nodes but it *does* eliminate further propagation of constraints and, at the same time, is path-independent. The algorithm distinguishes between three kinds of elements: the initial mesh elements *(kind=0)* and elements resulting from vertical *(kind=2)* or horizontal *(kind=1)* refinements. Analogously *nref=1* or *nref=2* denotes

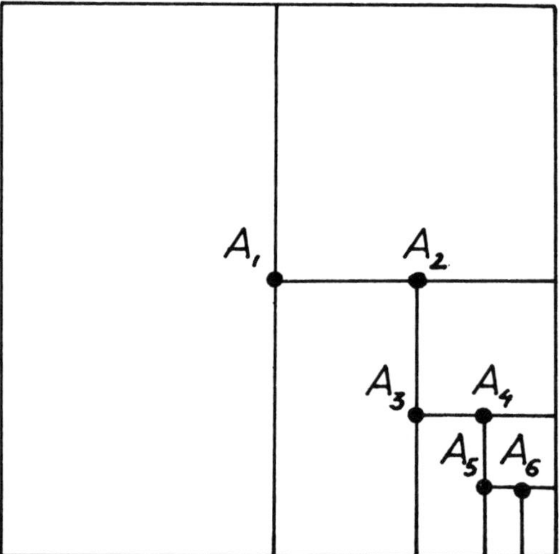

Figure 8.3 Example of a mesh showing failure of the *two-to-one* algorithm in avoiding multiple constrained nodes

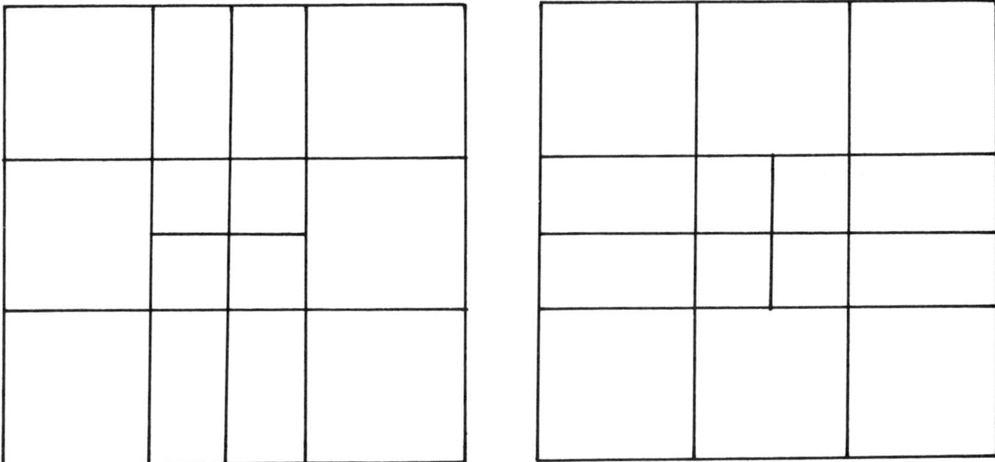

Figure 8.4   Refinement path dependency in anisotropic refinements

required horizontal or vertical refinement, respectively. We note also that within each of the three classes (kinds), elements are denumerated separately and therefore each element is identified using both its number *nel* and kind *kind*.

The proposed algorithm has the following form:

input: element number *Nel* and kind *Kind* (= 0,1,2)

       kind of refinement required *Nref* (= 1,2)

```
set  nel=Nel, kind=Kind, nref=Nref
10      continue
     if (kind = 0) go to 20
     if (nref = kind) then
           if all nodes of the element are active then
                 go to 20
           else
                 store the element on the waiting list
                 identify the element to be refined first and the kind of refinement
                 required,
                 i.e. set new nel, kind, nref
                 go to 10
           endif
     else
           if all nodes of the element's father are active then
                 go to 20
```

```
        else
                store the element on the waiting list
                identify the element to be refined first and the kind of refinement
                required,
                i.e. set new nel, kind and nref
                go to 10
        endif
    endif
20  break the (nel,kind) element
    if the waiting list is empty then
            stop
    else
            get the last element form the waiting list, i.e. take new nel,kind and nref
            from the list
            go to 10
    endif
```

A generalization of the algorithm to the 3-D FE meshes described in point 4 of Section 8.3 of the description of this project has been implemented. For all the details we refer to Demkowicz and Banaś (1993).

8.5.2 ASYMPTOTIC CONVERGENCE

The coupled, elastic scattering problem under consideration falls into the category of variational problems

$$\begin{cases} \text{Find } u \in V \text{ such that} \\ b(u,v) = l(v) \; \forall v \in V \end{cases} \tag{8.18}$$

where V is a Hilbert space, l is a continuous and antilinear functional on V and b is a continuous, sesquilinear form on V of the form

$$b(u,v) = a(u,v) + c(u,v) \tag{8.19}$$

where a is a (complex-) symmetric, V-coercive sesquilinear form and c is a compact sesquilinear form. Two other related problems of interest fall into the same cathegory: forced vibrations problems from structural mechanics and the model elastic scattering problem considered in Demkowicz et al. (1991(i)), Demkowicz et al. (1991(ii)).

Given a sequence of finite dimensional spaces $V_h \subset V$, approximating V, i.e.

$$\forall v \in V \quad \forall h \quad \exists v_h \in V_h \quad \text{such that} \quad v_h \to v \text{ as } h \to 0 \tag{8.20}$$

we construct the usual Galerkin approximation of (8.17) with the form

$$\begin{cases} \text{Find } u_h \in V_h \text{ such that} \\ b(u_h, v_h) = l(v_h) \; \forall v_h \in V_h \end{cases} \tag{8.21}$$

We record now, after Wendland (1988), two fundamental results

Theorem 1 (Babuska)

If

$$\sup_{v_h \neq 0} \frac{\mid B(u_h, v_h) \mid}{\parallel v_h \parallel} \geq \gamma \parallel u_h \parallel \tag{8.22}$$

then

$$\parallel u - u_h \parallel \leq (1 + M\gamma^{-1}) \parallel u - w_h \parallel \quad \forall w_h \in V_h \tag{8.23}$$

where M is the continuity constant for the form b.

Consequently, if γ_h is independent of h, then the approximation is said to be *optimal* as the approximation error $\parallel u - u_h \parallel$ is bounded by the best approximation error

$$\inf_{w_h \in V_h} \parallel u - w_h \parallel \tag{8.24}$$

The optimal constant

$$\gamma_h = \inf_{\parallel u_h \parallel = 1} \sup_{\parallel v_h \parallel = 1} \mid b(u_h, v_h) \mid \tag{8.25}$$

is called the *discrete LBB-constant* and provides information about the *conditioning* of the discrete problem.

Theorem 2 (Mikhlin)

If problem (8.17) has a unique solution then there exists an $h_0 > 0$ such that condition (8.21) holds for every $h < h_0$.

We say in such a case that the approximation is *asymptotically stable* and consequently *asymptotically optimal*. The practical question now is: *can we characterize h_0* ? The analysis presented in Demkowicz and Oden (1993(i)) for a particular forced vibration problem showed that the region of asymptotic stability (convergence) is reached once the eigenvalues neighbouring the wave number k^2 are sufficiently well resolved. More precisely, the corresponding *approximate* eigenvalues must be located on the same, appropriate, side of the k^2-border line as the exact ones.

A mathematical analysis of the general case has been given in Demkowicz (1993). Here we present a short summary of the results, referring for all the details to Demkowicz (1993).

1. Introducing operators $B, A, C : V \rightarrow V'$ and $B_h, A_h, C_h : V_h \rightarrow V_h'$ corresponding to bilinear forms b, a, c on V and V_h respectively, e.g.

$$b(u, v) = \; < Bu, v > \quad \forall v \in V \tag{8.26}$$

etc., we obtain

$$\begin{aligned}
\gamma_h^2 &= \; \inf_{\parallel u_h \parallel = 1} \parallel R_h^{-1} B_h u_h \parallel^2 \\
&= \; \inf_{u^T g u = 1} (g^{-1} b)^* g (g^{-1} b)
\end{aligned} \tag{8.27}$$

where R_h denotes the discrete Riesz operator, b and g are matrix representations of the forms $b(u, v)$ and the inner product $(u, v)_V$ (Gram matrix) with respect to a basis in V_h, and $*$ denotes the complex adjoint. It follows in particular that γ_h is equal to the *smallest* eigenvalue of $(g^{-1} b)^* g (g^{-1} b)$ and as such it is a perfectly *computable* quantity !

2. Replacing the original inner product in $V(V_h)$ with the equivalent one defined by the form $a(u, v)$, we identify Riesz operators R and R_h with operators A and A_h and introduce *compact* operators $K = A^{-1}C$ and $K_h = A_h^{-1}C_h$. Consequently

$$\gamma_h = \inf_{\|u_h\|=1} \| (I + K_h)u_h \| \tag{8.28}$$

where I is the identity operator.

3. Considering the compact operator K we conclude that

(a) $I + K$ invertible $\Rightarrow \inf_{\|u\|=1} \| (I + K)u \| =: \gamma > 0$
(b) K compact $\Rightarrow \| K - K_h \| \to 0$ as $h \to 0$
(c) $\gamma_h \geq \gamma - \| K - K_h \|$
(d) $\gamma_h \to \gamma$ as $h \to 0$

The third condition in particular implies that attaining the asymptotic stability region depends upon the value of the constant γ and the rate of convergence of $\| K - K_h \|$. Once a specific operator K is given, it may be possible to estimate the convergence rate precisely. The fourth condition indicates that, asymptotically in h, one cannot expect a better stability constant γ_h than for the infinite-dimensional problem.

4. The spectrum of the operator K, with a possible exception for 0, consists of a decreasing (in modulus) sequence of complex eigenvalues, converging to zero if the sequence is infinite

$$| \lambda_1 | \geq | \lambda_2 | \geq \ldots \geq | \lambda_n | \; (\to 0) \tag{8.29}$$

One may try in a way similar to that in the example presented in Demkowicz and Oden (1993(i)) describe the convergence of K_h to K, and consequently reaching the asymptotic stability region, in terms of convergence of approximate eigenvalues λ_i^h to the exact ones λ_i. A number of simple observations follows

(a) $\gamma_h \leq \min_{\lambda_h \in \mathrm{Sp}(K_h)} | 1 + \lambda_h |$
(b) If the operator K_h is normal then the equality holds
    $\gamma_h = \min_{\lambda_h \in \mathrm{Sp}(K_h)} | 1 + \lambda_h |$
(c) For a self-adjoint operator K_h all eigenvalues λ_h are real
(d) For an arbitrary operator K_h
    $\gamma_h^2 = \min_{\lambda_h \in \mathrm{Sp}(S_h)} | 1 + \lambda_h |$
    where $S_h = K_h^* + K_h + K_h^* K_h$ is self-adjoint.

The case described in Demkowicz and Oden (1993(i)) falls into the third cathegory. It is clear that in all cases, for getting γ_h close to γ, the resolution of eigenvalues of K (or S_h in the general case) close to unity, is crucial.

Acknowledgement: The work of the second author has been supported by the Office of Naval Research under Contract N00014-89-J-1451. The work of the first author has been supported in part by by the Polish National Science Foundation (Komitet Badań Naukowych), under Contract 1548/3/91 and in part by the Office of Naval Research under Contract N00014-89-J-1451.

REFERENCES

Demkowicz L., Oden J.T., Rachowicz W. and Hardy O. (1989), "Toward a Universal *h-p* Adaptive Finite Element Strategy. Part1: Constrained Approximation and Data Structure", *Computer Methods in Applied Mechanics and Engineering* **77** (1-2), 79-112;

Demkowicz L., Oden J.T., Ainsworth M. and Geng P. (1991(i)), "Solution of Elastic Scattering Problems in Linear Acoustics Using *h-p* Boundary Element Methods", *Journal of Computational and Applied Mathematics* **36** , 29-63;

Demkowicz L., Karafiat A. and Oden J.T. (1991(ii)) "Solution of Elastic Scattering Problems in Linear Acoustics Using h-p Boundary Element Method", *Computer Methods in Applied Mechanics and Engineering*,**101**, pp. 251-282 (Proceedings of Second Workshop on Reliability and Adaptive Methods in Computational Mechanics, Cracow, October 1991, eds. L.Demkowicz, J.T.Oden and I.Babuska);

Demkowicz L. and Oden J.T. (1993(i)), "Elastic Scattering Problems in Linear Acoustics Using an h-p Boundary/Finite Element Method", in **Adaptive Finite and Boundary Element Methods**, eds. C.A.Brebbia and M.H.Aliabadi, Computational Mechanics Publications;

Demkowicz L. (1992), "Geometrical Modeling Package", *TICOM REPORT* 92-06, The University of Texas at Austin, Austin TX 78712

Demkowicz L. and Oden J.T.(1993(ii)), "Recent Progress on Application of *hp*-Adaptive BE/FE Methods to Elastic Scattering", submitted to *International Journal for Numerical Methods in Engineering*;

Demkowicz L. and Banaś K. (1993), "3D hp Adaptive Package", *SAM Report 2/1993*, Section of Applied Mathematics, Cracow University of Technology, March 1993;

Demkowicz L. (1993), "Some Results on Asymptotic Convergence in Finite Element Method", *SAM Report 3/1993*, Section of Applied Mathematics, Cracow University of Technology, April 1993;

Hsiao G.C. (1990), "The Coupling of Boundary Element and Finite Element Methods", *ZAMM, Z. Angew. Math. Mech.*, **70** (6), T493-T503;

Junger M.C. and Feit D. (1986), **Sound, Structures and Their Interaction**, MIT Press, Cambridge, Massachusetts, 2nd ed.

Oden J.T. ,Patra A. and Feng Y. (1993), "An *h-p* Adaptive Strategy", in **Proceedings of Symposium on Adaptive, Multilevel and Hierarchical Computational Strategies**, ASME Winter Annual Meeting, Nov.8-13, 1992, Anaheim, C.A., to be published.

Wendland W.L. (1988), "On Asymptotic Error Estimates for Combined BEM and FEM", in **Finite Element and Boundary Element Techniques From a Mathematical and Engineering Point of View**, E.Stein & W.I.Wendland (eds), CISM Lecture Notes 301, Voline, Springer-Verlag, Vien-New York, pp.273-333.

Chapter 9

A Posteriori Error Estimates for Stokes and Navier Stokes Equations

Mark Ainsworth

Mathematics Department, Leicester University, Leicester LE1 7RH, U.K.

9.1 INTRODUCTION

Even a cursory glance through the proceedings of past MAFELAP meetings reveals the intense research activity towards the development of adaptive (h, p or hp) finite element methods. The critical ingredient of any adaptive algorithm is the ability to assess and control the accuracy of the discretization, usually by means of an *a posteriori* error estimate. Techniques for self-adjoint problems are well established and plentiful. However, for problems with constraints (*e.g.* incompressible Stokesian flow) or non-linear problems (*e.g.* Navier Stokes equations) there is considerably less scope, especially if one is interested in hp algorithms. The paper describes recent progress towards the development of effective error indicators and estimates for hp approximation of such problems.

9.1.1 THE STOKES PROBLEM

Let $\Omega \subset \mathbb{R}^n$ ($n = 2$ or 3) be a bounded domain with smooth boundary $\partial\Omega$. The spaces X and M are defined to be

$$X = [H_0^1(\Omega)]^n \quad \text{and} \quad M = \left\{ q \in L_2(\Omega) : \int_\Omega q \, \mathrm{d}\mathbf{x} = 0 \right\} \tag{9.1}$$

where $H_0^1(\Omega)$ and $L_2(\Omega)$ are the usual Sobolev and Lebesgue spaces respectively. For simplicity it is assumed that homogeneous Dirichlet conditions are specified on the whole of $\partial\Omega$. For given data $\mathbf{f} \in [L_2(\Omega)]^n$, the Stokes problem is to find a pair $(\mathbf{u}, p) \in X \times M$ such that:

$$-\operatorname{div} \underline{\sigma}(\mathbf{u}, p) = \mathbf{f} \text{ in } \Omega \tag{9.2}$$

$$\operatorname{div} \mathbf{u} = 0 \text{ in } \Omega \tag{9.3}$$

and

$$\mathbf{u} = 0 \text{ on } \partial\Omega \tag{9.4}$$

where $\mathbf{u}$ is the velocity field, p is the pressure and $\underline{\sigma}$ is the stress tensor defined by

$$\sigma_{ij}(\mathbf{u}, p) = \nu \, \partial u_j / \partial x_i - p \, \delta_{ij} \tag{9.5}$$

The Mathematics of Finite Elements and Applications
Edited by J. R. Whiteman © 1994 John Wiley & Sons Ltd

and $\nu > 0$ is the (constant) viscosity. The weak formulation of (9.2)–(9.4) is obtained by introducing the bilinear forms $B : X \times X \to \mathbb{R}$ and $b : X \times M \to \mathbb{R}$ defined by:

$$B(\mathbf{u}, \mathbf{v}) = \int_\Omega \nu \operatorname{grad} \mathbf{u} : \operatorname{grad} \mathbf{v} \, d\mathbf{x} \tag{9.6}$$

and

$$b(\mathbf{u}, q) = - \int_\Omega q \operatorname{div} \mathbf{v} \, d\mathbf{x} \tag{9.7}$$

The Stokes problem may then be rewritten as:

Find $(\mathbf{u}, p) \in X \times M$ such that for any $(\mathbf{v}, q) \in X \times M$

$$B(\mathbf{u}, \mathbf{v}) + b(\mathbf{v}, p) + b(\mathbf{u}, q) = L(\mathbf{v}) \tag{9.8}$$

where $L : X \to \mathbb{R}$ is the linear functional

$$L(\mathbf{v}) = \int_\Omega \mathbf{f} \cdot \mathbf{v} \, d\mathbf{x} \tag{9.9}$$

It will be assumed that the bilinear forms are continous

$$|B(\mathbf{u}, \mathbf{v})| \le C_B \| \mathbf{u} \|_X \| \mathbf{v} \|_X \tag{9.10}$$

$$|b(\mathbf{v}, q)| \le C_b \| \mathbf{v} \|_X \| q \|_M \tag{9.11}$$

and that $b(\cdot, \cdot)$ satisfies the inf-sup condition

$$\sup_{\mathbf{v} \in X} \frac{|b(\mathbf{v}, q)|}{\| \mathbf{v} \|_X} \ge \alpha_b \| q \|_M \tag{9.12}$$

Under these conditions it may be shown (Girault and Raviart (1986)) that there exists a unique solution to the Stokes problem (9.8).

9.1.2 NORM ON $X \times M$

The canonical norm on the product space $X \times M$ is

$$\| (\mathbf{e}, E) \|_{X \times M} = \left\{ \| \mathbf{e} \|_X^2 + \| E \|_M^2 \right\}^{1/2} \tag{9.13}$$

However, we shall introduce an alternative choice. The pair $(\mathbf{\Phi}, \psi) \in X \times M$ is uniquely defined by the condition

$$B(\mathbf{\Phi}, \mathbf{v}) + (\psi, q) = B(\mathbf{e}, \mathbf{v}) + b(\mathbf{v}, E) + b(\mathbf{e}, q) \tag{9.14}$$

for all $(\mathbf{v}, q) \in X \times M$. The new norm is then given by

$$\| (\mathbf{e}, E) \|_* = \left\{ \| \mathbf{\Phi} \|_X^2 + \| \psi \|_M^2 \right\}^{1/2} \tag{9.15}$$

In fact this norm is equivalent to the norm in equation (9.13):

Theorem 1   *There exist positive constants k_1 and k_2 depending only on C_B, C_b and α_b such that for any $(e, E) \in X \times M$*

$$k_1 \parallel (e, E) \parallel^2_* \leq \parallel (e, E) \parallel^2_{X \times M} \leq k_2 \parallel (e, E) \parallel^2_* \tag{9.16}$$

Proof See Ainsworth and Oden (1993(ii)).

9.2 FINITE ELEMENT DISCRETIZATION

The domain Ω is partitioned into a collection $\mathcal{P} = \{K\}$ of subdomains in the usual manner. The subdomains may be simplices, hexahedra or a mixture of the two. Following the standard finite element approach, piecewise polynomial (possibly of non-uniform degree) subspaces $\hat{X}$ and $\hat{M}$ of X and M are established on the partition. In particular, the approximation to the pressure space may comprise of functions which are discontinuous between the subdomains. Usually, the approximation will satisfy an inf-sup condition but this is not strictly necessary for our discussion. The finite element discretization of the Stokes problem (9.8) is

Find $(\hat{\mathbf{u}}, \hat{p}) \in \hat{X} \times \hat{M}$ such that for any $(\hat{\mathbf{v}}, \hat{q}) \in \hat{X} \times \hat{M}$

$$B(\hat{\mathbf{u}}, \hat{\mathbf{v}}) + b(\hat{\mathbf{v}}, \hat{p}) + b(\hat{\mathbf{u}}, \hat{q}) = L(\hat{\mathbf{v}}) \tag{9.17}$$

with discretization error defined to be

$$(e, E) = (\mathbf{u}, p) - (\hat{\mathbf{u}}, \hat{p}) \tag{9.18}$$

The objective is now to estimate the global error measured in the $*$-norm $\parallel (e, E) \parallel_*$ defined in equation (9.15). At first sight, it may seem bizarre to estimate the error in such a norm. However, there is no natural energy type norm for the Stokes problem as, for example, in the case of linear elasticity. The use of the $*$-norm is reminiscent of the the *symmetrizer* norm used by Oden *et al.* (1990).

Associated with the partition $\mathcal{P}$ are local versions of the bilinear and linear forms $B_K : X_K \times X_K \rightarrow \mathbb{R}$ and $b_K : X_K \times M_K \rightarrow \mathbb{R}$

$$B_K(\mathbf{u}, \mathbf{v}) = \int_K \nu \, \mathbf{grad}\, \mathbf{u} : \mathbf{grad}\, \mathbf{v} \, \mathrm{dx} \tag{9.19}$$

$$b_K(\mathbf{v}, q) = - \int_K q \, \mathbf{div}\, \mathbf{v} \, \mathrm{dx} \tag{9.20}$$

where X_K and M_K denote local spaces of functions consisting of the restrictions of elements of X and M to a single subdomain. Of course, the product space formed from local spaces X_K need not coincide with the original space X owing to discontinuities across the interfaces. However, the following hold

$$X \subset X(\mathcal{P}) := \prod_{K \in \mathcal{P}} X_K \text{ and } M = M(\mathcal{P}) := \prod_{K \in \mathcal{P}} M_K \tag{9.21}$$

We shall be concerned with the space of continous linear functionals τ acting on the product space $X(\mathcal{P}) \times M(\mathcal{P})$ which vanish on the subspace $X \times M$. Let $\mathbb{H}(\mathbf{div}, \Omega)$ denote the space

$$\mathbb{H}(\mathbf{div}, \Omega) = \left\{ \underline{A} \in L_2(\Omega)^{n \times n} : \mathbf{div}\,\underline{A} \in L_2(\Omega)^n \right\} \tag{9.22}$$

with the usual norm. The following subspace of $\mathbb{H}(\mathbf{div}, \Omega)$ will be of interest

$$\Upsilon = \left\{ \underline{A} \in \mathbb{H}(\mathbf{div}, \Omega) : \oint_{\partial\Omega} \mathbf{n} \cdot \underline{A} \cdot \mathbf{v}\,\mathrm{d}s = 0 \quad \forall \mathbf{v} \in X \right\} \tag{9.23}$$

where $\mathbf{n}$ denotes the unit, outward normal on the boundary $\partial\Omega$. The following result characterizes the functionals τ mentioned above.

Theorem 2   *A continuous linear functional τ on the space $X(\mathcal{P}) \times M(\mathcal{P})$ vanishes on the subspace $X \times M$ if and only if there exists $\underline{A} \in \Upsilon$ such that*

$$\tau\left[(\mathbf{v}, q)\right] = \sum_{K \in \mathcal{P}} \oint_{\partial K} \mathbf{n}_K \cdot \underline{A} \cdot \mathbf{v}_K\,\mathrm{d}s \tag{9.24}$$

where $\mathbf{n}_K$ is the unit, outward normal on ∂K.

Proof See Ainsworth and Oden (1993(ii)).

9.3 A POSTERIORI ERROR ANALYSIS

Let K and J be subdomains sharing a common interface $\Gamma_{KJ} = \partial K \cap \partial J$. Generally, the approximation to the true flux $\mathbf{n} \cdot \underline{\sigma}(\mathbf{u}, p)$ on Γ_{KJ} is discontinuous. An averaged approximate flux on the interface is denoted by

$$\langle \mathbf{n}_K \cdot \underline{\sigma}(\hat{\mathbf{u}}, \hat{p}) \rangle = \alpha_{KJ}(s)\,\mathbf{n}_K \cdot \underline{\sigma}_K + \alpha_{JK}(s)\,\mathbf{n}_K \cdot \underline{\sigma}_J \tag{9.25}$$

where $\underline{\sigma}_K = \underline{\sigma}(\hat{\mathbf{u}}|_K, \hat{p}|_K)$ is the approximate stress in the subdomain K. The diagonal matrices $\alpha_{KJ}(s)$ used in the weighted combination are required to satisfy the condition

$$\alpha_{KJ}(s) + \alpha_{JK}(s) \equiv I \text{ on } \Gamma_{KJ} \tag{9.26}$$

If one chooses $\alpha_{KJ} = I/2$ then a simple averaging results. However, later an alternative strategy is discussed.

With the aid of (9.8), (9.14) and (9.17) one obtains for all $(\mathbf{v}, q) \in X \times M$:

$$B(\boldsymbol{\Phi}, \mathbf{v}) + (\psi, q) = L(\mathbf{v}) - B(\hat{\mathbf{u}}, \mathbf{v}) - b(\mathbf{v}, \hat{p}) - b(\hat{\mathbf{u}}, q) \tag{9.27}$$

This may in turn be reformulated as a minimization principle

$$-\frac{1}{2}\,\|\,(\mathbf{e}, E)\,\|_*^2 = \mathcal{J}(\boldsymbol{\Phi}, \psi) \le \mathcal{J}(\mathbf{v}, q) \quad \forall (\mathbf{v}, q) \in X \times M \tag{9.28}$$

where

$$J(\mathbf{v}, q) = \frac{1}{2} \parallel \mathbf{v} \parallel_X^2 + \frac{1}{2} \parallel q \parallel_M^2 - L(\mathbf{v}) + B(\hat{\mathbf{u}}, \mathbf{v}) + b(\mathbf{v}, \hat{p}) + b(\hat{\mathbf{u}}, q) \qquad (9.29)$$

The Lagrangian functional $\mathcal{L} : X(\mathcal{P}) \times M(\mathcal{P}) \times \Upsilon \to \mathbb{R}$ is defined by

$$\mathcal{L}(\mathbf{v}, q, \tau) = J(\mathbf{v}, q) - \tau\,[(\mathbf{v}, q)] \qquad (9.30)$$

It is readily observed that

$$\sup_{\tau \in \Upsilon} \mathcal{L}(\mathbf{v}, q, \tau) = \begin{cases} J(\mathbf{v}, q), & \text{if } (\mathbf{v}, q) \in X \times M; \\ +\infty, & \text{otherwise.} \end{cases} \qquad (9.31)$$

Therefore,

$$\begin{aligned}
-\frac{1}{2} \parallel (e, E) \parallel_*^2 &= \inf_{(\mathbf{v},q) \in X \times M} J(\mathbf{v}, q) \\
&= \inf_{(\mathbf{v},q) \in X(\mathcal{P}) \times M(\mathcal{P})} \sup_{\tau \in \Upsilon} \mathcal{L}(\mathbf{v}, q, \tau) \\
&= \sup_{\tau \in \Upsilon} \inf_{(\mathbf{v},q) \in X(\mathcal{P}) \times M(\mathcal{P})} \mathcal{L}(\mathbf{v}, q, \tau) \\
&\geq \inf_{(\mathbf{v},q) \in X(\mathcal{P}) \times M(\mathcal{P})} \mathcal{L}(\mathbf{v}, q, \hat{\tau}) \qquad (9.32)
\end{aligned}$$

for any $\hat{\tau} \in \Upsilon$. Using Theorem 2 it can be shown (Ainsworth and Oden (1993(ii))) there exists $\hat{\tau} \in \Upsilon$ for which the Lagrangian is decoupled into a sum of local functionals. With this choice one obtains

$$-\frac{1}{2} \parallel (e, E) \parallel_*^2 \geq \inf_{(\mathbf{v},q) \in X(\mathcal{P}) \times M(\mathcal{P})} \sum_{K \in \mathcal{P}} \{ J_K(\mathbf{v}) + j_K(q) \} \qquad (9.33)$$

where

$$J_K(\mathbf{v}) = \frac{1}{2} B_K(\mathbf{v}, \mathbf{v}) - L_K(\mathbf{v}) + B_K(\hat{\mathbf{u}}, \mathbf{v}) + b_K(\mathbf{v}, \hat{p}) - \oint_{\partial K} \langle \mathbf{n}_K \cdot \underline{\sigma}(\hat{\mathbf{u}}, \hat{p}) \rangle \cdot \mathbf{v}\, ds \qquad (9.34)$$

and

$$j_K(q) = \frac{1}{2} \int_K q^2\, d\mathbf{x} + b_K(\hat{\mathbf{u}}, q) \qquad (9.35)$$

The significant difference between (9.33) and (9.28) is that in (9.33) the bounding terms are entirely local whereas in (9.28) the coupling occurs through the regularity of the global spaces X and M. This localization is critical since it means that the cost of performing error estimation is small relative to the overall solution procedure.

The minimizer of j_K is found *a priori* to be $\mathbf{div}\,\hat{\mathbf{u}}$. However, the minimizer of J_K is approximated by solving the local error residual problem

Find $\mathbf{w} \in X_K$ such that for all $\mathbf{v} \in X_K$

$$B_K(\mathbf{w}_K, \mathbf{v}) = L_K(\mathbf{v}) - B_K(\hat{\mathbf{u}}, \mathbf{v}) - b_K(\mathbf{v}, \hat{p}) + \oint_{\partial K} \langle \mathbf{n}_K \cdot \underline{\sigma}(\hat{\mathbf{u}}, \hat{p}) \rangle \cdot \mathbf{v}\, ds \qquad (9.36)$$

As it stands there is a slight difficulty with this procedure. Namely, owing to the Riesz-Schauder theory the problem (9.36) has a solution if and only if the data is orthogonal to the kernel of the bilinear form $B_K(\cdot,\cdot)$. Physically, this means that the load data has to be in equilibrium. Here, the parameters used in the flux averages (9.25) may be used to adjust the data on every subdomain in order to satisfy equilibrium. At first it appears that a global computation is necessary to determine the appropriate parameters. In fact, it can be shown (Ainsworth and Oden (1992(i-ii)), Ainsworth *et al.* (1993)) that the whole process can be localized reducing the computational costs considerably. An alternative procedure is to simply reformulate the local problem on the quotient space $X_K/\ker(B_K)$. In such a case one is at liberty to choose a simple averaged flux, however one then loses the upper bound property proclaimed by the following:

Theorem 3   *Let $\mathbf{w}_K$ denote the solutions of the local error residual problems (9.36). Then the error in the finite element approximation to the Stokes problem (9.8) may be bounded by*

$$\| (\mathbf{e}, E) \|_*^2 \leq \sum_{K \in \mathcal{P}} \left\{ B_K(\mathbf{w}_K, \mathbf{w}_K) + \int_K (\,\mathbf{div}\,\hat{\mathbf{u}})^2 \,\mathrm{d}\mathbf{x} \right\} \tag{9.37}$$

The efficiency of the resulting estimator can be established under additional conditions, the most important of which is a type of *Saturation Assumption* (c.f. Bank and Weiser (1985), Bank and Welfert (1991)):

There exists a higher degree piecewise polynomial approximation $(\mathbf{u}^*, p^*)$ to $(\mathbf{u}, p)$ on the partition $\mathcal{P}$ such that for some constant β

$$\| \mathbf{u} - \mathbf{u}^* \|_X^2 + \| p - p^* \|_M^2 + \sum_{K \in \mathcal{P}} \oint_{\partial K} h_K \langle \mathbf{n}_K \cdot \underline{\sigma}(\mathbf{u} - \mathbf{u}^*, p - p^*) \rangle^2 \,\mathrm{d}s$$

$$\leq \beta^2 \left\{ \| \mathbf{e} \|_X^2 + \| E \|_M^2 \right\} \tag{9.38}$$

Theorem 4   *Suppose that the assumption (9.38) holds. Then there exists a constant C such that*

$$\sum_{K \in \mathcal{P}} \left\{ B_K(\mathbf{w}_K, \mathbf{w}_K) + \int_K (\,\mathbf{div}\,\hat{\mathbf{u}})^2 \,\mathrm{d}\mathbf{x} \right\} \leq C \, \| (\mathbf{e}, E) \|_*^2 \tag{9.39}$$

Proof See Ainsworth and Oden (1993(ii))

9.3.1 *LOCAL ERROR INDICATORS*

If one is satisfied with merely identifying the elements in which the discretization needs to be refined then local error indicators can be derived as follows. Applying Green's identity to the right hand side of (9.36) and using (9.25) leads to

$$B_K(\mathbf{w}_K, \mathbf{v}) = \int_K (\mathbf{f} + \mathbf{div}\,\underline{\sigma}(\hat{\mathbf{u}}, \hat{p})) \cdot \mathbf{v} \,\mathrm{d}\mathbf{x} - \oint_{\partial K} \alpha[\mathbf{n} \cdot \underline{\sigma}(\hat{\mathbf{u}}, \hat{p})] \cdot \mathbf{v} \,\mathrm{d}s \quad \forall \mathbf{v} \in X_K \tag{9.40}$$

where $[\,\cdot\,]$ denotes the jump in the flux on the interfaces

$$[\mathbf{n}\cdot\underline{\sigma}(\hat{\mathbf{u}},\hat{p})] = \mathbf{n}_K\cdot\underline{\sigma}_K + \mathbf{n}_J\cdot\underline{\sigma}_J \text{ on } \Gamma_{KJ} \tag{9.41}$$

Supposing that the data has been equilibrated as above shows that both sides vanish for any choice of $\mathbf{v} = \overline{\mathbf{w}}$ in the null space of $B_K(\cdot,\cdot)$. Consequently (9.40) yields

$$B_K(\mathbf{w}_K,\mathbf{w}_K) = \int_K (\mathbf{f} + \mathbf{div}\,\underline{\sigma}(\hat{\mathbf{u}},\hat{p}))\cdot(\mathbf{w}_K - \overline{\mathbf{w}})\,\mathrm{d}\mathbf{x} - \oint_{\partial K} \alpha[\mathbf{n}\cdot\underline{\sigma}(\hat{\mathbf{u}},\hat{p})]\cdot(\mathbf{w}_K - \overline{\mathbf{w}})\,\mathrm{d}s \tag{9.42}$$

An application of the Trace Theorem and the Cauchy-Schwarz Inequality leads to the estimate

$$B_K(\mathbf{w}_K,\mathbf{w}_K) \le C\left[\|\mathbf{f} + \mathbf{div}\,\underline{\sigma}(\hat{\mathbf{u}},\hat{p})\|_{L_2(K)} + h_K^{-1/2}\|\alpha[\mathbf{n}\cdot\underline{\sigma}(\hat{\mathbf{u}},\hat{p})]\|_{L_2(\partial K)}\right]$$
$$\|\mathbf{w}_K - \overline{\mathbf{w}}\|_{L_2(K)} \tag{9.43}$$

Choosing $\overline{\mathbf{w}}$ to be the best $L_2(K)$-approximation to $\mathbf{w}_K$ and noting that the null space of $B_K(\cdot,\cdot)$ consists of all rigid body motions finally yields the local error indicator

$$\epsilon_K = \left[h_K\|\mathbf{f} + \mathbf{div}\,\underline{\sigma}(\hat{\mathbf{u}},\hat{p})\|_{L_2(K)} + h_K^{1/2}\|\alpha[\mathbf{n}\cdot\underline{\sigma}(\hat{\mathbf{u}},\hat{p})]\|_{L_2(\partial K)}\right] \tag{9.44}$$

Theorem 5 *Let ϵ be as above. Then there exists a constant C such that the error in the finite element approximation to the Stokes problem may be bounded by*

$$\|(\mathbf{e},E)\|_{X\times M}^2 \le C\sum_{K\in\mathcal{P}}\left\{\epsilon_K^2 + \int_K (\mathbf{div}\,\hat{\mathbf{u}})^2\,\mathrm{d}\mathbf{x}\right\} \tag{9.45}$$

Proof This follows from the above manipulations and Theorems 1 and 3.
Once again it is possible to make the choice $\alpha = 1/2$ to obtain a simpler scheme. The indicator thereby obtained is similar to that of Verfurth (1989) and Baranger and Amri(1991).

9.3.2 GENERALIZATION TO NAVIER STOKES EQUATIONS

Let $c : X \times X \times X \to \mathbb{R}$ denote the trilinear form

$$c(\mathbf{u};\mathbf{v},\mathbf{w}) = \int_\Omega \mathbf{u}\cdot\mathbf{grad}\,\mathbf{v}\cdot\mathbf{w}\,\mathrm{d}\mathbf{x} \tag{9.46}$$

then the incompressible Navier Stokes equations are
Find a pair $(\mathbf{u},p) \in X \times M$ such that for all $(\mathbf{v},q) \in X \times M$

$$B(\mathbf{u},\mathbf{v}) + b(\mathbf{v},p) + b(\mathbf{u},q) = L(\mathbf{v}) - c(\mathbf{u};\mathbf{u},\mathbf{v}) \tag{9.47}$$

The issue of existence and uniqueness of the infinite dimensional problem itself requires further assumptions. Let $V = \{\mathbf{v} \in X : \mathbf{div}\,\mathbf{v} = \mathbf{0}\}$ and define

$$\mathcal{N} = \sup_{\mathbf{u},\mathbf{v},\mathbf{w}\in V} \frac{c(\mathbf{u};\mathbf{v},\mathbf{w})}{|\mathbf{u}|_{1,\Omega}|\mathbf{v}|_{1,\Omega}|\mathbf{w}|_{1,\Omega}} \text{ and } \|\mathbf{f}\|_* = \sup_{\mathbf{v}\in V} \frac{(\mathbf{f},\mathbf{v})}{|\mathbf{v}|_{1,\Omega}} \tag{9.48}$$

The chief obstacle in extending the *a posteriori* error analysis for the Stokes case is the generalization of Theorem 1.

Theorem 6 *If $\|\mathbf{f}\|_* < \nu^2/\mathcal{N}$ then there exists a unique solution to the Navier Stokes problem. Moreover, suppose that the error in the finite element approximation $(\mathbf{e}, E)$ tends to zero, then the equivalence in equation (9.16) holds asymptotically.*

Proof See Girault and Raviart (1986) and Oden *et al.* (1993).

The result concerning the equivalence of norms is weaker than Theorem 1 in that $(\mathbf{e}, E)$ are no longer completely arbitrary. However, the analysis leading to the actual error estimator for the Navier Stokes equations is similar to the Stokes case, giving corresponding local error residual problems

Find $\mathbf{w} \in X_K$ such that for all $\mathbf{v} \in X_K$

$$B_K(\mathbf{w}_K, \mathbf{v}) = L_K(\mathbf{v}) - c_K(\hat{\mathbf{u}}; \hat{\mathbf{u}}, \mathbf{v})$$
$$- B_K(\hat{\mathbf{u}}, \mathbf{v}) - b_K(\mathbf{v}, \hat{p}) + \oint_{\partial K} \langle \mathbf{n}_K \cdot \underline{\sigma}(\hat{\mathbf{u}}, \hat{p}) \rangle \cdot \mathbf{v}\, \mathrm{d}s \qquad (9.49)$$

One again has the option of deriving local error indicators as in the Stokes case. Of course the interior residual appearing in (9.44) must be modified appropriately to account for the additional non-linear term.

REFERENCES

Ainsworth M and Oden J.T., (1992(i)), A procedure for a posteriori error estimation for h–p finite element methods. *Comp. Meth. Appl. Mech. Eng.* **101**: pp.73–96.

Ainsworth M. and Oden J.T., (1992(ii)), A Posteriori Error Estimators for Second Order Elliptic Systems: Part 2. An Optimal Order Process for Calculating Self Equilibrating Fluxes. *Comp. Math. Applic.* (To appear).

Ainsworth M. and Oden J.T. , (1993(i)), A Unified Approach to A Posteriori Error Estimation Based on Element Residual Methods. *Numer. Math.* (To appear).

Ainsworth M. and Oden J.T., (1993(ii)), A Posteriori Error Estimators for Stokes' Problem. (Submitted).

Ainsworth M. ,Oden J.T. and Wu W., (1993), A Posteriori Error Estimators for h-p Finite Element Approximations in Linear Elastostatics. *J. Appl. Numer. Meth.* (To appear).

Bank R.E. and Weiser A., (1985), Some a posteriori error estimators for elliptic partial differential equations. *Math. Comp.* **44**: pp.283–301.

Bank R.E. and Welfert B.D., (1990), A posteriori error estimators for the Stokes equations: a comparison. *Comp. Meth. Appl. Mech. Eng.* **82**: pp.323–340.

Bank R.E. and Welfert B.D., (1991), A Posteriori Error Estimates for the Stokes Problem. *SIAM J. Num. Anal.* **28**: pp.591–623.

Baranger J. and Amri H. El, (1991), Estimateurs A Posteriori D'Erreur Pour Le Calcul Adaptatif D'Ecoulements Quasi -Newtoniens. $M^2 AN$. **25**: pp.31–48.

Girault V. and Raviart P.A., (1986), Finite Element Methods for Navier Stokes Equations. Springer Series in Computational Mathematics. **5**: Springer-Verlag.

Oden J.T. , Demkowicz L. , Rachowicz W. and Westermann T.A., (1990), A posteriori error analysis in finite elements: the element residual method for symmetrizable problems with applications to compressible Euler and Navier Stokes equations. *Comp. Meth. Appl. Mech. Eng.* **82**: pp.183–203.

Oden J.T. , Wu W. and Ainsworth M., (1993), An A Posteriori Error Estimate for Finite Element Approximations of the Navier Stokes Equations. (Submitted).

Verfurth R., (1989), A posteriori error estimators for the Stokes equations. *Numer. Math.* **55**: pp.309–325.

Verfurth R., (1992), A posteriori error estimators for the Stokes equations II. Non-conforming discretizations. *Numer. Math.* **60**: pp.235–249.

Chapter 10

A Zooming-Technique Using a Hierarchical *hp*-Version of the Finite Element Method

Ernst Rank

Num. Meth. und Informationsverarbeitung, Universität Dortmund,
Fachbereich Bauwesen, August-Schmidt Strasse,
4600 Dortmund-Eichlinghofen, Germany

10.1 INTRODUCTION

The *hp*-version of the finite element method (e.g. Guo and Babuska (1986)) combines local mesh refinement with an increase of the polynomial order of the shape functions. It has been shown theoretically and in many numerical examples (e.g. Holzer *et al.* (1990)), that exponential rate of convergence in the energy norm and excellent accuracy can be obtained for linear elliptic boundary value problems. The *hp*-version has also been applied successfully to more general problems like reaction diffusion equations or nonlinear Navier Stokes equations (Demkowicz *et al.* (1990)). Recently a variant of the *hp*-version as a combination of a high order approximation with a domain decomposition method has been suggested by the author (Rank (1992)) . This '*hp − d*-version' is similar to the 's-version' (Fish (1992)) using a superposition of independent finite element meshes. The suggested approach is also related to domain decomposition methods analysed by Bramble *et al.* (1990) and to the multi-level splitting method of Yserentant (1986). The basic idea can be explained as follows. In a first step of the analysis a pure *p*-version approximation is performed on a coarse finite element mesh. Controlled by user interaction or by an a posteriori error estimation the coarse mesh is then covered partially by a geometrically independent fine mesh. On this second mesh a low order approximation is performed and the global approximation is defined as the *hierarchical sum* of the *p*-approximation on the coarse mesh and the *h*-approximation on the fine mesh. Global continuity of the finite element solution can be guaranteed by imposing homogeneous conditions at the fine mesh boundary. The hierarchical nature of the approximation also reflects in the structure of the linear equation system arising and can be used in an efficient solution algorithm.

In the next section algorithmic details are addressed, and it is shown how to apply this domain decomposition method for multiscale problems. The *p*-version is used to model the large scale solution behaviour, the *h*-version approximation being coupled consistently for simulation on the microscale. In the last section numerical examples

The Mathematics of Finite Elements and Applications
Edited by J. R. Whiteman © 1994 John Wiley & Sons Ltd

are presented showing the ability of the method to efficiently and accurately model problems with a scale ratio of more than four orders of magnitude.

10.2 THE hp DOMAIN DECOMPOSITION

As our model problem we will consider a reaction-diffusion equation

$$-\nabla(D(x)\nabla u) + ku = f \; in \; \Omega_1 \tag{10.1}$$

with appropriate boundary conditions. A subdomain $\Omega_2 \subset \Omega_1$ with boundary Γ_2 as in Figure 10.1 is defined and we will assume that a typical scale of Ω_2 is much smaller than the diameter of Ω_1. Ω_2 will cover local features important for an accurate finite element simulation on the global domain Ω_1. To mention a few, these features could be concentrated loads, locally inhomogeneous material , sharp reaction and shock fronts or localization zones. In the following a finite element space will be constructed allowing for a globally C^0-continuous approximation on a composite mesh schematically shown in Figure 10.2.

Let $T_1 = \{t_{1,i}, i \in I_1\}$ be a regular triangulation of Ω_1 which is chosen so that there is a subset $I_{2,1} \subset I_1$ with $\Omega_2 \subset \cup\{t_{1,i}, \; i \in I_{2,1}\}$.

Let $T_2 = \{t_{2,i}, \; i \in I_2\}$ be a regular triangulation of Ω_2 which is chosen so that for every $t_{2,i}$ in T_2 there is *exactly one* j in $I_{2,1}$ with $t_{2,i} \subset t_{1,j}$ i.e. elements of T_2 do not intersect edges of the triangulation T_1.

On Ω_2 we choose now S_{h,Ω_2} as a 'standard' finite element space over T_2 and a subspace

$$S^0_{h,\Omega_2} := \{v \in S_{h,\Omega_2} | \; v = 0 \; on \; \Gamma_2\} \tag{10.2}$$

Typically, we use for S_{h,Ω_2} an approximation of low order finite elements, for example $p = 1$ or $p = 2$. Similary, we define a p-version approximation space S_{p,Ω_1} over the triangulation T_1 with a subspace

$$S^0_{p,\Omega_1} = S_{p,\Omega_1} \setminus S^0_{h,\Omega_2} \tag{10.3}$$

The global approximation space is now defined as

$$S_{h,p} = S^0_{h,\Omega_2} \oplus S^0_{p,\Omega_1} \tag{10.4}$$

i.e. every function in $S_{h,p}$ can be represented by the sum of a (global) p-version function and a (local) h-version function. To be precise let $\{N_i^h \; | \; i \in I_h\}$ be the set of shape functions of the h-approximation on the fine mesh, $\{N_i^p \; | \; i \in I_p\}$ the set of shape functions of the p-approximation on the coarse mesh. With

$$u_p = \sum_{i \in I_p} x_i^p N_i^p = \mathbf{x}^{pT}\mathbf{N}^p \; and \; u_h = \sum_{i \in I_h} x_i^h N_i^h = \mathbf{x}^{hT}\mathbf{N}^h \tag{10.5}$$

and real coefficient vectors $\mathbf{x}^p$ and $\mathbf{x}^h$, every $u = u_h + u_p \in S_{h,p}$ is continuous by construction. The weak formulation for a hierarchical FE-approximation of (10.1), using

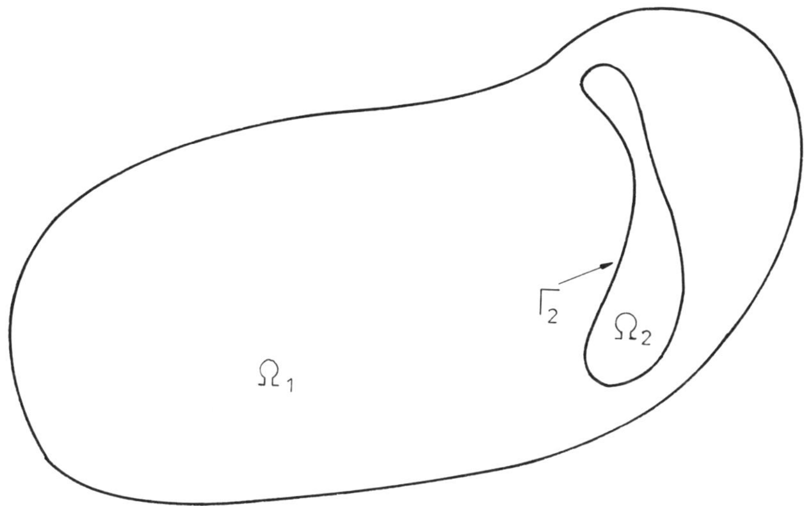

Figure 10.1 : Domains and boundaries

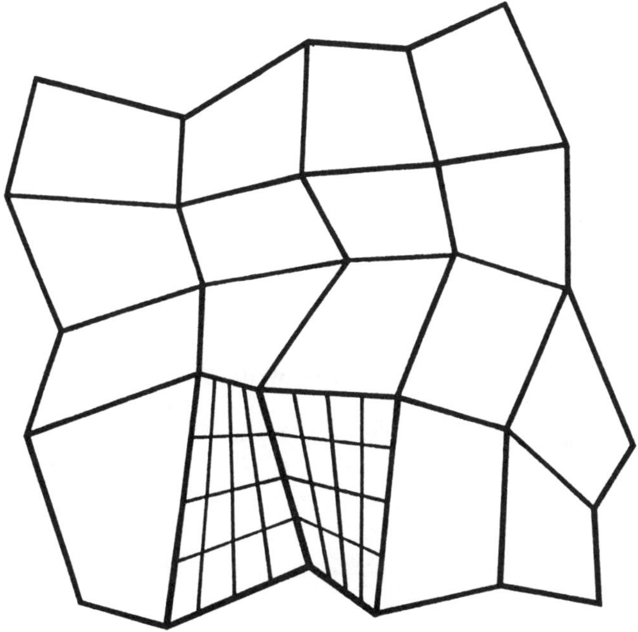

Figure 10.2 : Composed mesh

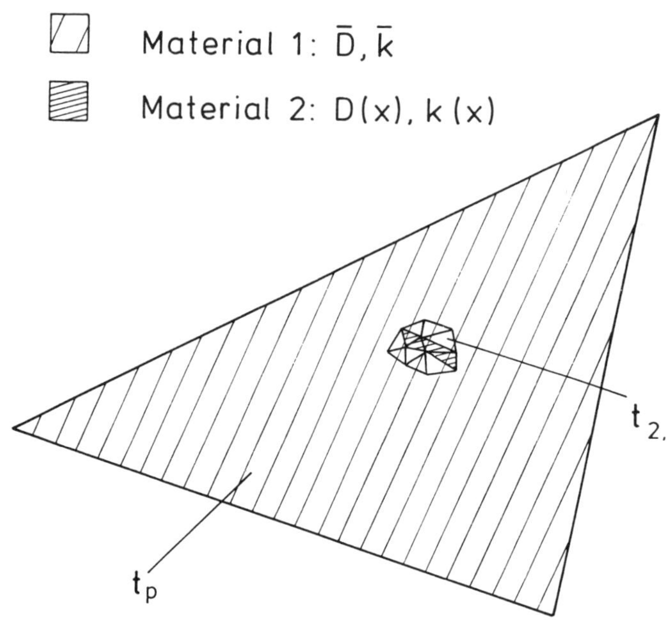

Figure 10.3 Element with overlay

$$B(u,v) = \int_{\Omega_1} D(x)\nabla u \nabla v \; + \; kuv \; d\Omega \quad and \quad F(v) = \int_{\Omega_1} fv \; d\Omega \qquad (10.6)$$

as bilinear form and load functional can now be formulated as:

Find $u_{FE} = u_h + u_p \in S_{h,p}$ *so that for every test function* $v_p \in S_{p,\Omega_1}^0$ *and* $v_h \in S_{h,\Omega_2}^0$

$$B(u_p + u_h, v_p) = F(v_p) \quad and \quad B(u_p + u_h, v_h) = F(v_h) \qquad (10.7)$$

In matrix form the weak formulation can be written as

$$\mathbf{A}\,\mathbf{x} = \begin{pmatrix} \mathbf{A}_{pp} & \mathbf{A}_{ph} \\ \mathbf{A}_{ph}^t & \mathbf{A}_{hh} \end{pmatrix} \begin{pmatrix} \mathbf{x}^p \\ \mathbf{x}^h \end{pmatrix} = \begin{pmatrix} f_p \\ f_h \end{pmatrix} \qquad (10.8)$$

with submatrices

$$\mathbf{A}_{pp} = (a_{ij})_{i,j \in I_p} \; , \; \mathbf{A}_{hh} = (a_{ij})_{i,j \in I_h} \; , \; \mathbf{A}_{hp} = (a_{ij})_{i \in I_h, j \in I_p} \; , \; a_{ij} = B(N_i, N_j) \; (10.9)$$

The righthand side of (10.8) is defined analoguously. A question to be addressed next is the computation of element matrix and load vectors for locally inhomogeneous material. Consider a situation schematically shown in Figure 10.3, a single element

t_p being partially overlaid by the refined domain Ω_2. Only the shaded part in Ω_2 is assumed to have material coefficients $D(x), k(x)$, differing from 'global' material constants $\overline{D}$ and $\overline{k}$ in the rest of t_p. Let $N_i \in S^0_{p,\Omega_1}$ and $N_j \in S_{h,p}$ be functions having support in t_p. The coefficient $a_{ij}^{t_p}$ of the element matrix of t_p is given as

$$
a_{ij}^{t_p} := \int_{t_p} D(x)\,\nabla N_i\,\nabla N_j \,+\, k(x)N_iN_j\,dx = \int_{t_p} \overline{D}\,\nabla N_i\,\nabla N_j \,+\,\overline{k}N_iN_j\,dx\,+
$$

$$
+ \sum_{t_{2,l}\subset\Omega_2\cap t_p} \int_{t_{2,l}} D(x)\,\nabla N_i\nabla N_j \,+\, k(x)N_iN_j\,dx
$$

$$
- \sum_{t_{2,l}\subset\Omega_2\cap t_p} \int_{t_{2,l}} \overline{D}\,\nabla N_i\nabla N_j \,+\, \overline{k}N_iN_j\,dx
$$

$$(10.10)$$

Thus the element matrix of t_p can be computed by a *composed* integration rule taking into account the locally different material in Ω_2 in the following manner.

Algorithm

<table>
<tr><td colspan="2" align="center">Loop over all p-elements</td></tr>
<tr><td></td><td align="center">Compute element matrices with
material $\overline{D}$, $\overline{k}$</td></tr>
</table>

<table>
<tr><td colspan="2" align="center">Loop over all h-elements</td></tr>
<tr><td></td><td align="center">Compute composed element
matrices with material $D(x)$, $k(x)$
(i.e. h-d.o.f.s and partial
domain-integrals of p-d.o.f.s)</td></tr>
<tr><td></td><td align="center">compute partial p-element-matrices
(corresponding to domain of
h-element) with material
$-\overline{D}$, $-\overline{k}$</td></tr>
</table>

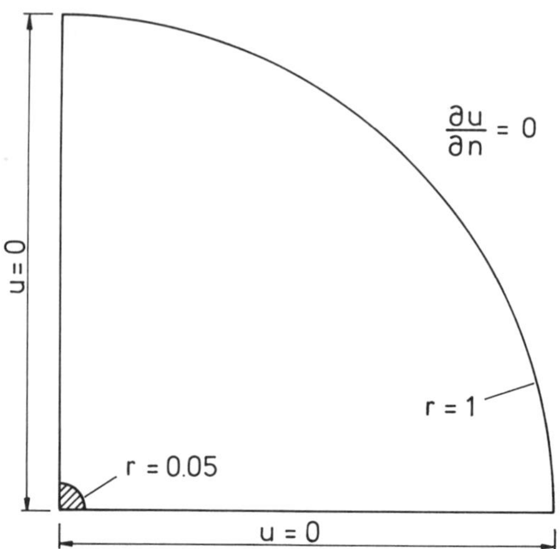

Figure 10.4 : domain of computation for example 1

10.3 NUMERICAL EXAMPLES

The domain of computation of the first example is given in Figure 10.4, showing two concentric quarter-circles with radius 1 for Ω_1 and 0.05 for Ω_2. In Ω_2 diffusion and reaction coefficients are chosen to be $D = .05$ and $k = 10000$; in $\Omega_1 \backslash \Omega_2$ $D = 1$ and $k = 0$. Homogeneous Dirichlet boundary conditions are set along the x- and y-axes, natural boundary conditions are used along the circular arc. In order to be able to compute an exact solution the righthand side of (10.1) is set to

$$f = -D(r)\frac{16}{r^2} g(r) \sin(4\theta)$$

$g(r)$ being the solution of an auxiliary 1D-problem

$$D(r)(u_{rr} - \frac{1}{r}u_r) - k(r)u = 0 \ , \ u(0) = 0 \ , \ u(1) = 1$$

The exact solution decays like a modified Bessel function of the first kind of order 0 for $r < 0.05$ with an exponential decay length of approximately 2/1000, yielding an interface layer on a microscopic scale. Figure 10.5 shows the composed mesh with the (global) circularly oscillating solution. The local solution near the interface is given in Figure 10.6.

A computation with 370 degrees of freedom in the fine mesh and 81 degrees of freedom for polynomial degree $p = 4$ in the coarse mesh yields a pointwise error distribution as shown in Figure 10.7 for the global scale. The maximal error (plotted in dark) is 0.065 compared to a maximal solution value of 1. The error distribution around the interface is shown in Figure 10.8 with a maximal error of 0.025 compared to a maximal solution value of 0.35 in the plotted window.

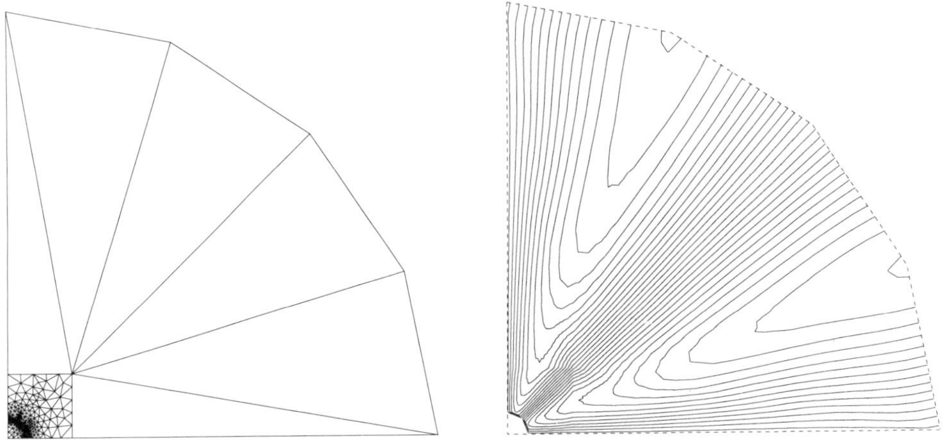

Figure 10.5 : Composed mesh and global solution

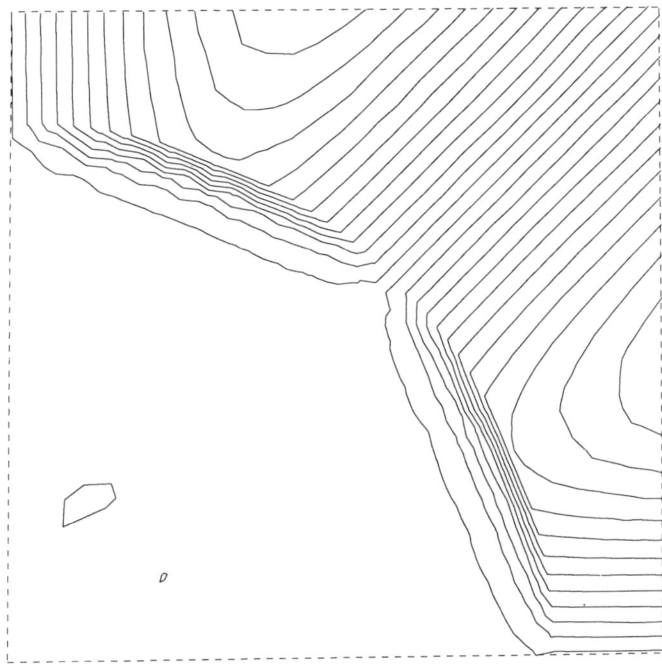

Figure 10.6 : local solution

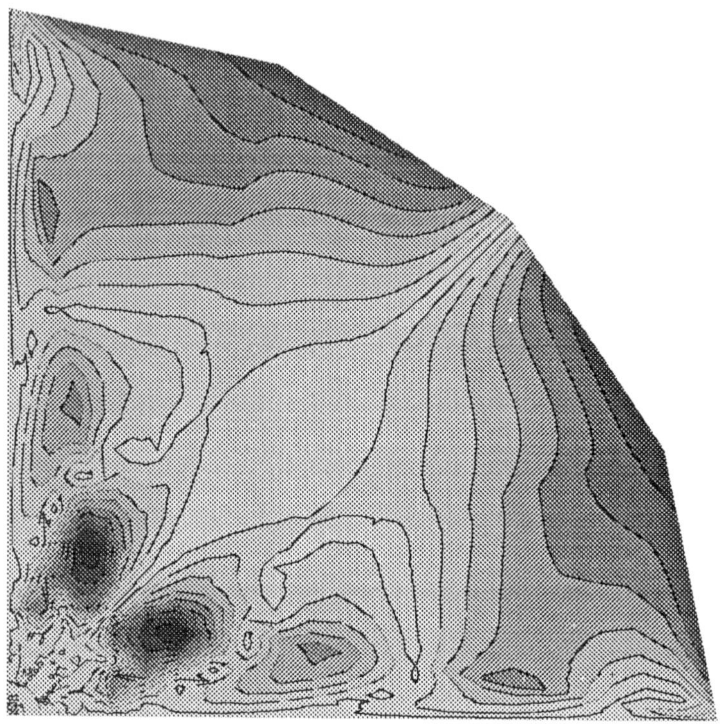

Figure 10.7 : Error on macroscopic scale

In the second example $k = 0$ $D = 1$ everywhere, boundary conditions are set as indicated in Figure 10.9, showing in the top part the geometrically refined mesh of the macroscopic scale and in the bottom part an enlarged area around the origin with a local heat source (plotted dark). $u = 0$ for $-1000 \leq x \leq -1$, $y = 0$ and $u = 1$ for $0 \leq y \leq 1000$, $x = 1000$.

The global mesh for the p-version is refined geometrically towards the origin. A contour plot of the global solution is shown in Figure 10.10. Figure 10.11 gives the finite element solution on the microscopic scale near the origin, showing the solution at the transition from fine mesh to coarse mesh to be continuous as it is guaranteed by construction of the method. It should also be noted that this example shows well the interaction of local and global solution. Neither the microscopic nor the macroscopic behaviour could be modelled accurately without taking into account the solution on the other scale.

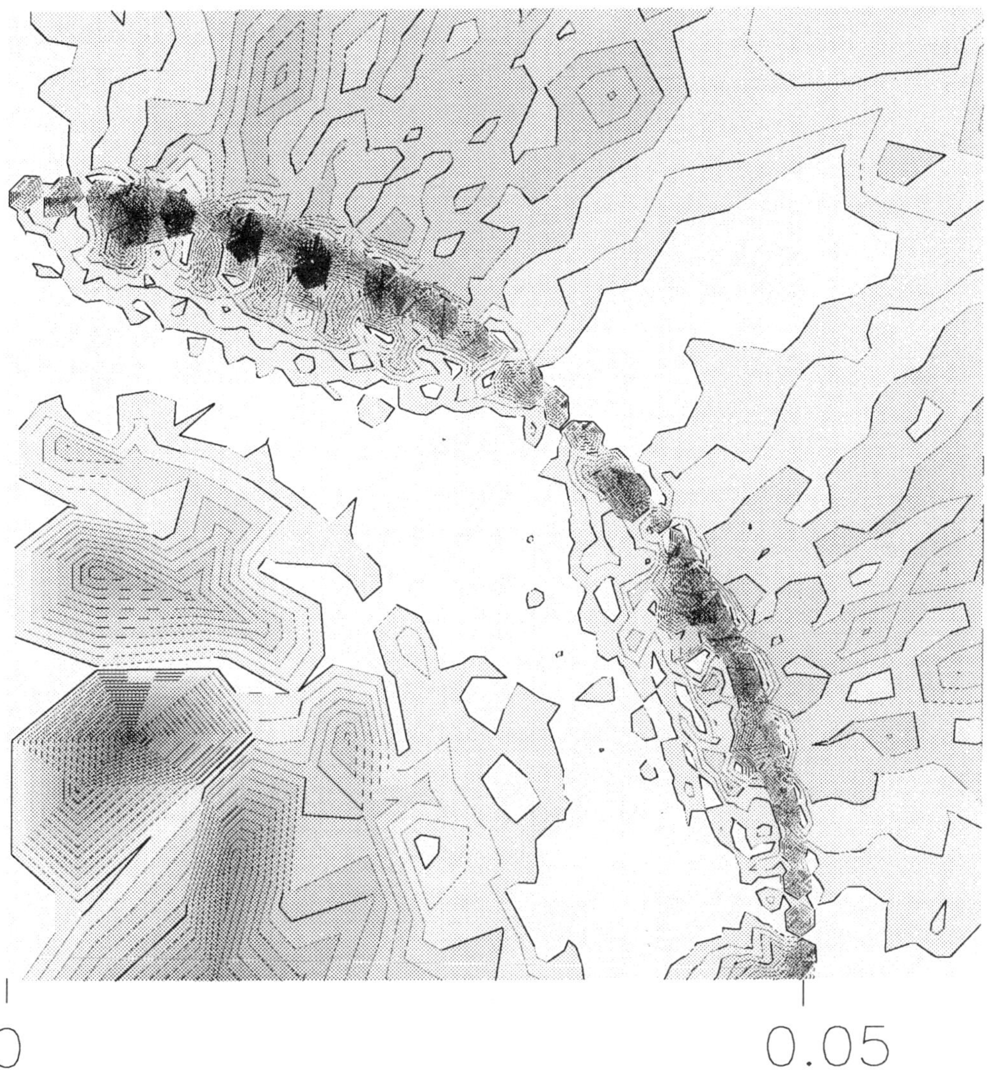

Figure 10.8 : error on microscopic scale

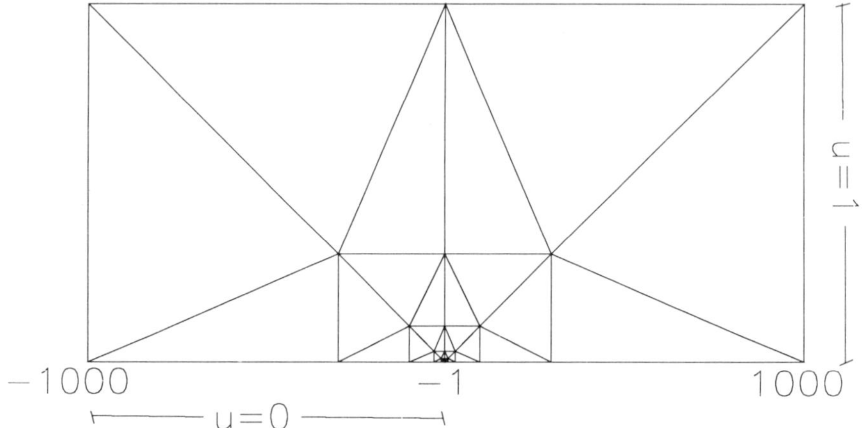

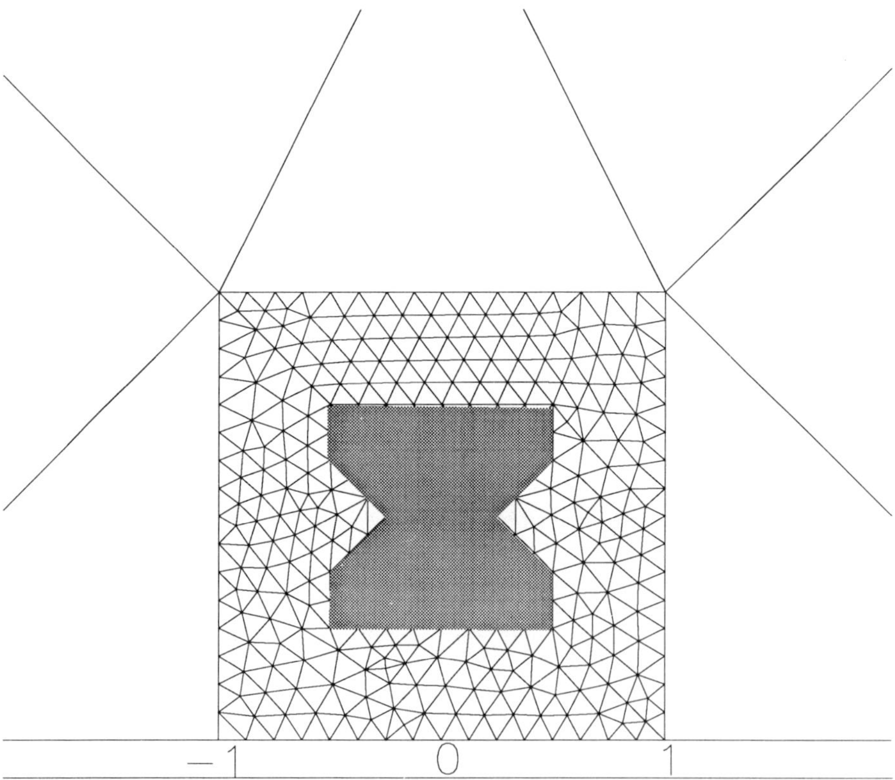

Figure 10.9 : Mesh, boundary conditions and local heat source for example 2

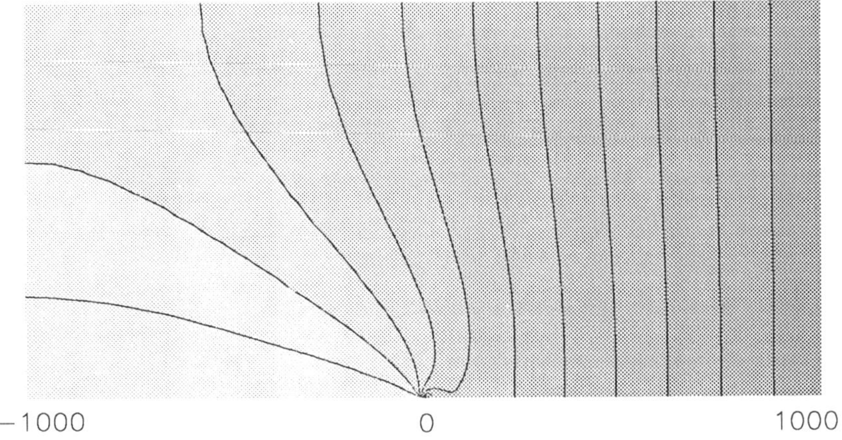

Figure 10.10 Solution on macroscopic scale

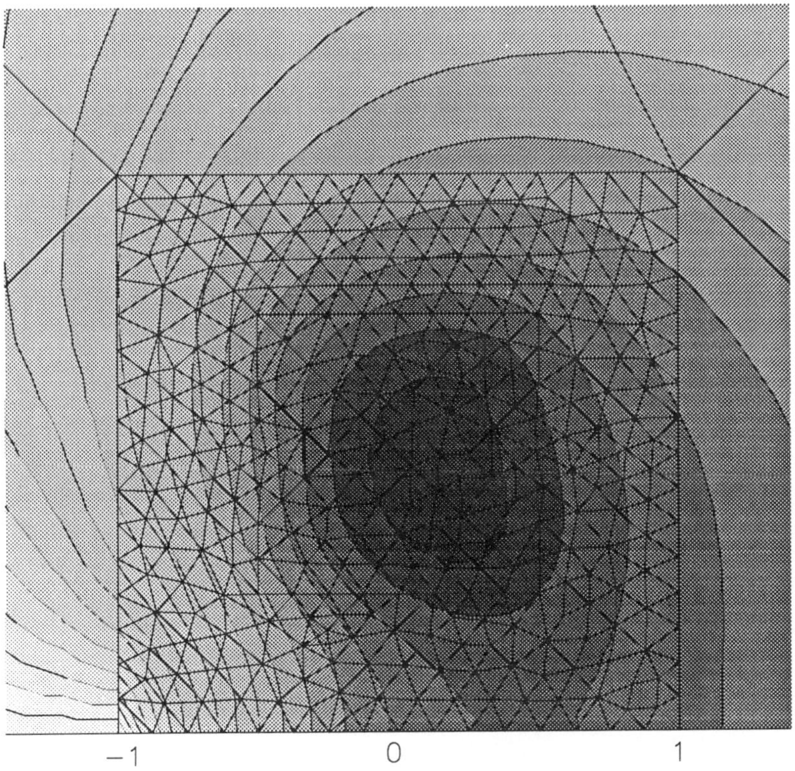

Figure 10.11 Solution near origin

REFERENCES

Rank E. (1992), Adaptive remeshing and h-p domain decomposition. *Comp. Meth. in Appl. Mech. Eng.*, **101**: 299–313.

Fish J. (1992, The s-version of the finite element method. *Computers and Structures*, **43** No. 3: 539-547.

Guo B. and Babuska I. (1986), The h-p version of the finite element method. *J. Comput. Mech.*, **1**: 21-42.

Holzer S., Rank E. and Werner H. (1990), An implementation of the hp-version of the finite element method for Reissner-Mindlin plate problems. *Int. J. for Numer. Meth. in Eng.*, **30**: 459-471.

Demkowicz L., Oden J.T. and Rachowicz W. (1990), A new finite element method for solving compressible Navier-Stokes equations based on an operator splitting method and h-p adaptivity, *Comp. Meth. in Appl. Mech. Eng.*, **84**: 275-376.

Bramble J., Ewing R., Parashkevov R. and Pasciak R. (1992), Domain decomposition methods for problems with partial refinement. In :*Proceedings of the Copper Mountain Conference on Iterative Methods, Book 4 of 4, April 1-5, 1990, Copper Mountain, Colorado; and SIAM J. Scientific and Statistical Computing* 13(1).

Yserentant H. (1986), On the multi-level splitting of finite element spaces. *Numer. Math.*, **49**: 379-412

Chapter 11

A Substructured Elastic-Plastic Fracture Analysis Program for Parallel Processing on Transputer Networks

H. van Lengen, B. May, F.-G. Buchholz, H. A. Richard

Institute of Applied Mechanics
University of Paderborn, Pohlweg 47-49
D-W4790 Paderborn, Germany

11.1 INTRODUCTION

This paper presents the elastic-plastic Finite-Element toolset PLASTPARA for complex fracture analysis on parallel transputer equipment. The PLASTPARA toolset consists of the following parts:

1. The PREDIVIDER, a program for automatic domain decomposition,
2. Finite-Element programs for 2D and 3D elastic or elastic-plastic problems,
3. miscellaneous fracture mechanic modules.

The programs have been implemented on different transputer systems using various FORTRAN Compilers. First implementations under the operating system HELIOS were realised on a PC-host based transputer system (6 T800 procs.) and on the Supercluster SC320 (1-320 procs.). The actual implementation was realised under PARIX on the GCel (32-1024 procs.) of PC^2 (Paderborn Centre of Parallel Computing). The *efficiency* of the program will be discussed through the analysis of three problems of structural mechanics.

In recent decades the finite-element-method has been acknowledged as a powerful and excellent computational tool for the analysis of large and complex problems in engineering science. However using conventional sequential computers, the calculation time for any computation rises dramatically with increasing size and complexity of the problem. Simultaneously the costs of the calculation also rise comparably. In order to reduce the calculation time and the associated costs, use can be made of different computer architectures, which continue to undergo progressive development. Important advances have been made in the development of multiple-instruction, multiple-data (MIMD) multi-processing systems, which consist of networks of programmable processors. One of these processors is the transputer, which is an inexpensive fast processor which has access to 4-16MB RAM. It is connected through

The Mathematics of Finite Elements and Applications
Edited by J. R. Whiteman © 1994 John Wiley & Sons Ltd

a maximum of four links to other transputers in a network which can very easily be expanded.

These new parallel architectures have revolutionised the design of computer algorithms, and promise to have significant influence on algorithms for structural engineering computations.

The Evaluation of the performance of a parallel program always has to be done with care. The first factor usually given is the overall speedup:

$$\text{Speedup} = \frac{\text{user time for one processor}}{\text{user time for n processors}}. \tag{11.1}$$

This gives an idea of the performance of a parallel program compared to a sequential one. A second factor is the *efficiency*, which is usually given by the ratio of the speedup to the number of processors.

$$Efficiency = \frac{\text{Speedup}}{\text{number of processors}}. \tag{11.2}$$

However, this only gives a partial idea of the program performance. Indeed, as part of the solution is performed sequentially, it is difficult to assess only the *efficiency* of the parallelisation. Another coefficient defines the maximum theoretical speedup for a program where a fraction f of the work is made in parallel.

$$\text{theor. Speedup} = \frac{\text{number of processors}}{\text{f} + (1 - \text{f}) \cdot \text{number of processors}} \tag{11.3}$$

The *efficiency* of the parallel part of the program or in other words the load balancing between processors, is given by the ratio of the speedup of the program to the theoretical speedup.

11.2 PARALLEL SUBSTRUCTURE TECHNIQUE

A straightforward approach to parallel computing is the substructure technique. This has been well known and used for many years in the FE-analysis of large engineering problems. Use of the substructure technique requires the subdivision of the whole FE-model into smaller sub regions with two classes of variables (Figure 11.1): the internal and the external variables related to the nodes of main net level. Using such substructured models most steps of the FE calculation can be done without synchronisation on different processors. The substructures can be calculated by parallel processing up to the step of reducing all variables of the substructures to external ones by Gaussian elimination. In the next step a Main Net Solver will assemble and solve these subequations and compute the displacements at main net level. In the last step the internal displacements are calculated in parallel at subnet level. The implicit parallelism of this method in combination with the frontal solution algorithm provides an excellent basis for the utilisation of modern hardware developments for massively parallel computing.

11.2.1 FRONTAL SOLUTION METHOD

The first version of PLASTPARA uses the Frontal Solution Technique at main and subnet levels. The idea of the Frontal Solution Technique first presented by Irons (1970) is the simultaneous assembly and elimination of the degrees of freedom in the stiffness matrix. Use of the technique at subnet level leads automatically to the reduced element stiffness at main node level. All reduced subnet stiffness matrices can be treated as hyper element stiffness matrices and a Hyper Frontal Solver can compute the results at main net level. The results at the internal nodes can be calculated by parallel backward substitution at subnet level.

This approach works with a sequential main net solver which is integrated into the master processor code. The frontal solvers are designed as "in core"-versions to avoid data transfer to the host computer and between the processors. It is clear that this algorithm is not scaleable because the maximum number of substructures (= max. number of processors) is given by the number of elements. The speedup for this marginal case would be very poor, because the sequential main solver would do exactly the work of the non-parallel frontal solver. This algorithm can be used efficiently only if the number of substructures is small compared to the number of elements and the amount of non-parallel work is also small.

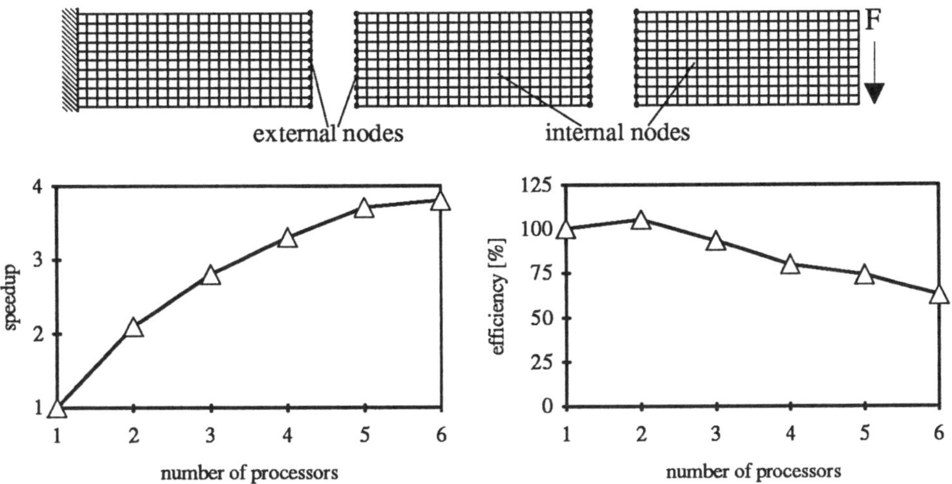

Figure 11.1 Speedup and *Efficiency* for a bending specimen

The first example shows speedup and the *efficiency* for the elastic plastic deformation of a 2D bending specimen with 900 elements and 3901 nodes. In this case PLASTPARA was implemented in TOPEXPRESS FORTRAN under the operating system HELIOS on a PC-host based transputer system containing 6 processors. Whilst Figure 11.1 shows very good *efficiency*, greater than 100% for the calculation with 2 processors, we also see that for this example 6 processors represent the upper limit for efficient use of the parallel substructuring technique. The increasing sequential calculation time leads to reduced *efficiency* for a greater number of processors. Other reasons for the decreasing of the *efficiency* are the bad load balancing between the meshes given by the different number of main nodes per substructure, and some special facits of the plastic calculations. A higher number of main nodes in the substructure at the middle of the specimen leads up to a higher frontwidth and so to a high calculation time for this subnet. During the elastic-plastic calculations a new element stiffness matrix must not be built up on all subnets for all loadsteps as this forces bad load balancing for such loadsteps.

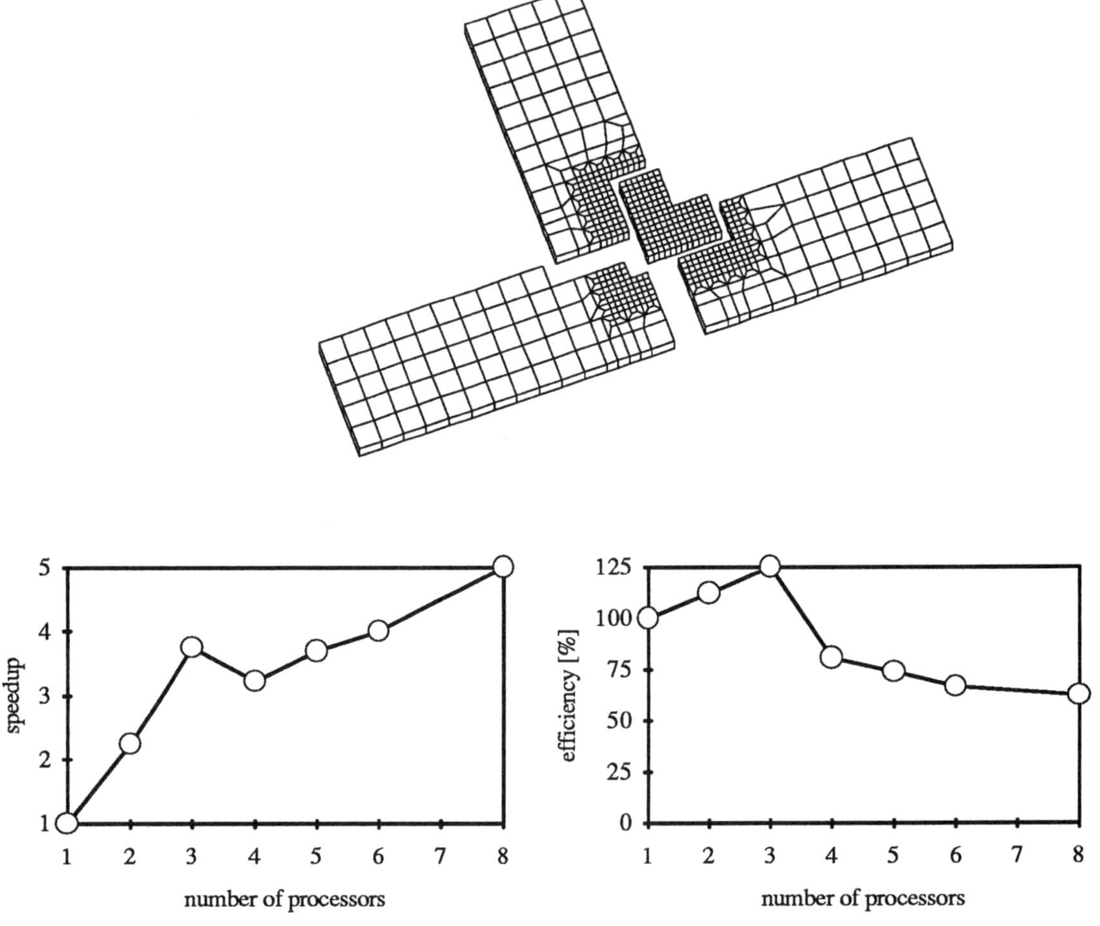

Figure 11.2   Speedup and *Efficiency* for a T-junction

The second example, calculated on the SC320 using MEIKO FORTRAN under HELIOS, shows the results for a 3D calculation of a T-junction problem. Very good *efficiency* of 125% for 3 processors but very bad speedup for 4 processors is obtained. With 5 to 6 six processors the speedup rises to 4. The bad speedup found with 4 processors can be explained by an unfavourable domain decomposition shown in Figure 11.2. Because one of the four substructures has a greater number of main nodes, and hence a higher frontwidth than the other ones, we get bad load balancing at subnet level for this case.

The example shown in Figures. 11.3 and 11.4 was calculated with PLASTPARA compiled with ACE FORTRAN under the operating system PARIX on the GCel. For this problem a greater number of processors can be used efficiently. With up to 12 processors we obtain very good speedup. In those cases where the speedup in this example is smaller than the number of processors the load balancing given by the domain decomposition was not optimal.

11.2.2 PARALLEL ITERATIVE MAIN NET SOLVER

To avoid the difficulties described above and caused by increasing sequential calculation time at main net level, a parallel main net solver due to Law (1986) was adapted to the Program. In this approach the iterative main solver works without building up the main net equation system. Iterative solvers work on every reduced substructure equation system with the method of conjugate gradients. During each iteration step a residual vector and two scalar values must be transmitted between neighbouring substructures. It is clear that for an increasing number of substructures and an increasing number of main nodes the number of iterations will rise. Because the number of main nodes per substructure can often be kept constant, while the number of substructures is increased, the main net calculation time can be kept approximately constant. The load balancing for this type of main net solver depends only on the number of main nodes per substructure.

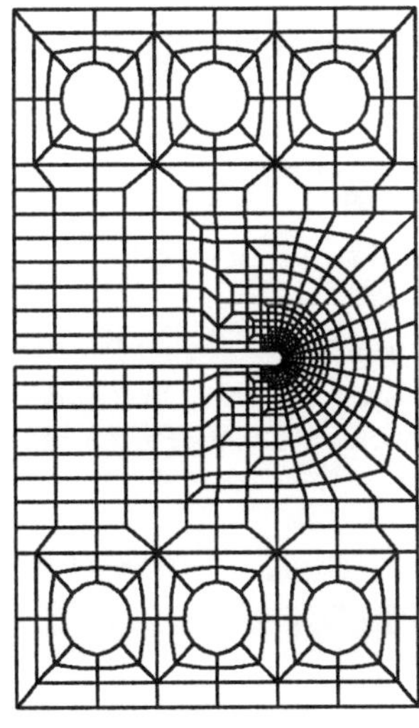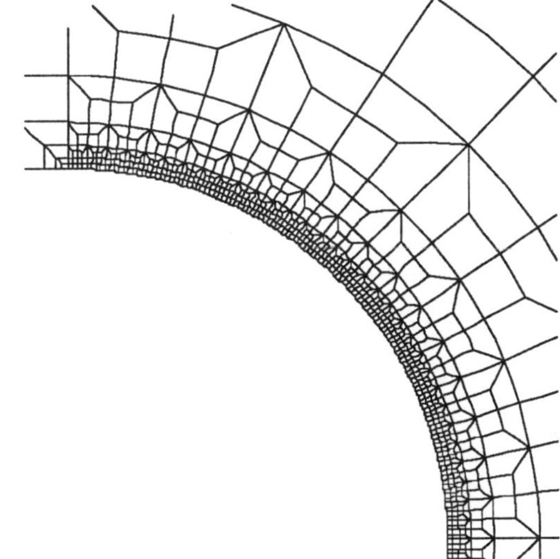

Figure 11.3   FE mesh for the U-notch problem

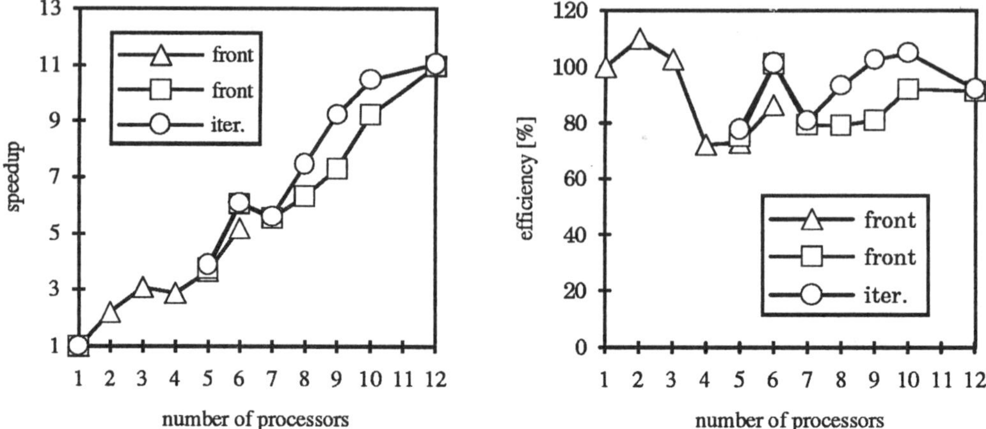

Figure 11.4 Speedup and *Efficiency* of the U-notch problem

To demonstrate the advantage of the parallel main net solver, Figure 11.4 shows the comparison of frontal and iterative main net solvers for the U-notch problem given in Figure 11.3. From Figure 11.4 we see that in this example the iterative solver makes better speedup possible. In some cases the *efficiency* reaches values higher than 100%.

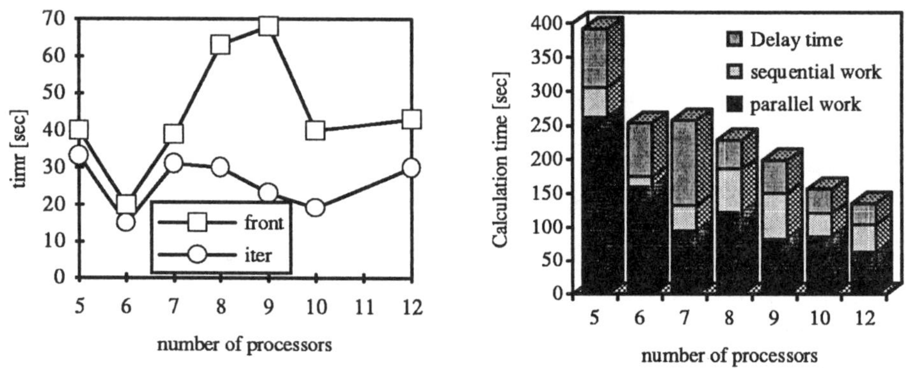

Figure 11.5 Time for main net solving and part of sequential work

Figure 11.5 shows the increasing time for main net solving using a sequential frontal solver in comparison to a parallel iterative conjugate gradient solver. The right hand diagram shows the fraction of sequential and parallel work for the PLASTPARA version with the sequential frontal solution technique.

11.3 AUTOMATIC DOMAIN DECOMPOSITION

The examples presented in Sections 11.2.1 and 11.2.2 show the relations between *efficiency* and the quality of the domain decomposition. The load balancing between the processors is determined by the following subnet parameters: number of elements, number of nodal points, front- or bandwidth and the number of main nodes. Simultaneous optimisation of all these parameters is very difficult and often not possible. On one hand Figure 11.4 shows that by good substructuring *efficiencies* greater than 100% can be achieved. By contrast the bad decomposition for seven processors leads to lower speedup than for six. These examples show that a manually performed domain decomposition for the parallel computation of large and complicated finite element meshes is not easy and can take much effort. For practical applications of the parallel substructuring technique fast automatic mesh division is required, because the time saved by the parallel FEM must not be wasted on laborious manual decomposition. We thus make use of the domain decomposition Program PREDIVIDER developed by Bürger and Buchholz (1992) which can be split into the following steps:

- Performance of a dividing approach using automatically chosen start elements
- Decomposition optimisation based on the load balancing parameters
- Main node minimisation
- Graphic control and manual modification possibilities
- Net numbering optimisation
- Optimisation of the element or nodal point numbering to minimise frontwidth or skyline
- Main node reduction by quadrilateral element dividing (Figure. 11.6)

Figure 11.6 explains the optimisation features mentioned above. In order to minimise the main nodes, element chains (I) and element islands (II) are cancelled by the PREDIVIDER. Elements such as (III), requiring the same number of main nodes even if they belong to mesh 1 or 2, will be shifted to the smaller net. The mesh sizes (number of elements or number of nodes) can be adjusted by shifting a row of neighbouring elements (IV) to the smaller net (here net 2). By repeating all these actions the domain decomposition is optimised, and at the end of this process the user can check the result via a graphic screen and, if necessary, add some manual optimisations. Last but not least the program offers the facility of triangulation of all quadrilateral elements at the interfaces. By this method the number of main nodes can be reduced as shown in Figure 11.6.

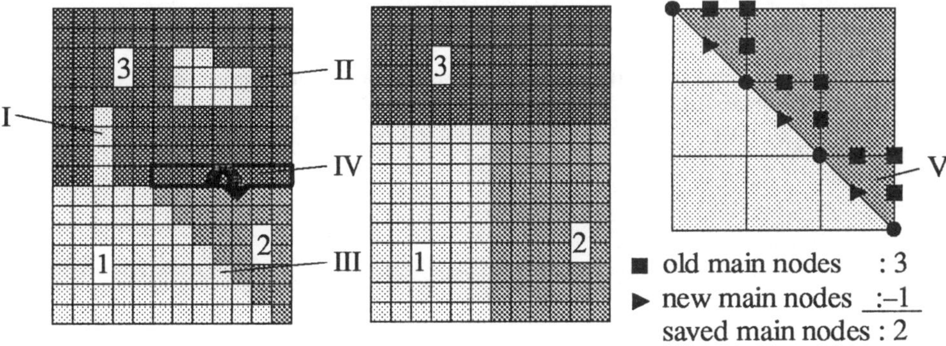

Figure 11.6   Decomposition optimisation by the PREDIVIDER

Eventually the PREDIVIDER will send the decomposition result direct to the FE program and also a 3D version of the program will be produced.

The *efficiency* of this domain decomposer can be seen from the speedup for the U-notch example shown in Figure 11.4.

11.4 CONCLUDING REMARKS

The numerical examples presented in this paper demonstrate the suitability of substructuring techniques for parallel finite element computing. Thus efficient and portable FE-software can be produced that requires only a few non standard FORTRAN statements. Because the Substructure Technique in combination with the frontal solution method distributes not only the numerical operations to different processors but also reduces the number of operations, very good *efficiency* (~ 100 %) can be achieved. The disadvantages are the increasing work for the solution of the main node equation system and the dependence of the load balancing on the decomposition result. Optimal decomposition for nonlinear 3D FE calculations cannot easily be obtained and so an efficient domain decomposer is necessary.

Research on iterative subnet level solvers, similar to the main net solver presented above but including the internal degrees of freedom of the substructures, shows that the load balancing depends mainly on the number of nodes in the substructures. Future work will show whether the substructuring technique in combination with this solution technique will achieve better speedup and an easier domain decomposition for a very large number of processors.

REFERENCES

Bürger M. and Buchholz F.-G. (1992), On a Substructure Finite Element- and Fracture Analysis Program for Parallel Processing on Transputer Networks in Personal Computers. In: *Proceedings of the International Conference on Parallel Computing and Transputer Applications (PACTA 92)*, 1280–1289.

Coulon D., Piret G. and Geradin M. (1992), Parallel Design of linear Equation Solvers for Finite Element Systems. In: , *Proceedings of the first European Conference on Numerical Methods in Engineering (ECCOMAS)*, 1280–1289.

Farhat C. (1988), A simple and efficient automatic FEM domain decomposer. *Computers and Structures*, **28**, 579-602.

Irons M.B. (1970), A Frontal Solution Program for Finite Element Analysis. *International Journal for Numerical Methods in Engineering.*, 2: 5–32.

Law K.H. (1986), A Parallel Finite Element Solution Method. *Computers and Structures*, **23**: 845–858.

Owen D.R.J. and Goncalves O.J.A. (1982), Substructuring Techniques in Material Nonlinear Analysis. *Computers and Structures*, **15**: 205–213.

Chapter 12

An Efficient Iterative Parallel Finite Element Computational Method

K.P. Wang and J.C. Bruch, Jr.

Dept. of Mechanical and Environmental Engineering,
University of California, Santa Barbara, California, 93106 U.S.A.

12.1 INTRODUCTION

In 1987, Zienkiewicz and Zhu (1987) presented a new error estimator. This scheme evaluates the global energy norm error and also estimates the local error. Burkley *et al.* (1991) and Burkley and Bruch (1991) have utilized the scheme on free boundary seepage problems. Their procedure will be used in this study for the adaptive mesh finite element analysis.

When refining a finite element mesh according to the results of the error estimation, the computation costs will increase when solving the finite element system based on the new mesh. However, advanced computer technologies make computations faster, more efficient, and less expensive. Multi-processor computers have been developed in order to fulfill the demand for fast solutions and the economical computation requirement. Hypercube, Ncube, Alliant, Transputer, and Connection Machine are several examples of parallel computers. A considerable number of studies utilizing these multi-processor computers have been made during the past decade. See, for example, Farhat (1990) and the references therein. Herein, a distributed memory parallel computer (Hypercube concurrent computer) is used to perform a study of utilizing parallel computation with the adaptive mesh finite element analysis.

12.2 ADAPTIVE MESH FINITE ELEMENT ANALYSIS

Error estimation and local mesh refinement are two major concepts of adaptive mesh finite element analysis. In addition, a mesh refinement algorithm is required to perform remeshing after obtaining the error of a finite element system. The error estimate decides how the computed results deviate from the exact solution. Local mesh refinement testing is performed to determine how the mesh is to be refined. The remeshing algorithm is then used to automatically generate a refined mesh according to the error obtained.

The error estimation procedure used herein was introduced by Zienkiewicz and Zhu (1987). It allows an accurate assessment of errors while remaining so simple that it

The Mathematics of Finite Elements and Applications
Edited by J. R. Whiteman © 1994 John Wiley & Sons Ltd

can readily be implemented as a post-processor involving minimal computation. This computationally simple error estimator with the modification introduced by Burkley *et al.* (1991) is used in an existing finite element code in the solution of a free surface flow through an earth dam using a parallel computer.

12.2.1 MESH GENERATION AND REFINEMENT STRATEGY

A mesh generator and a mesh refiner are required to perform the finite element analysis and the post-processing of the mesh refinement. For the purpose of this study isoceles right triangle elements are adopted and to generate the initial mesh, a simple mesh generator and refiner for these elements is implemented. The concept behind the mesh generation and mesh refinement is simple: divide an element into two by refining across its longest side.

In this study, incompatible elements are not allowed. Therefore, a recursive process proposed by Rivara (1984, 1987) is used to avoid having such elements. For the refining process, if the longest side of a refining element (target element) is shared with another element (neighbour element), the neighbour element must be refined first. However, once the shared side is not the longest side of the neighbour element, the neighbour element is refined first into two elements, then the new element sharing the same longest side with the target element is refined. Since the same situation may happen to the neighbour element, the refinement process will be propagated until no element shares the longest side of the element being refined, or the shared side is the longest side for both elements. Therefore, refining an element may cause a recursive refinement in the neighbourhood of that element. Since triangles are being divided into half, this procedure will cause elements to be similar triangles without there being any incompatible elements.

Accordingly, to create an initial mesh block we start with two large triangular elements in a square or rectangle. Then these two large elements can be subdivided (refined) into four smaller triangular elements. Thus, a mesh can be obtained by repeating this process until the area of every element is smaller than a common area criterion. Similarly, for the refinement of the mesh after the error estimate, the same procedure is applied. In this stage, the desired area for each element calculated from the error estimate process is used as the new area criterion for an element.

12.3 PARALLEL ITERATIVE SCHEME

To utilize parallel computation for the adaptive mesh finite element analysis, a parallel numerical scheme is required. A highly efficient parallel iterative scheme that has recently been developed by Wang and Bruch (1993) for a linear static finite element system is used for this analysis. This parallel iterative scheme is based on a domain decomposer that subdivides the computation domain into several subdomains and then calculates, in parallel, the solutions in all subdomains. This domain decomposer is capable of automatically dividing a finite element mesh into subdomains to guarantee load balance. After completing the decomposition, all subdomains are mapped onto the same number of processors of the parallel computer. Each processor treats the assigned subdomain as an independent sub-finite element system with information exchanged with its neighbouring processors. With this new parallel iterative scheme, these sub-finite element systems are solved fully in parallel as an independent finite element system. Also, the inherently sequential Gauss-Seidel and SOR schemes are altered into a fully parallel iterative scheme.

12.3.1 LOAD BALANCE AND DOMAIN DECOMPOSER

The optimal performance of a parallel computer can be achieved by evenly subdividing the computation load into smaller blocks of computation tasks and then distributing these blocks onto processors of a parallel computer. Thus, all processors can start the computations at the same time and can finish the computations at about the same time. This is the computation load balancing requirement without which some processors will be idle while the others are computing. This degrades the performance of a parallel computer. Therefore, the goal of the domain decomposer is to subdivide the finite element mesh into several smaller subdomains which have the same number of elements and nodes since linear triangles are being used.

This study uses the domain decomposer proposed by Wang and Bruch (1993). Their domain decomposer creates horizontal linked subdomains (column-wise subdividing). Vertical linked subdomains (row-wise subdividing) can be obtained by interchanging the x and y information. The same number of elements and approximately the same number of nodes were obtained for all the subdomains. Thus, the load balancing requirement is satisfied by using this domain decomposer.

12.3.2 FORMULATION OF PARALLEL ITERATIVE METHOD

Wang and Bruch (1993) also proposed parallel iterative Gauss-Seidel and SOR iterative schemes. Conventional Gauss-Seidel and SOR iteration schemes need to be performed sequentially. Reordering the equations could alter these two schemes into fully parallel iterative schemes. Wang and Bruch (1993) used this intuitive idea and implemented it on a free boundary seepage problem. Speedups were obtained exceeding the theoretical speedup.

After a computation domain is subdivided into subdomains, the problem domain boundary remains a boundary and the interfaces of a subdomain become new boundaries. Thus, the computation of values at interior mesh points for one subdomain is uncoupled from the other subdomains. Also, the computation of values at interface mesh points of an interface is uncoupled from the other interfaces. The iterative schemes use a combination of newly computed values and old values at mesh points surrounding a mesh point to compute the new value at that mesh point. Therefore, the iterative process can be performed for the interior mesh points of a subdomain using the old values at interface mesh points. Moreover, the values at interface mesh points can be updated using the newly computed values at interior mesh points by the iterative process. An example and explanation are given in Wang and Bruch (1993).

Accordingly, this parallel iterative scheme simply reorders the computing sequence such that the values at interior mesh points are computed first, then those at the interface mesh points are computed. With this parallel scheme, all processors can compute concurrently for the new values at the interior mesh points. Also, all processors update the values at mesh points on the interface in parallel.

12.4 NUMERICAL IMPLEMENTATION

As a numerical implementation, a free surface seepage problem is studied (Bruch
(1980)). The free surface seepage problem is modelled by the governing equations and
boundary conditions shown in Fig. 12.1. The relevant dimensions are taken to be:
$a = 40$, $y_1 = 10$ and $y_2 = 3$. In Fig. 12.1, D is the region $ABEF$ and Ω is the seepage
region $ABC_{\bar{f}}F$. The location of the curve $FC_{\bar{f}}$ is unknown *a priori*. The variable
ω is the Baiocchi transformation of the extended potential function. The detailed
derivations of these equations are given in Bruch (1980).

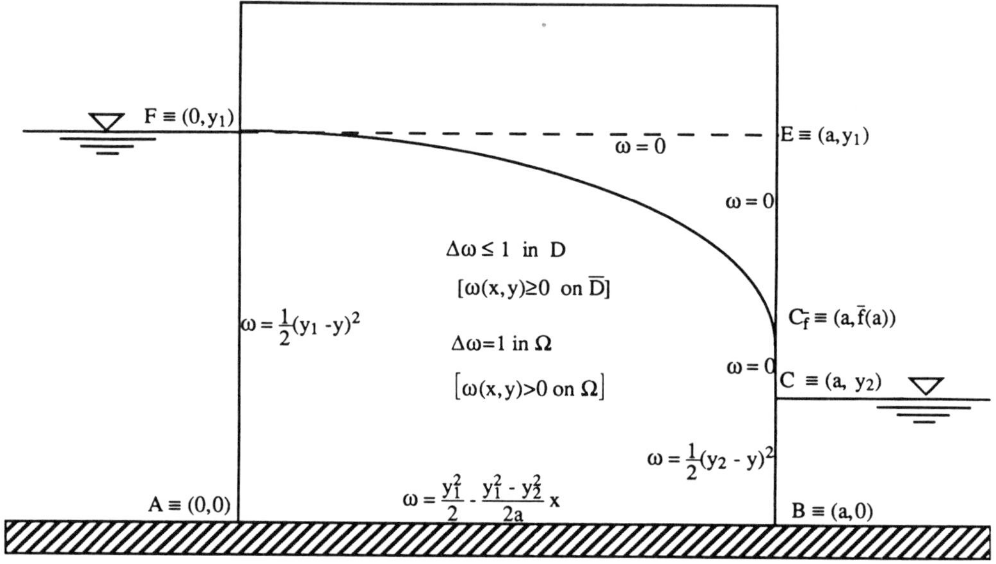

Figure 12.1 Free Boundary Seepage Problem

The problem shown in Fig. 12.1 can be written as a "complementarity system" and
its corresponding variational inequality formulation. From the latter a minimization
problem can be formulated:

$$J(\omega) \le J(u), \tag{12.1}$$

where

$$J(u) = a(u, u) + 2(f, u) \quad , \quad u \in H^1(D), \tag{12.2}$$

and u satisfies the Dirichlet boundary conditions, with $a(u, v)$ a bilinear form,
continuous, symmetric, positive definite on and f is given, i.e. $a(u, u) = \iint_D \nabla u \cdot \nabla u \, dx \, dy$ and $(f, u) = \iint_D fu \, dx \, dy$. For this example problem $f = 1$.

The functional J has one and only one minimum in a closed convex set. The
minimum is found using the following algorithm:

$$u_i^{(n+1/2)} = -\frac{1}{a_{ii}} \left(\sum_{j=1}^{i-1} a_{ij} u_j^{(n+1)} + \sum_{j=i+1}^{np} a_{ij} u_j^{(n)} + f_i \right), \tag{12.3}$$

$$u_i^{(n+1)} = P_i\left(u_i^{(n)} + \alpha\left(u_i^{(n+1/2)} - u_i^{(n)}\right)\right) = \max\left(0, u_i^{(n)} + \alpha\left(u_i^{(n+1/2)} - u_i^{(n)}\right)\right),$$

(12.4)

where $a_{ij} = a(N_i, N_j)$, $f_i = (f, N_i)$, N_i is the canonical basis, P_i is the projection on the convex set, $i = 1, ..., np$, np is the number of nodal points and α is the relaxation factor. Linear triangular elements, as already mentioned, will be used in the discretization. It should be noted that the projection operation in the numerical scheme must be applied during the iteration process. It cannot be applied after the iteration process has been completed since if it were, an incorrect solution would be obtained. For the numerical results given herein, the SOR relaxation factor was 1.85, while the stopping error criterion was 10^{-4} for the maximum absolute difference between iterates at a mesh point. This free boundary seepage problem will be solved by adaptive mesh finite element analysis with parallel computation. The procedure for the adaptive mesh finite element analysis with parallel computation is as follows:

1. Create an initial mesh.
2. Solve the problem by the parallel iterative scheme.
3. Perform, in parallel, error estimate evaluations. If the error calculated from this pass of the scheme is satisfied, then output the results and exit.
4. If the error criterion is not met, calculate, in parallel, new elemental areas for each element.
5. Refine the mesh and go to Step 2.

Note that Step 3 can be performed in parallel since it is an element by element calculation with global summations. Similarly, Step 4 is executed in parallel since it is an element by element calculation.

The measurement of the speedup is given by:

$$S_p = T_1/T_p,$$

(12.5)

where T_p is the time required to execute the algorithm using p processors and T_1 is the time required to execute the same program on a single processor. It should be noted that when the computations are performed on the p processors their sequence has been changed in that the computations on the interfaces are performed after those in the subdomains. Also, T_p is selected from the processor that requires the most time.

12.5 RESULTS AND DISCUSSION

12.5.1 ADAPTIVE MESH FINITE ELEMENT ANALYSIS

Cases were tested using different stages of adaptive mesh finite element analysis. A criterion that sets the final desired percentage error less than 0.03 was chosen for all cases. Refinement of a case was stopped when the calculated percentage error $^0\eta$ (see Burkley and Bruch (1991) for the meaning of $^0\eta$) was smaller than the desired percentage error η_{max} in a pass. A pass is one run of the solution scheme for a given system of equations. In addition, parallel computation was used in each case. A discussion will be given of the effectiveness of the adaptive mesh finite element analysis. Also, the speedup will be presented for the parallel computation.

The cases tested were:

1. Start with 297 nodes and 512 elements and set the η_{max} value to decrease from 0.07 to 0.02.
2. Start with 297 nodes and 512 elements and set η_{max} to always be 0.03.
3. Use a coarser mesh with 149 nodes and 256 elements and set the η_{max} value to decrease from 0.09 to 0.02.
4. Start with 149 nodes and 256 elements and set η_{max} to always be 0.03.
5. Test case without the adaptive mesh procedure. Set η_{max} to always be 0.03. Start from a coarse mesh and go to a fine mesh.

For the adaptive mesh finite element analysis, there are two refinement schemes for the h-refinement. The first scheme refines the mesh gradually. η_{max} decreases with the refinement procedure. Cases 1 and 3 use this scheme. The second scheme, however, is to refine the mesh according to a fixed η_{max}. Cases 2 and 4 are examples using the second scheme. For the non-adaptive mesh procedure, Case 5, all elements are refined uniformly until a final η_{max} less than 0.03 is reached.

Tables 12.1 to 12.5 present the results for the five cases tested. The percentage error $^0\eta$ is shown along with the time required to obtain the solution. A first observation is that, for all five cases, $^0\eta$ decreases when applying the refinement procedure. This shows that the errors arising from the finite element formulation are being reduced. However, there are differences in the results for these five cases.

One difference among these five cases is the number of nodes after refinement. As shown in Tables 12.1 to 12.5, after refinement, the final number of nodes is less for Cases 1 and 3. The more nodes there are the more computation time required. This is seen when comparing Cases 1 and 3 with Cases 2 and 4, respectively. Case 2 has 55% more nodes than Case 1; and Case 4 has 79% more nodes than Case 3. Thus, the computation time required to obtain the solution is smaller for Cases 1 and 3.

However, a contradiction is observed when comparing Cases 1 and 3 against Case 5. Although the number of nodes is less than for Case 5, Cases 1 and 3 require more computation time to obtain the solution. This violates the assumption that the smaller the number of nodes, the smaller the computation time. An inspection of the bandwidths of the linear systems for these cases gives an explanation. For the final mesh, the bandwidth is 104 for Case 1, and is 140 for Case 3. However, for Case 5, the bandwidth is only 35. Thus more operations per iteration are expected for Cases 1 and 3. Since the SOR iteration method is used to solve the linear system, a multiplication of non-diagonal terms with the solution vector is performed for each relevant degree of freedom. Cases 1 and 3 have approximately 3 and 4 times, respectively, the bandwidth of Case 5. Therefore, for Cases 1 and 3, more multiplications than for Case 5 are needed; and thus more computation is expected.

A bandwidth reduction process can be used to resolve this problem. However, even if the bandwidth is reduced effectively, extra computation time is required for the bandwidth reduction process. Thus the goal of using the local refinement process to reduce the number of nodes such that computation time could be reduced is not achieved here. The idea of reducing the bandwidth after the the local mesh refinement process will be considered in a future study.

Another difference is the effectiveness of the refinement process. It is intuitive that Cases 2 and 4 will be more effective than Cases 1 and 3, respectively. The percentage error $^0\eta$ drops rapidly in Cases 2 and 4. However, due to the over-refinement of some elements Cases 2 and 4 tend to generate more nodes than Cases 1 and 3, respectively. As stated previously, this increases the computation time. On the other hand, Cases 1 and 3 are more effective than Case 5. Case 5 refines all elements in the mesh such that more nodes are generated. The percentage error $^0\eta$ for Case 5 decreases slower than the percentage errors in Cases 1 and 3. Furthermore, Case 3 is more effective than Case 1. Therefore, despite the deficiency of the solution requirement stated previously, Case 3 is the best refinement scheme.

12.6 PARALLEL COMPUTATION

In addition to the adaptive mesh finite element analysis, parallel computation was used to fulfill the computation requirement after the mesh refinement. As shown in Tables 12.1 to 12.5, after refinement, the computation requirement is considerable. The study of the parallel computation was accomplished on an iPSC/2 D5 Hypercube Concurrent Computer (32 processors available).

To implement the parallel algorithm of Wang and Bruch (1993), the computation domain was subdivided into 2, 4, 8, 16, or 32 subdomains and then mapped onto the same number of processors. Figure 12.2 shows the computation domain subdivided into 8 subdomains for the last pass for Case 1. However, when 32 processors were employed, elements adjacent to more than two subdomains exist. This requires extra data transferring to nodes on these elements. This is not supported by the current parallel programs used. Therefore, only 16 processors were used to investigate the effectiveness of the parallel computation.

A direct observation from Tables 12.2 and 12.4 is that distributed memory multiprocessors computers can be used when the size of a problem is too large to be solved on a single processor computer. There is more memory required to store the new mesh information due to increasing the number of nodes and elements of the refined mesh. The memory required to store the stiffness matrix will also be increased if the number of nodes is increased after the refinement. Thus, once the memory required exceeds the memory limit of a computer, the external storage media (hard disk) is used as a virtual memory and swaps information in and out the RAM of the computer. This yields extra disk I/O time and slows down the solution process. When a multi-processor computer is used, the large amount of memory required can be split and then distributed onto the distributed memory of the multi-processor computer.

Another advantage of using the parallel computer is speed. The time required to solve a problem on a single processor computer is fixed. Multi-processor computers are used to split the problem and use more than one processor to solve the problem concurrently in order to speedup the solution. The theoretical speedup equals the number of processors used. Figure 12.3 shows the speedup vs. number of processors used for passes 3 and 4 of Case 1. As shown, speedups more than the ideal speedup were obtained. As noted in Tables 12.6 and 12.7, the number of iterations is smaller when using multi-processors. In addition, the number of nodes used to achieve more than

pass	1	2	3	4
no. of nodes	297	549	1427	2968
η_{max}	0.07	0.05	0.03	0.02
$^0\eta$	0.08	0.06	0.036	0.0197
time (sec)	4.86	25.65	175.14	691.58

Table 1. Results for Case 1.

pass	1	2	3
no. of nodes	297	1893	4604
η_{max}	0.03	0.03	0.03
$^0\eta$	0.08	0.031	0.0184
time (sec)	4.87	305.01	

Table 2. Results for Case 2.

pass	1	2	3	4
no. of nodes	149	446	976	2563
η_{max}	0.09	0.06	0.03	0.02
$^0\eta$	0.13	0.0681	0.0385	0.022
time (sec)	2.58	27.61	124.36	794.96

Table 3. Results for Case 3.

pass	1	2	3
no. of nodes	149	1561	4598
η_{max}	0.03	0.03	0.03
$^0\eta$	0.13	0.0359	0.0186
time (sec)	2.58	304.85	

Table 4. Results for Case 4.

pass	1	2	3	4	5
no. of nodes	149	297	553	2129	4257
η_{max}	0.03	0.03	0.03	0.03	0.03
$^0\eta$	0.13	0.08	0.065	0.032	0.019
time (sec)	2.58	4.87	22.59	220.30	322.78

Table 5. Results for Case 5.

pass	1	2	3	4
1 proc.	70	89	99	87
2 proc.	70	88	95	85
4 proc.	69	87	95	86
8 proc.	70	83	94	85
16 proc.	71	80	90	82

Table 6. Iteration numbers for Case 1.

pass	1	2	3	4
1 proc.	80	89	88	88
2 proc.	80	89	86	88
4 proc.	79	87	84	88
8 proc.	75	86	83	87
16 proc.			77	84

Table 7. Iteration numbers for Case 3.

Insert Tables 12.1–12.7 on this page

Figure 12.2 Domain Subdividing for Pass 4 of Case 1.

ideal speedup is reduced to 1427 in Case 1 and 976 in Case 3. Similar phenomenon were presented in Wang and Bruch (1993) with 8353 nodes. This is due to the uncoupling of the system of equations caused by this parallel algorithm and the fact that the boundary data is being fed into the solution region faster in the parallel computations.

ACKNOWLEDGMENTS

This material is based upon work supported by the National Science Foundation under Award No. ECS-9006516. The authors would also like to thank the Cornell Theory Center's Advanced Computing Research Institute for providing time on their Hypercube Concurrent Computer (iPSC/2).

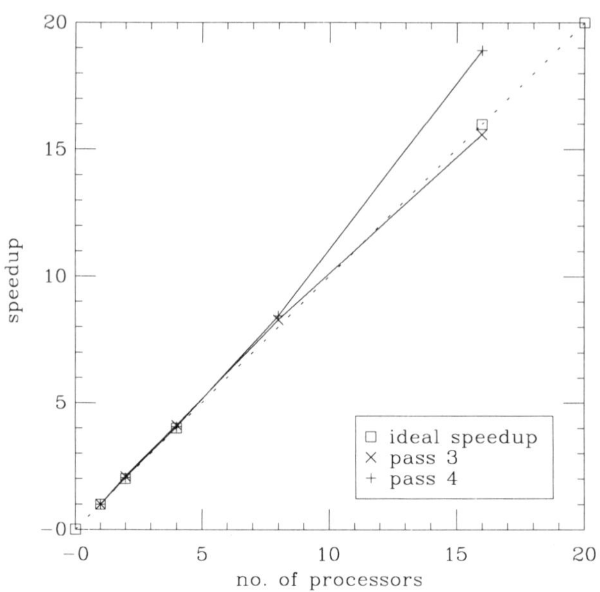

Figure 12.3 Speedup for Case 1.

REFERENCES

Bruch J.C. Jr. (1980), A Survey of Free Boundary Value Problems in the Theory of Flow Through Porous Media: Variational Inequality Approach. *Advances in Water Resources*, Part I, **3**: 65–80.

Burkley V.J., Bruch J.C. Jr., and Zienkiewicz O.C. (1991), Adaptive Meshes Used in Solving a Free Surface Seepage Problem. *The Mathematics of Finite Elements and Applications*, VII, edited by J.R. Whiteman, Academic Press, 101–110.

Burkley V.J. and Bruch J.C. Jr. (1991), Adaptive Error Analysis in Seepage Problems. *Int. J. Num. Meth. Engng.*, **31**: 1333–1356.

Farhat C. (1990), Which Parallel Finite Element Algorithm for Which Architecture and Which Problem? *Eng. Comp.*, **7**: 186–195.

Rivara M.C. (1984), Algorithms for Refining Triangular Grids Suitable for Adaptive and Multigrid Techniques. *Int. J. Num. Meth. Engng.*, **20**: 745–756.

Rivara M.C. (1987), A Grid Generator Based on 4-Triangles Conforming Mesh-Refinement Algorithms. *Int. J. Num. Meth. Engng.*, **24**: 1343–1354.

Wang K.P. and Bruch J.C. Jr. (1993), A Highly Efficient Iterative Parallel Computational Method for Finite Element Systems. *Eng. Comp..* **10**: 195–204.

Zienkiewicz O.C. and Zhu J.Z. (1987), A Simple Error Estimator and Adaptive Procedure for Practical Engineering Analysis. *Int. J. Num. Meth. Engng.*, **24**: 337–357.

Chapter 13

Application of a Variational Method for Computing Smooth Stresses, Stress Gradients, and Error Estimation in Finite Element Analysis

A. Tessler*, H. R. Riggs**, and S. C. Macy***

   * NASA Langley Research Center, Computational Mechanics Branch M/S 240 Hampton, Virginia 23681-0001, U.S.A.
  ** University of Hawaii at Manoa, Civil Engineering Department, Honolulu, Hawaii 96822, U.S.A.
 *** Lockheed Engineering and Sciences Company, Hampton, Virginia 23666, U.S.A.

13.1 INTRODUCTION

The qualitative deficiencies of stress (strain) results obtained from displacement finite element analysis are well established. For example, in elasticity problems and first-order shear-deformable plate theories, the element displacement field is C^o-continuous thus ensuring continuity of displacements along element boundaries; the stresses, however, are only C^{-1}-continuous. (i.e., discontinuous along element boundaries). The need for continuous stresses at interelement boundaries arises, for example, in stress contouring and *residual equilibrium* error estimation. Also, the practically important stresses along structural and component boundaries — where stress maxima often occur — are generally not as accurate as stresses in the interior.

For those cases in which the stresses are smooth, the stresses converge with the order $O(h^{p-m+1})$, where m is the highest-order derivative in the weak form, p is the degree of complete polynomial in the displacement interpolation and h is the mesh parameter, which may be taken as the diameter of the largest element in the mesh. A few optimal locations within the element interior are known to provide more accurate stresses than elsewhere in the element. At these locations, the stresses converge even faster with order $O(h^{p-m+2})$. These points have been identified for some elements and their properties investigated (e.g., Barlow (1976,1989); MacKinnon and Carey (1989); Krizek and Neittaanmaki (1987)).

Recently, Zienkiewicz and Zhu (1992) have proposed an effective stress–recovery procedure which is focused on extending the accuracy of the optimal stress points to the entire domain, as well as obtaining a C^o stress field. In their procedure, a *patch* of elements is taken around each vertex node; that is, each node which is not

The Mathematics of Finite Elements and Applications
Edited by J. R. Whiteman © 1994 John Wiley & Sons Ltd

a midside node. A local, discrete, least-squares approximation of the stress field is then determined based on the finite element stresses at optimal points within the patch (typically Gauss points). The same interpolation functions are used for the smoothed stresses as are used for the finite element displacements. They have shown that the recovered nodal stresses are superconvergent. The procedure does require certain *special cases*. At midside nodes which are involved in two patches, stresses are averaged to obtain a single-valued recovered stress. At certain boundary nodes, it is not possible to define a patch, and the recovered stress must be obtained by extrapolation from a nearby patch. Additional potential difficulties appear to exist for meshes which involve elements of different order. In the transition regions where adjacent elements have different order interpolation functions, the approach to define the interpolation for the stresses is not clear.

Another smoothing finite element method was developed by Tessler *et al.* (1987) for recovering globally smooth strain fields from experimental Moiré measurements using an approach analogous to that of Hinton and Irons (1968). Tessler and Freese (1991) discussed the application of this method to stress and stress-gradient recovery in two-dimensional finite element analysis. Tessler *et al.* (1993) proposed a general variational form for the smoothing analysis applicable to three-, two- and one-dimensional problems and established a theoretically-based smoothing discretization strategy which relies on superconvergent, optimal-point stresses. They demonstrated that the accuracy of superconvergent optimal stresses of the original finite element analysis can be extended over the entire domain.

The smoothing formulation of Tessler and co-workers—which henceforth will be referred to as Smoothing Element Analysis (SEA) — is based on a variational principle which combines the discrete least-squares and penalty-constraint functionals in a single variational form. The independent field variables are the smoothed quantity and n orthogonal gradients, with n denoting the dimension of the physical domain of the problem. The penalty-constraint term is included to ensure slope continuity across element interfaces of the smoothed quantity. The field variables are C^o-continuous; however, the smoothed quantity (stress) approaches C^1 continuity when a large value of the penalty number is applied. Thus, a single-valued C^1 continuous stress field is obtained over the entire domain (or whatever region of the domain is smoothed).

Because of improved accuracy, the smooth stress field can serve as a reference stress in the class of *post-processing* error estimators (Zienkiewicz and Zhu (1987)). In addition, within the same smoothing analysis, SEA is capable of recovering accurate stress gradients . This enables effective computations of *residual* based error estimators which require the knowledge of stress gradients (e.g., Babuska and Rheinboldt (1981); Ohtsubo and Kitamura (1992); Baehmann *et al.* (1992)). Such error estimators are particularly meaningful from an *engineering* perspective since they serve as direct measures of the error in the governing equations which are being approximated. Moreover, accurate stress gradients are required in the computation of transverse shear and normal stresses in the analysis of laminated composite plates and shells (Tessler and Freese (1991); Tessler and Saether (1991)).

This paper reviews the theoretical basis of SEA and discusses its application in the areas of stress smoothing, *post-processing* error estimation, stress-gradient recovery, and computation of *residual equilibrium* error estimators.

13.2 VARIATIONAL PRINCIPLE

Let $\hat{\sigma}_q = \sigma(\mathbf{x}_q)$ ($\mathbf{x}_q = (x_q^1, x_q^2, x_q^3)$, $q = 1, 2, .., N$) represent a set of discrete data defined on the domain $\Omega = \{(x^1, x^2, x^3) \in \mathbb{R}^3\}$, where $\mathbf{x}_q$ is the position vector of the data point relative to a three-dimensional orthogonal frame of reference. For present purposes, the discrete data $\hat{\sigma}_q$ are represented by a set of optimal stresses obtained by a finite element analysis which is characterized by the mesh parameter h.

It is desired to represent $\hat{\sigma}_q$ by a smooth, scalar-valued continuous function $\sigma(\mathbf{x})$. This may be accomplished through a minimization technique in which the domain Ω is discretized into n_{el} *smoothing* finite elements such that $\Omega \equiv \bigcup_{e=1}^{n_{el}} \Omega^e$, each element having a characteristic size h_s and possessing n^e points of stress data within its domain, Ω^e (i.e., $N \equiv \sum_{e=1}^{n_{el}} n^e$). Within each element, piecewise continuous approximations for the stress, σ, and orthogonal stress gradients, $\theta_i (i = 1, 2, 3)$, are assumed independently and denoted as σ^{h_s} and $\theta_i^{h_s}$, where the superscript h_s refers to the association of these variables with a smoothing finite element mesh. The variational statement for the discretized problem involves minimization of a functional Φ (i.e., $\delta \Phi = 0$) given as

$$\Phi = \sum_{e=1}^{n_{el}} \Phi^e,$$

$$\Phi^e = \sum_{q=1}^{n^e} \left[\hat{\sigma}_q - \sigma^{h_s}(\mathbf{x}_q)\right]^2 + \lambda \int_{\Omega^e} \left[(\sigma_{,1}^{h_s} - \theta_1^{h_s})^2 + (\sigma_{,2}^{h_s} - \theta_2^{h_s})^2 + (\sigma_{,3}^{h_s} - \theta_3^{h_s})^2\right] d\Omega \quad (13.1)$$

where λ is a *penalty number*, and $\sigma_{,i} \equiv \frac{\partial}{\partial x_i}(\sigma)$ $(i = 1, 2, 3)$. The first term in (13.1) represents a discrete least-squares functional with the term

$$\epsilon_q = \hat{\sigma}_q - \sigma^{h_s}(\mathbf{x}_q) \quad (13.2)$$

expressing the smoothing error or difference between the discrete stress, $\hat{\sigma}_q$ and the smooth stress state, $\sigma^{h_s}(\mathbf{x}_q)$. The second term in (13.1) is a penalty constraint functional which, for $\lambda \to \infty$, yields the following element-level constraint relations

$$\sigma_{,i}^{h_s} \to \theta_i^{h_s} \ (i = 1, 2, 3) \ in \ \Omega^e \quad (13.3)$$

When conditions (13.3) are fulfilled, $\theta_i^{h_s}(i = 1, 2, 3)$ represent the first-order gradients of σ^{h_s} with respect to the orthogonal coordinates. Equation (13.3) can also be interpreted as the limiting conditions of C^1 degree of smoothness for σ^{h_s}.

In practice, λ has to be sufficiently large in order to enforce conditions (13.3), yet it should not be excessively large to cause ill-conditioning or *locking* of the *smooth* solution. The optimal choice for λ will differ depending on the interpolations used for a given type of element and, therefore, it should be established for a particular element on the basis of numerical studies. It should also be noted at this point that whenever a vectorial or tensorial (stress) field is being smoothed, each component is smoothed independently, that is (13.1)–(13.3) apply to each component. Equations (13.1) through (13.3) are valid for a general three-dimensional case and are readily reducible to two- and one-dimensional cases by respectively omitting terms associated with subscript 3 or subscripts 2 and 3.

13.3 ELEMENT SHAPE FUNCTIONS AND MATRIX EQUATIONS

In (13.1), the gradients of σ^{h_s} and $\theta_i^{h_s}$ do not exceed order one; hence, the smoothing element shape functions for these field variables need only be C^o-continuous. Certain restrictions on the suitable choice of shape functions are due to the constraints (13.3). In a similar formulation for plate theory, constraints of this type are known to cause detrimental *locking* effect when conventional isoparametric interpolations are used. When σ^{h_s} is interpolated with a polynomial one degree higher than the $\theta_i^{h_s}$ variables using *anisoparametric* interpolations, the *locking* effect is alleviated or completely eliminated. (See Tessler *et al.* (1993) and references therein.)

For the three-dimensional case, the anisoparametric interpolations for σ^{h_s} and $\theta_i^{h_s}$ can be expressed in matrix form as

$$\sigma^{h_s} = \mathbf{P}\sigma + \mathbf{Q}_1\theta_1 + \mathbf{Q}_2\theta_2 + \mathbf{Q}_3\theta_3 \equiv \mathbf{N}\mathbf{d}^e, \quad \theta_i^{h_s} = \mathbf{P}\theta_i \quad (i = 1, 2, 3) \qquad (13.4)$$

where $\sigma^T = \{\sigma_1, \sigma_2, ..., \sigma_{n_{dof}}\}$ and $\theta_i^T = \{\theta_{i1}, \theta_{i2}, ..., \theta_{in_{dof}}\}$ are the element vectors of nodal degrees-of-freedom (dof) with n_{dof} denoting the number of element nodes; $\mathbf{P}$ and $\mathbf{Q}_i$ are row-vectors containing shape functions, with the $\mathbf{Q}_i$-functions being one order higher than $\mathbf{P}$.

Substituting (13.4) into (13.1) and performing the first variation with respect to the nodal dof (i.e., $\delta\Phi = 0$) results in a standard system of linear algebraic equations

$$\mathbf{K}\mathbf{d} = \mathbf{F} \qquad (13.5)$$

where $\mathbf{d}$ is the vector containing nodal dof in the discretization associated with the smoothing variable, σ^{h_s}, and its orthogonal gradients, $\theta_i^{h_s}$; further, $\mathbf{K}$ is a symmetric, banded square matrix similar to the stiffness matrix in the finite element method; $\mathbf{F}$ is a *consistent load* vector involving the $\hat{\sigma}_q$ values and element shape functions. The element-level counterpart of (13.5) is

$$\mathbf{K}^e\mathbf{d}^e = \mathbf{F}^e \qquad (13.6)$$

where

$$\mathbf{K}^e = \mathbf{K}_\varepsilon^e + \mathbf{K}_\lambda^e = \sum_{q=1}^{n^e} \mathbf{N}_q^T \mathbf{N}_q + \lambda \int_{\Omega^e} \mathbf{B}^T \mathbf{B} d\Omega, \quad \mathbf{F}^e = \sum_{q=1}^{n^e} \hat{\sigma}_q \mathbf{N}_q^T \qquad (13.6a)$$

where $\mathbf{N}_q \equiv \mathbf{N}(\mathbf{x}_q)$ and

$$\mathbf{B} = \begin{bmatrix} \mathbf{P}_{,1} & \mathbf{Q}_{1,1} - \mathbf{P} & \mathbf{Q}_{2,1} & \mathbf{Q}_{3,1} \\ \mathbf{P}_{,2} & \mathbf{Q}_{1,2} & \mathbf{Q}_{2,2} - \mathbf{P} & \mathbf{Q}_{3,2} \\ \mathbf{P}_{,3} & \mathbf{Q}_{1,3} & \mathbf{Q}_{2,3} & \mathbf{Q}_{3,3} - \mathbf{P} \end{bmatrix} \qquad (13.6b)$$

Equation (13.5) must be solved for each component of the stress tensor. However, since $\mathbf{K}^e$ depends only on the element shape functions evaluated at the stress locations and does not involve values of stress, each stress component corresponds to a different load case only. Thus, regardless of the number of smoothed quantities, $\mathbf{K}$ need only be assembled and factored once. The solution for $\mathbf{d}$ in (13.5) can be readily obtained if $\mathbf{K}$ is positive-definite; this aspect is assured if the number of independent dof in the smoothing mesh does not exceed the number of input stresses, N. The appropriate smoothing discretization strategy which guarantees positive-definiteness of $\mathbf{K}$ and ensures optimal stress-recovery convergence is discussed next.

13.4 ISO- AND SUPER-SMOOTHING DISCRETIZATIONS

To produce *optimal* SEA discretizations, Tessler *et al.* (1993) proposed an *order of accuracy* argument which may be arrived at from the following considerations. The finite element stresses in terms of the exact stresses, σ^{exact}, can be written as

$$\sigma^h = \sigma^{exact} + O(h^r) \tag{13.7}$$

where the exponent r is related to the interpolation order of the finite elements and the definition of stress. In elasticity problems without singularities, the stress is related to the first gradients of displacement and at the optimal stress points $r = p+1$, where p is the order of the complete polynomial in the displacement interpolation. Similarly, the smoothed stresses can be represented in terms of the finite element stresses as

$$\sigma^{h_s} = \sigma^h + O(h_s^{p_s+1}) \tag{13.8}$$

where p_s is the order of the complete polynomial in the smoothing element. Introducing (13.7) into (13.8) and subsequently allowing the error of the smoothing analysis and the error of the finite element analysis to be of the same order results in

$$h_s = h^\alpha, \quad \alpha = \frac{r}{p_s + 1} \tag{13.9}$$

If optimal-stress points are used, then the exponent in (13.9) becomes

$$\alpha = \frac{p+1}{p_s + 1} \tag{13.10}$$

and σ^{h_s} takes the form

$$\sigma^{h_s} = \sigma^{exact} + O(h^{p+1}) \tag{13.11}$$

That is, the smoothed stresses over the entire domain should have at least the same order of accuracy as the optimal finite element stresses.

Equation (13.9) (with the use of (13.10)) is the condition for achieving the super-convergent accuracy defined in (13.11). However, there exists an additional $h_s - h$ relation which needs to be satisfied in order to ensure positive-definiteness of $\mathbf{K}$. For $\mathbf{K}$ to be positive-definite, the number of discrete stresses used in the smoothing analysis, N, should be at least as large as the number of independent degrees-of-freedom in the smoothing element model, N^{dof}, i.e.,

$$N \geq N^{dof} \tag{13.12}$$

Considering for the sake of demonstration an uniformly discretized one-dimensional domain of unit length $\Omega \in [0,1]$, the requirement (13.12) can be readily expressed as

$$h_s \geq \frac{p/h - 2}{p_s - 1} \quad (p_s \geq 2) \tag{13.13}$$

Thus, for the smoothing solution to be possible, the optimal $h_s - h$ relation (13.9) must also satisfy (13.13).

Equation (13.10) also points towards the functional dependence between p and p_s. For $\alpha \leq 1$ or $p_s \geq p$, the smoothing mesh is coarser than or identical to the finite element mesh; the former situation is said to be *Super-smoothing* whereas the latter is *Iso-smoothing*. Similarly, for $\alpha > 1$ or $p_s < p$, the smoothing mesh is finer than the finite element mesh, this situation is referred to as *Sub-smoothing*. The impracticality of *Sub-smoothing* is obvious — the smoothing mesh is more complicated than the underlying finite element analysis mesh and additional *nonoptimal* stresses need to be invoked in order to fulfill the positive-definiteness requirement, with this latter aspect affecting the optimal stress recovery of SEA. (Refer also to results in Tessler *et al.* , (1993).) Both the *Super-smoothing*$(p_s > p)$ and *Iso-smoothing* $(p_s = p)$ strategies appear to be computationally viable, with the latter having a practical advantage of utilizing the same mesh as the underlying finite element analysis mesh.

13.5 SMOOTH STRESSES AND ERROR ESTIMATION: EXAMPLE PROBLEM

A two-point boundary value problem representing a fixed-fixed, unit-length elastic rod imbedded in an elastic foundation and subjected to a distributed force is a convenient one-dimensional test case to evaluate the effectiveness of SEA. The differential equation of equilibrium and boundary conditions in terms of the axial displacement, u, are given as (Zienkiewicz and Zhu, (1992))

$$-u_{,xx} + u = f \quad \text{in} \ \ \Omega = (0,1) \tag{13.14}$$

$$u(0) = u(1) = 0$$

in which $\sigma \equiv u_{,x}$. The function f is chosen such that the exact solution is given by

$$u = x^2 - \frac{\sinh 4x}{\sinh 4} \tag{13.15}$$

To test the proposed discretization strategy, the smoothing analyses are carried out using both quadratic (two-node, $p_s = 2$) and cubic (three-node, $p_s = 3$) *anisoparametric* one-dimensional elements that are analogous to the Timoshenko beam elements (Tessler and Dong, (1981)). Herein, only the shape functions for the quadratic smoothing element are illustrated. In terms of the natural coordinate ξ ($\xi \in [-1,1]$), the quadratic σ^{h_s} and linear $\theta_1^{h_s}$ interpolations are :

$$\sigma^{h_s} = \mathbf{P}\sigma + \mathbf{Q}_1\theta_1, \quad \theta_i{}^{h_s} = \mathbf{P}\theta_1 \tag{13.16}$$

where

$$\mathbf{P} = \frac{1}{2}\left[(1-\xi),\ (1+\xi)\right],\quad \mathbf{Q}_1 = \frac{h_s}{8}\left[(1-\xi^2),\ (\xi^2-1)\right] \tag{13.17}$$

The finite element stresses are obtained from uniform meshes along the span using linear ($p = 1$), quadratic ($p = 2$), and cubic ($p = 3$) displacement elements and *sampled* at the optimal Gauss points. The smoothing meshes are constructed on the basis of (13.9), (13.10), (13.13), and $p_s \geq p$; the range of $\lambda = 1 - 10^4$ produced virtually identical results.

Figure 13.1 shows convergence of the stress error norm defined as

$$\|e_\sigma\| \equiv \left[\int_0^1 (\sigma - \sigma^{exact})^2 dx\right]^{1/2} \tag{13.18}$$

This measure of error illustrates that stresses derived from the smoothing element analysis (SEA) are substantially more accurate than those derived from the finite element analysis (FEA). The smoothed stresses for the linear and quadratic elements (Figures 13.1(a) and 1(b), respectively) display convergence which is at least one order higher than the finite element stresses. When nonoptimal Gauss stresses are used in the SEA, as shown in Figure 13.1(b) for the case $(p, p_s) = (2, 2)$ where stresses were sampled according to the 3-point Gauss rule, the SEA accuracy is marginally superior; however, the rate of convergence is the same as the FEA. The *Super-smoothing* discretization for the case $(p, p_s) = (2, 3)$, although showing only slight improvements over the corresponding *Iso-smoothing* with $(p, p_s) = (2, 2)$, is achieved with fewer smoothing dof (Figure 13.1(b)).

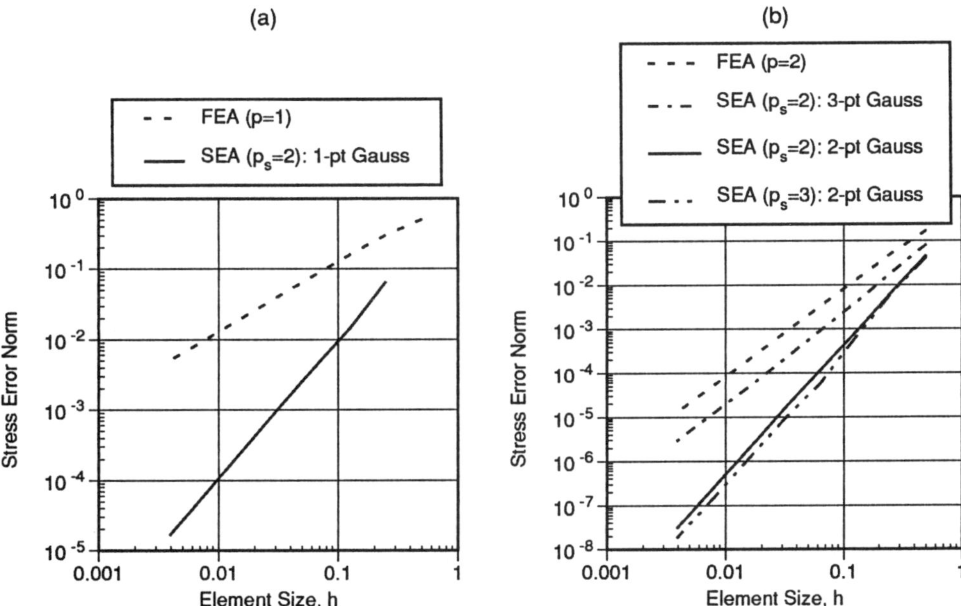

Figure 13.1 Stress error norm from finite element and smoothing element analyses.

Figure 13.2(a) depicts the distribution of the error in stress over the entire domain for the case of eight, linear finite elements and four, quadratic finite elements. The results show a dramatic increase in accuracy as a consequence of smoothing. The SEA stress distributions, nearly matching the values of optimal stresses at the optimal locations, extend the accuracy of the optimal stress points over the entire domain of finite element discretization.

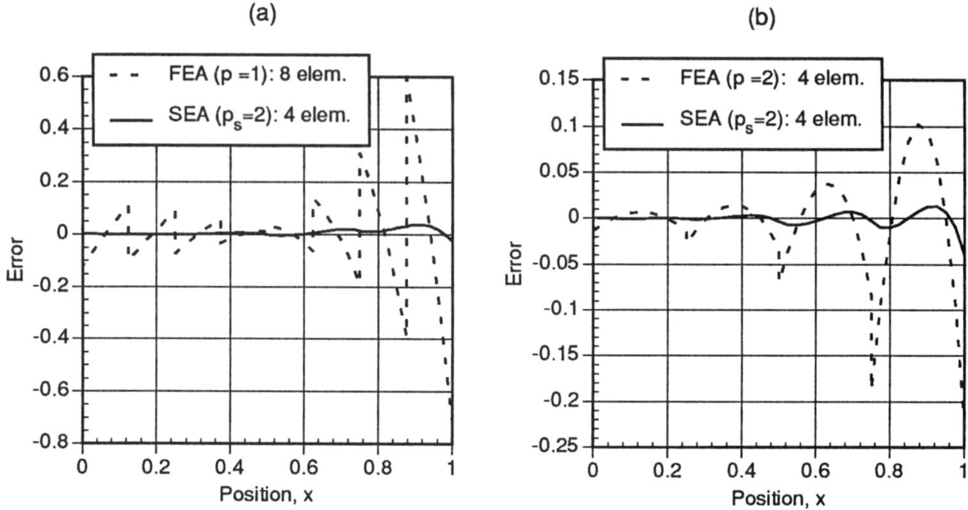

Figure 13.2 Distribution of stress error for discretizations of (a) eight linear and (b) four quadratic finite elements.

13.5.1 RESIDUAL ERROR ESTIMATOR

In addition to recovering smooth stresses, the smoothing approach also computes C^o-continuous stress gradients, θ_i. This enables a rather straightforward computation of residual error estimators. Thus, for the 1-D problem under consideration, the error norm is the L_2 norm of the residual equilibrium error r given by

$$\|e\| \equiv \left[\int_0^1 r^2 dx \right]^{1/2} \tag{13.19}$$

where r is the residual in the differential equation

$$r = \sigma_{,x} - u + f \tag{13.20}$$

The common difficulty with this type of error estimator using finite element stress results is that the stresses are discontinuous at element boundaries, and the gradient $\sigma_{,x}$ results in *jump* contributions to the error r. These contributions cannot be evaluated solely on an element basis. For simplicity, some investigators have ignored the contribution to the error from the discontinuous stresses (Baehmann *et al.* (1992)).

Figure 13.3(a) depicts convergence of $\|e\|$ which is computed by smoothing optimal stresses produced by elements of order $p = 1, 2, 3$ using the *Super-smoothing* strategy for $(p, p_s) = (1, 2)$, and *Iso-smoothing* for $(p, p_s) = (2, 2)$ and $(p, p_s) = (3, 3)$. Figure 13.3(b) shows the contribution to the total error norm in each element for two different meshes which have the same number of dof: (i) eight, $p = 1$ elements, and (ii) four, $p = 2$ elements using respectively *Super-* and *Iso-smoothing* meshes. These latter results illustrate that the estimator can be used in mesh adaptivity by showing which regions of the mesh should be refined.

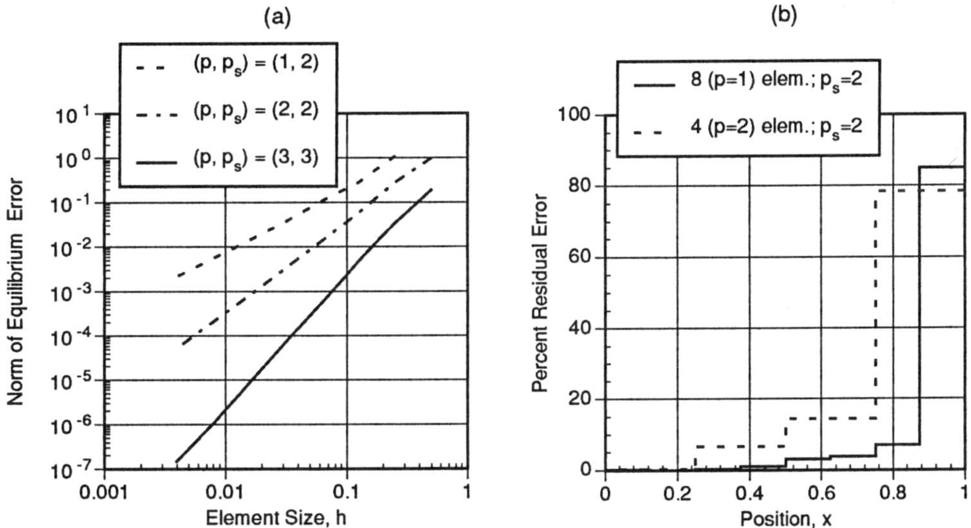

Figure 13.3 (a) L_2 norm of equilibrium error for smoothed $p = 1, 2, 3$ FEA results; (b) Distribution of residual equilibrium error over disretization domain.

13.6 CONCLUDING REMARKS

The Smoothing Element Analysis (SEA) method has been presented for one-, two-, and three-dimensional applications and its potential utility demonstrated for obtaining smooth stresses, stress gradients, and error estimation on a one-dimensional problem. The method is based on minimizing a functional involving discrete least-squares error with a penalty constraint that ensures smoothness of the stress field.

The smoothed stresses are C^1-continuous and are the same order of accuracy as the superconvergent optimal stresses of the original finite element analysis provided that interpolation functions for the smoothing element are of the same degree (*Iso-smoothing*) or greater (*Super-smoothing*) than those of the finite element. The overall accuracy corresponds to a finite element analysis of much higher refinement. Although an additional smoothing analysis is required, the approach appears to be much more efficient than refinement of the finite element mesh and subsequent reanalysis.

Other areas where SEA can be effectively applied include, but are not limited to, strain-field recovery based on experimental Moiré measurements, calculations of transverse shear and normal stresses in plate theory, improved potential energy estimation in frequency analysis, and adaptive mesh refinement.

REFERENCES

Babuska, I. and Rheinboldt, W. C. (1981), *A Posteriori* Error Analysis of Finite Element Solutions for One-Dimensional Problems. *SIAM Journal on Numerical Analysis*, **18**(3), 565–589.

Baehmann, P. L., Shephard, M. S. and Flaherty, J. E. (1992), *A Posteriori* Error Estimation for Triangular and Tetrahedral Quadratic Elements using Interior Residuals. *Int. J. Numer. Meths. Engrg.*, **34**, 979–996.

Barlow, J. (1976), Optimal Stress Locations in Finite Element Models. *Int. J. Numer. Meths. Engrg.*, **10**, 243P251.

Barlow, J. (1989), More on Optimal Stress Points and Reduced Integration, Element Distortions and Error Estimation. *Int. J. Numer. Meths. Engrg.*, **28**, 1487–1504.

Hinton, E. and Irons, B. (1968), Least Squares Smoothing of Experimental Data Using Finite Elements. *J. Strain*, **4**, 24–27.

Krizek, M. and Neittaanmaki, P. (1987), On Superconvergence Techniques. *Acta Applicandae Mathematicae*, **9**, 175–198.

MacKinnon, R. J. and Carey, G. F. (1989), Superconvergent Derivatives: A Taylor Series Analysis. *Int. J. Numer. Meths. Engrg*, **28**, 489–509.

Ohtsubo, H., and Kitamura, M. (1992), Numerical Investigation of Element-Wise *A Posteriori* Error Estimation in Two and Three Dimensional Elastic Problems. *Int. J. Numer. Meths. Engrg.*, **34**, 969–977.

Tessler, A., Freese, C., Anastasi, R., Serabian, S., Oplinger, D. and Katz, A. (1987), Least-Squares Penalty-Constraint Finite Element Method for Generating Strain Fields from Moiré Fringe Patterns. In: *Proceedings of SPIE: Photomechanics and Speckle Metrology*, **814**, 314–323.

Tessler, A. and Freese, C. (1991), A Global Penalty-Constraint Finite Element Formulation for Effective Strain and Stress Recovery. In: *Proceedings of Int. Conf. Comput. Engrg. Science, ICES Publications*, Melbourne, Australia, 1120–1123.

Tessler, A., and Saether, E. (1991), A Computationally Viable Higher-Order Theory for Laminated Composite Plates. *Int. J. Numer. Meths. Engrg.*, **31**, 1069–1086.

Tessler, A. and Dong, S. B. (1981), On a Hierarchy of Conforming Timoshenko Beam Elements. *Computers and Structures*, **14**(3P4), 335–344.

Tessler, A., Riggs, H. R., and Macy, S. C. (1993), A Variational Method for Finite Element Stress Recovery and Error Estimation. In: *Proceedings of 34th Structures, Structural dynamics and Materials Conference, AIAA Paper 93-1384*, La Jolla, California.

Zienkiewicz, O. C. and Zhu, J. Z. (1987), A Simple Error Estimator and Adaptive Procedure for Practical Engineering Analysis. *Int. J. Numer. Meths. Engrg.*, **24**, 337–357.

Zienkiewicz, O. C. and Zhu, J. Z. (1992), Superconvergent Patch Recovery and *a Posteriori* Error Estimates. Part 1: The Recovery Technique. *Int. J. Numer. Meths. Engrg.*, **33**, 1331–1364.

Chapter 14

A Parallel Numerical Model for Subsurface Contaminant Transport With Biodegradation Kinetics

Todd Arbogast and Mary F. Wheeler

Department of Computational and Applied Mathematics,
Rice University, Houston, Texas 77251-1892, U.S.A.

14.1 INTRODUCTION

In this paper we discuss the formulation of a simulator for groundwater flow and transport with biodegradation kinetics that has been developed at Rice University for massively parallel, distributed memory, message passing machines. The numerical procedures employed are a mixed finite element method for flow and the characteristics-mixed method for transport. Kinetics are treated by time splitting. The linear solvers are based on domain decomposition. Application of this procedure to a bioremediation problem as well as numerical experiments on the INTEL i860 and INTEL Delta are discussed. Results indicate that the procedure is theoretically mass conservative over each grid cell and is approximately so in implementation. Preliminary tests indicate that the procedure is robust and applicable to realistic groundwater problems. Moreover, the numerical model scales almost linearly with the number of processors even for fairly coarse grids.

Biorestoration techniques such as microbial biodegradation are part of an innovative, emerging technology for handling subsurface water contamination, see Borden and Bedient (1986), Bouwer and McCarthy (1983(i,ii)), Criddle *et al.* (1990), Hansen (1990), Semprini *et al.* (1991) and United States Department of Energy (1988). The microbial biodegradation process involves the stimulation of indigenous microorganisms to remove subsurface contaminants. U. S. Environmental Protection Agency studies (Thomas *et al.* (1987)) have shown that this type of restoration strategy can result in complete removal of contaminants, whereas other proposed restoration strategies have not proven as effective.

Biological decontamination is physically and chemically complex. It involves the transport and interaction of hydrocarbons, microbes, oxygen, nitrogen, and various chemical compounds as well as the movement of water within the aquifer. Numerical simulation of these processes is a critical step in understanding and designing biorestoration applications, see Arbogast *et al.* (1992), Chiang *et al.* (1991), Dawson *et al.* (1986-1987), Dawson and Wheeler (1992), Kindred and Celia (1989), Wheeler

The Mathematics of Finite Elements and Applications
Edited by J. R. Whiteman © 1994 John Wiley & Sons Ltd

et al. (1987), Wheeler and Dawson (1988) and Wood and Dawson (1993). Indeed, without computational science, in situ biodegradation of contaminants on a wide scale is impractical.

New parallel supercomputers, allowing simultaneous use of hundreds to thousands of processors, have greatly expanded the potential for building detailed models of these porous media processes. Parallel computing provides the capability of solving larger, more realistic and practical problems faster and more economically. This includes the ability to use an adequately refined discretization mesh, to incorporate complex chemical and physical effects associated with the transport of both hydrocarbons and organic contaminants in porous media, and to employ stochastic or conditional simulation. The latter is essential for simulating a realistic geologic aquifer, since much of the data needed to characterize it cannot be quantified accurately, and since often the chemical and physical processes are not well understood. Conditional simulation is ideally suited to parallel computation.

In this paper we will emphasize modeling of the flow and transport in the saturated zone. Extensions to multicomponent, multiphase models in the unsaturated zone are presently being considered, but because of page limitations, they will not be discussed herein.

The outline of the paper is as follows. In Section 14.2 we describe the governing flow and transport equations with biodegradation in a saturated porous medium. For simplicity, we assume linear sorption and aerobic conditions. More general kinetics such as Michaelis-Menton can be treated with the numerical techniques described in this paper.

In Section 3, we first describe the mixed finite element method for an elliptic flow equation, see Brezzi and Fortin (1991) and Raviart and Thomas (1977). We then formulate the characteristics-mixed method (Arbogast and Wheeler (1993(i))). Theoretical convergence estimates are stated for the characteristics-mixed method. Finally, we briefly discuss the time-splitting scheme employed for treating advection-diffusion-reaction problems.

In Section 4 we describe the parallel implementation of the procedure, and in Section 5 we present three dimensional, parallel, bioremediation simulation results. Conclusions and current directions on parallel implementation are given in Section 6.

14.2 CONTAMINANT TRANSPORT WITH BIODEGRADA-TION

The governing equations of transport with biodegradation in a saturated porous medium are described by a coupled, nonlinear advection-diffusion-reaction system consisting of m_s electron donors (substrates) and m_n electron acceptors or nutrients, and a system of m_x ordinary differential equations involving microbial mass. (Transport of microbes can be treated also if one assumes instead a system of advection-diffusion-reaction equations for the microbes.) The governing equation of flow is given by Darcy's Law and the continuity equation.

These equations defined in a one, two, or three dimensional porous medium Ω can be written as:

Electron Donor (Substrate)

$$\phi R_{S_i}\frac{\partial S_i}{\partial t} - \nabla \cdot (D\nabla S_i - \boldsymbol{u}S_i) = \phi\chi_{S_i} + g_{S_i}, \quad i = 1,\ldots,m_s, \tag{14.1}$$

Electron Acceptor

$$\phi R_{N_i}\frac{\partial N_i}{\partial t} - \nabla \cdot (D\nabla N_i - \boldsymbol{u}N_i) = \phi\chi_{N_i} + g_{N_i}, \quad i = 1,\ldots,m_n, \tag{14.2}$$

Microbial Mass
$$\phi\frac{\partial X_i}{\partial t} = \phi\chi_{x_i}, \quad i = 1,\ldots,m_x, \tag{14.3}$$

Darcy's Law and Continuity

$$\boldsymbol{u} = -K\nabla p \quad \text{and} \quad \nabla \cdot \boldsymbol{u} = f. \tag{14.4}$$

Here $D = D(\boldsymbol{u})$ is a hydrodynamic diffusion/dispersion tensor, ϕ is porosity, and R_i is a retardation factor for component i due to adsorption. The χ are possibly nonlinear kinetic terms which account for biodegradation of contaminants, utilization of nutrients, and growth and decay of microorganisms. The number and complexity of specific metabolic pathways or chemical reactions varies with the application. The source/sink terms f and g_i represent production and injection wells. The hydraulic conductivity K is a symmetric positive definite tensor. For convenience of exposition, we will assume that
$$\boldsymbol{u} \cdot \boldsymbol{\nu} = 0,$$
$$D\nabla N_i \cdot \boldsymbol{\nu} = D\nabla S_i \cdot \boldsymbol{\nu} = 0,$$

where $\boldsymbol{\nu}$ is the outward, unit, normal vector to $\partial\Omega$, the boundary of Ω.

14.3 THE NUMERICAL ALGORITHMS

Let $\Omega \subset \mathbb{R}^d$, $d = 1, 2$, or 3. Let $W = L^2(\Omega)$ denote the set of square integrable functions and $H(\Omega;\text{div}) = \{v \in (L^2(\Omega))^d \mid \nabla \cdot v \in L^2(\Omega)\}$. Let $V = H^0(\Omega;\text{div}) = \{v \in H(\Omega;\text{div}) \mid v \cdot \boldsymbol{\nu} = 0 \text{ on } \partial\Omega\}$.

For spatial discretization, we employ the lowest order Raviart-Thomas spaces (Raviart and Thomas (1977)) defined over a rectangular grid of Ω with maximal grid spacing $h > 0$. These finite dimensional spaces consist of $W_h \subset W$ equal to the set of discontinuous functions that are constant in each cell, and $\tilde{V}_h \subset H(\Omega;\text{div})$ equal to the set of vectors $v = (v_1, v_2, v_3)$ (if $d = 3$) such that v_i is continuous, piecewise linear over the grid in the ith direction and discontinuous, piecewise constant over the grid in the other two directions. We also need the subspace $V_h = \tilde{V}_h \cap V$.

We first describe the mixed finite element method for approximating (14.4). With $(\cdot,\cdot)$ denoting the $L^2(\Omega)$-inner product, we write (14.4) in variational form as

$$(K^{-1}\boldsymbol{u}, v) - (p, \nabla \cdot v) = 0, \qquad v \in V,$$

$$(\nabla \cdot \boldsymbol{u}, w) = (f, w), \qquad w \in W.$$

In the mixed finite element formulation, we seek the pair $(\boldsymbol{U}_h, P_h) \in V_h \times W_h$ satisfying

$$(K^{-1}\boldsymbol{U}_h, \boldsymbol{v}_h) - (P_h, \nabla \cdot \boldsymbol{v}_h) = 0, \qquad \boldsymbol{v}_h \in V_h, \tag{14.5a}$$

$$(\nabla \cdot \boldsymbol{U}_h, w_h) = (f, w_h), \qquad w_h \in W_h. \tag{14.5b}$$

In order to define the characteristics-mixed method for (14.1) and (14.2), we consider the following abstract transport equation for some component concentration c:

$$\phi \frac{\partial c}{\partial t} + \boldsymbol{u} \cdot \nabla c - \nabla \cdot D\nabla c = g \quad \text{in } \Omega, \ t > 0, \tag{14.6a}$$

$$D\nabla c \cdot \boldsymbol{\nu} = 0 \quad \text{on } \partial\Omega, \ t > 0, \tag{14.6b}$$

$$c(\boldsymbol{x}, 0) = c^0(\boldsymbol{x}) \quad \text{on } \Omega, \tag{14.6c}$$

where $\boldsymbol{u}$ is assumed given and c^0 is the initial concentration.

A numerical method for (14.6a)–(14.6c) that has been used with success is the modified method of characteristics (MMOC-Galerkin), see Arbogast *et al.* (1992), Chiang *et al.* (1991), Dawson *et al.* (1989), Dawson *et al.* (1986), Dawson *et al.* (1987), Dawson and Wheeler (1992), Douglas and Russell (1982), Ewing *et al.* (1983), Ewing *et al.* (1984), Russell *et al.* (1986), Russell and Wheeler (1983), Wheeler *et al.* (1987), Wheeler and Dawson (1988) and Wood and Dawson (1993). In this scheme, the time derivative and the advection term (i.e., the hyperbolic part of the equation, $\phi \partial c/\partial t + \boldsymbol{u} \cdot \nabla c$) are combined as a directional derivative along the characteristics, and then the equation is treated as in a Galerkin finite element method. Although this method allows one to use large time-step increments and to treat large Peclet numbers, the main drawback of the method is its nonconservative nature and the cost of tracing the characteristics from many points.

The characteristics-mixed method introduced by the authors, see Arbogast and Wheeler (1993(i)), retains all the advantages of MMOC-Galerkin in addition to being theoretically conservative. As in MMOC-Galerkin, a directional derivative is used to treat the hyperbolic part of (14.6a), but then a mixed finite element method is applied to treat the diffusion/dispersion.

Before defining the characteristics-mixed method, we write (14.6a)–(14.6c) in a saddle point weak formulation. Define $\tilde{V} = (L^2(\Omega))^d$, $\tilde{\boldsymbol{z}} = -\nabla c$, and the dispersive flux as $\boldsymbol{z} = -D\nabla c$. Then (14.6a) is expressed for $(c, \tilde{\boldsymbol{z}}, \boldsymbol{z}) \in W \times \tilde{V} \times V$ as

$$\left(\phi \frac{\partial c}{\partial t} + \boldsymbol{u} \cdot \nabla c, w\right) + (\nabla \cdot \boldsymbol{z}, w) = (g, w), \quad w \in W, \tag{14.7a}$$

$$(\tilde{\boldsymbol{z}}, \boldsymbol{v}) = (c, \nabla \cdot \boldsymbol{v}), \quad \boldsymbol{v} \in V, \tag{14.7b}$$

$$(\boldsymbol{z}, \boldsymbol{v}) = (D\tilde{\boldsymbol{z}}, \boldsymbol{v}), \quad \boldsymbol{v} \in \tilde{V}. \tag{14.7c}$$

To discretize this equation, we begin with the characteristic approximation. So let $\Delta t > 0$ be the time-step increment and $t^n = n\Delta t$ for $n = 1, 2, \ldots$ (Δt could vary with the time-step in practice). The characteristic trace-back along the velocity field $\boldsymbol{u}$ of a point $\boldsymbol{x} \in \Omega$ at time t^n to time t^{n-1} is approximately

$$\check{\boldsymbol{x}}(\boldsymbol{x}, t^{n-1}) = \boldsymbol{x} - \frac{\boldsymbol{u}(\boldsymbol{x}, t^n)}{\phi(\boldsymbol{x})}\Delta t. \tag{14.8}$$

As a consequence, the hyperbolic terms in (14.6a) at time t^n are

$$\phi\frac{\partial c^n}{\partial t} + \boldsymbol{u}^n \cdot \nabla c^n \approx \phi\frac{c^n - \check{c}^{n-1}}{\Delta t}, \tag{14.9}$$

wherein we use the notation $\psi^n = \psi(t^n)$ for functions of time and $\check{\psi}(\boldsymbol{x}) = \psi(\check{\boldsymbol{x}})$ for functions of space.

The characteristics-mixed algorithm can be stated as follows. First let $C^0 \in W_h$ approximate the initial concentration c^0. Then for each time level n, find $(C^n, \tilde{\boldsymbol{Z}}^n, \boldsymbol{Z}^n) \in W_h \times \tilde{V}_h \times V_h$ such that

$$\left(\phi\frac{C^n - \check{\tilde{C}}^{n-1}}{\Delta t}, w\right) + (\nabla \cdot \boldsymbol{Z}^n, w) = (g^n, w), \quad w \in W_h, \tag{14.10a}$$

$$(\tilde{\boldsymbol{Z}}^n, \boldsymbol{v}) = (C^n, \nabla \cdot \boldsymbol{v}), \quad \boldsymbol{v} \in V_h, \tag{14.10b}$$

$$(\boldsymbol{Z}^n, \boldsymbol{v}) = (D\tilde{\boldsymbol{Z}}^n, \boldsymbol{v}), \quad \boldsymbol{v} \in \tilde{V}_h, \tag{14.10c}$$

$$\tilde{C}^{n-1} = \mathcal{P}(C^{n-1}, \tilde{\boldsymbol{Z}}^{n-1}), \tag{14.10d}$$

where in (14.10d) we require the definition (14.11a)–(14.11b) below of the post-processing operator $\mathcal{P}$. (This more general form of the mixed method is analyzed in greater detail in Arbogast *et al.* (1993).) It is well known that in a single time-step of the mixed method, C^n is convergent to c^n only to the first order in h, and that we can improve this order through various post-processing techniques. Therefore, we apply the post-processing (14.10d) to the concentration at time t^{n-1} before taking the next time-step. In fact, we should view the post-processed quantity $\tilde{C}^n = \mathcal{P}(C^n, \boldsymbol{Z}^n)$ as our approximate concentration at time t^n. As a consequence, the theorem below shows that we obtain a better overall rate of convergence for the scheme.

We now define $\mathcal{P}$. Let $\tilde{W}_h \subset W$ denote our post-processing space. This space consists of discontinuous, piecewise linear functions defined over the grid. Define $\mathcal{P} : W \times \tilde{V} \to \tilde{W}_h$ for $(\omega, \boldsymbol{v}) \in W \times \tilde{V}$ by $\mathcal{P}(\omega, \boldsymbol{v}) = \tilde{\omega} \in \tilde{W}_h$, where over the grid cell R,

$$(\phi(\tilde{\omega} - \omega), w)_R = 0, \quad w \in W_h, \tag{14.11a}$$

$$(\nabla\tilde{\omega} + \boldsymbol{v}, \nabla\tilde{w})_R = 0, \quad \tilde{w} \in \tilde{W}_h. \tag{14.11b}$$

The notation $(\cdot, \cdot)_R$ means that the integration is restricted to R. Note that $w = 1$ gives material balance; in fact, mass is conserved cell by cell over the grid. The following convergence is attained by the scheme under reasonable hypotheses (see Arbogast and Wheeler (1993(i)) for the proof). The proof assumes the more usual form of the mixed method in which

$$(D^{-1}\boldsymbol{Z}^n, \boldsymbol{v}) - (C^n, \nabla \cdot \boldsymbol{v}) = 0, \quad \boldsymbol{v} \in V_h, \tag{14.12}$$

$$\tilde{C}^{n-1} = \mathcal{P}(C^{n-1}, \boldsymbol{Z}^{n-1}), \tag{14.13}$$

replaces (14.10b)–(14.10d), and also that $\mathcal{P} : W \times V \to \tilde{W}_h$, with

$$(D\nabla\tilde{\omega} + \boldsymbol{v}, \nabla\tilde{w})_R = 0, \quad \tilde{w} \in \tilde{W}_h \tag{14.14}$$

replacing (14.11b).

Theorem. *If (14.12)–(14.13) replaces (14.10b)–(14.10d) and (14.14) replaces (14.11b), then for h and Δt sufficiently small,*

$$\max_n \|\tilde{C}^n - c^n\| \le C\{h^{3/2} + \Delta t\}, \tag{14.15a}$$

$$\left\{ \sum_{n=1}^{N} \|D^{1/2}(\boldsymbol{Z}^n - z^n)\|^2 \Delta t \right\}^{1/2} \le C\{h + \Delta t\}, \tag{14.15b}$$

where $\|\psi\| = (\psi, \psi)^{1/2}$ is the $L^2(\Omega)$-norm.

Preliminary results by the authors indicate that the proof extends to the formulation (14.10a)–(14.10d), (14.11b), with (14.15b) replaced by

$$\left\{ \sum_{n=1}^{N} \|D^{1/2}(\tilde{\boldsymbol{Z}}^n - \tilde{z}^n)\|^2 \Delta t \right\}^{1/2} \le C\{h + \Delta t\}. \tag{14.16}$$

We remark that under special circumstances and if $\tilde{W}_h$ is replaced by the discontinuous, piecewise quadratic functions, then $h^{3/2}$ can be replaced by $h^{7/4}$ in the theorem.

14.4 PARALLEL IMPLEMENTATION

In our numerical formulation for modeling (14.1)–(14.4), we first approximate $\boldsymbol{u}$ by $\boldsymbol{U}$ by applying to (14.4) our mixed finite element procedure with the lowest order Raviart-Thomas space (1977) as described in the last section. In (14.5a) we employ the following quadrature rule for the ith component: the trapezoidal rule in the ith direction and midpoint rule in the other two directions. The resulting algorithm is equivalent to the cell-centered finite difference method, see Russell and Wheeler (1983), Weiser and Wheeler (1988). For a diagonal tensor K one obtains a finite stencil for pressure, five points if the dimension $d = 2$ and seven if $d = 3$.

A nonoverlapping parallel domain decomposition algorithm has been applied for solving the discrete system. The basic idea is to decompose the domain Ω into a number of subdomains Ω_j, one associated with each processor. We then have a series of boundary value problems similar to (14.4) in each Ω_j. We obtain a guess for the boundary data on the internal interfaces $\partial\Omega_j$, and solve the local problems in the subdomains. If the solutions and their normal derivatives match across subdomain boundaries, then the problem in Ω is solved. If not, then interface boundary data are updated and we iterate. In other words, in domain decomposition the solution process of the large global problem is decomposed into the repeated solution of numerous smaller, *independent* problems.

In such a domain decomposition algorithm, we must describe how we choose the guess for the internal interfaces boundary data. We use Method 2 of Glowinski and Wheeler (1988), which uses Dirichlet interface boundary data. From any convenient initial guess, the succeeding guesses are given as the solution of a positive definite, symmetric problem posed on the interfaces. There have been several interface problem solution techniques proposed (Cowsar *et al.* (1992(i)), Cowsar *et al.* (1992(ii))); however, the one used for the calculations reported in Section 5 is Balancing. This procedure was defined and analyzed by Cowsar, Mandel, and one of the authors of Cowsar *et al.* (1992(i)) and implemented by Cowsar. There are three components to multi-domain balancing. The first involves the solution of the subdomain problems with Dirichlet data provided on internal interfaces. The second involves the solution of subdomain problems with Neumann data. The third involves a global coarse grid problem with subdomains treated as "elements" to insure the well-posedness of the Neumann problems and to provide a mechanism of global exchange of information.

The number of iterations of the Balancing procedure is of the order $1 + \log(H/h)$, where the diameter of the global domain is $O(1)$, the diameter of the subdomain is $O(H)$, and the diameter of the cell is $O(h)$. Balancing scales almost linearly provided that care is taken treating the coarse grid problem. For massively parallel machines the coarse grid problem can be a costly bottleneck and a domain decomposition procedure needs to be applied to the coarse grid.

Since the approximate velocity U is discontinuous, there are certain numerical difficulties in obtaining good approximations to the characteristics (14.8). Therefore, we postprocess it into a continuous velocity $\tilde{U}$ in the space of piecewise discontinuous trilinear functions. This velocity field is actually more accurate than U because it exploits superconvergence, see Weiser and Wheeler (1988).

Given the velocity approximation $\tilde{U}$, the advection-diffusion-reaction system involving donor, acceptor and biological mass equations are approximated using a time splitting scheme. One global time step involves the following three sequential steps:

(i) Pure transport. For each electron donor or acceptor, characteristics are traced backwards in time to locate their origin at the previous time level. This may be done by taking small micro time steps. The trace-back points are joined to form a "twisted" grid. The time-step size is controlled so that this twisting is not excessive, and every grid cell is mapped to a distinct twisted grid cell. For each given cell, we average the postprocessed concentration from the previous time step over the corresponding "twisted" cell (i.e. we integrate and divide by the volume). This average is the transported concentration in the absence of reactions and diffusion/dispersion.

(ii) Reactions. The coupled system of reaction equations (i.e. (14.1)–(14.3) without the two divergence terms and without the g_i source terms) are approximated using a fourth order Runge-Kutta procedure. Initial conditions are the "twisted" cell averages from (i) for acceptors and donors, and the previous time step concentrations for the microbes. Many small time steps may be taken to improve the accuracy.

(iii) Diffusion/dispersion. The diffusion/dispersion step involves approximating a parabolic system for each donor or acceptor using initial data from (ii) and applying the mixed finite element method (14.10a)–(14.10d), with $C = S_i$ or $C = N_i$. In (14.10b) the same quadrature is used as described above for the flow. A tensor product trapezoidal rule is used in (14.10c). As in the case of the flow, a finite stencil is obtained for each component, nine points in two dimensions and nineteen in three. As in previous MMOC-Galerkin calculations (Ewing *et al.* (1983), Wheeler *et al.* (1987)), the discrete system is solved using a Jacobi preconditioned conjugate gradient algorithm. Respective concentrations and their spatial gradients $\tilde{Z}$ are thereby obtained. The latter are used to construct a higher order approximation to the concentrations in each grid cell as in (14.10d)–(14.11). Finally, a slope-limiting scheme is used to prevent overshoot and undershoot.

Using the previous time step as initial data, the global time step is repeated to some final time T.

The implementation of this time splitting reactive transport scheme in parallel was done by one of the authors, Ashokumar Chilakapati, and Doug Moore.

14.5 SOME BIOREMEDIATION RESULTS

The Hanford Site in Washington State in the United States occupies approximately 560 square miles of semiarid terrain and was selected in 1943 for producing materials (primarily plutonium) in support of the United States' World War II efforts. Chemical processes employed to recover and purify plutonium produced a waste stream containing actinide compounds as well as the typical aqueous and organic liquid industrial wastes. The primary organic contaminant carbon tetrachloride (CCl_4) totaled 637 to 1200 tons discharged. Today, plutonium production has ceased, and the primary mission has shifted to environmental restoration of the Hanford Site, see Hagood and Rohay (1991), Last *et al.* (1991).

Rice University and Pacific Northwest Laboratory (PNL) began a collaborative research effort in 1992 that involves laboratory, field, and simulation work directed toward validating remediation strategies, and includes both natural and *in situ* bioremediation. We discuss below some preliminary computational results based on some recent microbial CCl_4 destructive kinetics developed by Skeen *et al.* (1992) of PNL.

We assume that the domain is a $20\,\mathrm{m} \times 20\,\mathrm{m} \times 5\,\mathrm{m}$ aquifer and that the grid is $20 \times 20 \times 5$. The permeability is taken to be a scalar; a horizontal cross-section is shown in Figure 14.1. (In all the figures, horizontal cross-sections are taken at $z = 2.5\,\mathrm{m}$, and white is the highest value, black the lowest.) The model has six components: electron acceptors nitrate NO_3^-, nitrite NO_2^-, and acetate CH_3COO^-, donor CCl_4, microbes $C_5H_9O_3N$, and a nonreactive tracer. We also assume that the retardation factor R for acetate is 1.8. The chemical reactions for this system are:

$$8NO_3^- + 2CH_3COO^- + 2H^+ \rightarrow 4CO_2 + 8NO_2^- + 4H_2O,$$
$$8NO_2^- + 3CH_3COO^- + 11H^+ \rightarrow 6CO_2 + 4N_2 + 10H_2O,$$
$$7CH_3COO^- + 2NO_3^- + 9H^+ \rightarrow 4CO_2 + 6H_2O + 2C_5H_9O_3N,$$
$$13CH_3COO^- + 4NO_2^- + 17H^+ \rightarrow 6CO_2 + 10H_2O + 4C_5H_9O_3N,$$

and bioremediation is described by

$$\frac{d\,(CCl_4)}{dt} = \frac{-\mu\,(CCl_4)\,(\mathrm{microbes})}{1 + k_i(\,(NO_3^-) + (NO_2^-)\,)}.$$

The problem is a linear flow from the left to the right face of the aquifer with

Fig. 14.1 Horizontal cross-section of permeability field K.

pressure specified on the left and right faces and no flow or zero Neumann conditions specified on the remaining faces. The nutrient electron acceptors nitrate and acetate are introduced into the system on the left face for 20 days, with the goal of increasing the microbial population. We assume that the substrate CCl_4 is flowing through the system (i.e. it is continuously injected at its original concentration on the left face) so that any concentration reduction observed is due not to CCl_4 flowing out of the system, but rather to microbial degradation.

In Figure 14.2, horizontal cross-sections of a nonreactive tracer are shown at 2.5 days and at 5 days, respectively, and in Figure 14.3 a contour of the .5 concentration of the nonreactive tracer at 5 days. From Figures 14.1 and 14.2, we clearly observe the effect of the permeability; that is, the tracer moves into the high permeability zones. Figures 14.4–14.7 are horizontal cross-sections of carbon tetrachloride, microbial mass, acetate, and nitrate at 2.5 and 5 days respectively. One can observe the growth of the microbial mass when nitrate and acetate are both available, and the successful degradation of CCl_4.

The effects of retardation on acetate are clear from the figures by noting that acetate has not moved as fast as the nitrate or the tracer. Since all the assumed chemical reactions require acetate, this example illustrates the danger of a poor application of the technology: Some nitrate outruns the acetate and pollutes the aquifer (though nitrate is preferable to CCl_4!). A series of simulation studies could be used to identify the proper rates of nutrient injection.

Finally in Figure 14.8, we observe the performance of the Rice 3D Parallel Groundwater Reactive Flow and Transport Code (RPGW). Here, speed-up is plotted as a function of the number of processors, normalized to the time taken on two processors. The dotted line indicates the theoretical linear speed-up. Since the INTEL iPSC/860 and the INTEL DELTA have roughly the same speed, some of the experiments were run on only one machine; in particular the longer runs using two and four processors on the i860 were not repeated on the DELTA.

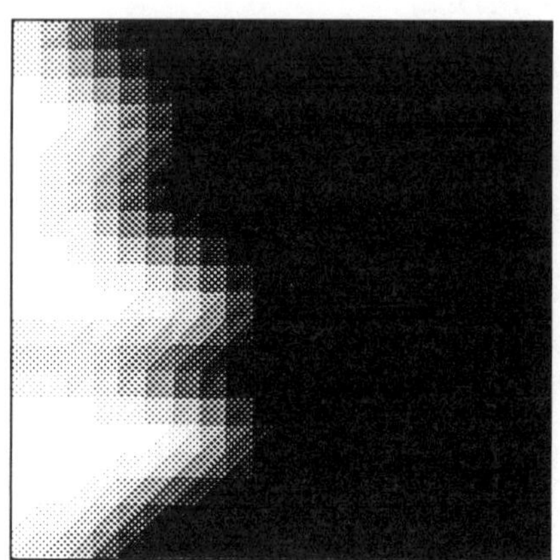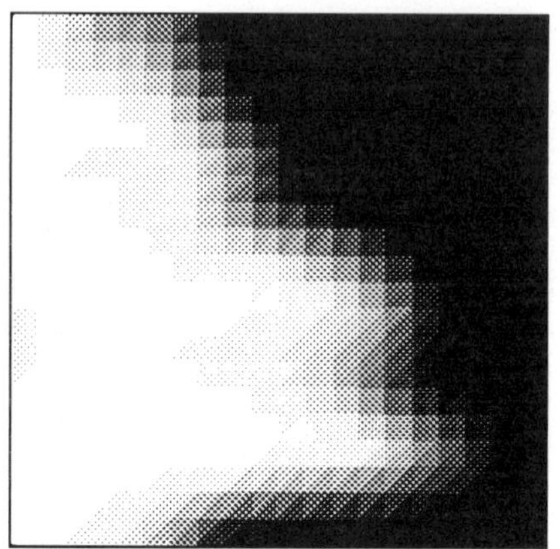

Fig. 14.2 Horizontal cross-section of nonreactive tracer at 2.5 and 5 days.

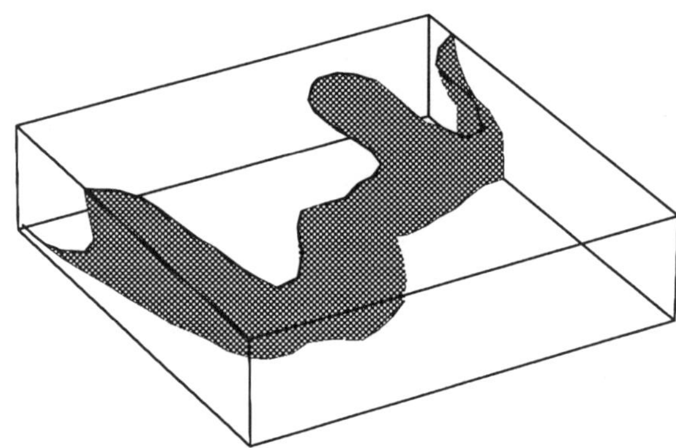

Fig. 14.3 Contour of nonreactive tracer concentration .5 at 5 days.

The reader should note that the subdomain grids are only $4 \times 4 \times 5$ when run on 25 processors, so then the surface to volume ratio is 80:80 (since the z-direction is not subdivided). This ratio is quite high, and it is proportional to the amount of interprocessor communication needed to solve the problem. The observed speedup is almost linear even up to this high level of surface to volume ratio.

Fig. 14.4 Horizontal cross-section of CCl$_4$ at 2.5 and 5 days.

Fig. 14.5 Horizontal cross-section of microbial mass at 2.5 and 5 days.

14.6 CONCLUSIONS AND FUTURE DIRECTIONS

Rice 3D Parallel Groundwater Reactive Flow and Transport is a parallel code under development entirely at Rice University. Its purpose is to simulate the flow and transport of reacting chemical species in the groundwater. This code is based on combining locally conservative schemes: a mixed finite element method for flow with a characteristics-mixed finite element method for transport. Computational experiments indicate that this approach is useful in solving grand challenge problems such as bioremediation and that the code scales almost linearly even on small problems.

We are presently adding general wells and boundary conditions to the transport

Fig. 14.6 Horizontal cross-section of acetate at 2.5 and 5 days.

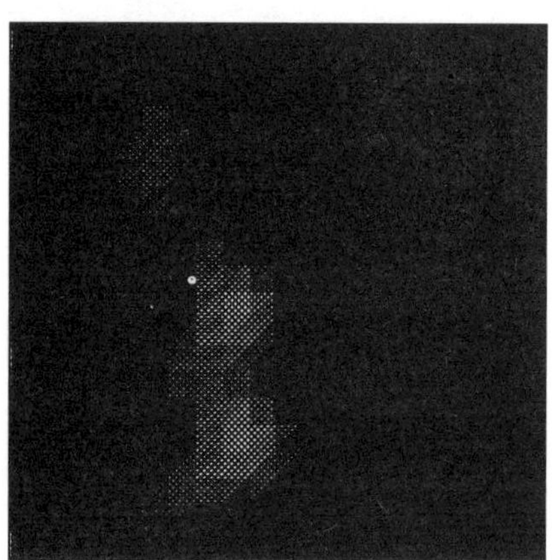 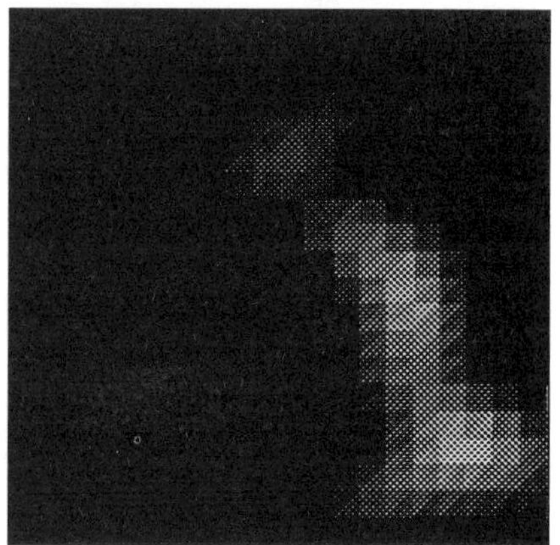

Fig. 14.7 Horizontal cross-section of nitrate at 2.5 and 5 days.

code as well as coupling this code to a three phase (gas, non-aqueous phase liquid, and aqueous) flow code. Future plans include the incorporation of meshes defined by fairly general geometry. In addition we plan to add more chemistry and microbiology as well as the capability of simulating fractured media. Since most environmental companies do not have access to large parallel machines, we plan to replace the present communications package PICL with PVM so that a collection of workstations can also be used.

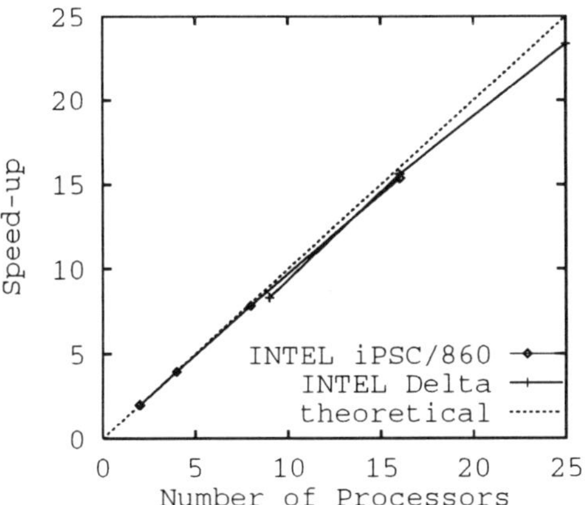

Fig. 14.8 Speed-up for RPGW.

ACKNOWLEDGEMENTS

The authors wish to acknowledge the Department of Energy who have supported this contract through two subcontracts: Oak Ridge National Laboratory (ORNL) as part of the Partnership in Computational Science (PICS) Consortium and Pacific Northwest Laboratory (PNL). The authors would also like to acknowledge the help of Kyle Roberson and Brian Wood of PNL in the formulation of the bioremediation problem as well as Perry Cheng, Doug Moore, Joe Warren, and Mark Wells of Rice for their help in computer visualization. Finally we wish to acknowledge the work of Ashokumar Chilakapati, Lawrence Cowsar, Clint Dawson, Frederic d'Hennezel, and Doug Moore in the development of the Rice Parallel Groundwater Reactive Flow and Transport Code.

REFERENCES

Arbogast, T., Chilakapati, A. and Wheeler, M.F., (1992), A Characteristic-Mixed Method for Contaminant Transport and Miscible Displacement. In: *Computational Methods in Water Resources IX, Vol. 1: Numerical Methods in Water Resources* (eds. T.F. Russell, R.E. Ewing, C.A. Brebbia, W.G. Gray, and G.F. Pinder), Computational Mechanics Publications, Southampton, U.K., 77-84.

Arbogast, T., Wheeler, M.F. and Yotov, I., (1993) Mixed Finite Element Methods with a Tensor Coefficient. In preparation.

Arbogast, T. and Wheeler, M.F., (1993(i)), A Characteristics-Mixed Finite Element Method for Advection Dominated Transport Problems. (To appear in *SIAM J. Numer. Anal.*)

Arbogast, T. and Wheeler, M.F., (1993(ii)), Characteristic Mixed Method Combined with Reactive Flow Along the Characteristics. In preparation.

Borden, R.C. and Bedient, P.B., (1986), Transport of Dissolved Hydrocarbons Influenced by Oxygen-Limited Biodegradation 1. Theoretical Development. *Water Resour. Res.*, **22**: 1973–1982.

Bouwer, E.J. and McCarthy, P.L., (1983(i)), Transformations of 1- and 2-Carbon Halogenated Aliphatic Organic Compounds Under Methanogenic Conditions. *Appl. Environ. Microbiol.*,

45: 1286–1294.

Bouwer, E.J. and McCarthy, P.L., (1983(ii), Transformations of Halogenated Organic Compounds Under Denitrification Conditions. *Appl. Environ. Microbiol.*, **45**: 1295–1299.

Brezzi, F. and Fortin, M., (1991), Mixed and Hybrid Finite Element Methods, Springer-Verlag, New York.

Chiang, C.Y., Dawson, C.N. and Wheeler, M.F., (1991), Modeling of *In-situ* Biorestoration of Organic Compounds in Groundwater. *Transport in Porous Media*, **6**: 667–702.

Cowsar, L.C., Mandel, J. and Wheeler, M.F., (1992(i)), Rice University Center for Research on Parallel Computation Technical Report CRPC-TR91154. Submitted for publication.

Cowsar, L.C., Weiser, A. and Wheeler, M.F., (1992(ii)), Multigrid and Domain Decomposition for Elliptic Equations. In: *Fifth International Symposium on Domain Decomposition Equations*, SIAM, Philadelphia, 376–385.

Criddle, C.S., DeWitt, J.T. and McCarty, P.L., (1990),. Reductive Dehalogenation of Carbon Tetrachloride by Escherichia Coli K-12. *Appl. Environ. Microbiol.*, **56**: 3247–3254.

Dawson, C.N., Russell, T.F. and Wheeler, M.F., (1989), Some Improved Error Estimates for the Modified Method of Characteristics. *SIAM J. Numer. Anal.*, **26**: 1487–1512.

Dawson, C.N., Wheeler, M.F., Nguyen, T.M. and Poole, S.W., (1986), Simulation of Hydrocarbon Biodegradation in Groundwater. *CRAY Channels*, **8**: 3 14–19.

Dawson, C.N., Wheeler, M.F., Nguyen, T.M. and Poole, S.W., (1987), Simulation of Subsurface Contaminant Transport with Biodegradation Kinetics. In: *Proceedings of Third International Symposium on Science and Engineering on CRAY Supercomputers*, Mendota Heights, Minnesota, 75–86.

Dawson, C.N. and Wheeler, M.F., (1992), Time-splitting Methods for Advection Diffusion-Reaction Equation Arising in Contaminant Transport. In: *Proceedings of ICIAM II*, SIAM, Philadelphia, 71–82.

Douglas, J. and Russell, T.F., (1982), Numerical Methods for Convection-Dominated Diffusion Problems Based on Combining the Method of Characteristics with Finite Element or Finite Difference Procedures. *SIAM J. Numer. Anal.*, **19**: 871–885.

Evans, J.C., Bryce, R.W., Bates, D.J. and Kemner, M.L., (1990), Hanford Site Ground-Water Surveillance for 1989, PNL-7396, Pacific Northwest Laboratory, Richland, Washington.

Ewing, R.E., Russell, T.F. and Wheeler, M.F., (1983), Simulation of Miscible Displacement Using a Mixed Method and a Modified Method of Characteristics. Paper SPE 1224, in: *Proceedings of the Seventh SPE Symposium on Reservoir Simulation, Society of Petroleum Engineers*, Dallas, 71–81.

Ewing, R.E., Russell, T.F. and Wheeler, M.F., (1984), Convergence Analysis of an Approximation of Miscible Displacement in Porous Media by Mixed Finite Elements and a Modified Method of Characteristics. *Comp. Meth. Appl. Mech. Eng.*, **47**: 73–92.

Glowinski, R. and Wheeler, M.F., (1988), Domain Decomposition and Mixed Finite Element Methods for Elliptic Problems. In: *Domain Decomposition Methods for Partial Differential Equations*, SIAM, Philadelphia, 144–172.

Hagood, M.C. and Rohay, V.J., (1991), 200 West Area Carbon Tetrachloride Expedited Response Action Project Plan, WHC-SD-EN-AP-046, Westinghouse Hanford Company, Richland, Washington.

Hansen, E.J., (1990), Transformation of Tetrachloromethane Under Denitrifying Conditions by a Subsurface Bacterial Consortium and its Isolates. M.S. Thesis, Washington State University, Pullman, Washington.

Kindred, J.S. and Celia, M.A., (1989), Contaminant Transport and Biodegradation 2. Conceptual Model and Test Simulations. *Water Resour. Res.*, **25**: 1149–1159.

Last, G.V., Lenhard, R.J., Bjornstad, B.N., Evans, J.C., Roberson, K.R., Spane, F.A., Amonette, J.E. and Rockhold, M.L., (1991), Characteristics of the Volatile Organic Compound-Arid Integrated Demonstration Site, PNL-7866, Pacific Northwest Laboratory, Richland, Washington.

Raviart, P.A. and Thomas, J.M., (1977), A Mixed Finite Element Method for Second Order Elliptic Problems. In: *Mathematical Aspects of the Finite Element Method*, Lecture Notes in Mathematics 606, Springer-Verlag, New York, 292–315.

Russell, T.F., Wheeler, M.F. and Chiang, C.Y., (1986), Large-Scale Simulation of Miscible

Displacement. In: *Proceedings of the SEG/SIAM/SPE Conference on Mathematical and Computational Methods in Seismic Exploration and Reservoir Modeling*, (ed. W.E. Fitzgibbon), SIAM, Philadelphia, 85–107.

Russell, T.F. and Wheeler, M.F., (1983), Finite Element and Finite Difference Methods for Continuous Flows in Porous Media. Chapter II, in: *The Mathematics of Reservoir Simulation*, (ed. R.E. Ewing), Frontiers in Applied Mathematics 1, SIAM, Philadelphia, 35-106.

Semprini, L., Hopkins, G.D., Janssen, D.B., Lang, M., Roberts, P.V. and McCarty, P.L., (1991), In-Situ Biotransformation of Carbon Tetrachloride Under Anoxic Conditions, EPA/2-90/060, U.S. EPA, Ada, Oklahoma.

Skeen, R.S., Roberson, K.R., Brouns, T.M., Petersen, J.N. and Shouche, M., (1992), In-situ Bioremediation of Hanford Groundwater. In: *Proceedings of the 1st Federal Environmental Restoration Conference*, Vienna, Virginia.

Thomas, J.M., Lee, M.D., Bedient, P.B., Borden, R.C., Canter, L.W. and Ward, C.H., (1987). Leaking Underground Storage Tanks: Remediation with Emphasis on *In Situ* Biorestoration. *Environmental Protection Agency* 600/2-87, 008.

United States Department of Energy, (1988), *Site-directed Subsurface Environmental Initiative, Five Year Summary and Plan for Fundamental Research in Subsoils and in Groundwater FY1989-1993*, April 1988, DOE/ER 03441, Office of Energy Research.

Weiser, A., and Wheeler, M.F., (1988), On Convergence of Block Centered Finite Differences for Elliptic Problems, *SIAM J. Numer. Anal.*, **25**: 351–375.

Wheeler, M.F., Dawson, C.N., Bedient, P.B., Chiang, C.Y., Borden, R.C. and Rifai, H.S., (1987), Numerical Simulation of Microbial Biodegradation of Hydrocarbons in Groundwater. In: *Proceedings of AGWSE/IGWMCH Conference on Solving Ground Water Problems with Models*, National Water Wells Association, 92–108.

Wheeler, M.F. and Dawson, C.N., (1988), An Operator-Splitting Method for Advection-Diffusion-Reaction Problems. In: *MAFELAP Proceedings*, (ed. J.R. Whiteman), 463–482, Academic Press, London.

Wood, B. and Dawson, C.N., (1993), Effects of Lag and Maximum Growth. In: *Contaminant Transport and Biodegradation Modeling*, to appear.

Chapter 15

Unstructured Grid Methods for High Speed Compressible Flows

K. Morgan*, J. Peraire**, J. Peiró*** and O. Hassan****

* Department of Civil Engineering, University College Swansea, Swansea, SA2 8PP
** Department of Aeronautical and Astronautics, Massachusetts Institute
of Technology, Cambridge, Massachusettes 02139, U.S.A.
*** Department of Aeronautics, Imperial College, London SW72BY, U.K.
**** CDR, Innovation Centre, University College, Swansea SA2 8PP, U.K.

15.1 INTRODUCTION

Over the past ten years, the views of the aerospace industry on the use of unstructured grid methods for the simulation of aerodynamic flows have changed considerably. In the early eighties, these methods were greeted with a large degree of suspicion. At that time, the main causes for concern were the perceived difficulty of generating unstructured grids of high quality and the question of the accuracy of the results of computations made on general unstructured grids. In the intervening period, researchers have made major strides in the areas of grid generation, accuracy and adaptivity so that, nowadays, most major aerospace companies have, or are working towards the acquisition of, an unstructured grid aerodynamics capability. In addition, it is already recognised by industrial users that the unstructured approach will have an important role to play in the development of any successful multi–disciplinary (i.e. fluid dynamics, electro–magnetics, structural mechanics, optimisation) analysis capability, see Shankar, Hall, Mohammadian and Rowell (1993).

In MAFELAP 1984, we described some early efforts on the development of a finite element based unstructured grid procedure for the solution of high speed aerodynamic flows see Zienkiewicz, Löhner and Morgan (1985). It is appropriate, therefore, that we should return to this subject at MAFELAP 1993 to review our current capability in this area and to demonstrate some of the characteristics of the approach which are now proving so attractive to industrial users.

We will begin by considering, briefly, unstructured tetrahedral grid generation, as this is of major concern in practical aerodynamic flow simulation, where the geometries involved are frequently extremely complicated in shape. A general framework for the development of algorithms for the solution of the flow equations on general unstructured triangular or tetrahedral grids will then be presented. The approach which will be followed employs an edge based representation of the grid. It will be shown that, as well as being a convenient mechanism for algorithm development, the

The Mathematics of Finite Elements and Applications
Edited by J. R. Whiteman © 1994 John Wiley & Sons Ltd

edge based representation has significant implications to the computational efficiency of the flow solvers, both in terms of computer time and memory requirements. Practical algorithms which have been utilised to date for the simulation of steady flows are generally based upon the use of explicit time stepping methods. In the structured grid context, more efficient methods have been developed which are based upon either explicit time stepping with multigrid acceleration (see Jameson (1983)) or implicit time stepping with line relaxation (see Chakravarthy (1984)). The implementation of such ideas within the unstructured grid environment is not straightforward and possible implementations are presented and the difficulties which need to be overcome are discussed. The paper concludes by considering the application of unstructured grid methods to the solution of truly unsteady problems.

15.2 UNSTRUCTURED GRID GENERATION

Our favoured approach to the problem of generating unstructured triangular or tetrahedral grids, over domains of arbitrary geometrical complexity, has been a form of the advancing front concept in which nodes and elements are simultaneously created (see Peraire *et al.* , (1987), Peraire *et al.* (1989)). This process begins with a discretisation of the computational boundaries and then the advancing front method is used to create elements within the computational domain. User specified grid parameters control the shape of the generated elements. The specification of the grid control function is accomplished by generating manually a coarse so–called background grid, of triangles or tetrahedra, to completely cover the computational domain. The values of the grid parameters are specified at the nodes of this grid and the functional variation of the parameters through the computational domain is obtained by linear interpolation over the background grid elements. The requirement of constructing an adequate background grid for complex two dimensional geometries and for most three dimensional geometries has proved to be a significant barrier to the successful use of the approach by the inexperienced user.

To alleviate these problems, the concept of the use of point and line sources has recently been added to the process of defining the variation of the grid parameters over the computational domain. With the location of the source specified, the nodal spacing δ defined by the source at location $\mathbf{x}$ is determined as

$$\delta(\mathbf{x}) = \begin{cases} \delta_1 & |\tilde{\mathbf{x}}| \leq R_1 \\ \delta_1 \exp\left(|\tilde{\mathbf{x}}| \, ln\left[2/(R_2 - R_1)\right]\right) & |\tilde{\mathbf{x}}| \geq R_1 \end{cases} \tag{15.1}$$

where $|\tilde{\mathbf{x}}|$ denotes the distance from $\mathbf{x}$ to the point or line source and δ_1, R_1 and R_2 are user–specified constants. The nodal spacing used by the grid generator at location $\mathbf{x}$ is taken as the minimum of the spacings defined at that location by the background grid and by all the active point and line sources. When simulating inviscid aerodynamic flows, it is known that suitable computational grids will be clustered in certain regions, for example near the leading and trailing edges of wings, while larger elements can be employed elsewhere. Grids of this type are readily generated by using a simple background grid supplemented with appropriate point and line sources. This is illustrated in Figure 15.1, which shows the surface grid generated on a typical wing when the background grid consists of a single tetrahedral element, supplemented by line sources defined to lie along the leading and trailing edges of the wing. The incorporation of this facility leads to a significant reduction in the time required to prepare the data necessary for the grid generation procedure.

15.3 EQUATIONS OF COMPRESSIBLE FLOW

The equations of interest here are the 3D unsteady compressible Euler equations, expressed in a cartesian coordinate system $Ox_1x_2x_3$ in the conservation form

$$\frac{\partial \mathbf{U}}{\partial t} + \frac{\partial \mathbf{F}^j}{\partial x_j} = 0 \qquad j = 1, 2, 3 \tag{15.2}$$

where the sumation convention is employed and where

$$\mathbf{U} = \begin{bmatrix} \rho \\ \rho u_1 \\ \rho u_2 \\ \rho u_3 \\ \rho\epsilon \end{bmatrix} \tag{15.3}$$

is the unknown vector of conservation variables. The inviscid flux vectors are defined by

$$\mathbf{F}^j = \begin{bmatrix} \rho u_j \\ \rho u_1 u_j + p\delta_{1j} \\ \rho u_2 u_j + p\delta_{2j} \\ \rho u_3 u_j + p\delta_{3j} \\ (\rho\epsilon + p)u_j \end{bmatrix} \tag{15.4}$$

Here p, ρ and ϵ denote the pressure, density and specific total energy of the fluid respectively, while the component of the fluid velocity in the direction x_j is denoted by u_j. The equation set is completed by the addition of the perfect gas equation of state

$$p = (\gamma - 1)\rho(\epsilon - 0.5u_m^2) \tag{15.5}$$

where γ denotes the ratio of the specific heats.

For the purpose of deriving algorithms for the solution of these equations in a region Ω bounded by a closed surface Γ, using general unstructured tetrahedral grids, it is convenient to start from the alternative weak variational formulation: find $\mathbf{U}(\mathbf{x}, t)$ such that

$$\int_\Omega \frac{\partial \mathbf{U}}{\partial t} W d\Omega = \int_\Omega \mathbf{F}^j \frac{\partial W}{\partial x_j} d\Omega - \int_\Gamma \bar{\mathbf{F}}^{(n)} W d\Gamma \tag{15.6}$$

for every suitable weighting function $\mathbf{W}(\mathbf{x})$ and every $t > 0$, subject to $\mathbf{U}(\mathbf{x}, 0) = \mathbf{U}_0(\mathbf{x})$. In equation (15.6), the quantity $\bar{\mathbf{F}}^{(n)}$ is defined by

$$\bar{\mathbf{F}}^{(n)} = \mathbf{F}^j n^j \tag{15.7}$$

                                        K. Morgan et al.

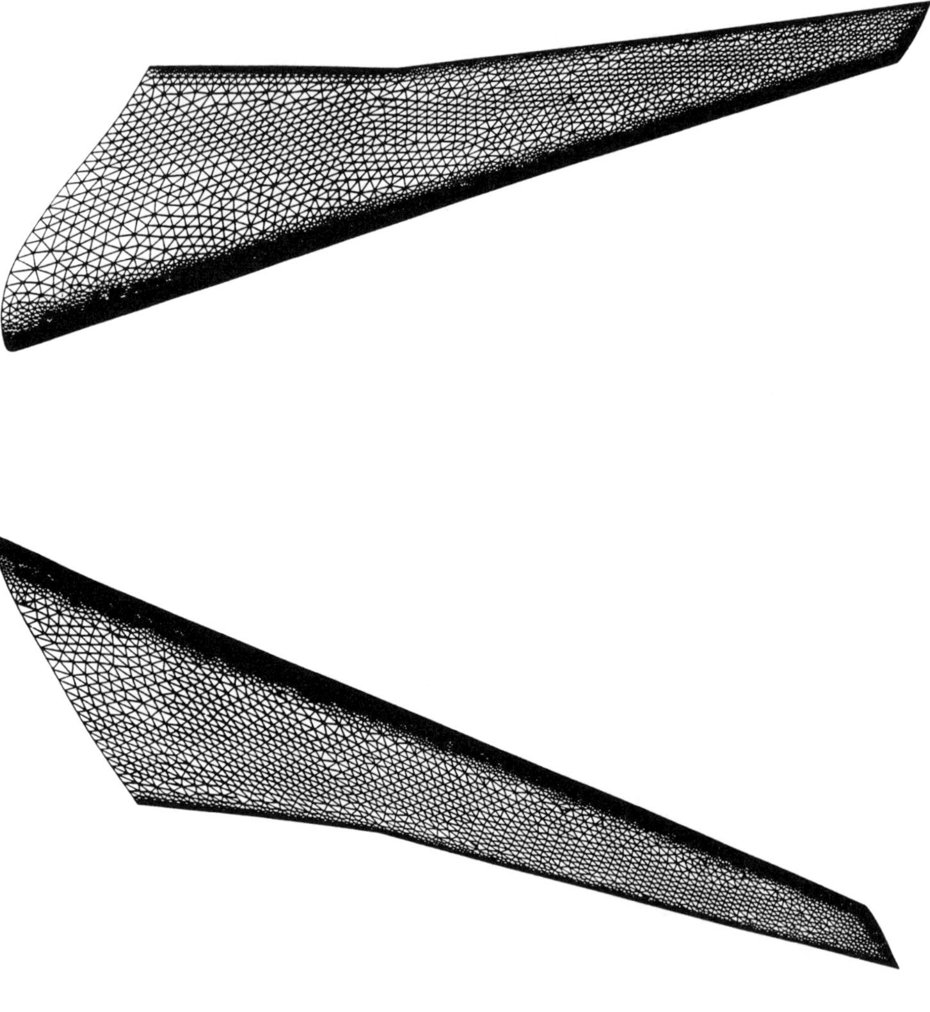

Figure 15.1 Discretisation of the surface of a wing

where $\mathbf{n} = (n_1, n_2, n_3)$ denotes the unit outward normal to Γ. The exact form of $\bar{\mathbf{F}}^{(n)}$ will depend on both the boundary condition being simulated and the local value of the solution.

15.4 UNSTRUCTURED GRID SOLUTION ALGORITHMS

The domain Ω is discretised using linear tetrahedral or triangular elements. In the standard finite element approach, a piecewise linear approximate solution $\mathbf{U}^*$ is constructed in the form

$$\mathbf{U}^* = \mathbf{U}_J N_J(\mathbf{x}) \tag{15.8}$$

where $N_J(\mathbf{x})$ denotes the piecewise linear finite element shape function associated with node J of the grid and $\mathbf{U}_J = \mathbf{U}_J(t)$ denotes the value of the approximate solution at node J. The nodal values are obtained from a standard Galerkin method, which is based upon the weak formulation of the problem in which the approximation is constructed by employing the shape functions themselves as the weighting functions (see Johnson, (1987)). Here, this is expressed as: find $\mathbf{U}^*$ such that

$$\int_\Omega \frac{\partial \mathbf{U}^*}{\partial t} N_I d\Omega = \int_\Omega \mathbf{F}^{j*} \frac{\partial N_I}{\partial x_j} d\Omega - \int_\Gamma \bar{\mathbf{F}}^{(n)} N_I d\Gamma \tag{15.9}$$

for every node I in the grid. The assumed form of $\mathbf{U}^*$ from equation (15.8) is inserted into this expression and the left hand side directly evaluated as

$$\int_\Omega \frac{\partial \mathbf{U}^*}{\partial t} N_I d\Omega = \mathbf{M}_{IJ} \frac{d\mathbf{U}_J}{dt} \tag{15.10}$$

where $\mathbf{M}_{IJ}$ denotes the entries in the standard consistent mass matrix. The right hand side terms in equation (15.9) are evaluated by making use of the local nature of the finite element shape functions. The process is illustrated here for the domain integral only. Suppose that each edge in the grid is numbered and that the information about the grid is given in terms of the nodes which are joined by these edges. Suppose also that edge S in the grid joins the interior node I with node S_1. If it is assumed that the fluxes are also linearly interpolated in terms of their nodal values i.e.

$$\mathbf{F}^{j*} \approx \mathbf{F}^j \left(\mathbf{U}_J\right) N_J(\mathbf{x}) = \mathbf{F}_J^j N_J(\mathbf{x}) \tag{15.11}$$

then it can be readily shown (see Peraire et $al.$ (1993)) that

$$\int_\Omega \mathbf{F}^{j*} \frac{\partial N_I}{\partial x_j} d\Omega = \sum_{S \epsilon I} C_{I,S_1}^j \left(\mathbf{F}_I^j + \mathbf{F}_{S_1}^j\right) \tag{15.12}$$

where C_{I,S_1}^j denotes the weight that must be applied to the sum of the edge nodal fluxes in order to obtain the contribution made to node I by the edge which joins nodes I and S_1. With an edge based representation for the grid, the discretised equation at an interior node I can therefore be written as

$$\mathbf{M}_{IJ} \frac{d\mathbf{U}_J}{dt} = \sum_{S \epsilon I} C_{I,S_1}^j \left(\mathbf{F}_I^j + \mathbf{F}_{S_1}^j\right) \tag{15.13}$$

It can be readily demonstrated from this expression that the flux at node I does not contribute to the discretised equation at node I, so that the application of the Galerkin method leads to a scheme which is central difference in character. In addition, the total contribution made by any interior edge in the grid is zero, so that the scheme can immediately be seen to be conservative in this sense.

To produce practical schemes for the simulation of compressible flows, we can employ the approach which is frequently adopted in 1D and replace the actual fluxes in equation (15.13) by appropriate consistent numerical fluxes. To achieve this in a convenient manner, we begin by defining the vector

$$\mathbf{C}_{I,S_1} = \left(C^1_{I,S_1}, C^2_{I,S_1}, C^3_{I,S_1}\right) \tag{15.14}$$

and the quantities

$$\mathcal{L}_{I,S_1} = \left|\mathbf{C}_{I,S_1}\right| \qquad \mathcal{S}^j_{I,S_1} = C^j_{I,S_1}/\mathcal{L}_{I,S_1} \tag{15.15}$$

Then, the equation at node I may be expressed as

$$\mathbf{M}_{IJ}\frac{d\mathbf{U}_J}{dt} = \sum_{S\varepsilon I} \mathcal{L}_{I,S_1}\mathbf{\Phi}_{I,S_1} \tag{15.16}$$

where

$$\mathcal{F}_I = \mathbf{F}^j_I \mathcal{S}^j_{I,S_1} \qquad \mathcal{F}_{S_1} = \mathbf{F}^j_{S_1} \mathcal{S}^j_{I,S_1} \tag{15.17}$$

are the actual fluxes, in the direction of the vector $\mathbf{C}_{I,S_1}$, associated with the two nodes belonging to the edge S and

$$\mathbf{\Phi}_{I,S_1} = \left(\mathcal{F}_I + \mathcal{F}_{S_1}\right) \tag{15.18}$$

denotes the actual flux associated with the edge S. Practical solution schemes for the compressible Euler equations on general unstructured meshes follow by replacing $\mathbf{\Phi}$ by a consistent $\tilde{\mathbf{\Phi}}$, which is constructed on the basis of 1D arguments.

15.4.1 CENTERED SCHEMES

Finite Element Lax–Wendroff Method

This is a simple finite element scheme for the solution of the compressible Euler equations on unstructured grids and is the finite element equivalent of the finite difference Lax–Wendroff method. For a problem involving only the spatial direction x_1, the Lax–Wendroff flux function is defined as (see Hirsch (1990))

$$\tilde{\mathbf{\Phi}} = \mathbf{\Phi} - \frac{\Delta t}{2h_S}\mathbf{A}_S\left(\mathbf{F}^1_{S_1} - \mathbf{F}^1_I\right) \tag{15.19}$$

where Δt is a time step, $\mathbf{A}_S = d\mathbf{F}^1/d\mathbf{U}$ is the Jacobian matrix of $\mathbf{F}^1$ and h_S is the length of the edge S. By direct extension to the multi–dimensional case, a Lax–Wendroff type scheme, for general tetrahedral grids, is written in the form

$$\mathbf{M}_{IJ}\Delta\mathbf{U}_J = \Delta t \sum_{S\varepsilon I} \mathcal{L}_{I,S_1}\tilde{\mathbf{\Phi}}^n_{I,S_1} \tag{15.20}$$

where

$$\tilde{\Phi} = \Phi - \frac{\Delta t}{2h_S} \mathbf{A}_S \left(\mathcal{F}_{S_1} - \mathcal{F}_I \right) \tag{15.21}$$

In equation (15.20),

$$\Delta \mathbf{U} = \mathbf{U}^{n+1} - \mathbf{U}^n \tag{15.22}$$

the superscript n denotes the time level, $\Delta t = t_{n+1} - t_n$ and $\mathbf{A}_S = d\mathcal{F}/d\mathbf{U}$. For the simulation of transient problems, equation (15.20) may be solved by explicit iteration (see Morgan and Peraire (1987)). For steady flow simulations, the consistent mass matrix $\mathbf{M}$ is replaced by the diagonal lumped mass matrix $\mathbf{M_L}$ to produce a truly explicit scheme.

A Multi–Stage Method
An approach which has proved to be extremely successful for the simulation of compressible transonic Euler flows is the use of a central difference discretisation in space together with the explicit addition of a stabilising scalar high order added artificial viscosity (see Jameson (1987)). For a problem involving only the spatial direction x_1, a suitable flux function for a scheme of this form is

$$\tilde{\Phi}_{I,S_1} = \Phi_{I,S_1} - d_4 \left| \lambda_S \right| \left[\mathbf{U}_{S_1} - \mathbf{U}_S - \left(h_S \frac{d\mathbf{U}^*}{d\eta} \right)_S \right] \tag{15.23}$$

where the coordinate η runs along the edge from node I to node S_1. In equation (15.23), d_4 denotes a user–specified constant and $\left| \lambda \right|$ is the maximum eigenvalue, in absolute value, of the Jacobian matrix $\mathbf{A}_S$. To implement this procedure, nodal gradients are calculated, using equation (15.12), in the form

$$[\mathbf{M_L}]_{II} \left. \frac{\partial \mathbf{U}^*}{\partial x_j} \right|_I = - \sum_{S \epsilon I} C^j_{I,S_1} \left(\mathbf{U}_I + \mathbf{U}_{S_1} \right) \tag{15.24}$$

with $j = 1$. The required edge gradients are obtained as the average of the edge nodal gradients. Expression (15.23) for the flux function can be shown to possess the property that the diffusion which is added is zero, on any grid, whenever $\mathbf{U}^*$ is linear. By direct extension, we employ the flux function of equation (15.23) directly for the multi–dimensional case on general tetrahedral grids.

The resulting equation can be advanced in time via the explicit multi–stage time stepping scheme

$$\mathbf{U}_I^{(0)} = \mathbf{U}_I^n$$
$$\mathbf{U}_I^{(k)} = \mathbf{U}_I^n + \alpha_k \Delta t \left[\mathbf{M_L}\right]_{II}^{-1} \sum_{S \epsilon I} \mathcal{L}_{I,S_1} \tilde{\Phi}_{I,S_1}^{(k-1)} \quad k = 1, \ldots, m \tag{15.25}$$
$$\mathbf{U}_I^{n+1} = \mathbf{U}_I^{(m)}$$

In an alterative implementation (see Swanson and Turkel (1990)), the scalar diffusion coefficient, $\left| \lambda_S \right|$, in equation (15.23) is replaced by a matrix coefficient, leading to the numerical flux function

$$\tilde{\Phi}_{I,S_1} = \Phi_{I,S_1} - d_4 \left| \mathbf{A}_S \left(\mathbf{U}_I, \mathbf{U}_{S_1} \right) \right| \left[\mathbf{U}_{S_1} - \mathbf{U}_S - \left(h_S \frac{d\mathbf{U}^*}{d\eta} \right)_S \right] \tag{15.26}$$

where $\left| \mathbf{A}_S \left(\mathbf{U}_I, \mathbf{U}_{S_1} \right) \right|$ denotes the standard Roe matrix (see Roe (1981)) evaluated in the direction $\mathbf{C}_{I,S_1}$.

Shock Capturing
Additional explicitly added low order artificial viscosity is needed if either of the above schemes is to be used successfully for the simulation of flows containing discontinuities. By comparison with the one dimensional case, it is apparent that the addition of diffusion in the multi–dimensional situation can be simply achieved by the use of the flux function

$$\tilde{\tilde{\mathbf{\Phi}}}_{I,S_1} = \tilde{\mathbf{\Phi}}_{I,S_1} - d_2 \pi_S \left| \lambda_S \right| [\mathbf{U}_{S_1} - \mathbf{U}_I] \tag{15.27}$$

where d_2 is a user defined constant. To maintain the accuracy of the original schemes in the smooth regions of the flow, an edge pressure switch π_S has been built into the definition of this additional diffusion. This pressure switch is designed to be significant only near any flow discontinuities and to be very small in the smooth flow regions (see Peraire *et al.* (1987)). The alternative expression

$$\tilde{\tilde{\mathbf{\Phi}}}_{I,S_1} = \tilde{\mathbf{\Phi}}_{I,S_1} - d_2 \pi_S \left| \mathbf{A}_S \left(\mathbf{U}_I, \mathbf{U}_{S_1} \right) \right| [\mathbf{U}_{S_1} - \mathbf{U}_I] \tag{15.28}$$

in which the scalar diffusion coefficient in equation (15.27) is replaced by a matrix diffusion coefficient, may also be employed here.

Higher Resolution via Flux Corrected Transport
Flux corrected transport (FCT) is a technique introduced originally (see Boris and Book (1973), Zalesak (1979)) to produce high resolution monotone solutions by combining the results of a low order monotone scheme with those of a higher order scheme. Unstructured grid implementations have proved successful (see Parrott and Christie (1986), Morgan *et al.* (1988)), particularly for transient simulations. The approach only needs slight modification for implementation in the curent context. For example, for the Lax–Wendroff type scheme, let $\Delta \mathbf{U}^H$ be the basic (high order) solution increment and let $\Delta \mathbf{U}^L$ be the (low order) increment computed with the same algorithm but with the addition of the unswitched (i.e. $\pi_S = 1$) shock capturing diffusion of equation (15.27). If this added diffusion is denoted by $\mathbf{D}$, it follows that these increments are related by

$$\Delta \mathbf{U}^H = \Delta \mathbf{U}^L - \mathbf{D} \tag{15.29}$$

It is apparent from equation (15.27) that the diffusion $\mathbf{D}$ is formed as the sum of edge contributions i.e.

$$\mathbf{D} = \sum_{S \epsilon I} \mathbf{D}_{I,S_1} \tag{15.30}$$

The idea behind FCT is to limit the amount of artificial diffusion which is removed from the low order solution to produce the high order solution in equation (15.29) by imposing the requirement that the high order solution should remain monotone. This is achieved by firstly identifying maximum and minimum allowable values for the solution at node I. We define

$$\mathbf{U}^H = \mathbf{U}^n + \Delta \mathbf{U}^H \qquad\qquad \mathbf{U}^L = \mathbf{U}^n + \Delta \mathbf{U}^L \tag{15.31}$$

and the maximum and minimum allowable nodal values are determined as

$$\mathbf{U}_I^{max} = \max_{S \epsilon I} \left(\mathbf{U}_I^L, \mathbf{U}_{S_1}^L, \mathbf{U}_I^n, \mathbf{U}_{S_1}^n \right)$$
$$\mathbf{U}_I^{min} = \min_{S \epsilon I} \left(\mathbf{U}_I^L, \mathbf{U}_{S_1}^L, \mathbf{U}_I^n, \mathbf{U}_{S_1}^n \right) \tag{15.32}$$

Edge limiters ψ_S are determined so as to ensure that the increment computed according to

$$\Delta \mathbf{U} = \Delta \mathbf{U}^L - \sum_{S \epsilon I} \psi_S \mathbf{D}_{I,S_1} \tag{15.33}$$

is such that the new solution at each node lies within the range specified by these maximum and minimum values. In practice, the edge limiter is determined from the requirement that variables such as density, pressure and Mach number should remain monotone.

As well as producing high resolution solutions, the FCT procedure acts as a stabilising influence on centered schemes when they are applied in the hypersonic range. Thus, with the addition of FCT, the explicit multi–stage time stepping scheme has been employed to solve the problem of steady flow over a space shuttle at a Mach number of ten and at thirty degrees angle of attack. A view of the surface grid used in the simulation is shown in Figure 15.2a and the computed distribution of the Mach number contours can be seen in Figure 15.2b. The complete unstructured grid, for one half of the configuration, consisted of 1,462,189 elements and 269,780 nodes.

Upwind Methods

By similar direct extension of schemes developed for the solution of the compressible Euler equations in 1D, a first order upwind method for general unstructured grids can be constructed by the use of the flux function

$$\tilde{\mathbf{\Phi}}_{I,S_1} = \mathbf{\Phi}_{I,S_1} - \left| \mathbf{A}_S \left(\mathbf{U}_I, \mathbf{U}_{S_1} \right) \right| \left[\mathbf{U}_{S_1} - \mathbf{U}_I \right] \tag{15.34}$$

Higher order methods can be derived by employing MUSCL (see van Leer (1979)) or TVD concepts (see Harten (1983, 1984)).

In the MUSCL scheme (see Vahdati *et al.* (1992)), the gradient of the solution is computed at the nodes, as in equation (15.24), and the solution is linearly reconstructed for each edge. Slope limiting is applied to ensure that the reconstructed solution does not go above (below) the maximum (minimum) of the value of the solution at the nodes of all elements associated with the two nodes of the edge. With a reconstructed solution, the scheme employs the numerical flux function

$$\tilde{\mathbf{\Phi}}_{I,S_1} = \left(\mathcal{F}_I^+ + \mathcal{F}_{S_1}^- \right) - \left| \mathbf{A}_S \left(\mathbf{U}_I^+, \mathbf{U}_{S_1}^- \right) \right| \left[\mathbf{U}_{S_1}^- - \mathbf{U}_I^+ \right] \tag{15.35}$$

where the superscripts $+$ and $-$ denote mid–edge states, which are determined from the reconstructed solution.

In a TVD implementation (see Lyra *et al.* (1993)), points S_2 and S_3 are located in the grid such that (a) S_2 and S_3 lie on the line of the edge S (b) S_2 and S_1 are equidistant from I (c) S_2 lies on the opposite side of the point I to the point S_1 (d) I and S_3 are equidistant from S_1 (c) S_3 lies on the opposite side of the point S_1 to the point I. The TVD flux is computed as (see Yee (1989))

$$\tilde{\mathbf{\Phi}}_{I,S_1} = \mathbf{\Phi}_{I,S_1} - \mathbf{R}^{-1} \left| \mathbf{\Lambda} \right| \mathbf{Q} \tag{15.36}$$

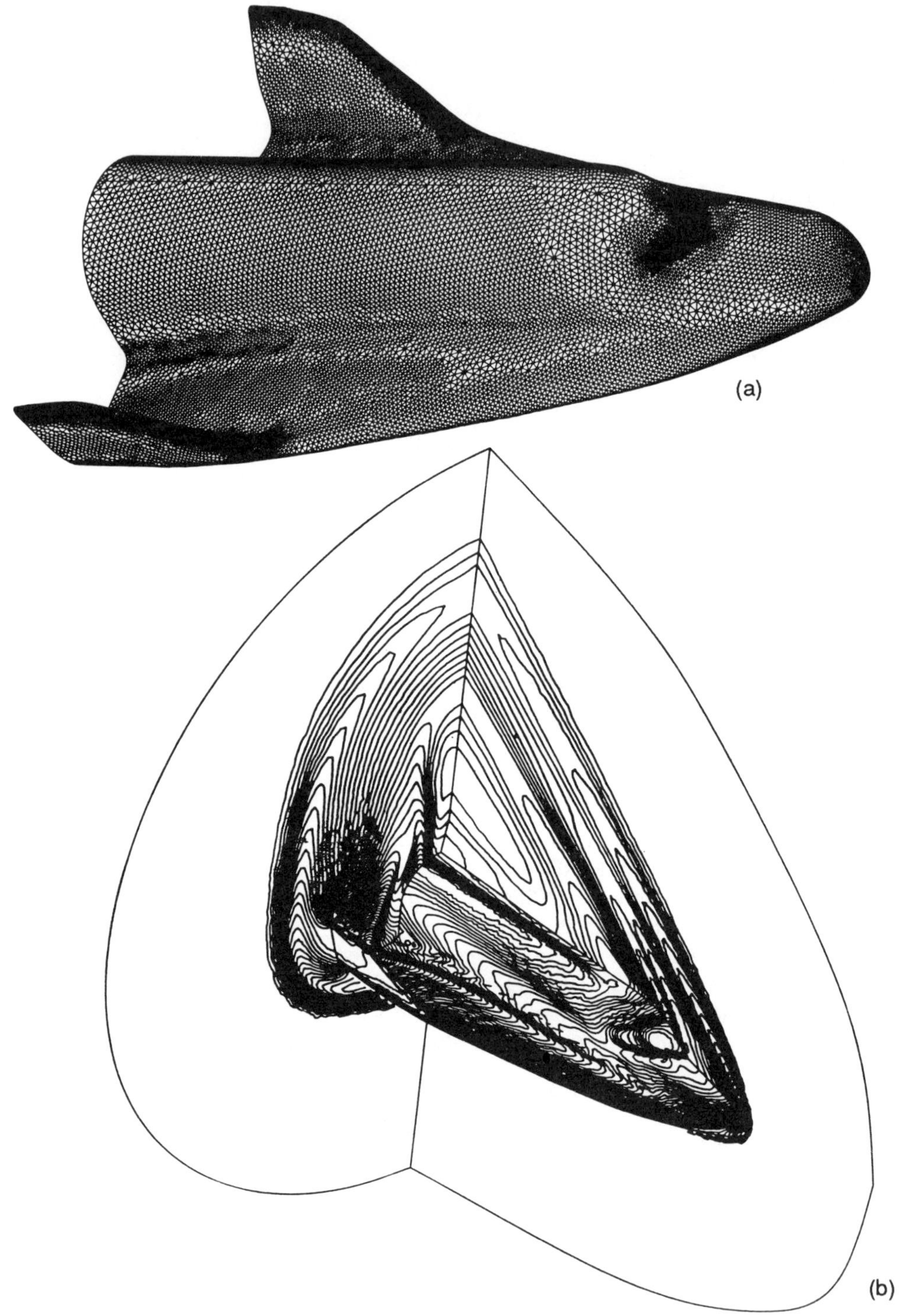

Figure 15.2 Flow over a shuttle at a Mach number of ten and thirty degrees angle of attack showing (a) surface discretisation (b) computed Mach number contours

where the Roe matrix is decomposed according to

$$\left| \mathbf{A}_S \right| = \mathbf{R}^{-1} \left| \mathbf{\Lambda} \right| \mathbf{R} \tag{15.37}$$

In this expression $\mathbf{R}$ denotes the matrix whose columns are the right eigenvectors of the Roe matrix and $\left| \mathbf{\Lambda} \right|$ is the diagonal matrix whose entries are the absolute values of the eigenvalues of $\mathbf{A}_S$. In equation (15.36),

$$\mathbf{Q} = \psi_{I,S_1} - \mathrm{minmod}\left(\psi_{S_2,I}, \psi_{I,S_1}, \psi_{S_1,S_3}\right) \tag{15.38}$$

where

$$\psi_{I,S_1} = \mathbf{R}\left(\mathbf{U}_{S_1} - \mathbf{U}_I\right) \tag{15.39}$$

With either of these schemes, the solution can be advanced in time to steady state by any suitable time stepping procedure.

15.4.2 GRID DATA STRUCTURE

For a general 3D grid of linear tetrahedral elements, the number of elements is approximately 5.5 times the number of nodes, the number of triangular faces is approximately twice the number of elements and the number of edges is approximately equal to the number of elements plus the number of nodes. Using these relations, it is possible to show that, in addition to being useful for the creation of flow algorithms, the adoption of the edge based representation for the unstructured grid has important computational implications.

Table 15.1 Grid Data Structures

Data Structure	Storage/Node	cpu/node
Element	25*	1.0*
	71.5	0.88
Face	58*	1.4*
	88	1.2
Edge	32.5	0.6

*With recomputation of the geometry

This is illustrated in Table 15.1, which compares the storage and cpu requirements for an explicit centered unstructured grid solution algorithm for the solution of the compressible Euler equations. The grid data structures which are considered in this table are the traditional finite element data structure, the face data structure, which is frequently favoured by the developers of finite volume algorithms, and the edge data structure, which has been considered in detail above. It can be observed that the computer code based upon the edge data structure for the grid is the most efficient in terms of cpu requirements and is close to being optimal in terms of its storage demands.

15.5 CONVERGENCE ACCELERATION TECHNIQUES

The solution of complex steady 3D flows on unstructured grids is most commonly
achieved by explicit time stepping. Although this approach is robust, it can prove
expensive in terms of cpu requirements, as a large number of time steps is frequently
needed to produce a converged solution. There is therefore the requirement to
search for more effective solution approaches, especially with the perceived increasing
industrial demand for the development of a capability for the solution of large scale
viscous flow problems.

15.5.1 MULTIGRID METHODS

Multigrid acceleration coupled with an explicit multi–stage time stepping scheme has
proved to be an extremely effective method for the solution of the compressible Euler
equations on structured grids (see Jameson (1983)). The basic idea of the multigrid
method is to use the residuals computed on a given grid to drive a time marching
scheme on a coarse grid. The corrections to the unknowns computed on the coarser
grid are then added to the fine grid solution. The time marching scheme on the coarse
grid may itself be accelerated by the use of an even coarser grid and, in this way,
the multigrid concept can be extended to incorporated any number of coarse grids.
Advancing the solution on the coarse grids is computationally relatively inexpensive,
because of the reduced number of unknowns and the larger allowable time step size.
The necessary intergrid transfers are readily accomplished because of the nested nature
of the grid sequence.

The development of appropriate intergrid transfer operators is a complication that
has to be addressed when attempting to implement a similar multigrid acceleration
scheme within the unstructured grid environment (see Mavriplis (1988), Mavriplis
(1991), Peraire, Morgan and Peiró (1993)).

A multigrid approach has been applied to the simulation of the flow over a model
of a twin–engined aircraft. A sequence of three grids was employed and the surface
triangulations are shown in Figure 15.3.

The corresponding volume grids consisted of 787568, 262664 and 172066 tetrahedra.
The computed steady state distribution of the pressure contours on the surface of the
aircraft when the free stream Mach number is 0.81 and the angle of attack is 2.739
degrees is shown in Figure 15.4. The predicted surface pressure distributions have
been shown to be in good agreement with experimental observations (see Peraire,
Peiró and Morgan (1993)). For this example, the grid generation and the flow solution
were performed on an IBM RS 6000/530H workstation.

The sequence of grids was generated in approximately seven cpu hours and the flow
solution required almost eight cpu hours.

15.5.2 IMPLICIT SCHEMES

The use of implicit timestepping procedures in the context of any of the above
algorithms leads to an equation system of the form

$$\mathbf{K} \, \Delta\mathbf{U} = \mathbf{R} \tag{15.40}$$

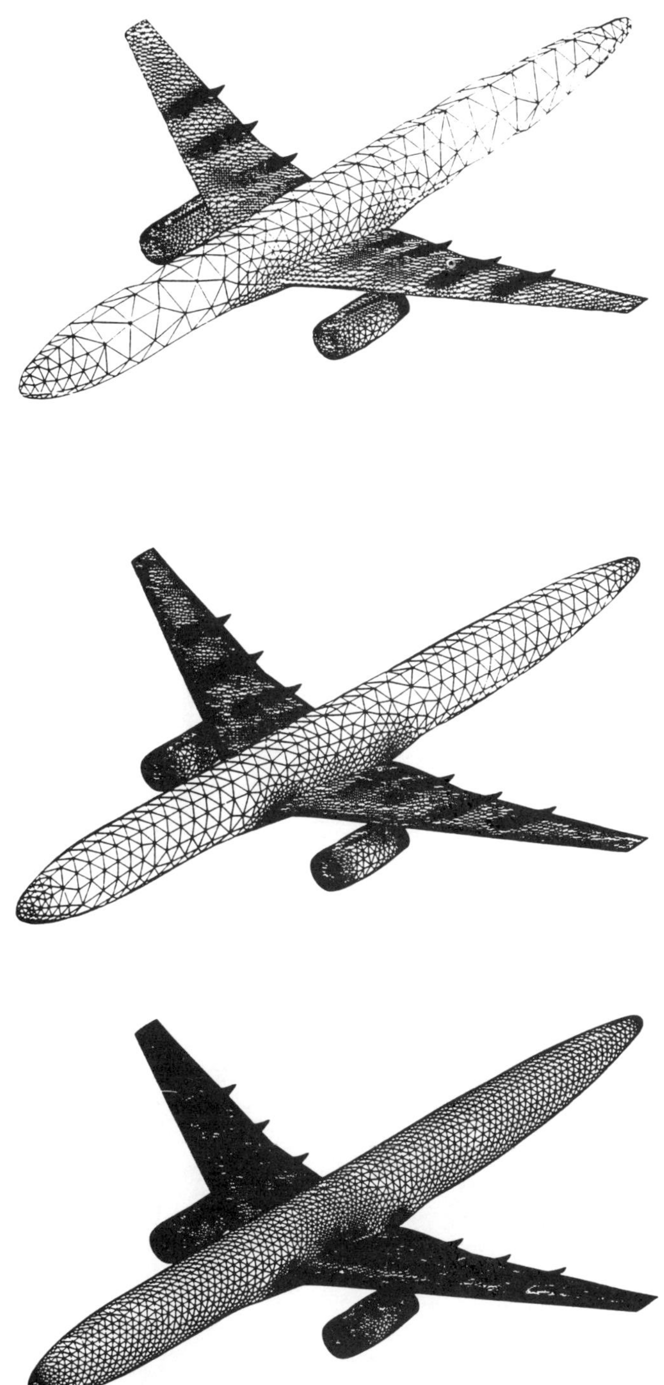

Figure 15.3 Inviscid flow over a twin–engined aircraft at a Mach number 0.81 and at an angle of attack of 2.739 degrees showing the surface triangulations of (a) the fine grid (b) the intermediate grid (c) the coarse grid.

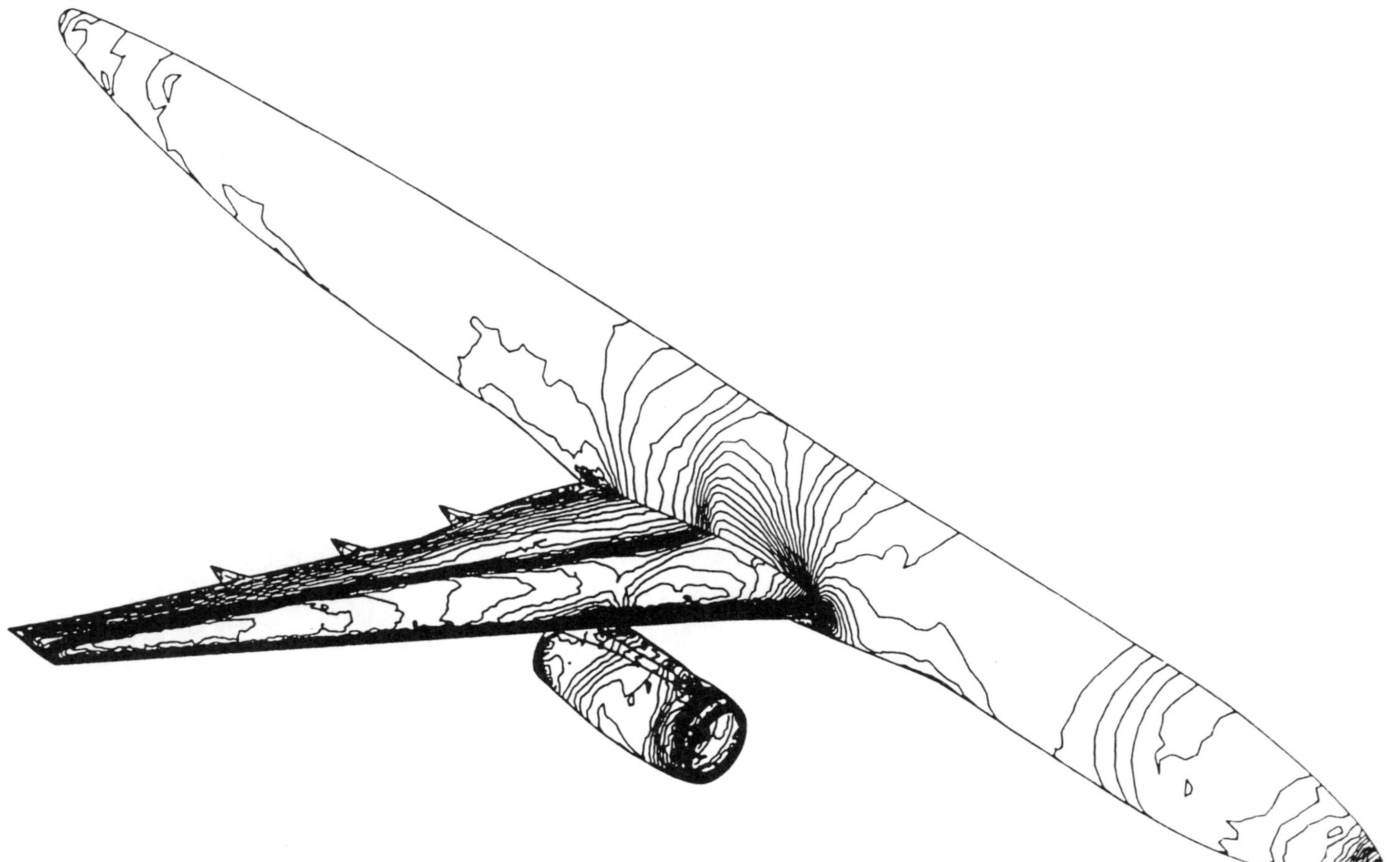

Figure 15.4 Inviscid flow over a twin–engined aircraft at a Mach number 0.81 and at an angle of attack of 2.739 degrees showing the computed surface pressure contour distribution.

which has to be solved at each time step. In many structured grid algorithms, the solution of equation systems of this form is achieved by line relaxation (see Chakravarty (1984), Thomas and Walters (1985)). The lines employed are easily identified by making use of the grid structure. However, with an unstructured grid, the identification of continuous lines which each pass through each node in the mesh once and only once is not a trivial problem. We have shown how this may be accomplished on a triangular mesh (see Hassan *et al.* (1991)) by using some of the ideas employed in the advancing front mesh generation method. Recently, this approach has been successfully extended to general tetrahedral meshes (see Hassan *et al.* (1993)).

When a number of such lines have been identified, the solution of equation (15.40) by a line relaxation procedure may be contemplated. The process begins by renumbering the nodes in the order implied by the line, so that the equation system takes the form

$$\hat{\mathbf{K}}\,\Delta\hat{\mathbf{U}} = \hat{\mathbf{R}} \tag{15.41}$$

and the matrix $\hat{\mathbf{K}}$ is decomposed and expressed as

$$\hat{\mathbf{K}} = \mathbf{D} + \left(\hat{\mathbf{K}} - \mathbf{D}\right) \tag{15.42}$$

where $\mathbf{D}$ is the block tri–diagonal of $\hat{\mathbf{K}}$. The solution of equation (15.41) is then sought via the iterative scheme

$$\mathbf{D}\,\Delta\hat{\mathbf{U}}^{(r+1)} = \hat{\mathbf{R}} - \left(\hat{\mathbf{K}} - \mathbf{D}\right)\Delta\hat{\mathbf{U}}^{(r)} \tag{15.43}$$

where r denotes the iteration count. For 3D simulations, a good solution strategy is to employ three lines, with a prescribed number of such iterations being used on each line during every time step. This approach has been used in the simulation of the inviscid flow over an ONERA M6 wing at a free stream Mach number of 0.84 and at 3.06 degrees angle of attack.

A detail of the surface mesh on the wing and on the symmetry plane is shown in Figure 15.5a. The complete volume mesh for this example contained 231,507 tetrahedral elements with 42,410 nodes. The solution was computed using both the standard explicit Lax–Wendroff type method described above and an equivalent implicit version. The implicit equation system was solved by using three lines, with one iteration along each line used during every time step. A view of one of these lines is given in Figure 15.5b.

In the implicit solution the Courant number was five times greater than that which was used with the explicit method. The computed pressure contour distribution on the wing surface is shown in Figure 15.6a. An indication of the numerical performance of the relaxation scheme can be gained from Figure 15.6b, which compares the convergence behaviour of the implicit and explicit schemes for this problem.

15.6 TRANSIENT PROBLEMS

Although the time dependent form of the equations has been employed in this paper, the main topic of interest has been the simulation of steady flows where the solution

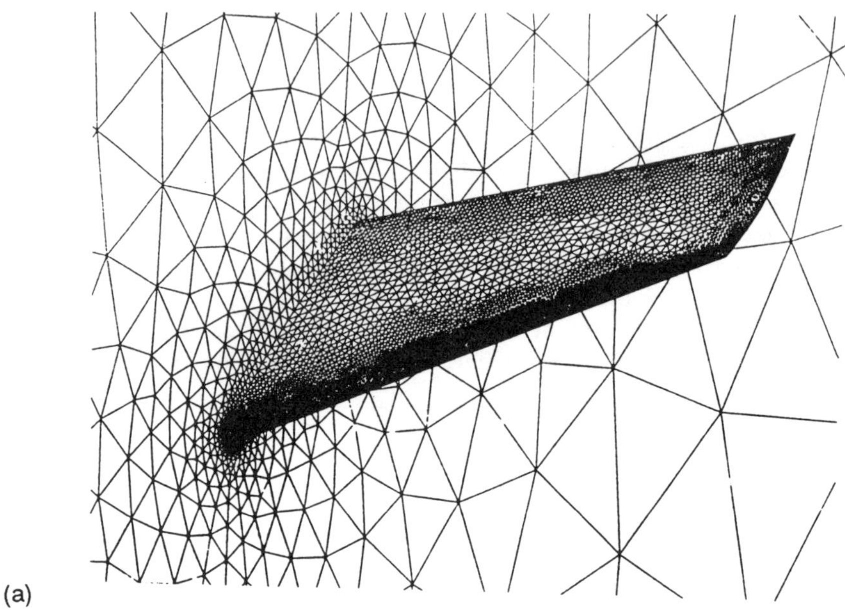

(a)

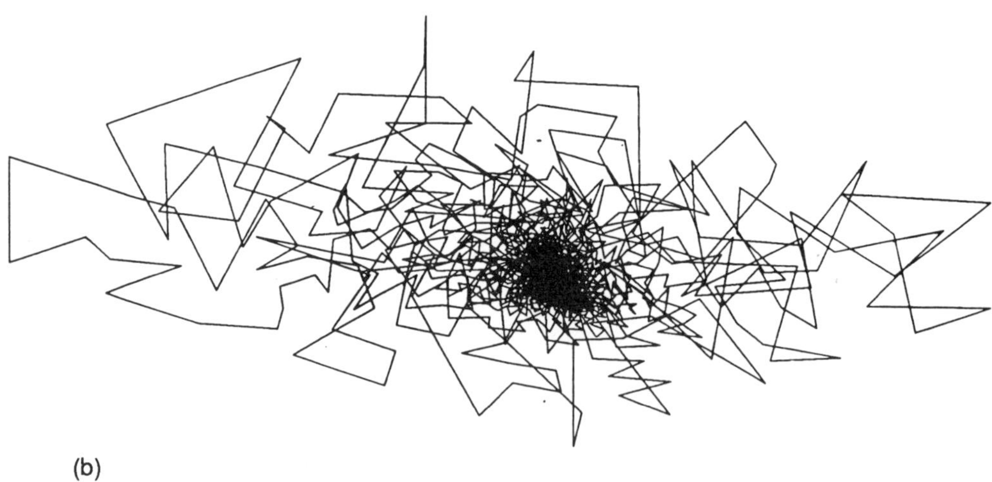

(b)

Figure 15.5 Inviscid flow over an ONERA M6 wing at a Mach number 0.84 and 3.06 degrees angle of attack showing (a) surface discretisation (b) one of the lines used for relaxation.

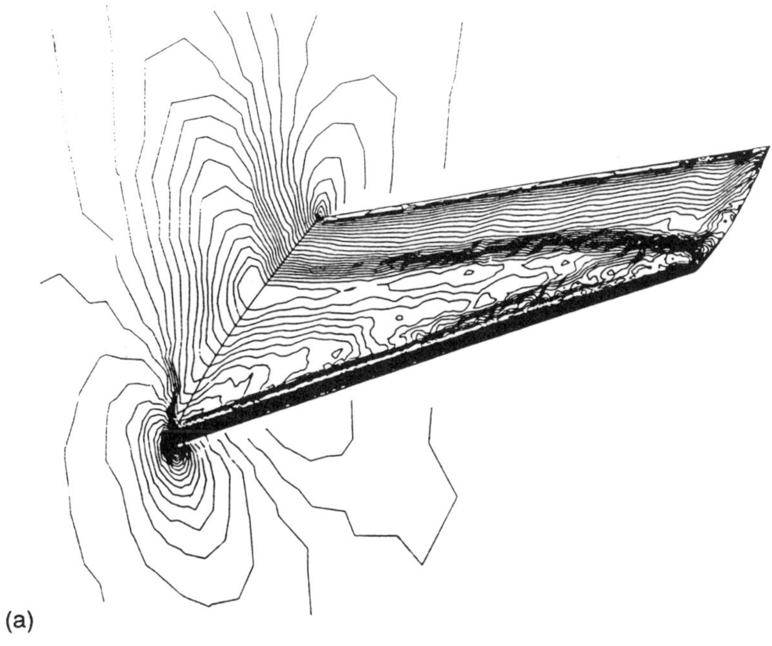

(a)

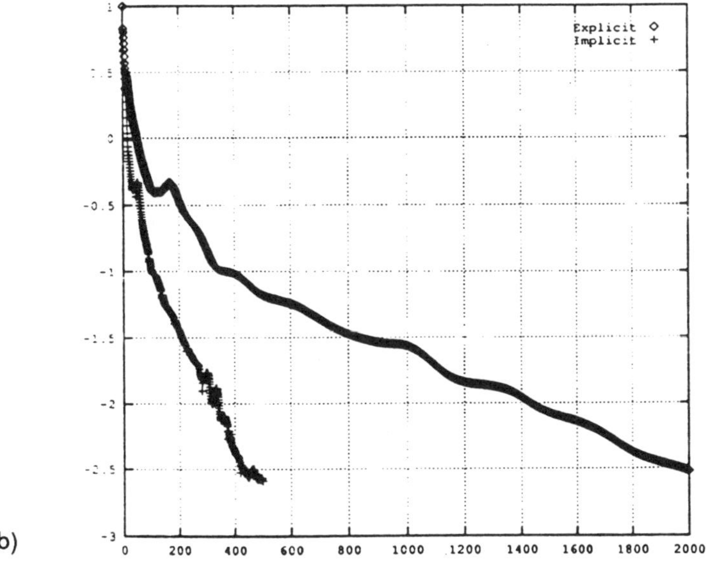

(b)

Figure 15.6 Inviscid flow over an ONERA M6 wing at a Mach number 0.84 and 3.06 degrees angle of attack showingn (a) computed pressure contours (b) convergence of the implicit and explicit schemes.

has been obtained via a false transient. We conclude by considering two applications in which the computation of the development of the true transient is important. In the first example the problem is solved by explicit time stepping while an implicit method is used for the second example.

15.6.1 ELECTROMAGNETIC SCATTERING

We consider the solution of the Maxwell's equations governing the scattering of an electromagnetic wave by an obstacle which is a perfect conductor. Although the governing equations are now different, Maxwell's equations can be expressed in a conservation form when a solution in the time domain is attempted (see Taflove (1992)) and the solution techniques described above can again be employed. The equations are suitably scaled and considered in the form

$$\frac{\partial \mathbf{E}}{\partial t} = curl\ \mathbf{H}$$
$$\frac{\partial \mathbf{H}}{\partial t} = -curl\ \mathbf{E}$$

(15.44)

where $\mathbf{E}$ and $\mathbf{H}$ denote the electric and magnetic field vectors respectively. By considering the individual components of these vector equations, it follows that this pair of equations can be expressed in the conservation form of equation (15.2), where now the unknown vector is

$$\mathbf{U} = \begin{pmatrix} E_1 \\ E_2 \\ E_3 \\ H_1 \\ H_2 \\ H_3 \end{pmatrix}$$

(15.45)

and the flux vectors are defined according to

$$\mathbf{F}^1 = \begin{pmatrix} 0 \\ H_3 \\ -H_2 \\ 0 \\ -E_3 \\ E_2 \end{pmatrix} \qquad \mathbf{F}^2 = \begin{pmatrix} -H_3 \\ 0 \\ H_1 \\ E_3 \\ 0 \\ -E_1 \end{pmatrix} \qquad \mathbf{F}^3 = \begin{pmatrix} H_2 \\ -H_1 \\ 0 \\ -E_2 \\ E_1 \\ 0 \end{pmatrix}$$

(15.46)

The total $\mathbf{E}$ and $\mathbf{H}$ fields are regarded as being composed of incident and scattered components. The incident fields satisfy the above equations and, hence, so do the scattered fields. The approach which has been adopted has then been to solve Maxwell's equations for the scattered field only. At the surface of the scatterer, the boundary condition is the vanishing of the tangential component of the total $\mathbf{E}$ field and the normal component of the total $\mathbf{H}$ field. The infinite region surrounding the scatterer is represented by a finite computational domain. The correct boundary condition which should be imposed at the computational far field boundary is that the scattered field consist only of outgoing waves. This condition is achieved by a simple filtering technique (see Givoli (1992)) which involves the explicit addition of diffusion to the governing equations in the vicinity of the far field boundary.

The equations are advanced in time, using the explicit Taylor Galerkin type scheme (see Morgan *et al.* (1993)), until 'steady' conditions are achieved. Far field scattering data is obtained from the results of the time domain computation by employing a near–field to far–field transformation (see Balanis (1989)).

This computational approach has been validated by considering the simulation of scattering of a plane TE wave by an infinite circular cylinder. The incident wave is such that $\beta L = 3\pi$, where β is the phase constant and L is the cylinder radius. The computational domain and the mesh employed, consisting of 68,612 elements and 34,632 nodes, are shown in Figure 15.7a. A detail of the computed amplitude of the scattered H_z field is shown in Figure 15.7b and the exact and computed distributions of the radar cross section are compared in Figure 15.7c.

15.6.2 FLOW OVER AN OSCILLATING WING

The spatial discretisation methods for the Euler equations which have been considered above lead to a discretised equation system

$$\frac{d\mathbf{U}}{dt} = \mathbf{R}\,(\mathbf{U})$$

(15.47)

Here we use an implicit three–level scheme in time (see Jameson (1992))

$$\frac{3\,\mathbf{U}^{n+1} - 4\,\mathbf{U}^n + \mathbf{U}^{n-1}}{2\,\Delta t} = \mathbf{R}\left(\mathbf{U}^{n+1}\right)$$

(15.48)

to advance the solution. Equation (15.48) can be re–arranged to give

$$0 = \mathbf{R}^*\left(\mathbf{U}^{n+1}\right) + \mathbf{S}\left(\mathbf{U}^n, \mathbf{U}^{n-1}\right)$$

(15.49)

where

$$\mathbf{R}^*\left(\mathbf{U}^{n+1}\right) = \mathbf{R}\left(\mathbf{U}^{n+1}\right) - \frac{3}{2\,\Delta t}\mathbf{U}^{n+1}$$

(15.50)

and

$$\mathbf{S}\left(\mathbf{U}^n, \mathbf{U}^{n-1}\right) = \frac{2\,\mathbf{U}^n}{\Delta t} - \frac{\mathbf{U}^{n-1}}{2\Delta t}$$

(15.51)

Equation (15.49) can then be solved by using an explicit multi–stage time stepping scheme to obtain the steady state solution of the equation

$$\frac{d\mathbf{U}^{n+1}}{dt} = \mathbf{R}^*\left(\mathbf{U}^{n+1}\right) + \mathbf{S}\left(\mathbf{U}^n, \mathbf{U}^{n-1}\right)$$

(15.52)

Again, as mentioned above, multigrid acceleration can be employed to accelerate the convergence of the explicit time stepping.

The numerical performance of the proposed procedure is demonstrated by considering the transient development of the flow over an oscillating ONERA M6 wing. In the computation, the wing is held fixed and its sinusoidal oscillation is simulated by modifying the boundary condition at the wing surface.

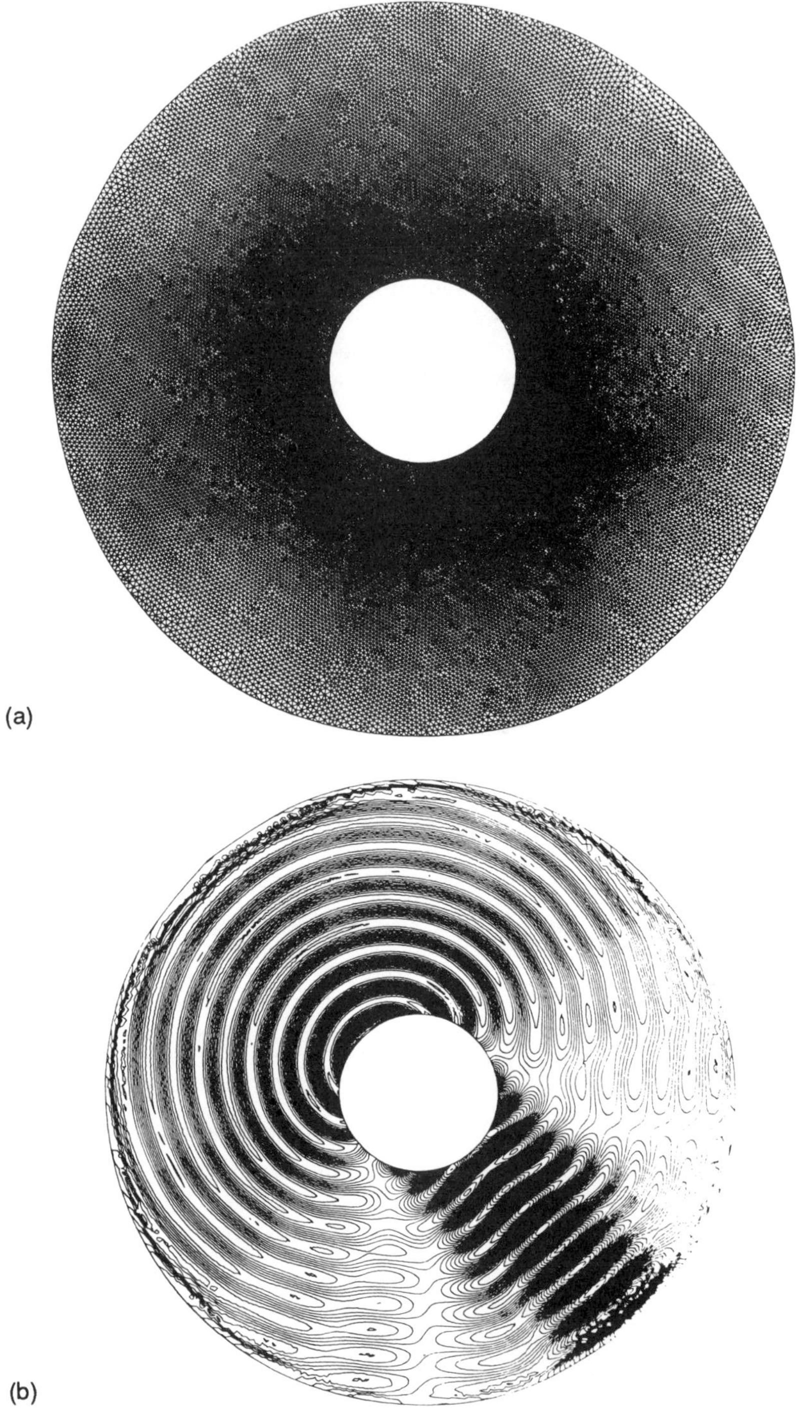

(a)

(b)

Figure 15.7a and b Scattering of a plane TE wave by an infinite circular cylinder showing (a) computational domain and the mesh (b) detail of the scattered H_z field .

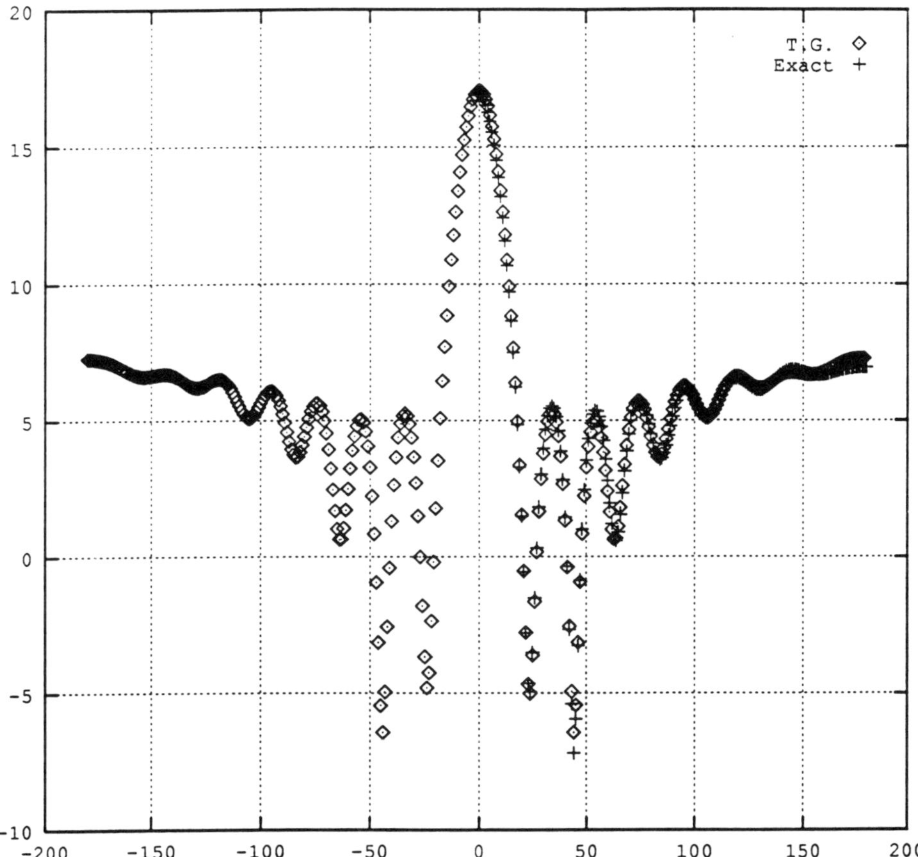

Figure 15.7c Comparison between the exact and numerically determined radar cross section.

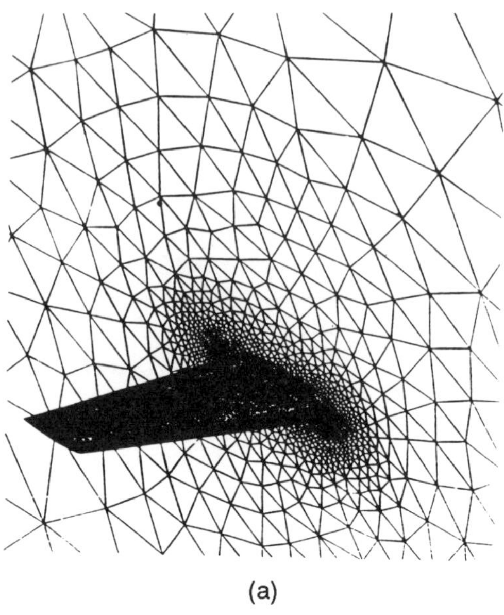

(a)

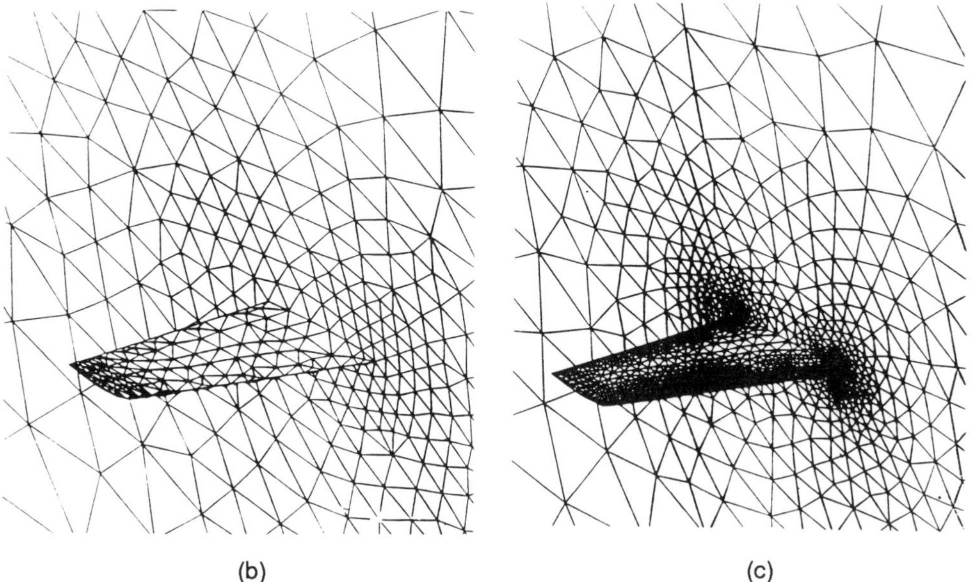

(b)   (c)

Figure 15.8 Oscillation of an ONERA M6 wing showing the surface discretisation of (a) the finest grid (b) an intermediate grid (c) the coarsest grid

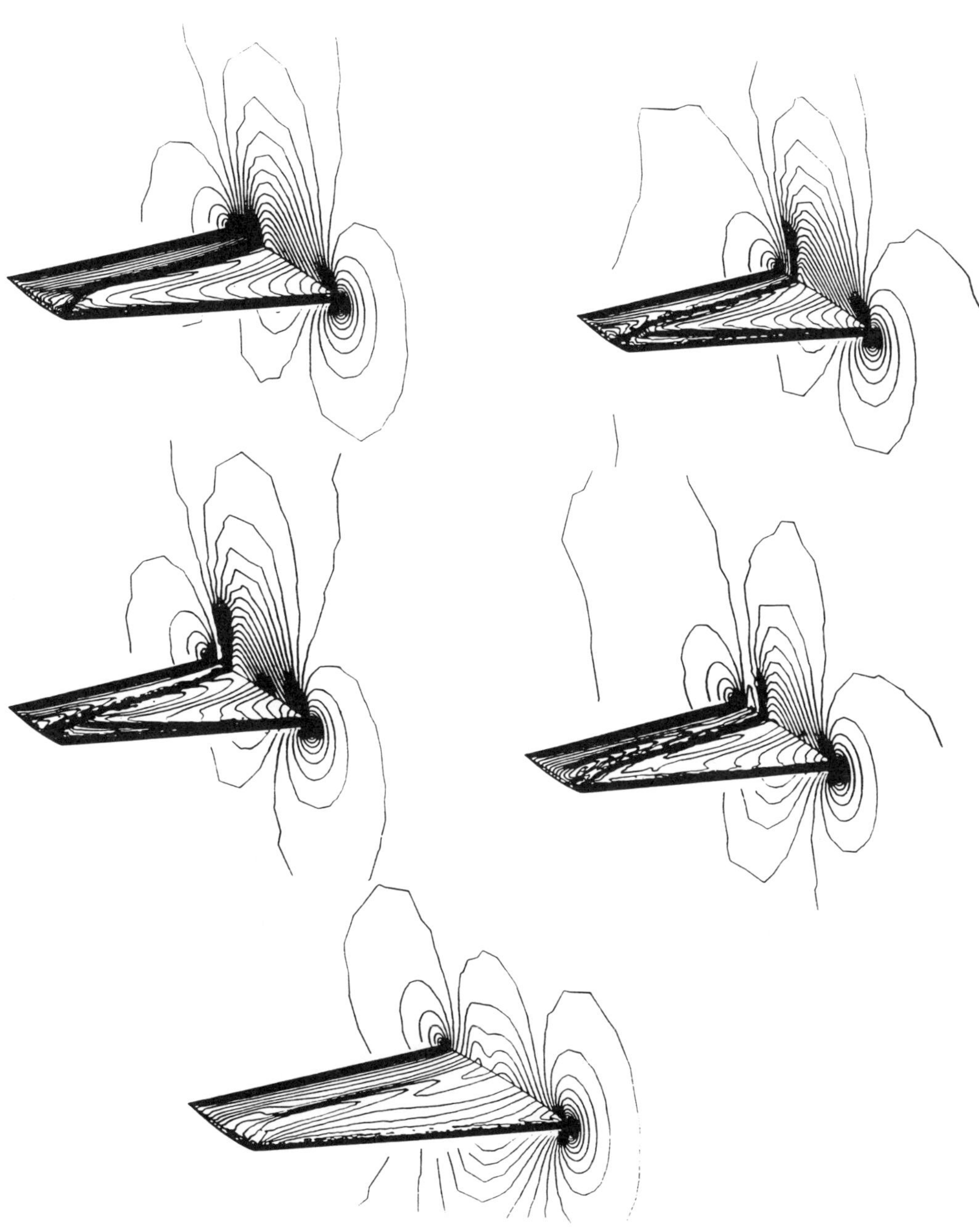

Figure 15.9 Oscillation of an ONERA M6 wing showing the distribution of the pressure contours on the upper surface of the wing during the initial portion of the cycle

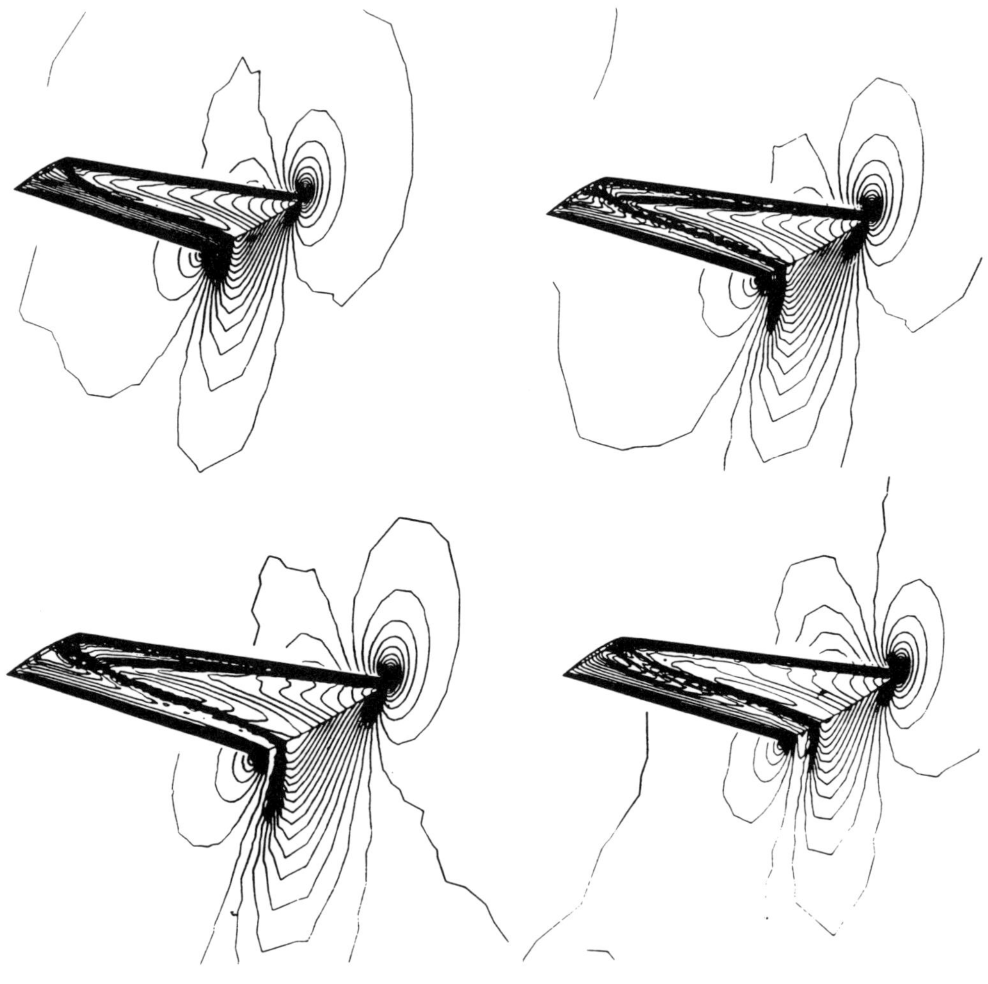

Figure 15.10 Oscillation of an ONERA M6 wing showing the distribution of the pressure contours on the lower surface of the wing during the latter portion of the cycle

This approach can be expected to be valid for oscillations of small amplitude. Here the simulation involved an amplitude of four degrees and 40 multigrid cycles were used per timestep with 24 timesteps taken per cycle. A sequence of five meshes was employed, with the surface discretisations employed being indicated in Figure 15.8.

The distribution of the pressure on the upper surface of the wing during the initial portion of the cycle is shown in Figure 15.9. The distribution of the pressure on the lower surface of the wing during the remainder of the cycle is shown in Figure 15.10.

15.7 CONCLUSIONS

We have reviewed our current unstructured mesh capability for the solution of inviscid aerodynamic flows. It has been shown that the procedures developed can also be used to simulate electromagnetic scattering problems in the time domain. The limits placed on the paper size mean that important developments in adaptive methods for aerodynamic flows have not been included. The advancement in the technology in this area since MAFELAP 1984 has been striking. However, the major problem of viscous flow simulation remains to be resolved. Although many algorithms for the solution of compressible viscous high speed flows have been proposed, no technique for the generation of adequate unstructured viscous grids on general 3D configurations has yet emerged. Given the world-wide level of activity in this area, it can be expected that this hurdle will be overcome before too long and that this major problem area will then be opened for simulation with unstructured grid techniques.

ACKNOWLEDGEMENTS

The authors acknowledge the partial support provided for the work on computational aerodynamics by the Aerothermal Loads Branch of the NASA Langley Research Center, under NASA Research Grant NAGW–1809 (Technical Monitor : Kim S. Bey) and the Dryden Flight Research Facility of the NASA Ames Research Center, under NASA Research Grant NAGW-2962 (Technical Monitor : Dr. K. Gupta). The work on computational electromagnetics has been supported by the Department of Computational Engineering, British Aerospace plc, Sowerby Research Centre and by the U.K. Science and Engineering Research Council under Research Grant GR/G59240. J. Peiró acknowledges the support of the U.K. Science and Engineering Research Council under Grant GR/H 46183.

REFERENCES

Balanis, C.A. (1989), *Advanced Engineering Electromagnetics*. Wiley.

Boris, J.P. and Book, D.L. (1973), Flux Corrected Transport I: SHASTA, A Fluid Transport Algorithm that Works. *Journal of Computational Physics* 11, 38–69.

Chakravarthy, S.R. (1984), Relaxation Methods for Unfactored Implicit Upwind Schemes. *AIAA Paper*, 84–0165.

Givoli, D. (1992), *Numerical Methods for Problems in Infinite Domains*. Elsevier.

Harten, A. (1983), High Resolution Schemes for Hyperbolic Conservation Laws. *Journal of Computational Physics* **49**, 357–393.

Harten, A. (1984), On a Class of High Resolution Total Variation Stable Finite Difference Schemes. *SIAM Journal of Numerical Analysis* **21**, 1–23.

Hassan, O., Morgan, K. and Peraire, J. (1991), An Implicit Finite Element Method for High Speed Flows. *International Journal for Numerical Methods in Engineering*, **32**, 183–205.

Hassan, O., Probert, E.J., Morgan, K. and Peraire, J. (1993), Line Relaxation Methods for the Solution of 2D and 3D Compressible Flows. *AIAA Paper*, 93–3366.

Hirsch, C. (1990), *Numerical Computation of Internal and External Flows*. Wiley.

Jameson, A. (1983), Solution of the Euler Equations for Two Dimensional Transonic Flow by a Multigrid Method. *Applied Mathematics and Computation*, **13**, 327–355.

Jameson, A. (1987), Successes and Challenges in Computational Aerodynamics. *AIAA Paper*, 87–1184.

Jameson, A. (1992), Time Dependent Calculations Using Multigrid, with Applications to Unsteady Flows Past Aerofoils and Wings. *AIAA Paper 91–1596*, 1991.

Johnson, C. (1987), *Numerical Solution of Partial Differential Equations by the Finite Element Method*. Cambridge University Press.

Lyra, P.R.M., Morgan, K., Peraire, J. and Peiró, J. (1993), Unstructured Grid FEM/TVD Algorithm for Systems of Hyperbolic Conservation Laws. In: *Proceedings of the Conference on Numerical Methods for Laminar and Turbulent Flow, Edited by C. Taylor et al.* , Pineridge Press, 1408–1420, 1993.

Mavriplis, D. (1988), Multigrid Solution of the Two Dimensional Euler Equations on Unstructured Triangular Meshes. *AIAA Journal* **26**, 824–831.

Mavriplis, D. (1991), Three Dimensional Unstructured Multigrid for the Euler Equations. *AIAA Paper*, 91–1549.

Morgan, K., Hassan, O. and Peraire, J. (1993), The Solution of Maxwell's Equations in the Time Domain Using Unstructured Grids. In: *Proceedings of the Conference on Numerical Methods in Laminar and Turbulent Flow, Edited by C. Taylor et al.* , Pineridge Press, 1382-1393.

Morgan, K. and Peraire, J. (1987), Finite Element Methods for Compressible Flows. In: *von Karman Institute for Fluid Dynamics Lecture Series 1987–04*, Brussels, 1–114.

Morgan, K., Peraire, J. and Löhner, R. (1988), Adaptive Finite Element Flux Corrected Transport Techniques for CFD. In: *Finite Elements—Theory and Applications, Edited by D.L. Dwoyer, M.Y. Hussaini and R.G. Voigt*, Springer–Verlag, 165–175.

Parrott, A.K. and Christie, M.A. (1986), FCT Applied to the 2–D Finite Element Solution of Tracer Transport by Single Phase Flow in a Porous Medium. In: *Numerical Methods for Fluid Dynamics II, Edited by K.W. Morton and M.J. Baines*, Clarendon Press, 609–619.

Peraire, J., Morgan, K. and Peiró, J. (1989), Unstructured Finite Element Mesh Generation and Adaptive Procedures for CFD. In: *Application of Mesh Generation to Complex 3–D Configurations*, AGARD Conference Proceedings No. 464, 18.1–18.12.

Peraire, J., Morgan, K. and Peiró, J. (1993), Multigrid Solution of the 3D Compressible Euler Equations on Unstructured Tetrahedral Grids. *International Journal for Numerical Methods in Engineering* **36**, 1029–1044.

Peraire, J., Morgan, K., Peiró, J. and Zienkiewicz, O.C. (1987), An Adaptive Finite Element Method for High Speed Flows. *AIAA Paper*, 87–0559.

Peraire, J., Peiró, J. and Morgan, K. (1993), Finite Element Multigrid Solution of Euler Flows Past Installed Aero–Engines. *Computational Mechanics* **11**, 433-451.

Peraire, J., Vahdati, M., Morgan, K. and Zienkiewicz, O.C. (1987), Adaptive Remeshing for Compressible Flow Computations. *Journal of Computational Physics*, **72**, 449–466.

Roe, P. (1981), Approximate Riemann Solvers, Parameter Vectors and Difference Schemes. *Journal of Computational Physics* **43**, 357–372.

Shankar, V., Hall, W.F., Mohammadian, A. and Rowell, C. (1993), Algorithmic Aspects and Trends in Computational Electromagnetics. *AIAA Paper*, 93–0367.

Swanson, R.C. and Turkel, E. (1990), On Central Difference and Upwind Schemes. *ICASE Report*, 90–44.

Taflove, A. (1992), Re–Inventing Electromagnetics: Supercomputing Solutions of Maxwell's Equations via Direct Time Integration on Space Grids. *AIAA Paper*, 92–0333.

Vahdati, M., Morgan, K. and Peraire, J. (1992), Computation of Viscous Compressible Flows Using an Upwind Algorithm and Unstructured Meshes. In: *Computational Mechanics in Aerospace Engineering, Edited by S.N. Atluri*, AIAA Progress in Astronautics and Aeronautics Series Volume 146, 479–505.

van Leer, B. (1979), Towards the Ultimate Conservative Difference Scheme. V. A Second Order Sequel to Godunov's Method. *Journal of Computational Physics* **32**, 101–136.

Yee, H.C. (1989), A Class of High–Resolution Explicit and Implicit Shock-Capturing Methods. In: *von Karman Institute for Fluid Dynamics Lecture Series 1989–04*, Brussels, 1–216.

Zalesak, S.T. (1979), Fully Multidimensional Flux Corrected Transport Algorithms for Fluids. *Journal of Computational Physics* **31**, 335–362.

Zienkiewicz, O.C., Löhner, R. and Morgan, K. (1985), High Speed Inviscid Compressible Flow by the Finite Element Method. In: *The Mathematics of Finite Elements and Applications— MAFELAP 1984, Edited by J.R. Whiteman*, Academic Press, 1–25.

Chapter 16

Simulation of Dense Non-aqueous Phase Fluid Flow in Porous Media using Collocation Finite Elements

Joseph F. Guarnaccia and George F. Pinder

College of Engineering and Mathematics
University of Vermont, Burlington, Vermont 05405, U.S.A.

16.1 INTRODUCTION

Herein is developed a computationally efficient and robust numerical model capable of simulating the movement of heavier-than-water organic liquids, commonly referred to as Dense Non-Aqueous Phase Liquids (DNAPL). The model addresses aquifer contamination from the point of emplacement of the DNAPL to the point of dissolution of the DNAPL residual, all under saturated conditions. The DNAPL enters the aquifer by displacing the water, and because it is heavier than water, it tends to sink through the aquifer until it encounters an impenetrable barrier. As the slug of DNAPL moves through the aquifer, it leaves behind an immobile residual held in place by capillary forces. Some of the components of the DNAPL residual slowly dissolve into the flowing groundwater and are transported downstream resulting in widespread aquifer pollution.

The equations describing fluid-flow and contaminant transport in the above-referenced problem can be written as a system of coupled partial-differential equations which include an elliptic-like pressure equation, a parabolic-like saturation equation, and a system of parabolic-like contaminant species-transport equations. An efficient method of solving this system of equations is to solve the pressure, saturation, and contaminant-species equations sequentially, such that not only is one dependent variable solved for at any one time, but also that different, equation-dependent, solvers can be applied to maximize efficiency.

This paper deals specifically with a solver written for general linear or nonlinear parabolic equations which relies on a parallel domain-decomposition method. In this approach interfacial coupling is achieved by enforcing continuity of the function and its normal derivative.

The Mathematics of Finite Elements and Applications
Edited by J. R. Whiteman © 1994 John Wiley & Sons Ltd

16.2　EQUATION DEVELOPMENT

The simultaneous flow of two semi-miscible incompressible fluid phases, one wetting (w, e.g. water) and one non-wetting (n, e.g. DNAPL), under isothermal conditions, in a non-deforming porous medium can be simulated using a set of four coupled nonlinear equations (Pinder and Abriola (1986)): two phase (α) mass balance equations,

$$\epsilon \frac{\partial}{\partial t}\left(S_w \rho^w\right) - \nabla \cdot \left[\frac{\mathbf{k}\, k_{rw}}{\mu^w}(\nabla P^w - \rho^w \mathbf{g})\right] = \rho^w q^w + \widehat{\rho}^w \qquad (16.1)$$

$$-\epsilon \frac{\partial}{\partial t}\left(S_w \rho^n\right) - \nabla \cdot \left[\frac{\mathbf{k}\, k_{rn}}{\mu^n}(\nabla P^w + P_c'\nabla S_w - \rho^n \mathbf{g})\right] = \rho^n q^n - \widehat{\rho}^w \qquad (16.2)$$

which describe the relative distribution and flow field of the water and DNAPL respectively, and two dissolved constituent (i, α) mass-balance equations,

$$\frac{\partial}{\partial t}\left(\epsilon S_\alpha \rho_i^\alpha\right) + \nabla \cdot \left(\epsilon S_\alpha \rho_i^\alpha \mathbf{v}^\alpha\right) - \nabla \cdot \left(\epsilon S_\alpha \mathbf{D}^\alpha \nabla \rho_i^\alpha\right) = \rho_i^\alpha q^\alpha + \widehat{\rho}_i^\alpha \qquad (16.3)$$

which describe the distribution of dissolved contaminants in the phases. This would include DNAPL contaminants dissolved in the water phase (o,w) and a water contaminant dissolved in the DNAPL phase (w,n). Equations (16.1), (16.2) are written in terms of the dependent variables, water phase saturation, S_w, and water phase pressure, P^w. This is achieved by noting that the phase saturations sum to unity, i.e. $S_w + S_n = 1$ and the multiphase extension of Darcy's law applies, where $\mathbf{k}$ is the intrinsic permeability tensor, $k_{r\alpha}$ is the relative permeability, μ^α is the phase viscosity, ρ^α is the phase mass density, and $\mathbf{g}$ is the gravity vector. In addition one can relate S_w to the capillary pressure $P_c(S_w) = P^n - P^w$ when both phases are mobile, and where $P_c' = dP_c/dS_w$. Equations (16.3) are nonlinear convection-diffusion equations written in terms of the dependent variable mass concentration, ρ_i^α. Here ϵ is porosity, $\mathbf{v}^\alpha$ is the mass average Darcy velocity for the α phase, and $\mathbf{D}^\alpha$ is a velocity dependent hydrodynamic dispersion tensor. The right-hand sides of equations (16.1), (16.2), and (16.3) include sources and sinks of mass, where q^α is an external source if it is positive (e.g. an injection well), and $\widehat{\rho}_i^\alpha$ is an inter-phase mass exchange term. The total mass flux into the α phase is given by:

$$\widehat{\rho}^\alpha = \widehat{\rho}_o^\alpha + \widehat{\rho}_w^\alpha$$

and, for this case,

$$\widehat{\rho}^n = -\widehat{\rho}^w.$$

16.3　NUMERICAL DISCRETIZATION

Equations (16.1), (16.2), and (16.3) are dicretized in time using an implicit backward differencing scheme. The equations are linearized by the typically robust and relatively simple to implement Picard iteration technique, wherein the functions of the dependent variables are dated at the known iteration level and successively updated.

The dependent variables in equations (16.1), (16.2), P^w and S_w, and in equation (16.3), ρ_i^α, are approximated in space using a linear combination of bicubic Hermite polynomial basis functions. For two-dimensional problems this results in four degrees of freedom at each node for each variable, i.e. the function, the two first-order derivatives and the cross-derivative. The other variables, which are space dependent, are approximated using bilinear Lagrange basis functions. The nonlinear terms, $k_{r\alpha}$ and P_c' in equations (16.1), (16.2) and S^α, $\mathbf{v}^\alpha$ and $\mathbf{D}^\alpha$ in equation (16.3), are evaluated at the node and interpolated into the element using bilinear Lagrange basis functions. Because the mass exchange term represents the main coupling between the flow and the transport equations (a function of all the dependent variables), it is evaluated using Hermite interpolated dependent variables.

The discretized equations are reduced to a set of linear algebraic equations by employing the orthogonal collocation method: a method of weighted residuals wherein the weighting function is the Dirac delta function. This is equivalent to driving the residual to zero at specified points in the domain which are denoted as collocation points (orthogonal collocation results when the four Gauss quadrature points are used as collocation points).

16.4 PARALLEL SOLUTION COLLOCATION (PSC)

Figure 16.1 shows an example of a collocation finite-element mesh (a) highlighting the numbering scheme for degrees of freedom (boxed numbers) and collocation point equations (open numbers). Boundary conditions have been removed from the system. The degree of freedom numbering is for one dependent variable, and the block tri-diagonal structure (b) results from the numbering scheme selected. This non-symmetric matrix can be solved using direct or iterative solution techniques.

Consider the following iterative solution algorithm wherein the block tri-diagonal system matrix (see Figure 16.1) is diagonalized, thereby reducing the solution to a series of independent subproblems which can be computed in parallel. The algorithm is efficiently described by viewing it as a domain decomposition scheme wherein independent subdomain problems are solved and interface relaxation conditions are imposed. Several similar algorithms have recently been put forth. For example see (Funaro *et al.* (1988)), (Marini (1989)), (Borgers (1989)), and (Lai (1992)).

The first step is to idealize the grid in Figure 16.1 as a series of one-dimensional strips of elements. Thus, the global domain is decomposed into a series of subdomains. Figure 16.2 depicts an example of this idealization where the grid in Figure 16.1 has been decomposed into four horizontal subdomains. The dashed interfaces represent the imposed boundaries. Note that a vertical partitioning is equally feasible. The subdomains are linked together using an iterative interface relaxation solution technique, wherein continuity of the function and its normal derivative across the imposed interface is enforced.

A procedural summary of the interface relaxation iteration scheme for the parabolic equation solver is described as follows. Interface continuity is achieved by specifying a Dirichlet condition (D) on one side of the interface and a Neumann condition (N) on

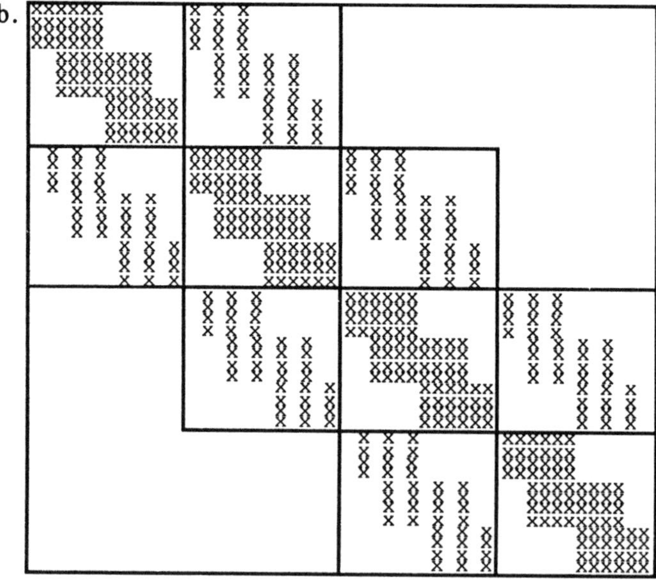

Figure 16.1 Collocation finite element mesh.

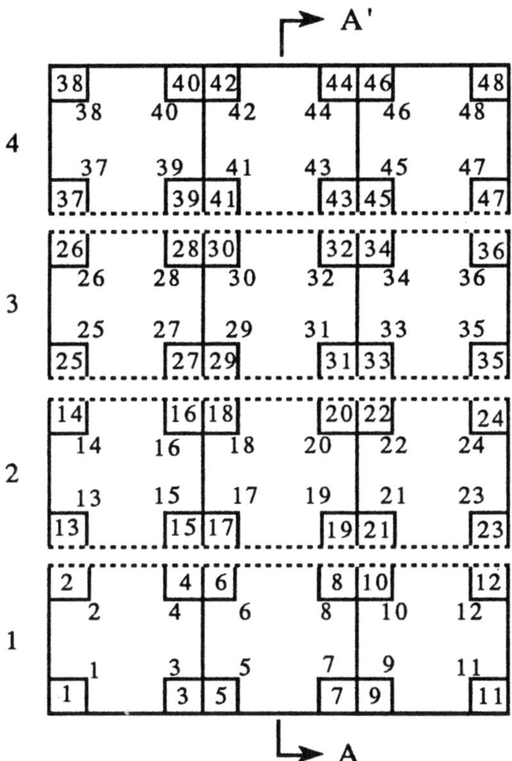

Figure 16.2 Example of a collocation finite-element grid which has been partitioned into four horizontal subdomains.

Figure 16.3 Example of D-N condition ordering used to solve the matrix equation associated with the transport solver.

the other (referred to here as a D-N condition). Figure 16.3 depicts one scenario for D-N condition ordering. In general D-N condition ordering is arbitrary. Therefore, the first step is to define subdomain partitions and D-N condition ordering (for example see Figure 16.2 and 16.3 respectively). This, in turn, defines the numbering scheme of the nodal unknowns, where, in this example, odd numbered unknowns in Figure 16.2 are associated with Dirichlet interface conditions and even numbered unknowns are associated with Neumann interface conditions. This is a direct result of the Hermite interpolation, wherein the function and its tangential derivative are associated with Dirichlet conditions, and the normal and cross derivatives are associated with Neumann conditions.

The next step is to define an interface relaxation parameter, θ, where $0 < \theta < 1$. The choice for the value of θ is important to the performance of the algorithm. To date only a heuristic analysis has been performed, and in general the choice is problem dependent. For the contaminant transport problems considered in this study, values of θ between 0.60 and 0.85 yield convergent iteration schemes.

Given that this setup information is held constant during the iteration procedure, consider now the general collocation system of equations to be solved:

$$A^M U^{M+1} = f^M$$

which, if written for the problem domain and numbering scheme in Figure 16.1, can be represented by the following symbolic matrix equations:

$$\begin{bmatrix} a_{11} & a_{12} & 0 & 0 \\ a_{21} & a_{22} & a_{23} & 0 \\ 0 & a_{32} & a_{33} & a_{34} \\ 0 & 0 & a_{43} & a_{44} \end{bmatrix}^M \begin{bmatrix} u_1 \\ u_2 \\ u_3 \\ u_4 \end{bmatrix}^{M+1} = \begin{bmatrix} f_1 \\ f_2 \\ f_3 \\ f_4 \end{bmatrix}^M \qquad (16.4)$$

where the system matrix A is represented as tri-block diagonal with sub-matrices a (as in Figure 16.1), the vectors U and f are decomposed into sub-vectors u_i and f_i, $i = 1, .., 4$, where each sub-vector corresponds to that part of the global vector associated with each subdomain $i = 1, .., 4$ (Figure 16.2). It can be seen by comparing equation (16.4) with Figure 16.1 and 16.2 that the main diagonal blocks of A, $a_{i,i}$, are associated with the unknowns in subdomain i, located in u_i, while the off-diagonal blocks in A, $a_{i, i\pm1}$, are associated with the imposed boundary information in adjacent subdomains $i \pm 1$, located in vectors $u_{i\pm1}$.

Given the global solution at iteration level M, U^M, evaluation of U^{M+1} is a three step procedure:

1. solve odd sweeps (i=1,3):

$$a_{i,i}^M u_i^* = f_i^M - a_{i,i\pm1}^M u_{i\pm1}^M$$

relax solution:

$$u_i^{M+1^*} = \theta u_i^* + (1-\theta)u_i^M$$

update global solution:

$$u_i^{M+1^*} \Rightarrow U^{M+1^*}$$

2. solve even sweeps $(i = 2, 4)$:

$$a_{i,i}^{M+1^*} u_i^{**} = f_i^{M+1^*} - a_{i,i\pm1}^{M+1^*} u_{i\pm1}^{M+1^*}$$

relax solution:

$$u_i^{M+1} = \theta u_i^{**} + (1 - \theta) u_i^{M+1^*}$$

update global solution:

$$u_i^{M+1} \Rightarrow U^{M+1}$$

3. check convergence:

$$\left\| U^{M+1} - U^M \right\|_\infty \leq \epsilon$$

The solution of each subdomain problem involves the direct solution of each main diagonal block system matrix. Note that the fully two-dimensional system matrix need not be constructed, thus allowing substantial savings in storage requirements.

16.5 IMPLEMENTATION AND DISCUSSION

A heuristic analysis of the PSC algorithm has been conducted using numerical experiments to determine a relationship between the relaxation weighting parameter, θ, defined previously, and the number of iterations required to obtain a solution to a given problem. The analysis is based on the following simplifying conditions:

- the nearly elliptic pressure equation is solved using a direct-solution algorithm;
- the relaxation weight is held constant for the duration of each simulation;
- the same weight is used to relax both the Dirichlet and Neumann interface conditions.

Consider the problem depicted in Figure 16.4. DNAPL is injected at the point sources indicated at a rate of 0.02 cm^3/s. Fluid and soil properties [granular soil] are given as follows (cgs units): $\rho^n = 1.4657$, $\rho^w = 1.0$, $\mu^w/\mu^n = 2.0$, $\epsilon = 0.3$, $\mathbf{k} = 5 \times 10^{-7}$. The initial condition on water pressure is that it is in static equilibrium, and the aquifer is initially free of contaminants. Applying the PSC algorithm, the grid is partitioned into vertical subdomains (subdomains are parallel to the Dirichlet boundary) with Dirichlet conditions applied along the upstream side of each subdomain and Neumann conditions applied along the downstream side. Figure 16.5 provides the solutions of the DNAPL saturation distribution and dissolved contaminant concentration plume at time = 1000 seconds.

Figure 16.6 shows the relationship between the relaxation weight and the total number of iterations required to generate a saturation solution at time 1000s using a maximum Δt of 10s. The optimal constant relaxation weight was approximately 0.65. Using this weight, the average number of iterations per time step was 5.0. Figure 16.7 shows the relationship between the relaxation weight and the total number of iterations required to generate the concentration solution at time 1000s using a maximum Δt

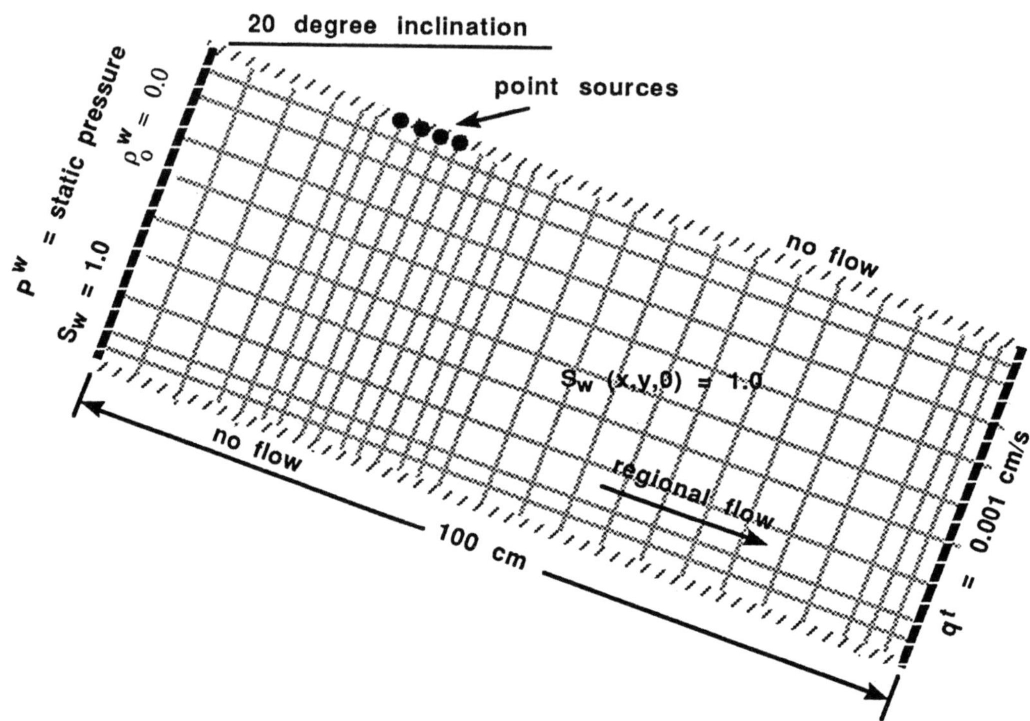

Figure 16.4 Model problem definition.

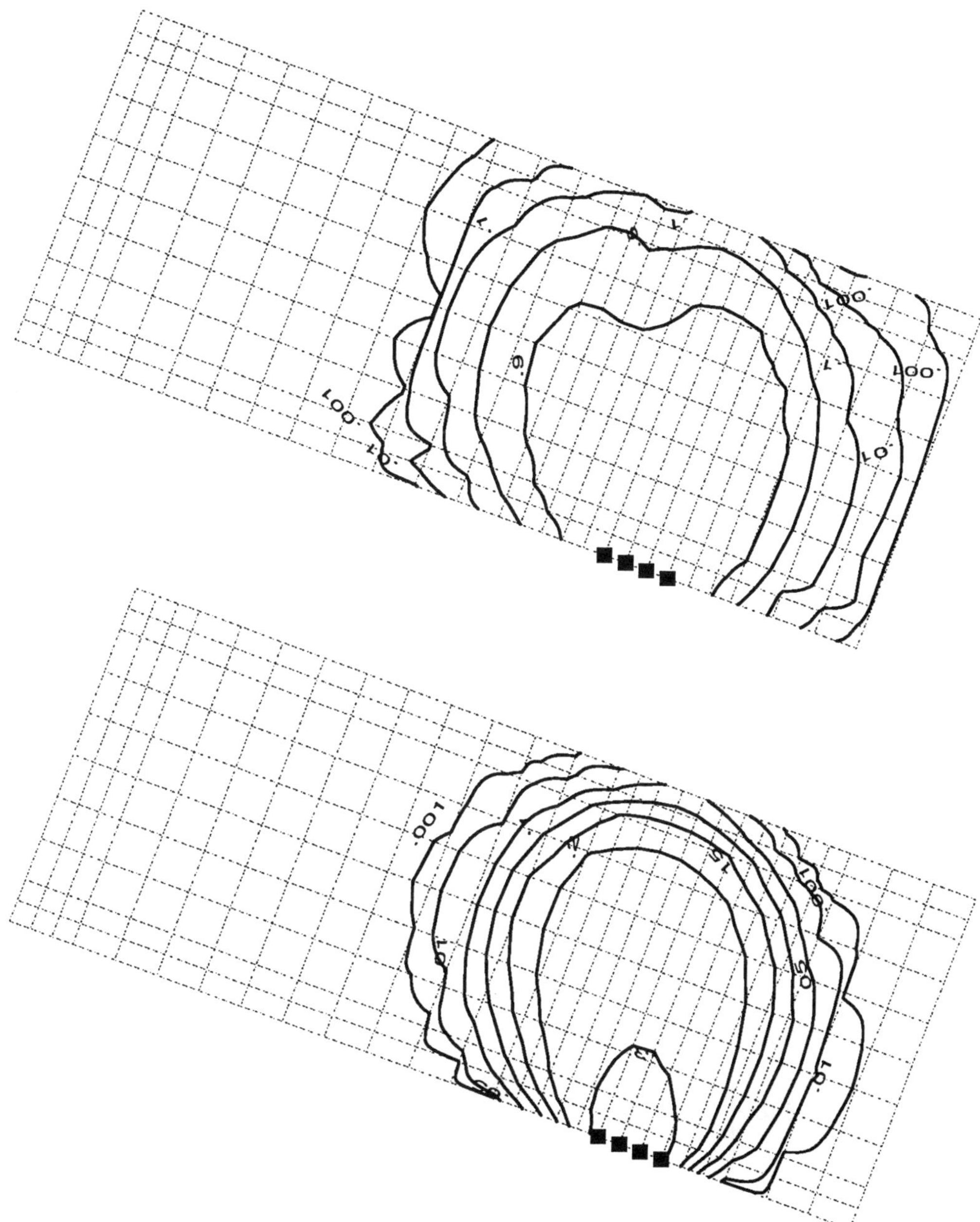

Figure 16.5 Solution at time 1000s. (a) the DNAPL saturation solution (contours are 0.001, 0.01, 0.05, 0.1, 0.14, 0.2, 0.3). (b) dissolved DNAPL concentration normalized by the solubility limit (contours are 0.001, 0.01, 0.1, 0.5, 0.9).

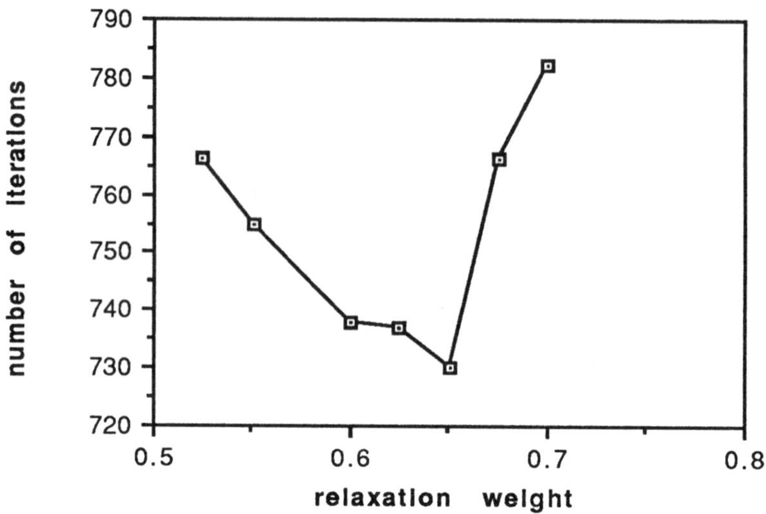

Figure 16.6   Relationship between the relaxation weight and the total number of iterations needed to generate a saturation solution from static conditions at time 0s to time 1000s (maximum time step is 10s). The optimal weight is approximately 0.65.

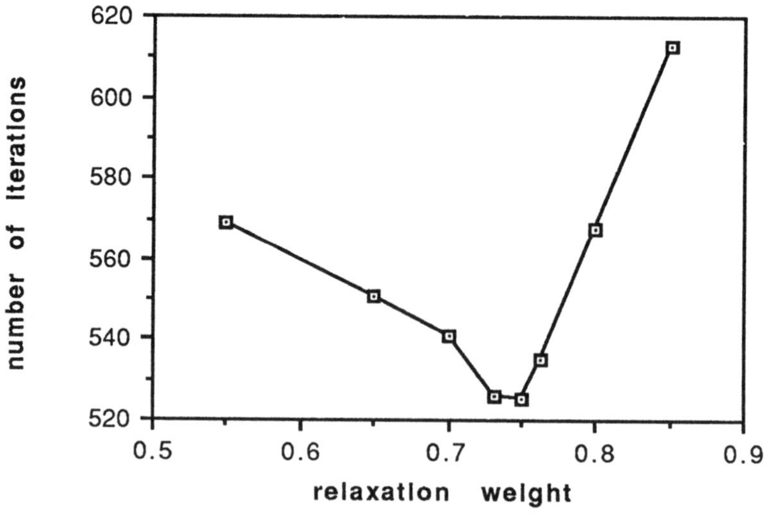

Figure 16.7   Relationship between the relaxation weight and the total number of transport equation iterations needed to generate a solution at time 1000s (maximum time step 10s). The optimal weight for this case is approximately 0.75.

of 10s. The optimal constant relaxation weight was about 0.75. Using this weight the average number of iterations per time step was 3.9. Note that an optimal constant relaxation weight exists and is different for each equation. In addition, there is a broad range of values which admit a convergent solution.

ACKNOWLEDGMENT

The authors wish to acknowledge the financial support of the U.S. Environmental Protection Agency (EPA) assistance I.D. No CR-818955 and CR-820499. Although the research described in this article has been funded by the U.S. EPA, it has not been subjected to agency review and therefore does not necessarily reflect the views of the Agency and no official endorsement should be inferred.

REFERENCES

Demkowicz L., Oden J.T., Rachowicz W. and Hardy O. (1989), Toward a Universal *h-p* Adaptive Finite Element Strategy. Part1: Constrained Approximation and Data Structure, *Computer Methods in Applied Mechanics and Engineering* **77** (1-2), 79-112;

Pinder, G.F. and L.M. Abriola, (1986), On the Simulation of Nonaqueous Phase Organic Compounds in the Subsurface, Water Resources Research, **22** (9), 109-119.

Funaro, D., Quarteroni, A., and P. Zanolli, (1988), An Iterative Procedure with Interface Relaxation for Domain Decomposition Methods, SIAM J. Numerical Analysis, **25**(6), 1213-1236.

Lai, C-H, (1992), A Nonoverlapped Domain Decomposition for a Class of Convection-diffusion Problems, Applied Math. Modelling, **16**, 101-106.

Borgers, C., (1989), The Neumann-Dirichlet Domain Decomposition Method with Inexact Solvers on the Subdomains, *Numer. Math.*, **55**, 123-136.

Marini, L.D. and A. Quateroni, (1989), A Relaxation Procedure for Domain Decomposition Methods using Finite Elements, *Numer. Math.*, **55**, 599-618.

Chapter 17

A Complementary Volume Approach for Modeling Three-Dimensional Navier-Stokes Equations Using Dual Delaunay/Voronoi Tessellations

J.C. Cavendish*, C.A. Hall** and T.A. Porsching***

* Mathematics Department, General Motors Research Laboratories
Warren, Michigan 48090-9055, U.S.A.
** Department of Mathematics and Statistics
University of Pittsburgh, Pittsburgh, Pennsylvania 15260, U.S.A.

17.1 INTRODUCTION

In this paper we extend to three-dimensions a covolume approach proposed in
Cavendish *et al.* (1993), Hall (1985), Hall *et al.* (1991) and (1992) for modeling the flow
of a viscous incompressible fluid in an open bounded domain. The approach integrates
three components: (i) **Unstructured grid generation**, (ii) **Covolume equation
generation**, and (iii) **Dual variable reduction**.

17.2 UNSTRUCTURED GRID GENERATION

At MAFELAP V (1984) we described an algorithm for the computer generation of
Delaunay tetrahedral finite element meshes for bounded three-dimensional domains
Cavendish *et al.* (1984) and (1985).

Closely associated with a *Delaunay mesh* of tetrahedra is a dual geometric construct
called the *Voronoi tessellation*. Each point, p, in the tetrahedral mesh is surrounded
by a union of disjoint tetrahedra, say T_p. Appropriate straight-line connections of the
circumcentres of these tetrahedra (see Cavendish *et al.* (1984) and (1985), Finney
(1979) for details) produces the edges of a convex polytope, V_p, containing p in its
interior. The Voronoi polytope, V_p, has the property that any point in its interior
is closer to p than it is to any other point in the mesh. Consequently, the faces of
V_p are composed of bounded planes that bisect tetrahedral edges connecting p to its
neighbouring points in the mesh. On the other hand, an edge on the surface of V_p
is perpendicular to a triangle face contained in T_p and that edge passes through the

circumcentre of that triangle face. As a result of all this, V_p and T_p share a mutual orthogonality which we summarise as follows:

- Faces of V_p are perpendicular bisectors of edges of tetrahedra in T_p.
- Edges of V_p are perpendicular to triangle faces in T_p and pass through tetrahedral and triangle facial circumcentres in T_p.
- Corners on the surface of V_p are circumcentres of tetrahedra in T_p.

Visualization and manipulation of three-dimensional tetrahedral meshes and Voronoi polytopes is difficult at best. To help with this, and to lay the foundation for covolume equation generation in Section 17.4, we consider some simple illustrations. These illustrations consider a single point p_1, a single set of surrounding tetrahedra, T_{p_1}, and a simple associated Voronoi polytope, V_{p_1}, all extracted from a Delaunay tetrahedral mesh of a large set of points containing p_1. Figure 17.1a displays the Voronoi polytope associated with p_1. Note that in general, the corners of ∂V_{p_1} (e.g., **A** in Figure 17.1a) are shared by three faces on ∂V_{p_1} (∂ signifies boundary). This means that such a corner is also a corner of three abutting Voronoi polytopes, $V_{p_2}, V_{p_3}, V_{p_4}$. Connecting p_1, p_2, p_3, p_4 forms one of the Delaunay tetrahedra in T_{p_1}. Figure 17.1b illustrates T_{p_1} (in wire frame mode) and V_{p_1} (in shaded mode) simultaneously. From Figure 17.1a and 17.1b the mutual orthogonality of T_{p_1} and V_{p_1} is evident. For example, edge $\overline{p_1 p_2}$ is perpendicular to and is bisected by face ABCDE of the polytope V_{p_1}. Also, edge $\overline{BC}$ of V_{p_1} is perpendicular to and passes through the circumcentre of face $\overline{p_1 p_2 p_4}$.

In Figure 17.2a we show a Delaunay tetrahedral mesh of an automobile catalytic converter (3241 tetrahedra); Figure 17.2b displays a rendering of the associated dual Voronoi tessellation containing 792 polytopes (a subset of only 77 polytopes interior to the flow region is shown for clarity).

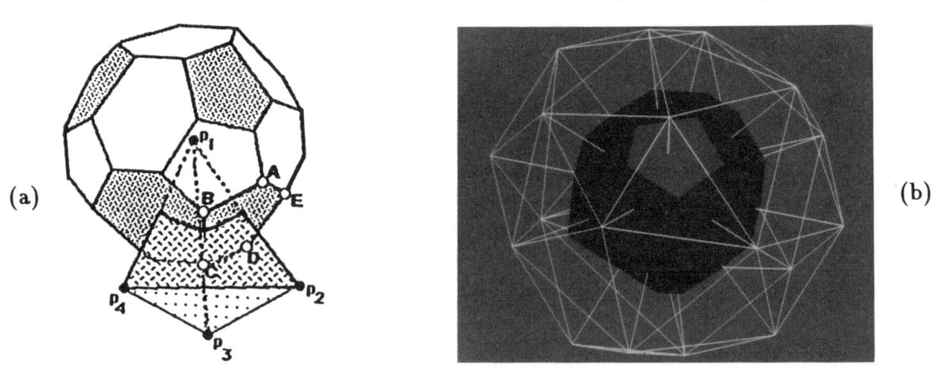

Figure 17.1 (a) Polytope V_{p_1} and one tetrahedron, (b) T_{p_1} and Voronoi polyhedron V_{p_1}

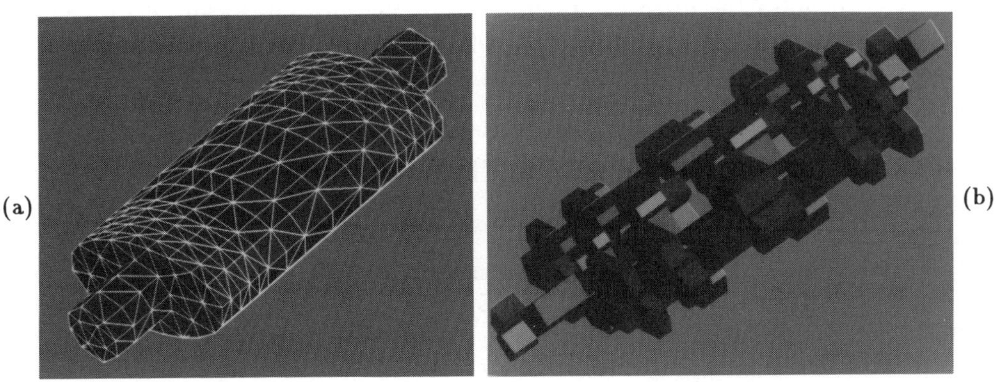

Figure 17.2 (a) Delaunay tetrahedral mesh, (b) Voronoi tessellation

17.3 THE INCOMPRESSIBLE FLOW PROBLEM

We consider a bounded flow domain Ω in $\mathbb{R}^3$ with boundary $\partial\Omega$. The continuum problem is to find a velocity vector $\mathbf{q} = (u(x,y,z,t), v(x,y,z,t), w(x,y,z,t))^T$ and a scalar pressure field $p(x,y,z,t)$ which satisfy the continuity equation

$$\nabla \cdot \mathbf{q} = 0, \qquad\qquad (x,y,z) \in \Omega, t > 0 \qquad (17.1)$$

and the vector-valued conservation of momentum equation

$$\frac{\partial \mathbf{q}}{\partial t} - \nu\nabla^2\mathbf{q} + (\mathbf{q}\cdot\nabla)\mathbf{q} + \nabla p = \mathbf{F}(\mathbf{x},t), \qquad (17.2)$$

where $\mathbf{x} = (x,y,z)$, ν is the kinematic viscosity of the fluid and $\mathbf{F}(\mathbf{x},t)$ is a prescribed source term. We assume boundary conditions of the form

$$p(\mathbf{x},t) = p_b(\mathbf{x}), \qquad\qquad \mathbf{x} \in \partial\Omega_1, t > 0 \qquad (17.3)$$

and

$$\mathbf{q}(\mathbf{x},t) = \mathbf{q}_b(\mathbf{x}), \qquad\qquad \mathbf{x} \in \partial\Omega_2, t > 0 \qquad (17.4)$$

where $\partial\Omega = \partial\Omega_1 \bigcup \partial\Omega_2$ and $\partial\Omega_1 \bigcap \partial\Omega_2 = \phi$. Further, we assume initial conditions of the form

$$\mathbf{q}(\mathbf{x},0) = \mathbf{q}_0(\mathbf{x}), \qquad\qquad p(\mathbf{x},0) = p_0(\mathbf{x}), \; \mathbf{x} \in \Omega. \qquad (17.5)$$

Following the development in Hall *et al.* (1991) and (1992), if $\mathbf{n}$ is a unit vector in $\mathbb{R}^3$, then taking the dot product of $\mathbf{n}$ with the vector differential equation (17.2) produces a *scalar* version of the momentum equation

$$\frac{\partial}{\partial t}(\mathbf{n}\cdot\mathbf{q}) + \nu\,\mathbf{curl}(\mathbf{curl}\,\mathbf{q}\,)\cdot\mathbf{n} + \parallel\mathbf{q}\parallel\frac{\partial}{\partial\mathbf{q}}(\mathbf{q}\cdot\mathbf{n}) + \frac{\partial\mathbf{p}}{\partial\mathbf{n}} = \mathbf{n}\cdot\mathbf{F} \qquad (17.6)$$

In (17.6) $\frac{\partial}{\partial \mathbf{q}}$ denotes the directional derivative in the direction $\mathbf{q}$ and $\| \cdot \|$ denotes Euclidean length. It is (17.1) and (17.6) to which we apply the covolume technique in Section 17.4.

17.4 COVOLUME DISCRETIZATION OF THE FLOW PROBLEM

In the three-dimensional setting, we integrate (17.1) over each tetrahedral control volume, $\hat{\Omega}_j$ (see Figure 17.3), apply the 3D Divergence Theorem and approximate the normal velocity component $\mathbf{n} \cdot \mathbf{q}$ by velocity components normal to the faces of the tetrahedra and passing through facial circumcentres. The result is

$$0 = \int\int\int_{\hat{\Omega}_j} \nabla \cdot \mathbf{q} \, dV = \int\int_{\partial\hat{\Omega}_j} \mathbf{q} \cdot \mathbf{n} \, d\sigma \simeq \sum_{i=1}^{4} u_i(t) A_i (\mathbf{n} \cdot \mathbf{n}_i), \qquad (17.7)$$

where $\mathbf{n}$ is the outward normal vector to $\partial\hat{\Omega}_j$, $\mathbf{n}_i$ is a preassigned unit normal to the ith face of $\partial\hat{\Omega}_j$, A_i is the area of the ith face of $\hat{\Omega}_j$, $u_i(t) = \mathbf{q} \cdot \mathbf{n}_i \mid_{P_i}$ and P_i is the circumcentre of the ith face of $\hat{\Omega}_j$. Note that $(\mathbf{n} \cdot \mathbf{n}_i) = \pm 1$. If there are N_T tetrahedra in the tessellation of Ω, then there are N_T such discrete continuity equations which we write as

$$\mathbf{A} \mathbf{D}_1 \mathbf{u}(t) = \mathbf{b} \qquad (17.8)$$

where $\mathbf{u}(t)$ is an N_F-vector of normal velocity components with ith entry $u_i(t)$, and N_F is the number of triangular faces in the tetrahedral mesh for which the normal velocity is unknown. The diagonal matrix D_1 has its ith entry equal to A_i. The vector $\mathbf{b}$ contains known boundary data (that is, velocities specified in (17.4)). The $N_T \times N_F$ matrix $\mathbf{A} \mathbf{D_1}$ is called the *discrete divergence operator* associated with the tetrahedral Delaunay tessellation. We note also that the entries of the matrix $\mathbf{A}$ are ± 1 or 0.

The momentum equation (in scalar form defined by (17.6)) is approximated at the circumcentre of each face of the N_F faces of the tetrahedral mesh bearing an unknown normal velocity component as follows: with reference to Figure 17.3 and starting with facial circumcentre node P_1 associated with face **ABC**, we take $\mathbf{n} = \mathbf{n}_1$ in (17.6) where $\mathbf{n}_1$ is the unit vector through P_1 normal to face **ABC**.

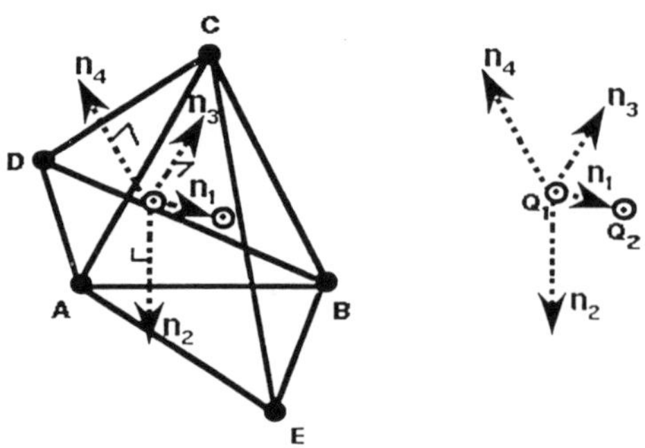

Figure 17.3 Tetrahedral control volume $\hat{\Omega}_j$ with vertices **A,B,C,D**

Then, we define the following discretizations:

I. Temporal Term: $\frac{\partial}{\partial t}(\mathbf{n} \cdot \mathbf{q})$

$$\frac{\partial}{\partial t}(\mathbf{n}_1 \cdot \mathbf{q}) \mid_{P_1} = \frac{d}{dt}u_1(t) \tag{17.9}$$

II. Pressure Term: $\frac{\partial p}{\partial \mathbf{n}}$

$$\frac{\partial p}{\partial \mathbf{n}_1} \simeq \frac{p_2(t) - p_1(t)}{\| Q_2 - Q_1 \|} \tag{17.10}$$

where p_1 approximates pressure at the circumcentre, Q_1 of tetrahedron **ABCD** and p_2 approximates pressure at circumcentre Q_2 of tetrahedron **ABCE** (see Figure 17.3).

III. Viscous Term: $\mathbf{curl(curl\ q\ )} \cdot \mathbf{n}$

If $A_{\mathbf{ABC}}$ is the area of triangular face **ABC**, then

$$\mathbf{curl(curl\ q\ )} \cdot \mathbf{n_1} \simeq \frac{1}{A_{\mathbf{ABC}}} \int\!\!\int_{\mathbf{ABC}} (\mathbf{curl\ curl\ q}) \cdot \mathbf{n_1} d\sigma. \tag{17.11}$$

Let **T** be a unit vector tangent to $\partial(\mathbf{ABC})$ oriented in the counter clockwise direction with respect to $\mathbf{n}_1$.

Then, by Stokes' Theorem

$$\int\!\!\int_{\mathbf{ABC}} (\mathbf{curl\ curl\ q}) \cdot \mathbf{n}_1 d\sigma = \int_{\partial(\mathbf{ABC})} \mathbf{curl q} \cdot \mathbf{T} ds$$

$$= \int_{\mathbf{AB}} \mathbf{curl\ q} \cdot \mathbf{T_{AB}} ds + \int_{\mathbf{BC}} \mathbf{curl\ q} \cdot \mathbf{T_{BC}} ds + \int_{\mathbf{CA}} \mathbf{curl\ q} \cdot \mathbf{T_{CA}} ds \tag{17.12}$$

where $\mathbf{T_{AB}}, \mathbf{T_{BC}}$ and $\mathbf{T_{CA}}$ are the unit tangent vectors to the sides of triangle face **ABC** (see Figure 17.4). Because of the mutual orthogonality of Delaunay tetrahedral and Voronoi polyhedral tessellations, these tangent vectors are also normal vectors to the faces (planes bisecting edges **AB, BC** and **CA**) of three abutting Voronoi polyhedra. Also,

$$\int_{\mathbf{BC}} \mathbf{curl\ q} \cdot \mathbf{T_{BC}} ds \simeq h_{\mathbf{BC}}(\mathbf{curl\ q})_{\frac{\mathbf{B+C}}{2}} \cdot \mathbf{T_{BC}} \tag{17.13}$$

where $h_{\mathbf{BC}}$ is the length of edge **BC**.

Next, if $\mathbf{F_{BC}}$ is the Voronoi polyhedral face corresponding to the plane bisecting **BC**, then

$$(\mathbf{curl\ q})_{\frac{\mathbf{B+C}}{2}} \cdot \mathbf{T_{BC}} \simeq \frac{1}{Area(\mathbf{F_{BC}})} \int\!\!\int_{\mathbf{F_{BC}}} \mathbf{curl\ q} \cdot \mathbf{T_{BC}} d\sigma \tag{17.14}$$

Because $\mathbf{T_{BC}}$ is normal to $\mathbf{F_{BC}}$, a second application of Stoke's Theorem to (17.14) yields

$$\int\int_{\mathbf{F_{BC}}} \mathrm{curl}\,\mathbf{q}\cdot\mathbf{T_{BC}}d\sigma = \int_{\partial\mathbf{F_{BC}}} \mathbf{q}\cdot\mathbf{n}ds \tag{17.15}$$

where $\mathbf{q}\cdot\mathbf{n}$ is the component of flow tangential to the edges of $\mathbf{F_{BC}}$, and therefore, normal to certain triangle faces of the tetrahedral mesh. Finally, we have

$$\int_{\partial\mathbf{F_{BC}}} \mathbf{q}\cdot\mathbf{n}ds \simeq \sum_{i=1}^{k} u_i(t)h_i'(\mathbf{n}\cdot\mathbf{n}_i) \tag{17.16}$$

where h_i' are the lengths of the k (= 6 in Figure 17.4) edges bounding the polyhedral face $\mathbf{F_{BC}}$.

Combining (17.15) and (17.16) with equation (17.13) produces

$$\int_{\mathbf{BC}} \mathrm{curl}\,\mathbf{q}\cdot\mathbf{T_{BC}}ds \simeq \frac{h_{\mathbf{BC}}}{Area(\mathbf{F_{BC}})}\sum_{i=1}^{k} u_i(t)h_i'(\mathbf{n}\cdot\mathbf{n}_i) \tag{17.17}$$

Similarly, $\int_{\mathbf{CA}} \mathrm{curl}\,\mathbf{q}\cdot\mathbf{T_{CA}}ds$ and $\int_{\mathbf{AB}} \mathrm{curl}\,\mathbf{q}\cdot\mathbf{T_{AB}}ds$ can be approximated by linear combinations of normal velocity components, $u_i(t)$, assigned to the edges of polyhedral faces that are planes bisecting $\mathbf{CA}$ and $\mathbf{AB}$ respectively. Finally, (17.11) indicates that $\mathrm{curl}\,(\mathrm{curl}\,\mathbf{q})\cdot\mathbf{n}_1$ can be approximated by a sum of tangential flows along polyhedral edges (hence, components of flow, $u_i(t)$, normal to triangle faces of tetrahedral covolumes) corresponding to planes bisecting $\mathbf{AB},\mathbf{BC}$ and $\mathbf{CA}$.

IV Convective Term: $\|\mathbf{q}\|\frac{\partial}{\partial\mathbf{q}}(\mathbf{q}\cdot\mathbf{n})$

Step 1. Determine an approximation to the velocity at the vertices of each tetrahedron: Let p_i be a vertex of tetrahedron $\hat{\Omega}_j$ and V_{p_i} be the Voronoi polytope associated with p_i. Further, let T_{p_i} be the set of all tetrahedra that share vertex p_i and let S_{p_i} be the set of circumcentres of those faces of the tetrahedra in the set T_{p_i} that have p_i as a vertex. The point p_i is in the convex hull of S_{p_i}. The velocity vector, $\mathbf{q}$ at vertex p_i can be approximated (in several ways) as a linear combination of the velocities at the facial circumcentres in S_{p_i}; assuming the latter are known (see Section 17.5). For example, choose four points P_k in S_{p_i}, k = 1,2,3,4, such that the tetrahedron with vertices $\{P_k\}_{k=1}^{4}$ contains p_i (see Figure 17.5). The velocity $\mathbf{q}(p_i)$ is approximated by

$$\mathbf{q}(p_i) \simeq \sum_{k=1}^{4} \beta_{ik}\mathbf{q}(P_k) \tag{17.18}$$

where $\{\beta_{ik}\}_{k=1}^{4}$ are the barrycentric coordinates of the point p_i in the tetrahedron $P_1P_2P_3P_4$.

Step 2. Upwind differencing of the convective term: Let P_1 be a facial circumcentre where the convection term is to be discretized and let $\mathbf{q}(P_1)$ be the velocity vector at P_1 (calculated from a previous time step). Assume $\mathbf{n}_1$ is an outward normal to the triangular face $\mathbf{ABC}$ with circumcentre P_1. The upwind strategy then involves testing whether $\mathbf{q}(P_1)\cdot\mathbf{n}_1 \geq 0$; i.e. the flow across this face is out of tetrahedron $\hat{\Omega}_j$ in Figure

17.6. (If $\mathbf{q}(P_1) \cdot \mathbf{n}_1 < 0$, then consider the other abutting tetrahedron sharing face **ABC**.)

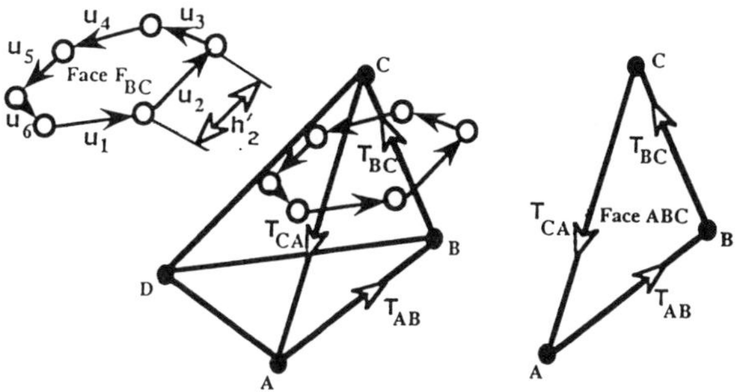

Figure 17.4 Tetrahedral face **ABC** and Voronoi face $\mathbf{F_{BC}}$

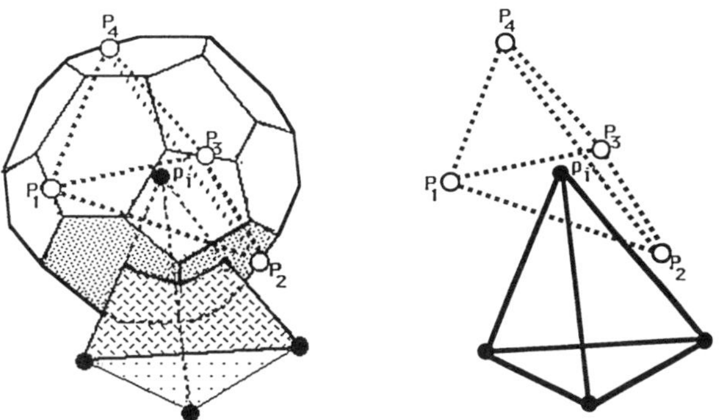

Figure 17.5 Voronoi poytope V_{p_i} associated with vertex p_i and an enclosing tetrahedron.

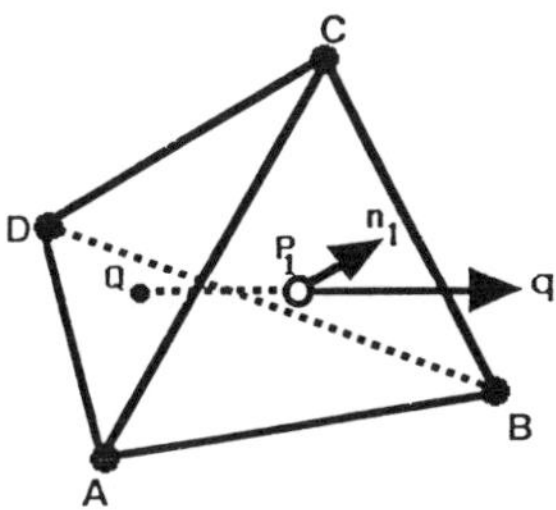

Figure 17.6 Upwind differencing on a tetrahedral mesh

Let Q be the point of intersection of the vector $-\mathbf{q}(P_1)$ with a face (say $\mathbf{ABD}$) of $\hat{\Omega}_j$. Then

$$\| \mathbf{q} \| \frac{\partial}{\partial \mathbf{q}}(\mathbf{q} \cdot \mathbf{n}_1) \simeq \| \mathbf{q}(P_1) \| \frac{\mathbf{q}(P_1) \cdot \mathbf{n}_1 - \mathbf{q}(Q) \cdot \mathbf{n}_1}{\| P_1 Q \|} \tag{17.19}$$

The velocity $\mathbf{q}(Q)$ is determined by linear interpolation of the velocity at the vertices $\mathbf{A,B,D}$, the latter having been determined in **Step 1**. If the facial circumcentre P_1 is not on the face $\mathbf{ABC}$, then replace P_1 in the above discussion by the centroid of that face.

Combining (17.9)-(17.12), (17.13) and (17.19), we obtain a semi-discrete momentum equation of the form

$$\frac{d\mathbf{u}}{dt} + Q(\mathbf{u})\mathbf{u} = D_2^{-1} A^T \mathbf{p} + \mathbf{b} \tag{17.20}$$

where $Q(\mathbf{u})$ is an $N_F \times N_F$ matrix containing couplings associated with the discretization of the viscous and convective terms, the latter depending on $\mathbf{u}$. The $N_F \times N_F$ diagonal matrix D_2 has as its ith entry, the distance between the two circumcentres of the two tetrahedra sharing the ith triangular face in the mesh; see equation (17.10). The vector $\mathbf{p}$ of pressures is $N_T \times 1$ and the matrix A^T is the transpose of the $N_T \times N_F$ matrix in (17.8).

Equations (17.8) and (17.20) form a differential algebraic equation (DAE) system of $(N_T + N_F)$ equations in $(N_T + N_F)$ unknowns; the N_T pressures and the N_F velocity components. This system can be solved using any one of a variety of DAE solvers to obtain a numerical solution of the fluid dynamics system (17.1) and (17.6).

17.5 RECONSTRUCTION OF THE VELOCITY FIELD

Inherent in the covolume method is the need to reconstruct the velocity (vector) field $\mathbf{q}$, given components of $\mathbf{q}$ at various points in the flow region, Cavendish *et al.* (1993), Hall *et al.* (1991), and (1992), Nicolaides (1989(i-ii)) and (1992). In Hall *et al.* (1993), a reconstruction procedure for triangulations was presented that reproduces *linear* flow fields exactly. We now extend this procedure to three-dimensional tetrahedral meshes.

We introduce a local coordinate system on each of the tetrahedral faces. The system is centred at the circumcentre P_i of the ith face and $\mathbf{n}_i$ is the preassigned unit normal to the ith face. The remaining coordinate directions, $\mathbf{s}_i$ and $\mathbf{t}_i$ are chosen in the plane of the ith face so that $\mathbf{n}_i$, $\mathbf{s}_i$ and $\mathbf{t}_i$ form a (non orthogonal) right handed system. For example, a convenient choice for $\mathbf{s}_i$ and $\mathbf{t}_i$ would be unit vectors in the directions $\overline{\mathbf{AC}}$ and $\overline{\mathbf{CB}}$, in Figure 17.7.

Given a tetrahedron, Ω_1, let $\Omega_j, j = 2, 3, 4, 5$ be the four tetrahedra sharing a face with Ω_1 (see Figure 17.8). Let $\{P_i\}_{i=1}^{16}$ be the 16 facial circumcentres of these four neighbouring tetrahedra. At P_j the velocity $\mathbf{q}$ is approximated by

$$\mathbf{q}_j(t) = u_j(t)\mathbf{n}_j + v_j(t)\mathbf{s}_j + w_j(t)\mathbf{t}_j \tag{17.21}$$

where u_j is known and we seek values for v_j and $w_j, j = 1, 2, ..., 16$. These 32 components are chosen so as to reproduce an arbitrary linear flow field (in $\cup_{j=1}^{5}\Omega_j$) of the form

$$\mathbf{q}(t) = \mathbf{a}_0(t) + \mathbf{a}_1(t)x + \mathbf{a}_2(t)y + \mathbf{a}_3(t)z \tag{17.22}$$

where the four 3-vectors $\mathbf{a}_i(t)$ are fixed but arbitrary. Thus, there are a total of 44 unknowns. Equating (17.21) and (17.22) at each of the 16 circumcentres yields 48 scalar equations in 44 unknowns. We have yet to investigate the rank of this system, but our experience in the two-dimensional case suggests that it is of full rank. As such least squares can be used for its solution.

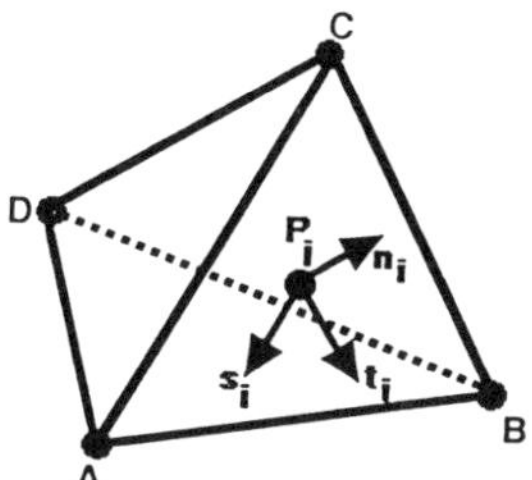

Figure 17.7 A local coordinate system

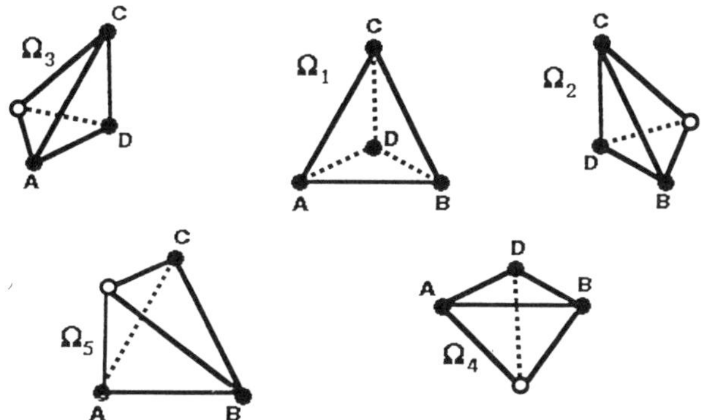

Figure 17.8 A tetrahedron Ω_1 and its four neighbours

17.6 THE DUAL VARIABLE METHOD

In this section we describe a variable reduction technique which replaces the $N_F + N_T$ dimensional system (17.8),(17.20) by an equivalent system of dimension $N_F - N_T$. This network method (called the *dual variable method*) was presented in Hall (1985) and Porsching (1985) for rectangular grids and in Cavendish *et al.* (1993), Hall *et al.* (1991), (1992) and (1993) for triangular grids.

In network terminology Berge and Ghouila-Houri (1965) the edges (*links*) and vertices (*nodes*) of the Voronoi polytopes form a directed network, Γ while those of the Delaunay tessellation constitute its dual, Γ^*. Each link of Γ carries a *flow* that is an approximation of the velocity component normal to the face of a tetrahedron, and each node of Γ carries a *state* that is an approximation of the pressure at the circumcentre of a tetrahedron. The matrix A in (17.8) is the *incidence matrix* of the network Γ, equations (17.8) are its node laws and equations (17.20) constitute the *link characteristics*.

The dual variables are states on the nodes of Γ^*, i.e. the vertices of the tetrahedra. If the flow region is simply connected, then the boundaries of the faces of the Voronoi polytopes are *elementary cycles* in Γ. (An elementary cycle is a closed path of links in the network such that in a complete traverse of the cycle a node is encountered exactly once.) In two dimensions an edge of a Voronoi polygon is shared by at most two polygons, and the boundaries of the Voronoi polygons form a basis of elementary cycles. However, in three dimensions, an edge of a Voronoi polytope may be common to 3 or more polytopes. Hence, the set of all elementary cycles that are boundaries of faces of polytopes may yield a dependent set. For example, in Figure 17.9, the elementary cycles corresponding to facial boundaries for a simple Voronoi polytope are indicated. Note that the elementary cycle on face $F_{p_1 p_2}$ is the negative of the sum of the other elementary cycles associated with the faces of this polytope

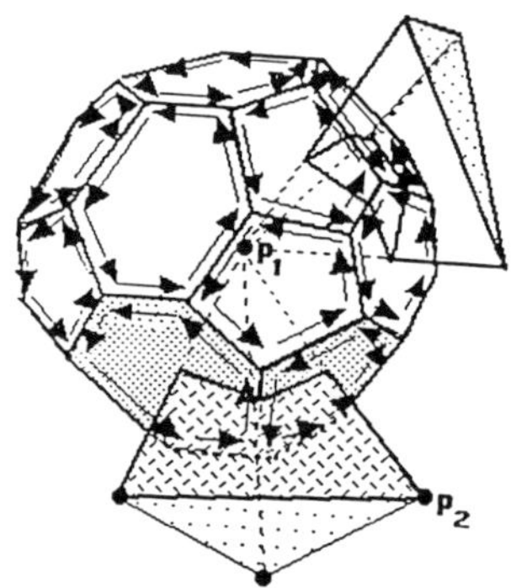

Figure 17.9 Elementary Cycles

The dimension of the elementary cycle basis is $N_F - N_T$; that is, the dimension of $ker(A)$ is $N_F - N_T$. ($ker(A)$ is the *kernel* or the *null* space of A: the set of linearly independent vectors $\{\mathbf{C_1}, \mathbf{C_2}, ..., \mathbf{C_{N_F-N_T}}\}$ all of which satisfy $\mathbf{AC_i} = 0$). In theory (Berge and Ghouila-Houri (1965), Theorem 7) it is straight forward to construct an $N_F \times (N_F - N_T)$ fundamental matrix $\mathbf{C}$, whose columns $\{\mathbf{C_i}\}_{i=1}^{N_F-N_T}$ form a basis for $ker(A)$.

Given the fundamental matrix $\mathbf{C}$ such that $\mathbf{AC} = \mathbf{0}$, we proceed to reduce the size of the primitive system as follows. If we let $\mathbf{D_1 u_0}$ be any particular solution of the discrete continuity equation (17.8), then $(D_1 \mathbf{u} - D_1 \mathbf{u_0})$ is in $ker(A)$. Such a particular solution can be obtained using the notion of a spanning tree for Γ, Berge and Ghouila-Houri (1965). Hence $D_1 \mathbf{u} - D_1 \mathbf{u_0} = C\gamma$ for some vector of *dual variables*, $\gamma = (\gamma_1, \gamma_2, ..., \gamma_{N_F-N_T})^T$, and

$$\mathbf{u} = D_1^{-1}(C\gamma + D_1 \mathbf{u_0}) \tag{17.23}$$

Equation (17.23) expresses the N_F unknown normal components of velocity in terms of the $(N_F - N_T)$ dual variables. If we substitute (17.23) into (17.20), we have a system of N_F ordinary differential equations in the $(N_F - N_T)$ components of γ and the N_T components of $\mathbf{p}$. However, it is also possible to eliminate the pressures from this system since $C^T A^T = (AC)^T = 0$. Multiply (17.20) by $C^T D_2$ after substituting (17.23) into (17.20) to obtain

$$(C^T D_2 D_1^{-1} C)\frac{d\gamma}{dt} + C^T D_2 Q(\gamma)D_1^{-1}C\gamma = C^T D_2(\mathbf{b} - Q(\gamma)\mathbf{u}_0) \qquad (17.24)$$

This is a system of $N_F - N_T$ ordinary differential equations for γ.

Ignoring the faces of tetrahedra in common with $\partial\Omega$, where pressures may be specified, it is known Ewing *et al.* (1970) that N_T (the number of tetrahedra) is approximately $\frac{1}{2}N_F$ (the number of internal tetrahedral faces). Hence, the primitive system is of dimension $N_F + N_T \simeq 3N_T$, and the dual variable system of dimension $N_F - N_T \simeq N_T$, for a reduction factor of 3.

17.8 CONCLUSIONS

In Cavendish *et al.* (1993), Hall *et al.* (1991), (1992) and (1993) we presented an approach that integrates three computational components into a single algorithm for modeling the two-dimensional incompressible Navier-Stokes equations: (i) automatic Delaunay mesh generation, (ii) co-volume finite difference equation generation, and (iii) dual variable reduction of primitive systems. In this paper we have extended this approach to the modeling of the three-dimensional incompressible Navier-Stokes equations. The primitive discrete model results from the approximation of velocity components tangent to boundary edges on Voronoi polyhedral covolumes, and the pressure field is approximated at the corners of the polyhedral surface (the circumcentres of Delaunay tetrahedra).

Finally, as in two dimensions, the DAE's generated by the covolume method in three-dimensions are well suited for variable reduction by the dual variable method. The primitive DAE system can be replaced by an equivalent system of ODE's with dimension a factor of three smaller.

REFERENCES

Berge C. and Ghouila-Houri A. (1965), Programming, Games and Transportation Networks, Methuen, London.

Cavendish, J.C., Field D.A. and Frey W.H. (1985), An Approach to Automatic Three-Dimensional Finite Element Mesh Generation, *Int. J. Num. Meth. Engrg.* **21**: 329-347.

Cavendish, J.C., Field D.A. and Frey W.H. (1984), A Finite Element/Computer Aided Design Interface, *The Mathematics of Finite Elements and Applications V*, Academic Press.

Cavendish, J.C., Hall C.A. and Porsching T.A. (1993), Solution of Incompressible Navier-Stokes Equations on Unstructured Grids Using Dual Tessellations, *Int. J. Numer. Meth. Heat and Fluid Flow*, (to appear)

Ewing D.J., Fawkes A.J. and Griffiths J.R. (1970), Rules Governing the Number of Nodes and Elements in a Finite Element Mesh, *Int. J. Num. Meth. Engrg.* **2**: 597-601

Finney J.L. (1979), A Procedure for the Construction of Voronoi Polyhedra, *J. Comp. Phys.* 137-143.

Hall C.A. (1985), Numerical Solution of Navier-Stokes Problems by the Dual Variable Method, *SIAM J. Alg. and Disc. Meth.* **6**: 220-236.

Hall C.A., Cavendish J.C. and Frey W.H. (1991), The Dual Variable Method for Solving Fluid Flow Difference Equations on Delaunay Triangulations, *Computers and Fluids* **20**: 145-164.

Hall C.A., Porsching T.A. and Mesina G.L. (1992), On a Network Method for Unsteady Incompressible Fluid Flow on Triangular Grids, *Int. J. Numer. Meth. Fluids*, **15**: 1383-1406.

Hall C.A., Porsching T.A. and Hu P. (1993), Covolume-Dual Variable Method for Thermally Expandable Flow on Unstructured Triangular Grids, (submitted for publication).

Nicolaides R.A. (1989(i)), Flow Discretization by Complementary Volume Techniques, *Proc. 9th AIAA CFD Mtg*, Buffalo, N.Y., AIAA paper 89-1978.

Nicolaides R.A. (1989(ii)), Triangular Discretization for the Vorticity-Velocity Equations, *Proc. 7th Int. Conf. Finite Elements in Flow Problems*, Huntsville, Ala.

Nicolaides R.A. (1992), Direct Discretization of Planar div-curl Problems, *SIAM J. Numer. Analysis* **29**: 32-56.

Porsching T.A. (1985), A Network Model for Two-Fluid, Two Phase Flow, *Num. Meth. Part. Diff. Eqs.* **1**: 295-313

Watson D.F. (1981), Computing the n-dimensional Delaunay Tessellation with Applications to Voronoi Polytopes, *Computer Journal* **24**: 167.

Wu X. (1991), Analysis and Applications of the Covolume Method for the Navier-Stokes Equations, Ph.D. dissertation, Carnegie Mellon University.

Chapter 18

A Posteriori and A Priori Error Analysis of Finite Volume Methods

K.W. Morton and Endre Süli

Oxford University Computing Laboratory, 11 Keble Road,
Oxford, OX1 3QD, U.K.

18.1 INTRODUCTION

In recent years finite volume methods have been widely and successfully used in the numerical solution of conservations laws, such as those occurring in fluid dynamics and electromagnetics. The term *finite volume method* was coined by Jameson (1979) for discretizations of the full potential equation of gas dynamics $\nabla \cdot (\rho \nabla \phi) = 0$, where $\rho = \rho(|\nabla \phi|)$ is given by Bernoulli's equation. However the method itself was in use earlier by McDonald (1971) both for this problem, and for the neutron diffusion equation. Earlier still, the finite volume methodology was employed by MacNeal (1953) (see also Varga's classical book — Varga (1962)) to derive finite difference schemes for elliptic equations on irregular two-dimensional grids.

The basic idea behind the construction of finite volume methods is the exploiting of the divergence form of the differential equation by integrating it over a finite (control) volume and the use of Gauss' theorem to convert the result into surface integrals which are then discretized. Much of the practical attraction of this class of methods stems from their compactness and their effectiveness on irregular multi-dimensional meshes. Thus they are now the most widely used numerical methods in the aerospace industry for the solution of transonic gas dynamics equations.

Depending on the location of the point(s) within the cell where the unknowns are kept, finite volume methods can be classified as *cell centre* or *cell vertex* schemes. A familiar example of a cell vertex method for a hyperbolic equation on a rectangular grid is the box scheme which has been particularly popular among hydraulic engineers for the numerical solution of the shallow water equations and related problems with multiple time scales: the resolution of slow waves exploits the second order convergence of the box scheme and that of the fast waves its unconditional stability.

Despite the practical success of finite volume methods, their theoretical foundation is still unsatisfactory and their stability and accuracy properties are not well understood. The aim of this paper is to present an overview of some recent developments concerning finite volume approximations of hyperbolic problems, and elliptic problems with dominant hyperbolic behaviour. Operating within the framework of nonconforming Petrov-Galerkin methods, we develop the analysis of cell-vertex finite volume methods. A key ingredient in the *a priori* error analysis is a discrete Gårding inequality which ensures coercivity in a generalised sense. Hence we derive optimal error bounds in

The Mathematics of Finite Elements and Applications
Edited by J. R. Whiteman © 1994 John Wiley & Sons Ltd

various mesh-dependent norms and investigate the effects of mesh distortion on the accuracy of the scheme. For finite difference operators on $\mathbb{R}^n$ with non-negative symbols a discrete Gårding inequality was first established by Lax and Nirenberg (1966); they have also shown that if the norm of the symbol is bounded by unity then Lax-Richmyer stability can be deduced from the discrete Gårding inequality (see also Vaillancourt (1970)). More recently Michelson (1983) has developed the stability theory of dissipative and weakly dissipative finite difference approximations of multi-dimensional initial boundary value problems for strictly hyperbolic systems using weak and local discrete Gårding inequalities over algebras of pseudo- difference operators, and thereby extended the stability theory of Gustafsson, Kreiss and Sundström (1972) to higher dimensions. However, even on a simple uniform finite difference mesh, the cell vertex finite volume approximation of a hyperbolic initial boundary value problem in two space dimensions gives rise to a non-dissipative difference scheme, the box scheme, which cannot be handled within the theory of Lax-Nirenberg or Michelson.

In order to highlight the general approach to the analysis of cell vertex finite volume methods proposed here, we give a brief sketch of the argument employed in Section 18.4 for linear advection equations. The analysis is based on the work of Süli (1989,1991,1992) and Morton and Süli (1991); for elliptic equations analogous results are presented in Morton and Stynes (1991) and Morton, Stynes and Süli (1993).

Suppose that Ω is a bounded open set in $\mathbb{R}^2$ with a polygonal boundary which has been partitioned by a structured quadrilateral grid. The approximate solution u^h, of the partial differential equation $Lu = f$ in Ω (subject to appropriate boundary conditions on $\partial\Omega$) will belong to an associated trial space $\mathcal{U}^h$ consisting of continuous piecewise bilinear functions, and the test space $\mathcal{M}^h$ will consist of piecewise constant functions on the partition. Then, $u^h \in \mathcal{U}^h$ is defined by requiring that

$$B^h(u^h, p) = (f, p) \quad \forall p \in \mathcal{M}^h,$$

where $B^h(\cdot, \cdot) : \mathcal{U}^h \times \mathcal{M}^h \to \mathbb{R}$ is the sesquilinear form associated with the differential operator L and $(\cdot, \cdot)$ is the $L_2(\Omega)$-inner product. The key idea in the analysis is to construct a mapping $G : \mathcal{U}^h \to \mathcal{M}^h$ such that the following discrete Gårding inequality holds:

$$\Re\, B^h(v, Gv) \geq C\|v\|_h^2 \quad \forall v \in \mathcal{U}^h,$$

where C is a positive constant, independent of h, $\Re$ denotes the real part of a complex number, and $\|\cdot\|_h$ is a suitable discrete norm or seminorm on $\mathcal{U}^h$, sufficiently strong to ensure the stability of the method. If such a mapping G exists, it is necessarily injective and therefore it is invertible on its range. We note in passing that the invertibility of G becomes important in the implementation of the method in that the inverse mapping $G^{-1} : \mathcal{M}^h \to \mathcal{U}^h$ defines a correspondence between a control volume and associated degrees of freedom; we shall return to this point in the next section where we discuss some of the practical aspects of the scheme including the solution of the cell vertex equations. By extending the sesquilinear form $B^h(\cdot, \cdot)$ to a class of suitably smooth functions, which contains the solution u, so that $B^h(u, p) = (f, p)$ for all $p \in \mathcal{M}^h$, and denoting by $I^h u$ the continuous bilinear interpolant of u, we have that

$$C\|I^h u - u^h\|_h^2 \leq \Re\, B^h(I^h u - u^h, G(I^h u - u^h))$$
$$= \Re\, B^h(I^h u - u, G(I^h u - u^h)) + \Re\, B^h(u - u^h, G(I^h u - u^h))$$
$$= \Re\, B^h(I^h u - u, G(I^h u - u^h)).$$

The right-hand side can be bounded in terms of $\|I^h u - u^h\|_h$ and a norm of $I^h u - u$. The former can be absorbed into the left-hand side, leading to a bound on $\|I^h u - u^h\|_h$ in terms of $I^h u - u$. An application of the triangle inequality then yields the *a priori* error bound on $\|u - u^h\|_h$ in terms of the interpolation error $I^h u - u$.

In the final part of the paper, using a decomposition of the global error into a local part and a propagating component, we derive residual-based local *a posteriori* error bounds and indicate their relevance in the construction of error indicators and error estimators for adaptive mesh refinement algorithms.

18.2 CELL VERTEX SCHEME FOR COMPRESSIBLE FLOW PROBLEMS

18.2.1 CELL RESIDUALS

In a bounded open set $\Omega \subset \mathbb{R}^2$ with a Lipschitz continuous boundary, we consider the system of conservation laws for the unknown vector function w contained in $[BV(\Omega)]^2$:

$$\frac{\partial}{\partial x} f(w(x,y)) + \frac{\partial}{\partial y} g(w(x,y)) = 0 \tag{18.1}$$

subject to appropriate boundary conditions on w which we shall ignore for the time being; here f and g are continuously differentiable flux vectors. The generalised derivatives appearing in (18.1) are Radon measures and therefore the equality is to be understood in the sense of measures.

Let h denote the spatial discretization parameter which approximates zero and suppose that Ω has been partitioned by a finite family $\mathcal{T}^h$ of disjoint non-empty open sets K_α^h with Lipschitz continuous boundaries, and $\mathrm{diam}(K_\alpha^h) \leq h$ for all α. For the sake of simplicity we shall assume that each of these sets is an open convex quadrilateral.

According to Vol'pert's generalisation of the Gauss-Green formula to BV-functions (see Volpert (1967)) and noting that the inward trace of a BV-function on ∂K_α^h exists and is integrable with respect to the one-dimensional Hausdorff measure H_1 on ∂K_α^h (see Giusti (1984)), we have that

$$0 = \frac{1}{|K_\alpha^h|} \int_{K_\alpha^h} \frac{\partial}{\partial x} f(w(x,y)) + \frac{\partial}{\partial y} g(w(x,y)) dx dy$$
$$= \frac{1}{|K_\alpha^h|} \int_{\partial K_\alpha^h} (f(w), g(w))^+ \cdot \mathbf{n}\, dH_1, \tag{18.2}$$

where $(\cdot)^+$ signifies the inward trace operator on ∂K_α^h, $\mathbf{n}$ denotes the unit outward normal vector (defined H_1-almost everywhere) on ∂K_α^h and $|K_\alpha^h| = \mathrm{meas}(K_\alpha^h)$.

The cell vertex scheme is constructed from identity (18.2) by rewriting the contour integral on the right as a sum of four line integrals, each along a side of the quadrilateral K_α^h, and evaluating these line integrals approximately by the trapezium rule. Thus labelling the vertices of K_α^h by 1, 2, 3, 4, starting with the lower right corner and proceeding anticlockwise, we obtain

$$R_\alpha(w) = \frac{1}{2|K_\alpha^h|}\{(f(w_1) - f(w_3))\delta y_{24} + (f(w_2) - f(w_4))\delta y_{31}$$
$$-(g(w_1) - g(w_3))\delta x_{24} - (g(w_2) - g(w_4))\delta x_{31}\} \approx 0, \qquad (18.3)$$

where (x_i, y_i) denotes the coordinate of the vertex numbered i, $w_i = w(x_i, y_i)$, $\delta x_{ij} = x_i - x_j$ and $\delta y_{ij} = y_i - y_j$. The finite volume approximation of equation (18.1) is defined as follows:

find a mesh function w^h such that

$$R_\alpha(w^h) = 0 \quad \forall K_\alpha^h \in T^h.$$

This compact four-point approximation of the conservation law has many attractive properties, the most important being its accuracy on distorted quadrilateral grids. Morton and Paisley (1989) have shown that the truncation error, $R_\alpha(w)$, is $O(h^2)$ provided w is a smooth function and K_α^h is close to a parallelogram in a sense that will be made precise in Section 18.4 (cf. Hypothesis H2(i)). Second-order convergence on structured quadrilateral partitions, under similar regularity of the partition, has been shown by Süli (1989,1991,1992) for steady and unsteady linear hyperbolic problems using a finite element analysis.

Indeed, by noting that the trapezium rule integrates linear functions exactly, (18.3) can be restated as a Petrov-Galerkin finite element method. For this purpose, let us denote by $\mathcal{U}^h$ the set of all continuous bilinear vector functions on the partition T^h, and let $\mathcal{M}^h$ be the set of scalar piecewise constant functions on T^h. Regarding w^h as an element of $\mathcal{U}^h$, the finite volume method can be rewritten as follows:

find $w^h \in \mathcal{U}^h$ such that

$$B^h(w^h, p) \equiv \sum_{K_\alpha^h \in T^h} |K_\alpha^h|\, p|_{K_\alpha^h} \frac{1}{|K_\alpha^h|} \int_{\partial K_\alpha^h} I^h(f(w^h), g(w^h)) \cdot \mathbf{n}\, dH_1 = 0 \qquad (18.4)$$

for all $p \in \mathcal{M}^h$, where I^h is the interpolation projector onto $[\mathcal{U}^h]^2$. Alternatively, in terms of cell residuals, (18.4) can be written

find $w^h \in \mathcal{U}^h$ such that

$$B^h(w^h, p) \equiv \sum_{K_\alpha^h \in T^h} |K_\alpha^h|\, p|_{K_\alpha^h} R_\alpha(w^h) = 0 \quad \forall p \in \mathcal{M}^h. \qquad (18.5)$$

Implicit in the transition from (18.2) to (18.3) has been the assumption that $(f(w(\cdot)))^+$ and $(g(w(\cdot)))^+$ are continuous functions along the boundary of the cell.

If the solution is discontinuous then this is no longer true. The gross inaccuracies of the cell vertex scheme across shocks have been studied and overcome in Morton and Paisley (1989) by a shock fitting algorithm. In the present work we adopt a more conventional shock capturing approach by adding artifical dissipation in the vicinity of shocks. The details of this are described below.

18.2.2 NODAL RESIDUALS, DISTRIBUTION MATRICES AND ARTIFICIAL VISCOSITY

Having defined the cell-based residual $R_\alpha(w^h)$, it remains to describe the solution procedure to determine w^h.

The box scheme for the St. Venant equations can be marched forward in time by solving for the cell residuals by Newton's method applied to the nonlinear block-tridiagonal system obtained for the new time level. Similarly, for cell vertex approximations of some steady problems this is possible when there is a time-like direction (for example, in a supersonic flow past a body with no subsonic pockets at the leading edge). However for a general steady compressible flow, the system (18.1) is of mixed hyperbolic/elliptic type and the solution of the cell residual equations by marching in a particular direction is not possible. Worse still, in general it is not even possible to ensure that the number of unknowns equals the number of cell equations (see Morton (1991)).

In order to overcome this problem we adopt an iterative algorithm of setting combinations of residuals to zero. This is achieved by distributing each cell residual to the four corners of a cell which creates a node-based residual equation for each unknown. The system of nodal equations can then be solved iteratively by a relaxation method. This approach is a genuine generalisation of the marching schemes mentioned above (see Crumpton *et al.*(1991), Morton *et al.*(1991)).

For a cell K^h_α, the distribution of cell residual R_α to the nodes whose local numbers are $j = 1,...,4$ is effected by the distribution matrices $D_{\alpha,j}$. Thus any node j surrounded by cells K^h_α and corresponding residuals R_α, has a nodal nonlinear equation of the form

$$N_j(w^h) \equiv \frac{\sum_{\alpha \in N(j)} |K^h_\alpha|(D_{\alpha,j} R_\alpha(w^h) + A^2_{\alpha,j} + A^4_{\alpha,j})}{\sum^p_{\alpha \in B(j)} |K^h_\alpha|} = 0, \qquad (18.6)$$

where $A^2_{\alpha,j}$ and $A^4_{\alpha,j}$ are second- and fourth-order artificial dissipation terms; $N(j)$ is the set of element numbers surrounding node j, and $B(j) = N(j)$ when j is an interior node. When j is a boundary node $N(j)$ contains fewer than four elements whereas the cardinality of the set $B(j)$ is always 4; this is achieved by including ficticious element numbers into $B(j)$ corresponding to dummy elements outside the domain whose measure is equal to the average of all $|K^h_\alpha|$, $\alpha \in N(j)$. The iterative algorithm which is used to solve this is

$$w^{h,n+1}_j = w^{h,n}_j - \omega_j N_j(w^{h,n}), \quad n = 0, 1, ..., \qquad (18.7)$$

where ω_j is a local scalar relaxation parameter.

The simplest choice of the distribution matrices is $D_{\alpha,j} = I$, for each α and j, where I is the identity matrix; this would assign to each node j the area-weighted average of residuals in surrounding cells. By further developing this simple choice, more general distribution matrices can be devised to introduce some degree of upwinding into the scheme in a local manner.

For example, on a structured quadrilateral mesh where each node is surrounded by four cells, we may define the distribution matrices by the formula

$$D_{\alpha,j} = I - \nu_C \frac{\Delta t_\alpha}{|K_\alpha^h|}(\delta y_{j-1,j+1} A_\alpha - \delta x_{j-1,j+1} B_\alpha), \quad j = 1,...,4, \qquad (18.8)$$

with cyclic indexing, where A_α and B_α denote the Jacobi matrices of the flux vectors f and g evaluated at the centroid of K_α^h; ν_C is a global cell-based Courant number. These distribution matrices can also be derived from a generalised Lax-Wendroff or Taylor-Galerkin pseudo-timestepping procedure — see Morton (1988). The relaxation parameter is $\omega_j = \nu_N \Delta t_j = \nu_N \min(\Delta t_\alpha, \Delta t_\beta, \Delta t_\gamma, \Delta t_\delta)$, where ν_N is a global node-based Courant number.

The two Courant numbers ν_C and ν_N play different roles. As shown in Crumpton *et al.* (1991) and Morton *et al.* (1991) by a Fourier analysis of the iteration for a scalar linear hyperbolic equation, the node-based Courant number controls the phase speed of waves and hence the transport of errors to the boundary. Taking the cell-based residual ν_C greater than ν_N introduces some second order dissipation beyond the usual fourth-order one obtained with $\nu_N = \nu_C$.

There are two specific reasons for introducing artificial dissipation as well as the distribution matrices in (18.6). As indicated earlier, at shocks our definition of $R_\alpha(w^h)$ may have an $O(1)$ truncation error (see Morton and Paisley (1989)). In the opposite extreme of expanding flow there is an excess of unknowns which are not made up by the boundary conditions. The two differing artificial dissipation terms are required to deal with these two very different situations. In the first case the second-order dissipation models the physical viscosity to overcome the inadequacy of the mathematical model at the shock: shocks are detected and second-order dissipation is applied in their neighbourhood only. In the second case, a fourth-order dissipation term is required to select a smooth solution to the residual equations by cutting out spurious solutions not eliminated by the boundary conditions.

In order to incorporate the distribution matrices into the Petrov-Galerkin formulation (18.5) of the cell vertex scheme, let us define, for each node j, the matrix-valued piecewise constant test function

$$\mathcal{D}_j = \frac{\sum\limits_{\alpha \in N(j)} \chi_\alpha D_{\alpha,j}}{\sum\limits_{\alpha \in B(j)} (\chi_\alpha, 1)}, \qquad (18.9)$$

where χ_α is the characteristic function of the cell K_α^h. Adopting the notation

$$B^h(w^h, \chi_\alpha D_{\alpha,j}) = \int_\Omega \chi_\alpha D_{\alpha,j} R_\alpha(w^h) dx,$$

the node-based iteration scheme can be rewritten as

$$w_j^{h,n+1} = w_j^{h,n} - \omega_j B^h(w^{h,n}, \mathcal{D}_j) \quad \forall j, \quad n = 0, 1, \dots$$

At convergence of this iteration, and in the absense of any artificial dissipation, the nodal residuals $N_j(w^h)$ are driven to zero for all j. To deduce that the cell residuals $R_\alpha(w^h)$ have also converged to zero requires the invertibility of the mapping between cells and nodes through the correct choice of the distribution matrices and boundary conditions. This can be ensured for the scalar linear advection equation and for one- and two-dimensional systems of conservation laws by defining the distribution matrices through pure upwinding (see Morton *et al.* (1991)).

One-dimensional studies of the cell vertex scheme have included transonic inviscid problems (Morton *et al.* (1991)) and advection-diffusion problems as a model of viscous phenomena (Morton and Stynes (1991)). The present paper focuses on two-dimensional effects. The subsonic Euler equations can be decomposed into a pair of quantities (entropy and vorticity) convected along the velocity field lines and a pair of equations representing an elliptic system (the pressure waves). We wish to emphasize the hyperbolic effects, so the key problem is scalar advection in two dimensions.

18.3   SCALAR ADVECTION MODEL PROBLEMS

Let $\Omega = (0, 1)^2$ be the open unit square in $\mathbb{R}^2$ and suppose that $a = (a_1, a_2)$ is a vector-function defined on $\bar{\Omega}$ with continuously differentiable entries. Given $f \in L_2(\Omega)$, we consider the linear hyperbolic equation

$$\nabla \cdot (au) - f = 0 \quad \text{in} \quad \Omega$$
$$u = 0 \quad \text{on} \quad \partial_-\Omega,$$

where $\partial_-\Omega = \{\mathbf{u} \in \partial\Omega : a(\mathbf{x}) \cdot n(\mathbf{x}) < 0\}$ denotes the inflow boundary.

Consider three distinct model problems, which are representative of the general case:-

(a) Transversal flow. Suppose $a_1 \cdot a_2 \neq 0$ on $\bar{\Omega}$ (i.e. both a_1 and a_2 are of constant sign on $\bar{\Omega}$; say $a_1 > 0$ and $a_2 > 0$.)

(b) Compression Wave. Suppose $a_1 > 0$ on $\bar{\Omega}$, and $a_2(x, y) = (2y - 1)A(x, y)$, where $A > 0$ on $\bar{\Omega}$.

(c) Expansion Wave. Suppose $a_1 > 0$ on $\bar{\Omega}$, and $a_2(x, y) = (1 - 2y)A(x, y)$ where $A > 0$ on $\bar{\Omega}$.

Case (a) will be considered in more detail in the next section. In particular for a constant linear advection equation with $a_1 > 0$ and $a_2 > 0$ the cell vertex finite volume scheme will be shown to be stable and optimally accurate. At the heart of the analysis in this case is the fact that there is a one-to-one correspondence between unknowns and cell-residuals. For the general variable coefficient problem where the conditions of (a) hold such a correspondence still exists and the stability and convergence results established for the constant coefficient problem can be extended to this case (see Morton and Süli (1993)).

However, in the cases (b) and (c), the situation is radically different: in case (b) there is an excess of cell residuals compared with the number of unknowns, whereas in case (c) the converse is true. In order to overcome these two counting problems the basic cell vertex scheme has to be modified appropriately. At the previous conference one of us in Morton (1991) proposed the splitting and merging of cells to obtain a correct match in these cases. For the one-dimensional Euler equations this was shown in Morton *et al.* (1991) to work very successfully. It has also been implemented for the present model problems, but numerical experiments have not given particularly encouraging results. In any case, although such an approach was hoped to produce benchmark solutions, it is clear that it would be difficult to apply to systems of equations in several dimensions.

Hence we are driven back to the general algorithm of the previous section, summarised in (18.6) and (18.7). It is this that we need to analyse and, by that means, tune it to our purpose of driving the cell residuals as close to zero as possible. In case (b) of the compression wave no artificial dissipation terms are needed: since a shock cannot develop in this linear problem we do not need A^2, and, as we have already noted, A^4 is introduced to deal with spurious modes not controlled by the boundary conditions, which will apply in case (c) and not in this case. What we do need here are the distribution matrices, which in effect reduce the dimensionality of the test space by introducing the test functions of (18.9). In the analysis one has to show how these can be chosen so as to maintain the accuracy obtained in case (a). The problem is related to that of choosing the pressure space in a Stokes problem to satisfy the Babuška-Brezzi condition.

In the expansion wave model problem of case (c), the distribution matrices are not needed in principle but the artificial dissipation A^4 is important. Suppose we consider the problem as a constrained optimisation problem, in which we seek the smoothest solution that satisfies the cell residual equations, since these are too few to determine it uniquely. Suppose further that we use an iterative procedure to find this solution and that we take as a measure of smoothness $\|Mw^h\|_{l_2}$, where M is a second order difference operator. Then it can be shown that one arrives at an iteration of the form (18.7) with $A^2 = 0$ and the fourth-order dissipation arising from $M^T M w^h$.

These two cases will be considered in detail in Morton and Süli (1993), but for the present we return to the simpler case (a) whose analysis forms the foundation for the others.

18.4   THE ANALYSIS FOR LINEAR HYPERBOLIC EQUATIONS

Let Ω denote the open unit square $(0,1) \times (0,1)$ in $\mathbb{R}^2$. For $\mathbf{a} = (a_1, a_2)$, a two-component vector-function with continuously differentiable real-valued entries defined on $\overline{\Omega}$, we introduce the following subsets of $\partial\Omega$:

$$\partial_-\Omega = \{\mathbf{x} \in \partial\Omega \,:\, \mathbf{a}(\mathbf{x}) \cdot \mathbf{n}(\mathbf{x}) < 0\}, \quad \partial_+\Omega = \{\mathbf{x} \in \partial\Omega \,:\, \mathbf{a}(\mathbf{x}) \cdot \mathbf{n}(\mathbf{x}) \geq 0\},$$

where $\mathbf{n}(\mathbf{x})$ denotes the unit outward normal to $\partial\Omega$ at $\mathbf{x} \in \partial\Omega$. Suppose that b is a continuous complex-valued function defined on $\overline{\Omega}$, and let f be a square-integrable complex-valued function on Ω.

We consider the boundary value problem

$$\nabla \cdot (\mathbf{a}u) + bu = f \qquad \text{in } \Omega, \tag{18.10}$$

$$u = 0 \qquad \text{on } \partial_-\Omega. \tag{18.11}$$

In order to transform this problem into its weak formulation, we associate with $\mathbf{a}$ the space $H^1_-(\Omega)$ consisting of all those elements of the complex Sobolev space $H^1(\Omega)$ whose trace on the inflow boundary $\partial_-\Omega$ is zero, and define the sesquilinear form $B : H^1_-(\Omega) \times L_2(\Omega) \to \mathbf{C}$ associated with the hyperbolic operator appearing in equation (18.10) by

$$B(u, p) = (\nabla \cdot (\mathbf{a}u) + bu, p).$$

Now problem (18.10), (18.11) can be restated as follows: find $u \in H^1_-(\Omega)$ satisfying

$$B(u, p) = (f, p) \qquad \forall p \in L_2(\Omega). \tag{18.12}$$

In order to construct the finite volume approximation of (18.12), we consider $\mathcal{F} = \{T^h\}$, $h > 0$, a family of partitions $T^h = \{K^h_\alpha\}$ where each K^h_α is a convex quadrilateral. We assume that, with $T^h = \{K^h_\alpha\}$, $\alpha = 1, 2, ..., m_h$, $\overline{\Omega} = \bigcup_{\alpha=1}^{m_h} \overline{K}^h_\alpha$, and that each pair $\overline{K}^h_\alpha$, $\overline{K}^h_\beta$, $\alpha \neq \beta$, has either an entire side or a vertex in common, or has empty intersection.

Let $h_{K^h_\alpha}$ denote the diameter of K^h_α, and let $\rho_{K^h_\alpha}$ denote the maximum diameter of circles contained in K^h_α. We denote by $P_{K^h_\alpha}$ and $Q_{K^h_\alpha}$ the midpoints of the diagonals of K^h_α. The family $\mathcal{F}$ will be assumed to possess the following regularity properties:

Hypothesis H1

$\mathcal{F}$ is structured; i.e., for each $h > 0$, T^h is topologically equivalent to a rectangular partition of Ω.

Hypothesis H2

There exist two constants $c_0 \geq 0$ and $c_1 > 0$ independent of h such that for all $K^h_\alpha \in T^h$, $T^h \in \mathcal{F}$,

(i) $dist(P_{K^h_\alpha}, Q_{K^h_\alpha}) \leq c_0 m(K^h_\alpha)$,

(ii) $h_{K^h_\alpha} \leq c_1 \rho_{K^h_\alpha}$.

We assume that $h = max\{h_{K^h_\alpha} : K^h_\alpha \in T^h\}$ approximates zero.

The finite volume discretization of (18.12) is performed on a family of partitions satisfying hypotheses H1 and H2. We define the finite element spaces

$$\mathcal{U}^h = \{v \in H^1(\Omega) : v|_{K^h_\alpha} \in Q_1(K^h_\alpha), \ K^h_\alpha \in T^h\},$$

$$\mathcal{M}^h = \{p \in L_2(\Omega) : p|_{K^h_\alpha} \in Q_0(K^h_\alpha), \ K^h_\alpha \in T^h\},$$

and $\mathcal{U}^h_- = \mathcal{U}^h \cap H^1_-(\Omega)$. Here $Q_1(K^h_\alpha)$ (respectively, $Q_0(K^h_\alpha)$ is the set of all bilinear (respectively, constant) functions on the quadrilateral K^h_α. Let $I^h : (H^1_-(\Omega) \cap C(\overline{\Omega}))^2 \to (\mathcal{U}^h_-)^2$ be the interpolation projector onto $(\mathcal{U}^h_-)^2$. The discrete analogue of the sesquilinear form is given by

$$B^h(v, p) = (\nabla \cdot I^h(\mathbf{a}v) + I^h(bv), p) \qquad \forall v \in \mathcal{U}^h_- \ \forall p \in \mathcal{M}^h.$$

We define the finite volume approximation of (18.12) as follows: find $u^h \in \mathcal{U}_-^h$ satisfying

$$B^h(u^h, p) = (f, p) \qquad \forall p \in \mathcal{M}^h. \tag{18.13}$$

As a simple first instance which represents a centrally important class of problems, we consider the linear hyperbolic equation (18.10) with constant coefficients. Thus, in the rest of the Section b will be assumed to be a fixed complex number and $\mathbf{a}$ will be taken a fixed constant vector. We can also assume, without restricting generality, that both entries of $\mathbf{a}$ are positive, in which case the inflow boundary $\partial_-\Omega$ coincides with the intersection of $\partial\Omega$ with the coordinate axes. Our results can be extended, at the expense of some technical difficulties, to problems with variable coefficients, provided that the components of $\mathbf{a}$ are of constant sign, or more generally, to domains which can be decomposed into a finite number of rectangular subdomains over each of which the components of $\mathbf{a}$ have constant sign. The algorithmic aspects of the method for such a general case have been described in the previous section.

By virtue of H1, for each $\mathcal{T}^h \in \mathcal{F}$ there exists a pair of positive integers $M(h)$, $N(h)$, such that m_h, the cardinality of $\mathcal{T}^h$, equals $M(h)N(h)$, and with each $K_\alpha^h \in \mathcal{T}^h$ we can associate a pair (i, j), $0 \le i \le M(h) - 1$, $0 \le j \le N(h) - 1$. Thus we label K_α^h by the subscript ij and write K_{ij}^h instead. The vertices of K_{ij}^h will be denoted $\mathbf{x}_{ij}^h$, $\mathbf{x}_{i+1,j}^h$, $\mathbf{x}_{i+1,j+1}^h$, $\mathbf{x}_{i,j+1}^h$, starting with the lower-left corner and labelling anticlockwise.

For a partition $\mathcal{T}^h = \{K_{ij}^h \; : \; 0 \le i \le M(h) - 1, \, 0 \le j \le N(h) - 1\}$, let

$$\Omega_{kl}^h := \bigcup_{i=0}^{k-1} \bigcup_{j=0}^{l-1} \overline{K}_{ij}^h, \quad \partial_+\Omega_{kl}^h := \partial\Omega_{kl}^h \setminus \partial_-\Omega, \quad k = 1, ..., M(h), \; l = 1, ..., N(h).$$

Clearly $\Omega_{M(h),N(h)}^h = \overline{\Omega}$ and $\partial_+\Omega_{M(h),N(h)}^h = \partial_+\Omega$ for all h. For a continuous function v on $\overline{\Omega}$, we define the following local averages:

$$\mu v_{ij} = \frac{1}{m(K_{ij}^h)} \int_{K_{ij}^h} v \, dx,$$

$$\mu_1 v_{ij} = \frac{1}{|\mathbf{x}_{i+1,j}^h - \mathbf{x}_{ij}^h|} \int_{[\mathbf{x}_{ij}^h, \mathbf{x}_{i+1,j}^h]} v \, ds, \quad \mu_2 v_{ij} = \frac{1}{|\mathbf{x}_{i,j+1}^h - \mathbf{x}_{ij}^h|} \int_{[\mathbf{x}_{ij}^h, \mathbf{x}_{i,j+1}^h]} v \, ds.$$

With these notations, we introduce the mesh-dependent seminorms:

$$\|v\|_{l_2(\Omega_{kl}^h)}^2 = \sum_{i=0}^{k-1} \sum_{j=0}^{l-1} m(K_{ij}^h) |\mu v_{ij}|^2,$$

$$\|v\|_{l_2(\partial_+\Omega_{kl}^h)}^2 = \sum_{i=0}^{k-1} |\mathbf{x}_{i+1,l}^h - \mathbf{x}_{il}^h| \, |\mu_1 v_{il}|^2 + \sum_{j=0}^{l-1} |\mathbf{x}_{k,j+1}^h - \mathbf{x}_{kj}^h| \, |\mu_2 v_{kj}|^2,$$

$$\|v\|_{\dot{l}_2(\Omega_{kl}^h)}^2 = \sum_{i=0}^{k-1} \sum_{j=1}^{l-1} m(K_{ij}^h \cup K_{i,j-1}^h) |\mu_1 v_{ij}|^2 + \sum_{i=1}^{k-1} \sum_{j=0}^{l-1} m(K_{ij}^h \cup K_{i-1,j}^h) |\mu_2 v_{ij}|^2.$$

The choice of the inflow boundary conditions ensures that each of these is a norm on the linear space $\mathcal{U}_-^h$. In the definition of the norm $\| \cdot \|_{\dot{l}_2(\Omega_{kl}^h)}$ we adopt the convention that empty sums, corresponding to $l = 1$ or $k = 1$, are equal to zero.

Let χ_{kl} denote the characteristic function of Ω_{kl}^h, and let $P^h : L_2 \to \mathcal{M}^h$ be the orthogonal projector in $L_2(\Omega)$ onto $\mathcal{M}^h$. The stability of the finite volume method (18.13) is based on the following discrete Gårding inequality (see Süli (1989,1991,1992)).

Theorem 18.1 *Let* $c_a := min(a_1, a_2)/|\mathbf{a}|$, $c_3 := (\Re b/|\mathbf{a}|) - 4c_0$, *and assume that hypotheses H1 and H2 hold with* $c_2 := c_a - 8c_0 c_1^2(1 + 2c_0) > 0$. *Then, for all* k, l, $1 \leq k \leq M(h)$, $1 \leq l \leq N(h)$, *and all* $v \in \mathcal{U}_-^h$,

$$\Re B^h(v, \chi_{kl} P^h v) \geq c_3 |\mathbf{a}| \|v\|_{l_2(\Omega_{kl}^h)}^2 - \frac{1}{2} c_0 |\mathbf{a}| \|v\|_{l_2^\circ(\Omega_{kl}^h)}^2 + \frac{1}{2} c_2 |\mathbf{a}| \|v\|_{l_2(\partial_+ \Omega_{kl}^h)}^2.$$

Now we are ready to prove the stability of the cell vertex scheme.

Theorem 18.2 *Assume that H1 and H2 hold with* $c_3 := (\Re b/|\mathbf{a}|) - 4c_0 > 0$ *and* $c_2 = c_a - 8c_0 c_1^2(1 + 2c_0) > 0$, *then (18.13) has a unique solution* $u^h \in \mathcal{U}^h$, *and*

$$c_3 \|u^h\|_{l_2(\Omega)}^2 + c_2 \|u^h\|_{l_2(\partial_+ \Omega)}^2 \leq c_4 \|f\|_{l_2(\Omega)}^2, \tag{18.14}$$

where $c_4 = c_3^{-1} |\mathbf{a}|^{-2} \exp\{c_5 + c_5 \exp(c_5)\}$ *and* $c_5 = 8c_0 c_1^2(1 + 2c_0)/c_2$.

Proof. The existence of a unique solution follows from (18.14) by the Fredholm alternative. In order to establish (18.14), let us choose $p = \chi_{kl} P^h v$ in (18.13). Employing Theorem 18.1, we obtain:

$$c_3 |\mathbf{a}| \|u^h\|_{l_2(\Omega_{kl}^h)}^2 - c_0 |\mathbf{a}| \|u^h\|_{l_2^\circ(\Omega_{kl}^h)}^2 + c_2 |\mathbf{a}| \|u^h\|_{l_2(\partial_+ \Omega_{kl}^h)}^2 \leq \frac{1}{c_3 |\mathbf{a}|} \|f\|_{l_2(\Omega_{kl}^h)}^2$$

for $1 \leq k \leq M(h)$ and $1 \leq l \leq N(h)$. Moving the second term on the left to the right-hand side, recognising the structural similarity between $\|u^h\|_{l_2(\partial_+ \Omega_{kl}^h)}^2$ and its two-dimensional analogue $\|u^h\|_{l_2^\circ(\Omega_{kl}^h)}^2$, and applying a bivariate discrete Gronwall lemma (see Süli (1989,1991,1992) for details) we obtain (18.14). $\square$

On a tensor-product non-uniform mesh, H1 is automatically satisfied and, since each $K_{ij}^h \in \mathcal{T}^h$ is a rectangle, H2(i) also holds with $c_0 = 0$. In fact the dependence on c_1 can also be removed and stability can be shown without any further assumptions on the regularity of the mesh (see Süli (1989,1991,1992) for details).

Theorem 18.3 *Suppose that* $\mathcal{F} = \{\mathcal{T}^h\}$ *is a family of rectangular partitions of* $\overline{\Omega}$ *and assume that* $c_a := \min(a_1, a_2)/|\mathbf{a}| > 0$ *and* $c_3 := \Re b/|\mathbf{a}| > 0$. *Then*

$$c_3 \|u^h\|_{l_2(\Omega)}^2 + c_a \|u^h\|_{l_2(\partial_+ \Omega)}^2 \leq \frac{1}{c_3 |\mathbf{a}|^2} \|f\|_{l_2(\Omega)}^2.$$

The convergence of the cell vertex scheme and appropriate error bounds can now be deduced from these stability results.

Theorem 18.4 *Let* $c_3 := (\Re b/|\mathbf{a}|) - 4c_0 > 0$. *Assume that the family of partitions* $\mathcal{F} = \{\mathcal{T}^h\}$ *satisfies hypotheses H1 and H2 with* $c_2 := c_a - 8c_0 c_1^2(1 + 2c_0) > 0$, *and let* $u \in H^3(\Omega)$. *Then*

$$\|u - u^h\|_{l_2(\Omega)} + \|u - u^h\|_{l_2(\partial_+ \Omega)} \leq Ch^2 \|u\|_{H^3(\Omega)}, \tag{18.15}$$

where $C = C(c_0, c_1, c_2, |\mathbf{a}|, b)$ *and* $h = \max_{K \in \mathcal{T}^h} h_K$.

Proof. Let $\eta = u - I^h u$ and $\xi = I^h u - u^h$. Then $u - u^h = \eta + \xi$. We begin by estimating ξ. Clearly $B^h(\xi, p) = -(\nabla \cdot (\mathbf{a}\eta) + b\eta, p)$ for all p in $\mathcal{M}^h$. Thus ξ is the solution of problem (18.13) with f replaced by $-(\nabla \cdot (\mathbf{a}\eta) + b\eta)$, and by virtue of Theorem 18.2,

$$\|\xi\|_{l_2(\Omega)}^2 + \|\xi\|_{l_2(\partial_+\Omega)}^2 \leq C(c_0, c_1, c_2, |\mathbf{a}|, b)\|\nabla \cdot (\mathbf{a}\eta) + b\eta\|_{l_2(\Omega)}^2.$$

Using the triangle inequality,

$$\|u - u^h\|_{l_2(\Omega)} + \|u - u^h\|_{l_2(\partial_+\Omega)}$$
$$\leq C(c_0, c_1, c_2, |\mathbf{a}|, b)(\|\eta\|_{l_2(\Omega)} + \|\eta\|_{l_2(\partial_+\Omega)} + \|\nabla \cdot (\mathbf{a}\eta)\|_{l_2(\Omega)}). \qquad (18.16)$$

The first two terms on the right are easily estimated by using the Bramble-Hilbert lemma together with hypothesis H2:

$$\|\eta\|_{l_2(\Omega)} \leq C(c_0, c_1)h^2\|u\|_{H^2(\Omega)}, \qquad (18.17)$$

$$\|\eta\|_{l_2(\partial_+\Omega)} \leq C(c_0, c_1)h^2\|u\|_{H^2(\partial_+\Omega)}. \qquad (18.18)$$

The nontrivial part of the proof consists of estimating the last term on the right hand-side of (18.16). For this we refer to Süli (1991,1992) where it is shown that

$$\left|\frac{1}{m(K)} \int_K \nabla \cdot (\mathbf{a}\eta)dx\right| \leq C(c_0, c_1)|\mathbf{a}|h_K\|u\|_{H^3(K)},$$

and therefore

$$\|\nabla \cdot (\mathbf{a}\eta)\|_{l_2(\Omega)} \leq C(c_0, c_1)h^2|\mathbf{a}|\,\|u\|_{H^3(K)}, \qquad (18.19)$$

Substituting (18.17)–(18.19) into (18.16), we obtain (18.15). $\square$

For a family of tensor product non-uniform meshes second-order convergence can be shown without any additional regularity assumption on the mesh (see Süli (1989,1991,1992) and Morton and Süli (1991)).

Theorem 18.5 *Suppose that $\mathcal{F} = \{T^h\}$ is a family of rectangular partitions of $\overline{\Omega}$, $\Re b > 0$, and let $u \in H^3(\Omega)$. Then*

$$\|u - u^h\|_{l_2(\Omega)} + \|u - u^h\|_{l_2(\partial_+\Omega)} \leq C(|\mathbf{a}|, b)h^2|u|_{H^3(\Omega)}.$$

Here we have only considered steady hyperbolic problems. The analysis of the cell vertex scheme for a time-dependent linear hyperbolic equation is given in Süli (1989,1992). Alternative stability and error bounds without the restrictive assumption $(\Re b/\mathbf{a}) - 4c_0 > 0$ have been derived by Morton and Stynes (1991).

18.5   A POSTERIORI ERROR BOUNDS: THE HYPERBOLIC CASE

The accurate and efficient resolution of localised structures, such as shocks and boundary layers, in complex multi-dimensional fluid flow problems necessitates the use of locally refined adaptive computational grids. The construction of such grids should be guided by sharp *a posteriori* error estimates which ensure that the adaptive algorithm is both reliable (i.e. a norm of the error is guaranteed to be below a certain tolerance) and efficient (i.e. it does not produce a grid which is overly refined).

In this section, as a first step towards developing successful adaptive algorithms for the compressible Euler and Navier-Stokes equations we derive residual-based two-sided *a posteriori* error bounds for finite volume approximations of steady two-dimensional hyperbolic problems. *A posteriori* error estimates for the cell vertex finite volume method were first derived by Süli (1989), and implemented by Mackenzie *et al.* (1992) (see also Sonar (1992)). Further work in this direction, including numerical experiments, has been carried out in Mackenzie, Sonar and Süli (1993(i)) and Mackenzie, Süli and Warnecke (1993(ii)). The results presented below sharpen the estimates given in these papers.

Given a real two-component vector function $\mathbf{a} \in [C^1(\overline{\Omega})]^2$ with non-negative entries a_1 and a_2 such that $|a_1(x,y)| + |a_2(x,y)| \geq c_* > 0$, consider the model hyperbolic boundary value problem

$$Lu \equiv \nabla \cdot (au) + bu = f, \quad x \in \Omega,$$
$$u = 0, \quad x \in \partial_- \Omega$$

and its finite volume approximation. A key ingredient in the *a posteriori* error analysis is the finite element residual

$$r = \nabla \cdot (au^h) + bu - f;$$

defining the global error as $e = u - u^h$, it is clear that

$$\nabla \cdot (ae) + be = -r, \quad x \in \Omega,$$
$$e = 0, \quad x \in \partial_- \Omega.$$

In order to state our basic *a posteriori* error bound on the global error in terms of the finite volume residual we introduce appropriate norms for measuring these quantities.

Let D be a non-empty open subset of $\Omega = (0,1)^2$ with a Lipschitz continuous boundary ∂D which is almost everywhere non-characteristic for the hyperbolic operator L; we define

$$H(D) = \{v \in L^2(D) \ : \ \nabla \cdot (\mathbf{a}v) \in L_2(D)\}$$

equipped with the weighted graph norm

$$\|v\|_{H(D)} = \{\|mv\|^2_{L_2(D)} + \|\nabla \cdot (\mathbf{a}v)\|^2_{L_2(D)}\}^{1/2},$$

where $m \in C(\overline{\Omega})$ is a strictly positive weight function. For the sake of notational simplicity, we shall suppress m when referring to the space $H(D)$; the actual choice of the weight function will be clear from the context. We note that v belongs to $H(D)$ if and only if $\mathbf{a}v$ belongs to the Duvaut-Lions space

$$H(div, D) = \{w \in [L_2(D)]^2 \ : \ \nabla \cdot w \in L_2(D)\}.$$

The relevance of this relationship will be seen later when we consider the trace operator on $H(D)$.

For v in $L_2(\partial_\pm D)$, we define the (weighted) norm

$$\|v\|_{L_2(\partial_\pm D)} = \left[\int_{\partial_\pm D} |\mathbf{a} \cdot \mathbf{n}| \, |mv|^2 ds \right]^{1/2},$$

where, again, the weight m has been omitted from the notation $L_2(\partial_\pm D)$.

Theorem 18.6 *For each cell $K_{ij}^h \in T^h$, there exists a positive constant $c_1(K_{ij}^h) = \sqrt{2}(1 + \|be^{\alpha \cdot x}\|_{L_\infty(K_{ij}^h)})^2$, and a continuous positive weight function $m = e^{-\alpha \cdot x}$, such that*

$$\|e\|_{H(K_{ij}^h)}^2 + M^2 \|e\|_{L_2(\partial_+ K_{ij}^h)}^2 \le c_1^2 \|r\|_{L_2(K_{ij}^h)}^2 + M^2 \|e\|_{L_2(\partial_- K_{ij}^h)}^2,$$

where $M^2 = 1 + \|be^{\alpha \cdot x}\|_{L_\infty(K_{ij}^h)}^2$, with $\alpha = (\alpha_1, \alpha_2)$ chosen so that

$$\frac{1}{2}(\nabla \cdot \mathbf{a}) + \mathbf{a} \cdot \alpha + b \ge 1 \quad on \; K_{ij}^h.$$

Proof. Let g be the continuous function on the closure of K_{ij}^h defined by

$$g(x) = -b(x) + M^2 e^{-2\alpha \cdot x},$$

where $M^2 = 1 + \|be^{\alpha \cdot x}\|_{L_\infty(K_{ij}^h)}^2$ and α is chosen as in the statement of the theorem. We note that since the entries of $\mathbf{a}$ are non-negative and at least one of them is positive at each point of K_{ij}^h, such an α exists. Now

$$\frac{1}{2}(g + b)(\nabla \cdot \mathbf{a}) - \frac{1}{2}\mathbf{a} \cdot \nabla(g + b) + gb \ge e^{-2\alpha \cdot x},$$

so that

$$(\nabla \cdot (\mathbf{a}e) + be, \nabla \cdot (\mathbf{a}e) + ge)_{K_{ij}^h}$$

$$= \|\nabla \cdot (\mathbf{a}e)\|_{L_2(K_{ij}^h)}^2 + \frac{M^2}{2} \int_{\partial K_{ij}^h} (\mathbf{a} \cdot \mathbf{n})(g + b)|e|^2 ds$$

$$+ (\frac{1}{2}(g + b)(\nabla \cdot \mathbf{a}) - \frac{1}{2}\mathbf{a} \cdot \nabla(g + b) + gb, e^2)_{K_{ij}^h}$$

$$\ge \|\nabla \cdot (\mathbf{a}e)\|_{L_2(K_{ij}^h)}^2 + \|e^{-\alpha \cdot x}e\|_{L_2(K_{ij}^h)}^2$$

$$+ \frac{M^2}{2} \int_{\partial K_{ij}^h} (\mathbf{a} \cdot \mathbf{n})|e^{-\alpha \cdot x}e|^2 ds.$$

On the other hand,

$$(\nabla \cdot (\mathbf{a}e) + be, \nabla \cdot (\mathbf{a}e) + ge)_{K_{ij}^h} = (Le, \nabla \cdot (\mathbf{a}e) + ge)_{K_{ij}^h}$$

$$\le \|Le\|_{L_2(K_{ij}^h)}(\|\nabla \cdot (\mathbf{a}e)\|_{L_2(K_{ij}^h)} + \max_{x \in K_{ij}^h} |e^{\alpha \cdot x}g(x)| \|e^{-\alpha \cdot x}e\|_{L_2(K_{ij}^h)}).$$

Since

$$\max_{x \in K_{ij}^h} |e^{\alpha \cdot x} g(x)| \leq M^2 e^{-\alpha \cdot x} + \|be^{\alpha \cdot x}\|_{L_\infty(K_{ij}^h)}$$

$$\leq (1 + \|be^{\alpha \cdot x}\|_{L_\infty(K_{ij}^h)})^2,$$

by choosing m and c_1 as in the statement of the theorem, we deduce that

$$\|e\|_{H(K_{ij}^h)}^2 + \frac{M^2}{2}\|e\|_{L_2(\partial_+ K_{ij}^h)}^2$$

$$\leq \frac{1}{2}c_1^2\|Le\|_{L_2(K_{ij}^h)}^2 + \frac{1}{2}\|e\|_{H(K_{ij}^h)}^2 + \frac{1}{2}\|e\|_{L_2(\partial_- K_{ij}^h)}^2.$$

As $Le = -r$, this inequality implies the desired estimate. $\square$

The application of this theorem requires the knowledge of the error at the inflow boundary of a cell in order to bound the error in the cell and at its outflow boundary. Since the entries of the velocity field, $\mathbf{a}$, are of constant sign, this can be achieved by sweeping through the (structured) mesh from left to right and bottom to top. However this process can be expected to badly overestimate the global error. In order to obtain sharp local *a posteriori* error bounds, we shall carry out a localisation process based on decomposing the global error into its locally created component and its propagating part (see also Mackenzie, Süli and Warnecke (1993(iii))). Let us, therefore, define $\tilde{u}$ and $\tilde{u}^h$ as the solutions of the local boundary value problems

$$L\tilde{u} = f \quad \text{on } K_{ij}^h,$$

$$\tilde{u}|_{\partial_- K_{ij}^h} = u|_{\partial_- K_{ij}^h};$$

and

$$L\tilde{u}^h = f \quad \text{on } K_{ij}^h,$$

$$\tilde{u}^h|_{\partial_- K_{ij}^h} = u^h|_{\partial_- K_{ij}^h}.$$

Here and below we assume that ∂K_{ij}^h is a non-characteristic curve for the hyperbolic operator L. Because of the uniqueness of solutions to these boundary value problems we have that $\tilde{u} = u|_{K_{ij}^h}$ on K_{ij}^h. The function $\tilde{u}^h$ can be interpreted as the exact solution of the local boundary value problem subject to a boundary condition which is a perturbation of the exact boundary condition due to the accumulation of the numerical error upwind of the cell K_{ij}^h. The quantity $e_c = \tilde{u}^h - u^h$ is then the error generated by the scheme within the cell K_{ij}^h and will be referred to as the *cell error*. The complementary quantity $e_{tr} = u - \tilde{u}^h$ reflects the component of the error which has been created upwind of the cell K_{ij}^h and is just advected through the cell; we call this the *transmitted error*. Thus we have the following decomposition of the global error:

$$e|_{K_{ij}^h} = (u - u^h)|_{K_{ij}^h} = e_c + e_{tr}.$$

Clearly e_c and e_{tr} satisfy the following local boundary value problems:

$$Le_c = -r \quad \text{on } K_{ij}^h,$$
$$e_c = 0 \quad \text{on } \partial K_{ij}^h;$$

and

$$Le_{tr} = 0 \quad \text{on } K_{ij}^h,$$
$$e_{tr} = e \quad \text{on } \partial K_{ij}^h.$$

Now we have the following two-sided local *a posteriori* bound for the cell error.

Theorem 18.7 *For each cell* $K_{ij}^h \in T^h$, *there exist positive constants* $c_0(K_{ij}^h) = [\sqrt{2}(1 + \|be^{\alpha \cdot x}\|_{L_\infty(K_{ij}^h)})]^{-1}$ *and* $c_1(K_{ij}^h) = \sqrt{2}(1 + \|be^{\alpha \cdot x}\|_{L_\infty(K_{ij}^h)})^2$, *and a continuous positive weight function* $m = e^{-\alpha \cdot x}$, *such that*

$$c_0 \|r\|_{L_2(K_{ij}^h)} \leq \|e_c\|_{H(K_{ij}^h)} \leq c_1 \|r\|_{L_2(K_{ij}^h)},$$

where $\alpha = (\alpha_1, \alpha_2)$ *is chosen so that*

$$\frac{1}{2}(\nabla \cdot \mathbf{a}) + \mathbf{a} \cdot \alpha + b \geq 1 \quad \text{on } K_{ij}^h.$$

Proof. The upper bound follows by applying the previous theorem to the boundary value problem

$$Le_c = -r \quad \text{on } K_{ij}^h,$$
$$e_c = 0 \quad \text{on } \partial K_{ij}^h.$$

Noting that

$$\|r\|_{L_2(K_{ij}^h)} \leq \|\nabla \cdot (ae_c)\|_{L_2(K_{ij}^h)} + \|be_c\|_{L_2(K_{ij}^h)} \leq \frac{1}{c_0}\|e_c\|_{H(K_{ij}^h)},$$

the lower bound follows. $\square$

Employing the two-sided bound from this theorem we define a lower cell error indicator $\underline{\epsilon}_{ij}$ and a global lower estimate $\underline{\epsilon}$ by

$$\|e_c\|_{H(\Omega)}^2 = \sum_{i=1}^{M}\sum_{j=1}^{N} \|e_c\|_{H(K_{ij}^h)}^2 \geq \sum_{i=1}^{M}\sum_{j=1}^{N} c_0^2(K_{ij}^h)\|r\|_{L_2(K_{ij}^h)}^2 \equiv \sum_{i=1}^{M}\sum_{j=1}^{N} \underline{\epsilon}_{ij}^2 \equiv \underline{\epsilon}^2.$$

Analogously, we define an upper cell error indicator $\overline{\epsilon}_{ij}$ and a global upper estimate $\overline{\epsilon}$ by

$$\|e_c\|_{H(\Omega)}^2 = \sum_{i=1}^{M}\sum_{j=1}^{N} \|e_c\|_{H(K_{ij}^h)}^2 \leq \sum_{i=1}^{M}\sum_{j=1}^{N} c_1^2(K_{ij}^h)\|r\|_{L_2(K_{ij}^h)}^2 \equiv \sum_{i=1}^{M}\sum_{j=1}^{N} \overline{\epsilon}_{ij}^2 \equiv \overline{\epsilon}^2.$$

Thus,

$$\underline{\epsilon} \le \|e_c\|_{H(\Omega)} \le \overline{\epsilon}.$$

Further, it follows that the *effectivity index*, $\mathrm{Eff}(\eta_\Omega)$, of the error estimator $\eta_\Omega = \|r\|_{L_2(\Omega)}$, defined as the ratio of the estimated error to the true error, satisfies

$$\mathrm{Eff}(\eta_\Omega) \in [(\max_{i,j} c_1(K_{ij}^h))^{-1}, (\min_{i,j} c_0(K_{ij}^h))^{-1}].$$

In particular, when $b \equiv 0$, we have that $\mathrm{Eff}(\eta_\Omega) \in [1/\sqrt{2}, \sqrt{2}]$.

For the transmitted error, Theorem 18.6 implies the following upper bound.

Theorem 18.8 *For each cell* $K_{ij}^h \in \mathcal{T}^h$, *there exists a positive constant* $c_1(K_{ij}^h) = \sqrt{2}(1 + \|be^{\alpha \cdot x}\|_{L_\infty(K_{ij}^h)})^2$, *and a continuous positive weight function* $m = e^{-\alpha \cdot x}$, *such that*

$$\|e_{tr}\|_{H(K_{ij}^h)}^2 + M^2\|e_{tr}\|_{L_2(\partial_+ K_{ij}^h)}^2 \le M^2\|e\|_{L_2(\partial_- K_{ij}^h)}^2,$$

where $M^2 = 1 + \|be^{\alpha \cdot x}\|_{L_\infty(K_{ij}^h)}^2$, *with* $\alpha = (\alpha_1, \alpha_2)$ *chosen so that*

$$\frac{1}{2}(\nabla \cdot \mathbf{a}) + \mathbf{a} \cdot \alpha + b \ge 1 \quad on \ K_{ij}^h.$$

So far we have implicitly assumed that the solution of the boundary value problem is sufficiently smooth so that $\|r\|_{L_2(\Omega)}$ converges to zero as the mesh is refined. When this is not the case a weaker norm of the residual has to be used instead. In order to illustrate this point, let us consider the boundary value problem

$$\begin{aligned}
u_x + u_y &= 0, \quad (x,y) \in (0,1]^2, \\
u(0,y) &= 1, \quad 0 \le y \le 1, \\
u(x,0) &= 0, \quad 0 < x \le 1.
\end{aligned}$$

The generalised solution of this problem is a step function translated along the characteristic $y = x$, i.e. $u(x,y) = 1 - H(x - y)$ where H is the Heaviside function. The cell vertex finite volume approximation of this problem on a uniform mesh of step size h is easily seen to be a bilinear function with nodal values $u_{ij}^h = 1$ for $i \le j$ and 0 otherwise. A simple calculation reveals that

$$\|r\|_{L_2(\Omega)} = \underline{\epsilon} = \overline{\epsilon} = \left(\frac{1}{3h} - \frac{1}{6}\right)^{1/2} \asymp Const.h^{-1/2}.$$

The unbounded growth of the residual in the L_2-norm under mesh refinement indicates that it is an inappropriate error estimator for problems which have the type of discontinuity found in this example. This undesirable behaviour of the L_2-norm of the residual has also been observed numerically for the Euler equations of compressible gas dynamics in the presence of shock waves (see Sonar (1992))

In order to ensure that mesh refinement reduces the residual, we are led to measuring it in a norm which is weaker than L_2. Note that an L_p-norm with $1 \le p < 2$ is not weak enough: indeed, for the model problem with the propagating discontinuity described above, $\|r\|_{L_p(\Omega)} \asymp Const.h^{\frac{1}{p}-1}$, which does not converge to zero when $h \to 0$. The negative Sobolev norm, $\| \cdot \|_{H^{-1}(\Omega)}$, may seem to be appropriate; in fact lower error estimators for linear and nonlinear hyperbolic problems based on the negative Sobolev norm have already been successfully implemented in Sonar (1992), Mackenzie *et al.* (1993(i)) and Mackenzie *et al.* (1993(iii)). The numerical experiments with the nonsmooth problem presented in Mackenzie *et al.* (1993(iii)) indicate that a lower error estimator based on the H^{-1}-norm converges to zero like $O(h^{1/2})$ whereas the global error in the L_2-norm converges at the slower rate of $O(h^{1/3})$ (see also Theorem 4.3 in Chapter 5 of Brenner *et al.* (1975) and recall that the Heaviside function belongs to the Besov space $B_{2,loc}^{1/2,\infty}(\mathbb{R})$), so that the corresponding effectivity index decays to zero as the mesh is refined, which is clearly undesirable. Here we propose to measure the residual in the dual graph norm, which is intermediate (in the sense of function space interpolation) between the H^{-1} and the L^2 norms. In addition, we derive two-sided error bounds for the L_2-norm of the cell error in terms of this new norm of the residual. We begin by introducing the necessary notation.

Suppose that D is a non-empty open subset of Ω with a Lipschitz continuous boundary ∂D which is an almost everywhere non-characteristic curve for the hyperbolic operator L. Let us define the spaces $\mathcal{D}_+(D)$ and $\mathcal{D}_-(D)$ by

$$\mathcal{D}_\pm(D) = \{v \in C^\infty(D) \, : \, \text{supp } v \cap \partial_\pm D = \emptyset\}.$$

We define the Hilbert space $H_\pm(div; D)$ as the closure of $[\mathcal{D}_\pm(D)]^2$ in the norm of the Duvaut-Lions space $H(div; D)$. Mimicking the proof of the trace theorem for $H(div; D)$ which states that the normal trace operator γ_n is a continuous surjection from $H(div; D)$ onto $H^{-1/2}(\partial D)$, it can be shown that

$$\mathbf{a}v \in H_\pm(div; D) \Leftrightarrow v \in H(D) \text{ and } \gamma_n(\mathbf{a}v) = 0 \text{ in } H^{-1/2}(\partial_\pm D).$$

This identification allows us to define the Hilbert space

$$H_\pm(D) = \{v \in H(D) \, : \, \gamma_n(\mathbf{a}v) = 0 \text{ on } \partial_\pm D\}$$

equipped with the norm of $H(D)$; $H_\pm(D)$ is a dense subspace of $L_2(D)$. Let $H'_\pm(D)$ denote the dual space of $H_\pm(D)$ equipped with the norm

$$\|u\|_{H'_\pm(D)} = \sup_{\phi \in H_\pm(D)} \frac{|\langle u, \phi \rangle|}{\|\phi\|_{H(D)}},$$

where $\langle \cdot, \cdot \rangle$ is the duality pairing between $H'_\pm(D)$ and $H_\pm(D)$. The injection $u \in L_2(D) \mapsto (u, \cdot)_D \in H'_\pm(D)$ defines a natural embedding of $L_2(D)$ into $H'_\pm(D)$, so that for u in $L_2(D)$ the duality pairing in the definition of $\|u\|_{H'_\pm(D)}$ can be replaced by the $L_2(D)$-inner product $(\cdot, \cdot)_D$.

Theorem 18.9 *For each cell $K_{ij}^h \in \mathcal{T}^h$ there exist positive constants $c_2(K_{ij}^h)$ and $c_3(K_{ij}^h)$ such that*

$$c_2\|r\|_{H'_+(K_{ij}^h)} \leq \|e_c\|_{L_2(K_{ij}^h)} \leq c_3\|r\|_{H'_+(K_{ij}^h)}.$$

Proof. Because $e_c|_{\partial_- K_{ij}^h} = 0$ and if $\phi \in H_+(K_{ij}^h)$ then $\phi|_{\partial_+ K_{ij}^h} = 0$, we have that

$$\|r\|_{H'_+(K_{ij}^h)} = \|Le_c\|_{H'_+(K_{ij}^h)} = \sup_{\phi \in H_+(K_{ij}^h)} \frac{|(Le_c, \phi)_{K_{ij}^h}|}{\|\phi\|_{H(K_{ij}^h)}} = \sup_{\phi \in H_+(K_{ij}^h)} \frac{|(e_c, L^T\phi)_{K_{ij}^h}|}{\|\phi\|_{H(K_{ij}^h)}},$$

where L^T denotes the adjoint of the operator L, defined by

$$L^T\phi = -\mathbf{a} \cdot \nabla\phi + b\phi \quad \text{in } K_{ij}^h,$$
$$\phi = 0 \quad \text{on } \partial_+ K_{ij}^h.$$

Let ϕ_0 denote the (unique) solution of the local boundary value problem

$$L^T\phi = e_c \quad \text{in } K_{ij}^h,$$
$$\phi = 0 \quad \text{on } \partial_+ K_{ij}^h.$$

Equivalently, this can be rewritten as

$$\nabla \cdot (-\mathbf{a}\phi) + (b + \nabla \cdot a)\phi = e_c \quad \text{in } K_{ij}^h,$$
$$\phi = 0 \quad \text{on } \partial_+ K_{ij}^h,$$

so that the principal part of the operator is in divergence form. Thus, by repeating the argument presented in the proof of Theorem 18.6 with a, b replaced by $-a$, $b + \nabla \cdot a$, respectively, and the inflow and outflow boundaries of the cell K_{ij}^h interchanged, we find that

$$\|\phi_0\|_{H(K_{ij}^h)} \leq c_3 \|e_c\|_{L_2(K_{ij}^h)},$$

where $c_3 = \sqrt{2}(1 + \|(b + \nabla \cdot a)\exp(-\alpha \cdot x)\|_{L_\infty(K_{ij}^h)})$ and the weight function in the norm is $m(x) = \exp(\alpha \cdot x)$ with α chosen in such a way that

$$\frac{1}{2}(\nabla \cdot \mathbf{a}) + \mathbf{a} \cdot \boldsymbol{\alpha} + b \geq 1.$$

Thus,

$$\|r\|_{H'_+(K_{ij}^h)} \geq \frac{\|e_c\|^2_{L_2(K_{ij}^h)}}{\|\phi_0\|_{H(K_{ij}^h)}} \geq \frac{1}{c_3}\|e_c\|_{L_2(K_{ij}^h)},$$

which implies the upper bound

$$\|e_c\|_{L_2(K_{ij}^h)} \leq c_3 \|r\|_{H'_+(K_{ij}^h)}.$$

On the other hand,

$$|(e_c, L^T \phi)| \leq \|e_c\|_{L_2(K_{ij}^h)} (\|\nabla \cdot (\mathbf{a}\phi)\|_{L_2(K_{ij}^h)} + \|(b + \nabla \cdot \mathbf{a})\phi\|_{L_2(K_{ij}^h)})$$

$$\leq \sqrt{2}(1 + \|(b + \nabla \cdot a)\exp(-\alpha \cdot x)\|_{L_\infty(K_{ij}^h)})\|e_c\|_{L_2(K_{ij}^h)}\|\phi\|_{H(K_{ij}^h)}$$

$$\equiv c_3\|e_c\|_{L_2(K_{ij}^h)}\|\phi\|_{H(K_{ij}^h)}.$$

Hence, upon dividing both sides by $\|\phi\|_{H(K_{ij}^h)}$ and recalling the definition of the norm $\|\cdot\|_{H'_+(K_{ij}^h)}$, we arrive at the lower bound

$$c_2\|r\|_{H'_+(K_{ij}^h)} \leq \|e_c\|_{L_2(K_{ij}^h)}$$

with $c_2 = 1/c_3$. $\square$

The *a posteriori* error bounds derived here exploited the L_2-well posedness of mixed initial boundary value problems for scalar hyperbolic equations. Indeed, two-sided local *a posteriori* error estimates for Friedrichs systems have been derived in Mackenzie *et al.* (1993(ii)). The extension of these results to linear strictly hyperbolic systems could be developed similarly using the well-posedness theory of Kreiss (1970) and Rauch (1972).

REFERENCES

Brenner, P., Thomée, V. and Wahlbin, L. (1975), Besov spaces and applications to difference methods for initial value problems, *in* 'Lecture Notes in Mathematics', number 434, Springer Verlag, Heidelberg.

Crumpton, P.I., Mackenzie, J.A. and Morton, K.W. (1991), Cell vertex algorithms for the compressible Navier-Stokes equations, Technical Report NA91/12, Oxford University Computing Laboratory, 11 Keble Road, Oxford, OX1 3QD. Published in the Proceedings of the Thirteenth International Conference on Numerical Methods in Fluid Dynamics, Lecture Notes in Physics 414.

Giusti, E. (1984), Minimal surfaces and functions of bounded variation, *in* A. Borel, J. Moser and S.-T. Yau, eds, 'Monographs in Mathematics', Vol. 80, Birkhäuser, Basel.

Gustafsson, B., Kreiss, H.-O. and Sundström, A. (1972), Stability theory of difference approximations for mixed initial boundary value problems. II, *Mathematics of Computation* **26**(119), 649–686.

Jameson, A. (1979), Acceleration of transonic potential flow calculations on arbitrary meshes by the multiple grid method, Paper 79-1458, AIAA, New York.

Kreiss, H.-O. (1970), Initial boundary value problems for hyperbolic systems, *Communications on Pure and Applied Mathematics* **23**, 277–298.

Lax, P.D. and Nirenberg, L. (1966), On stability for difference schemes; a sharp form of Gårding's inequality, *Comm. Pure and Appl. Math.* (XIX), 473–492.

Mackenzie, J.A., Mayers, D.F. and Mayfield, A.J. (1992), Error estimates and mesh adaption for a cell vertex finite volume scheme, Technical Report NA92/10, Oxford University Computing Laboratory, 11 Keble Road, Oxford, OX1 3QD.

Mackenzie, J.A., Sonar, T. and Süli, E. (1993(i)), Adaptive finite volume methods for hyperbolic problems. Paper presented at MAFELAP 1993. This volume, Chapter 19.

Mackenzie, J.A., Süli, E. and Warnecke, G. (1993(ii)), A posteriori error analysis of Galerkin approximations of Friedrichs systems. In preparation.

Mackenzie, J.A., Süli, E. and Warnecke, G. (1993(iii)), A posteriori error estimates for the cell-vertex finite volume method. To appear in the Proceedings of the 9th GAMM-Seminar, Kiel, 1993.

MacNeal, R.H. (1953), An asymmetrical finite difference network, *Quart. Appl. Math.* (11), 295–310.

McDonald, P.W. (1971), The computation of transonic flow through two-dimensional gas turbine cascades, Paper 71-GT-89, ASME, New York.

Michelson, E. (1983), Stability theory of difference approximations for multidimensional initial-boundary value problems, *Mathematics of Computation* (40), 1–45.

Morton, K.W. (1988), Finite volume and finite element methods for the steady Euler equations of gas dynamics, *in* J.R. Whiteman, ed., 'The Mathematics of Finite Elements and Applications VI MAFELAP 1987', Academic Press, London, pp. 353–378.

Morton, K.W. (1991), Finite volume methods and their analysis, *in* J.R. Whiteman, ed., 'The Mathematics of Finite Elements and Applications VII MAFELAP 1990', Academic Press, London pp. 189–214.

Morton, K.W. and Paisley, M.F. (1989), A finite volume scheme with shock fitting for the steady Euler equations, *Journal of Computational Physics* **80**, 168–203.

Morton, K.W. and Stynes, M. (1991), An analysis of the cell vertex method, Technical Report NA91/07, Oxford University Computing Laboratory, 11 Keble Road, Oxford, OX1 3QD. Submitted for publication.

Morton, K.W. and Süli, E. (1991), Finite volume methods and their analysis, *IMA Journal of Numerical Analysis* **11**, 241–260.

Morton, K.W. and Süli, E. (1993), Analysis of cell-vertex methods for convection problems. In preparation.

Morton, K.W., Rudgyard, M.A. and Shaw, G.J. (1991), Upwind iteration methods for the cell vertex scheme in one dimension, Technical Report NA91/09, Oxford University Computing Laboratory, 11 Keble Road, Oxford, OX1 3QD. Submitted for publication.

Morton, K.W., Stynes, M. and Süli, E. (1993), Analysis of cell-vertex methods for convection-diffusion problems. In preparation.

Rauch, J. (1972), L_2 is a continuable initial condition for Kriess' mixed problems, *Communications on Pure and Applied Mathematics* **XXV**, 265.

Sonar, T. (1992), Strong and weak norm error indicators based on the finite element residual for compressible flow calculations, Technical Report 92/07, DLR, Göttingen.

Süli, E. (1989), Finite volume methods on distorted meshes: stability, accuracy, adaptivity, Technical Report NA89/6, Oxford University Computing Laboratory, 11 Keble Road, Oxford, OX1 3QD.

Süli, E. (1991), The accuracy of finite volume methods on distorted partitions, *in* J.R. Whiteman, ed., 'The Proceedings of the Conference on The Mathematics of Finite Elements and Applications VII MAFELAP', Academic Press, pp. 253–260.

Süli, E. (1992), The accuracy of cell vertex finite volume methods on quadrilateral meshes, *Mathematics of Computation* **59**(200), 359–382.

Vaillancourt, R. (1970), A simple proof of Lax-Nirenberg theorems, *Comm. Pure and Appl. Math.* (XXIII), 151–163.

Varga, R.S. (1962), *Matrix Iterative Analysis*, Automatic Computation, Prentice–Hall.

Volpert, A. (1967), The spaces BV and quasilinear equations, *Mat. Sb.* (73), 255–302. English transl. in *Math. USSR. Sb.*, 2, 225-267.

Chapter 19

Adaptive Finite Volume Methods for Hyperbolic Problems

J.A. Mackenzie*, T. Sonar** and E. Süli***

* Department of Mathematics, University of Strathclyde, Glasgow, G1 1XH, U.K.
** DLR, Institut für Theoretische Strömungsmechanik, D-3400 Göttingen, Germany
*** Computing Laboratory, University of Oxford, OX1 3QD, U.K.

19.1 INTRODUCTION

One of the major challenges today in the field of computational fluid dynamics is the efficient numerical solution of the partial differential equations describing complex multidimensional fluid problems. Though there is much knowledge and literature concerning the adaptive control of numerical algorithms for the solution of linear elliptic partial differential equations, the treatment of nonlinear hyperbolic or parabolic systems of partial differential equations as encountered in CFD applications is an open field.

Fundamental to the success of adaptive algorithms is the availability of sharp *a posteriori* error estimates. The use of such an estimate ensures that the adaptive algorithm is reliable, in the sense that a norm of the error is guaranteed to be below a given tolerance when the error estimate is below the tolerance. The use of sharp error estimates ensures that the adaptive algorithm is efficient in that it does not produce a grid which is overly refined.

In this paper we consider the idea of using L_2 and H^{-1} norms of the finite-element residual for the adaptive control in the computation of steady hyperbolic flow problems using finite-volume methods.

19.2 GOVERNING EQUATIONS

In this paper we concentrate on the use of adaptive finite volume methods for the solution of two model problems: two-dimensional linear advection and the two-dimensional compressible Euler equations.

The Mathematics of Finite Elements and Applications
Edited by J. R. Whiteman © 1994 John Wiley & Sons Ltd

19.2.1 TWO-DIMENSIONAL LINEAR ADVECTION

Let $\Omega = (0,1)^2$. We consider the boundary value problem:

$$\nabla.(\mathbf{a}u) + bu = f, \qquad \mathbf{x} \in \Omega, \tag{19.1}$$

$$u = g, \qquad \mathbf{x} \in \partial_-\Omega, \tag{19.2}$$

where $\partial_-\Omega = \{\mathbf{x} \in \partial\Omega : \mathbf{a}(\mathbf{x}).\mathbf{n}(\mathbf{x}) < 0\}$. To transform this into its weak formulation, we define the bilinear form $B : H^1(\Omega) \times L_2(\Omega) \to \mathbb{R}$ associated with the hyperbolic operator appearing in (19.1) by

$$B(u,p) = (\nabla.(\mathbf{a}u) + bu, p),$$

where $(\cdot,\cdot)$ denotes the L_2 inner product. The boundary value problem (19.1), (19.2) can then be reformulated as follows: find $u \in H_-^1(\Omega)$ satisfying

$$B(u,p) = (f,p) \qquad \forall p \in L_2(\Omega), \tag{19.3}$$

where $H_-^1(\Omega)$ consists of all functions in $H^1(\Omega)$ whose trace on the inflow boundary $\partial_-\Omega$ is equal to g.

19.2.2 TWO-DIMENSIONAL COMPRESSIBLE EULER EQUATIONS

Steady inviscid compressible fluid flow in two dimensions is governed by a system of equations of the form

$$\sum_{j=1}^{2} \partial_{x^j} \mathbf{f}_j(\mathbf{u}) = \mathbf{0}, \tag{19.4}$$

representing the conservation of mass, momentum in both Cartesian directions and energy. The state vector, $\mathbf{u}(\mathbf{x})$ and the flux functions, $\mathbf{f}_j(\mathbf{u})$, are given by

$$\mathbf{u} = (\rho, \rho v^1, \rho v^2, \rho E)^T \tag{19.5}$$

$$\mathbf{f}_j(\mathbf{u}) = (\rho v^j, \rho v^1 v^j + p\delta^{1j}, \rho v^2 v^j + p\delta^{2j}, \rho H v^j)^T, \tag{19.6}$$

where $\rho, v^1, v^2, p, E,$ and H denote density, velocity components in x^1- and x^2-directions, pressure, internal energy and enthalpy, respectively. The quantity $\delta^{ij} := \{1 \text{ if } i = j, 0 \text{ if } i \neq j\}$ is the usual Kronecker delta. To close system (19.4) an equation of state is needed. In the case of an ideal gas this equation is given by

$$p = (\gamma - 1)\rho \left(E - \frac{|v|^2}{2} \right), \quad v := (v^1, v^2)^T, \tag{19.7}$$

where γ denotes the ratio of specific heats. The enthalpy H can be expressed in terms of energy, pressure and density:

$$H = E + \frac{p}{\rho}.$$

A remarkable property of the Euler equations is the existence of a matrix $\mathbb{R}^2 \ni \mathbf{n} \xrightarrow{T} T(\mathbf{n}) \in \mathbb{R}^{4\times 4}$ such that

$$n_1 \mathbf{f}_1(\mathbf{u}) + n_2 \mathbf{f}_2(\mathbf{u}) = T^{-1}(\mathbf{n}) \mathbf{f}_1(T(\mathbf{n})\mathbf{u}). \tag{19.8}$$

For the exact form of this matrix and the proof of the above statement we refer to Spekreijse (1987).

19.3 THE FINITE VOLUME METHODS

19.3.1 THE CELL-VERTEX METHOD

The discretisation of (19.1) and (19.2) is performed on a partition $T^h = \{K_{ij}\}$ of convex quadrilaterals. With $h_{K_{ij}}$ denoting the diameter of K_{ij}, we shall assume that $h = \max\{h_{K_{ij}} : K_{ij} \in T^h\}$ is small. In order to introduce the relevant finite element spaces, we define the reference square $\hat{K} = (0,1)^2$, and denote by $F_{K_{ij}}$ the bilinear function which maps $\hat{K}$ onto K_{ij}. Let $Q_1(\hat{K})$ be the set of bilinear functions on $\hat{K}$ and $Q_0(\hat{K})$ the set of constant functions on $\hat{K}$; we define

$$\mathcal{U}^h = \{v \in H^1(\Omega) : \quad v = \hat{v} \circ F_{K_{ij}}^{-1}; \ \hat{v} \in Q_1(\hat{K}), \ K_{ij} \in T^h\}$$

and

$$\mathcal{M}^h = \{p \in L_2(\Omega) : \quad p = \hat{p} \circ F_{K_{ij}}^{-1}; \ \hat{p} \in Q_0(\hat{K}), \ K_{ij} \in T^h\},$$

as well as $\mathcal{U}_-^h = \mathcal{U}^h \cap H_-^1(\Omega)$. Let $I^h : (H_-^1(\Omega) \cap C(\overline{\Omega}))^2 \to (\mathcal{U}_-^h)^2$ be the interpolation projector onto $(\mathcal{U}_-^h)^2$. The discrete analogue of the bilinear form B is given by

$$B^h(v,p) = (\nabla . I^h(\mathbf{a}v) + I^h(bv), p) \qquad \forall v \in \mathcal{U}_-^h, \ \forall p \in \mathcal{M}^h.$$

The cell-vertex finite volume approximation of (19.3) is defined as follows: find $u^h \in \mathcal{U}_-^h$ satisfying

$$B^h(u^h, p) = (f, p) \quad \forall p \in \mathcal{M}^h. \tag{19.9}$$

19.3.2 UPWIND VERTEX-CENTRED METHOD

The discretisation of (19.4) is performed on a conforming triangulation $T^h = \{K_i\}$ of the domain Ω. The triangulation, T^h, will be called the primary network. If i is a node of the triangulation, then let $C_{h,i}$ denote the set of triangles having node i in common. Since any triangle $K \in T^h$ consists of three edges, $e_{K,k}$, $k = 1, 2, 3$, we define the set $E_{h,i} := \{e_{K,k} \mid K \in T^h, k \in \{1,2,3\}, e_{K,k} \text{ has node } i \text{ as vertex}\}$. Associated with each node i is a box σ_i, the boundary of which is defined by straight lines connecting the barycenters of the $K \in C_{h,i}$ with the midpoint of the edges $e_{K,k} \in E_{h,i}$. The union of all boxes is called a secondary network, denoted here as $\mathcal{S}^h$, and is the well-known barycentric subdivision of $\mathcal{T}^h$. The finite volume solution $\mathbf{u}^h$ is assumed to belong to the finite dimensional trial space

$$\mathcal{V}_0^h = \{\mathbf{v} \in [L_1(\Omega)]^4 : \; v_j|_{\sigma_i} \quad \text{is constant for all } \sigma_i \in \mathcal{S}^h, j = 1, \cdots, 4\}. \quad (19.10)$$

Let $N(i)$ denote the nodes which are direct neighbours of node i. Furthermore let $l_{ij}^k, k = 1, 2$, denote the two straight line segments separating boxes σ_i and σ_j. Testing (19.4) against piecewise constant functions over σ_i results in the discrete equation

$$[\mathcal{N}^w(\mathbf{u}^h)]_i = -\frac{1}{\text{meas}(\sigma_i)} \sum_{j \in N(i)} \sum_{k=1}^{2} \int_{l_{ij}^k} T^{-1}(n_{ij}^k)\mathbf{f}_1(T(n_{ij}^k)\mathbf{u}^h)\, ds, \quad (19.11)$$

where n_{ij}^k is the outer unit normal vector at l_{ij}^k. Trying to compute the integral leads to trouble since $\mathbf{u}^h$ is discontinuous across l_{ij}^k. We therefore introduce a numerical flux function $h(\mathbf{u}_i, \mathbf{u}_j; n_{ij}^k)$ consistent with the flux $\mathbf{f}_1$ in the sense that $\forall \mu \in \mathbb{R}^4$: $h(\mu, \mu, n_{ij}^k) = T^{-1}(n_{ij}^k)\mathbf{f}_1(T(n_{ij}^k)\mu)$. This then gives the discrete equation

$$[\mathcal{N}^w(\mathbf{u}^h)]_i = \frac{-1}{\text{meas}(\sigma_i)} \sum_{j \in N(i)} \sum_{k=1}^{2} h(\mathbf{u}_i, \mathbf{u}_j; n_{ij}^k)\, |\, l_{ij}^k\, |\,. \quad (19.12)$$

For the finite volume method presented here, the numerical flux function, h, is chosen to be that of Osher and Solomon (1982). Since the resulting scheme would be first order accurate, a MUSCL technique is applied to increase the order of the method. Details of the MUSCL technique and the iteration scheme used to solve the resulting system of non-linear equations are given in Sonar (1992(i)). To complete the description of finite-element spaces we introduce the space

$$\mathcal{W}_1^h = \{\mathbf{w} \in [H^1(\Omega)]^4 : \; w_j|_K \quad \text{is linear for all } K \in \mathcal{T}^h, j = 1, \cdots, 4\}. \quad (19.13)$$

The function spaces $\mathcal{V}_0^h$, and $\mathcal{W}_1^h$ are connected by means of the projection

$$\mathcal{W}_1^h \ni \mathbf{w}^h\big|_{K \in C_{h,i}} \xmapsto{\Pi_i} \Pi_i\, \mathbf{w}^h\big|_{K \in C_{h,i}} = \mathbf{w}(x_i^1, x_i^2) = \mathbf{u}^h\big|_{\sigma_i} \in \mathcal{V}_0^h. \quad (19.14)$$

This mapping clearly establishes a bijection between the spaces and therefore we view our solution as piecewise constant on the boxes while at the same time as continuous on $\overline{\Omega}$ and piecewise linear on the triangles $K \in \mathcal{T}^h$.

19.4 *A POSTERIORI* ERROR ESTIMATES

In order that our adaptive algorithms are reliable we have based our mesh refinement criteria on certain norms of the residual, r, of the finite volume approximations. For the linear scalar advection problem it can be shown that there exist positive computable constants c_0 and c_1 such that

$$c_0||r||_{H^{-1}(K)} \leq ||e_c||_{L_2(K)} \leq c_1||r||_{L_2(K)},$$

where e_c is the error locally generated in cell K (Mackenzie *et al.* (1993(i))). A *posteriori* error estimates have also recently been derived for Friedrichs systems (Mackenzie *et al.* (1993(ii))). The numerical evaluation of $||r||_{L_2(\Omega)}$ is obtained using quadrature. The difficulty in evaluating the H^{-1} norm of r lies in its definition:

$$||r||_{H^{-1}(K_{ij})} \equiv \sup_{\phi \in H_0^1(K_{ij})} \frac{|\int_{K_{ij}} r\phi \mathrm{d}\mathbf{x}|}{||\phi||_{H^1(K_{ij})}}.$$

For practical purposes we cannot test the residual against all functions in $H_0^1(K_{ij})$ and instead we use the approximation

$$||r||_{H^{-1}(K_{ij})} \approx \max_{l,m=1,2} \frac{|\int_{K_{ij}} r\phi_{lm} \mathrm{d}\mathbf{x}|}{||\phi_{lm}||_{H^1(K_{ij})}}, \tag{19.15}$$

where $\{\phi_{lm}\}_{l,m=1}^2$ is a set of four bilinear functions lying in $H_0^1(K_{ij})$. The particular functions used here are $\phi_{lm} = \hat{\phi}_{lm} \circ F_{K_{ij}}^{-1}$, where $\hat{\phi}_{lm} = \hat{\phi}_l(\hat{x}^1)\hat{\phi}_m(\hat{x}^2)$ and

$$\hat{\phi}_1(\hat{x}^1) = \begin{cases} \frac{2}{3}(\hat{x}^1 + 1), & -1 \le \hat{x}^1 \le \frac{1}{2}, \\ -2(\hat{x}^1 - 1), & \frac{1}{2} \le \hat{x}^1 \le 1, \end{cases}$$

$$\hat{\phi}_2(\hat{x}^1) = \begin{cases} 2(\hat{x}^1 + 1), & -1 \le \hat{x}^1 \le -\frac{1}{2}, \\ -\frac{2}{3}(\hat{x}^1 - 1), & -\frac{1}{2} \le \hat{x}^1 \le 1. \end{cases}$$

Clearly, since this approximation is a lower bound on the H^{-1} norm, we still have maintained reliability of the lower error estimate.

For the upwind method applied to the solution of the Euler equations we define the residual

$$\mathbf{r} = \sum_{j=1}^2 \partial_{x^j} \mathbf{f}_j(\mathbf{w}^h) = \nabla_{\mathbf{u}} \mathbf{f}_1(\mathbf{w}^h) \partial_{x^1} \mathbf{w}^h + \nabla_{\mathbf{u}} \mathbf{f}_2(\mathbf{w}^h) \partial_{x^2} \mathbf{w}^h$$

on each $K \in \mathcal{T}^h$. Since $\mathbf{w}^h$ is linear on K, the derivatives $\partial_{x^1} \mathbf{w}^h$ and $\partial_{x^2} \mathbf{w}^h$ are constant. However, the nonlinearity remains in the components of the Jacobian matrices. For this method we will also investigate the use of $||\mathbf{r}||_{H^{-1}}$ as a refinement criteria which is approximated here by

$$||\mathbf{r}||_{H^{-1}(K)} \approx \sum_{j=1}^4 \left\{ \max_{i=1,2,3} \frac{|\int_K r_j \phi_i dx|}{||\phi_i||_{H_0^1(K)}} \right\},$$

where $\{\phi_i, \ i = 1,2,3\}$ are three linear functions with compact support on $\overline{K}$.

19.5 NUMERICAL EXAMPLES

The first test case considered is scalar advection with the convective velocity field given by

$$\mathbf{a} = \left(\frac{(1+x^1)^{1+\beta}}{2\beta}, \ \frac{(1+x^2)^{1+\beta}}{2\beta} \right)^T,$$

which for $\beta > 0$ has both components strictly positive. With a forcing function given by

$$f = \left(\frac{(1+\beta)((1+x^1)^\beta + (1+x^2)^\beta)}{2\beta} + 1 \right) e^{-(1+x^1)^{-\beta} - (1+x^2)^{-\beta}}$$

and $b = 0$, the solution of (19.1) with boundary conditions

$$u(x^1, x^2)|_{\partial_- \Omega} = \begin{cases} e^{-(1+x^2)^{-\beta} - 1}, & x^1 = 0 \\ e^{-(1+x^1)^{-\beta} - 1}, & x^2 = 0 \end{cases}$$

is

$$u(x^1, x^2) = e^{-(1+x^1)^{-\beta} - (1+x^2)^{-\beta}}.$$

For simplicity, the problem will be approximated on a uniform grid. Figures 19.1(a) and (b) are of $\|e\|_{L_2(\Omega)}$ and the local error indicator, $\bar{\epsilon}_{ij} = c_1 \|r\|_{L_2(K_{ij})}$ on a 32×32 cell grid. Both of the integral norms have been approximated using one-point Gaussian quadrature. From these figures it is clear that there is a good correlation between the error indicator and the actual error. The agreement is quantified in Table 19.1 which shows second-order convergence of the approximate L_2-norm of the error and the one-point quadrature approximation of the global upper error estimate $\bar{\epsilon} = (\sum_{i,j=1}^{Ni,Nj} \bar{\epsilon}_{ij}^2)^{1/2}$. The third column of Table 19.1 shows an $O(h)$ convergence of the efficiency index $\bar{\theta} = \bar{\epsilon}/\|e\|_{L_2(\Omega)}$ of $\bar{\epsilon}$, to a value which is not much larger than unity. The fourth and fifth columns of Table 19.1 show the performance of the lower bound estimate $\underline{\epsilon} = (\sum_{i,j=1}^{Ni,Nj} (c_0 \|r\|_{H^{-1}(K_{ij})})^2)^{1/2}$. The lower efficiency index $\underline{\theta} = \|e\|_{L_2(\Omega)}/\underline{\epsilon}$ of $\underline{\epsilon}$ shows that the error estimate converges at the same rate as the error. The degree of underestimation of the error is not excessive and could probably be improved by testing the residual against a larger set of functions. More extensive numerical investigations of the *a posteriori* error estimates for the cell-vertex method are given in Mackenzie *et al.* (1993(i)).

To show the ability of the L_2 and H^{-1} norms of the residual in detecting refinement zones in inviscid compressible flow fields, we choose a typical transonic flow over the NACA0012 airfoil section, where $M_\infty = 0.8$ at an angle of attack of $1.25°$. The original grid and the Mach number obtained on that grid are shown in Figure 19.2(a). Figure 19.2(b) shows isolines of the quantities $\|\mathbf{r}\|_{H^{-1}(\Omega)}$ and $\|\mathbf{r}\|_{L_2(\Omega)}$ which show that the strong shock on the upper side as well as the weak shock on the lower side of the airfoil were detected. Adapting the grid three times using the refinement indicator $\|\mathbf{r}\|_{H^{-1}(\Omega)} > 10^{-5}$ and an isotropic point insertion technique leads to the grid and Mach number distribution as shown in Figure 19.2(c). An interesting fact is the refinement of the trailing edge and of the wake behind the upper shock. Here one would have expected large errors due to large grid cells. In contrast, the stagnation point region on the leading edge was not refined. The initial triangulation was already

Table 19.1 Performance of upper and lower error estimators for 2D linear advection with $\beta = 5$, where the integral norms have been approximated using one-point Gaussian quadrature

N	$\|e\|_{L_2(\Omega)}$	$\bar{\epsilon}$	$\bar{\theta}$	$\underline{\epsilon}$	$\underline{\theta}$
16	8.26E-4	1.84E-3	2.23	5.22E-5	15.8
32	2.06E-4	4.36E-4	2.10	1.26E-5	16.4
64	5.15E-5	1.05E-4	2.04	3.07E-6	16.8
128	1.29E-5	2.59E-5	2.02	7.58E-7	17.0
256	3.22E-6	6.44E-6	2.00	1.88E-7	17.1
512	8.05E-7	1.60E-6	1.99	4.69E-8	17.2

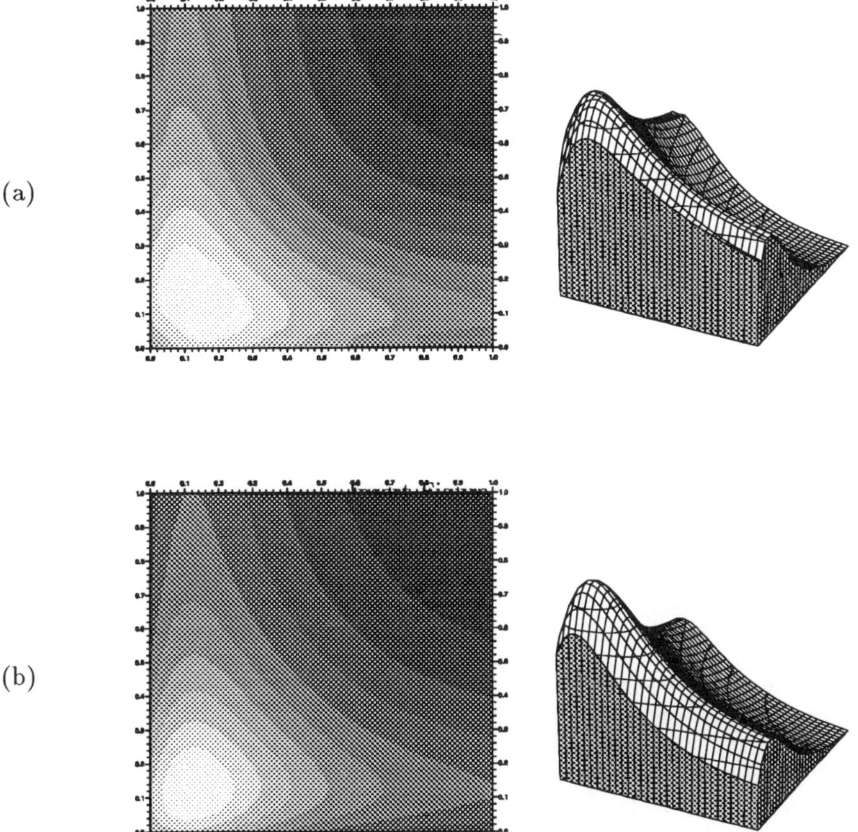

Figure 19.1 Contour and surface plots of (a) $\|e\|_{L_2(\Omega)}$, and (b) $\bar{e}_{ij}$ where $\beta = 5$ where the integral norms have been approximated using one-point quadrature

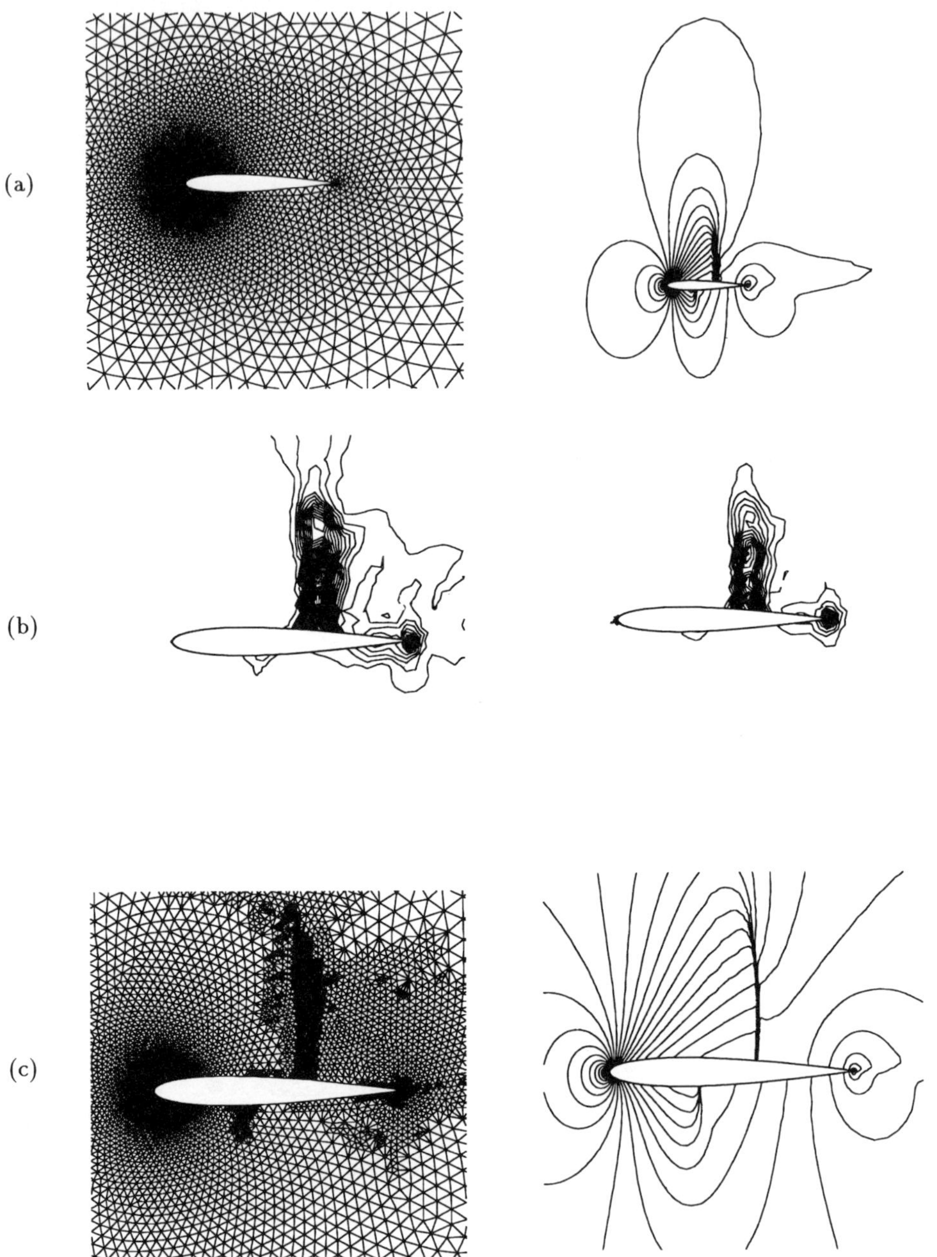

Figure 19.2 Plots of (a) original NACA0012 grid and the computed Mach number distribution, (b) isolines of $||r^h||_{H^{-1}}$ and $||r^h||_{L_2}$ and (c) the adapted grid and Mach number distribution after three refinement cycles.

very fine in this region such that a further refinement was not necessary. A comparison of the refinement regions indicated by gradients of flow variables such as pressure and entropy are given in Sonar (1992(ii)). It should be remembered however that these gradients provide no quantitative indication of the size of the error and in fact their values increase as the mesh is refined.

REFERENCES

Mackenzie J.A., Süli E. and Warnecke G. (1993(i)), *A posteriori* error estimates for the cell-vertex finite volume method. To appear in *Notes on Numerical Fluid Mechanics*, Vieweg Verlag.

Mackenzie J. A., Süli E. and Warnecke G. (1993(ii)), *A posteriori* error analysis for Galerkin approximations of Friedrichs systems. (In preparation).

Osher S. and Solomon F. (1982), Upwind Difference Schemes for Hyperbolic Conservation Laws. *Math. Comp.*, **38**, 339–374.

Sonar T. (1992(i)), On the Design of an Upwind Scheme for Compressible Flow on General Triangulations. To appear in *Num. Algorithms*.

Sonar T. (1992(ii)), Strong and Weak Norm Refinement Indicators Based on the Finite Element Residual for Compressible Flow Computation. To appear in *IMPACT Comput. Sci. Engrg.*

Spekreijse S.P. (1987), Multigrid Solution of the Steady Euler Equations. *Centrum voor Wiskunde en Informatica, Amsterdam.*

Chapter 20

Reliable Finite Volume Methods for Time-Dependent Partial Differential Equations

M. Berzins and J. Ware

School of Computer Studies,
University of Leeds, Leeds LS2 9JT, U.K.

20.1 INTRODUCTION

The accuracy of numerical solutions to time-dependent partial differential equations (p.d.e.s) depends on the spatial discretization method, the spatial mesh, the method of time integration and the timestep. In particular the spatial discretization method and positioning of the spatial mesh points should ensure that the spatial error is controlled to meet the user's requirements. It is then desirable to integrate the ordinary differential equation (o.d.e.) system in time with sufficient accuracy so that the temporal error does not corrupt the spatial accuracy or the reliability of the spatial error estimates. This paper describes the prototype of such an automated solver which has been developed from the framework laid down by Berzins *et al.*(1991). The components of this solver are the spatial and temporal error balancing approach of Berzins (1993), the discretization method of Ware and Berzins (1993) and the adaptive mesh algorithms of Berzins *et al.*(1993). This combination of error control strategies not only ensures that the main error present is that due to spatial discretization but offers the tantalising possibility of overall error control.
The approach used in the solver is described as follows. Section 2 describes the problem class and spatial discretization method. A method of lines approach is used which allows the global error to be decomposed into spatial and temporal errors in Section 3 and the spatio-temporal error balancing strategy to be described in Section 4. The performance of this strategy is illustrated in Section 5 while the full adaptive method is outlined in Section 6 and applied to a numerical Euler equations example.

20.2 FINITE VOLUME DISCRETIZATION METHOD

For ease of exposition consider the class of p.d.e.s in cartesian co-ordinates as

$$u_t \; = \; (\, f(u, u_x, u_y)\,)_x \; + \; (\, g(u, u_x, u_y)\,)_y \; , t \, \epsilon \, (0, t_e] \; , \; (x, y) \, \epsilon \, \Omega \qquad (20.1)$$

with appropriate boundary and initial conditions.

The Mathematics of Finite Elements and Applications
Edited by J. R. Whiteman © 1994 John Wiley & Sons Ltd

The spatial discretization algorithm has been developed for systems of equations of the form of (20.1). The algorithm uses upwind methods developed for the Euler equations for the convective parts of the fluxes f and g and uses centered discretization methods for the diffusive parts of the fluxes. Such discretizations are considered by many authors, e.g. Spekreise (1987) computes solutions free of oscillations using quadrilateral meshes. Ware and Berzins (1993) and Durlofsky *et al.*(1991) extend the methods to the use of unstructured triangular meshes. A finite volume approach is adopted to integrate equations (20.1) over a triangular element i. Solution values are assumed to be located at the centroids of the triangles. Applying the divergence theorem and a one point quadrature rule along the edges of the triangle gives

$$Area_i \dot{u}_i = -\sum_{j=1}^{3} [(f(u_{i,j}, (u_{i,j})_x, (u_{i,j})_y)\Delta y_j \quad (20.2)$$
$$-(g(u_{i,j}, (u_{i,j})_x, (u_{i,j})_y)\Delta x_j]$$

where $u_{i,j}$ is the solution value midway along the edge j , Δx_j is the change in the x co-ordinate in going from one end of the edge to the other , u_i is the solution value associated with the centroid of the i th triangle and the other values in the equation are similarly defined.

A standard first-order scheme uses only piecewise constant values inside each triangle; the evaluation of the convective fluxes midway along the edge involves the approximate solution of three one-dimensional Riemann problems in the direction of the normals to the edges of the triangle. This is done by using the standard scheme of Osher with a van Leer limiter.

The extension to a second order scheme on an unstructured triangular mesh is a cell-centered finite volume approach that uses a six triangle stencil around each edge, Ware and Berzins (1993), giving a ten triangle stencil for the discretization of the convective terms on each triangle.

Calculation of the derivative values used in the diffusive flux evaluations at the mid-points of an edge is by a control volume approach using the two triangles on either side of the edge. The divergence theorem is applied and mid-point quadrature used to evaluate the integrals of the solution along the edges of the two-triangle control volume. The mid-point solution values are calculated using linear interpolation based on the centroid values of the two triangle control volume plus the centroid value of one of the triangles external to the control volume. This interpolant has the advantege that it uses values pre-calculated for the convective part of the discretization. The derivative values at the mid-point of an edge thus involve contributions from six centroid solution values. Calculation of all the edge derivatives for a triangle thus involves the same ten point stencil as is used for the convective terms.

20.3   TIME INTEGRATION

The above spatial discretization scheme results in a system of differential equations, each of which is of the form of equation (20.2) . This system of equations can be written as the initial value problem:

$$\dot{U} = F_N\ (\ t,\ U(t)\ )\ ,\ U(0)\ given\ ,\qquad(20.3)$$

where the N dimensional vector, $U(t)$, is defined by

$$U(t) = (U(x_1,y_1,t)\ ,\ U(x_2,y_2,t),...,U(x_N,y_N,t)\ )^T\ .$$

The point (x_i,y_i) is the centroid of the ith triangle and $U(x_i,y_i,t)$ is a numerical approximation to $u(x_i,y_i,t)$. Numerical integration of (20.3) provides the approximation, $V(t)$, to the vector of exact p.d.e. solution values at the mesh points, $u(t)$. The global error in the numerical solution can be expressed as the sum of the spatial discretization error, $e(t) = u(t) - U(t)$, and the global time error, $g(t) = U(t) - V(t)$. That is,

$$\begin{aligned}
E(t) = u(t) - V(t) &= (u(t) - U(t)) + (U(t) - V(t))\\
&= e(t) + g(t).
\end{aligned}\qquad(20.4)$$

The Theta method code of Berzins and Furzeland (1992) used here selects functional iteration automatically for the non-stiff o.d.e.s resulting from the Euler equations. The numerical solution at $t_{n+1} = t_n + k$ where k is the time step size, as denoted by $V(t_{n+1})$ is defined by

$$V(t_{n+1}) = V(t_n) + (1-\theta)k\ \dot{V}(t_n) + \theta\ k\ F_N(t_{n+1}, V(t_{n+1}))\ ,\qquad(20.5)$$

in which $V(t_n)$ and $\dot{V}(t_n)$ are the numerical solution and its time derivative at the previous time t_n and the default value of θ is 0.55 .

In most time dependent p.d.e. codes either a CFL stability control is employed or a standard o.d.e. solver is used which controls the local error $l_{n+1}(t_{n+1})$ using local time error per step, (LEPS), control with respect to a user supplied accuracy tolerance, TOL, i.e.

$$\|\ l_{n+1}(t_{n+1})\ \| < TOL.\qquad(20.6)$$

or TOL is multiplied by the timestep k in a local time error per unit step (LEPUS) control. When controlling the LEPS it is difficult to establish a relationship between the accuracy tolerance, TOL , and the global time and space errors. In contrast, if the LEPUS is controlled then it is well known that the time global error is proportional to the tolerance. This suggests the use of LEPUS error control with a tolerance TOL that reflects the spatial error present.

20.4 BALANCING THE SPACE AND TIME ERRORS

Efficient time integration requires that the spatial and temporal errors are roughly the same order of magnitude. The need for spatial error estimates unpolluted by temporal error requires that the spatial error is the larger of the two errors. Lawson and Berzins (1992) have developed a strategy for parabolic equations which achieves this by controlling the local time error to be a fraction of the growth in the spatial discretization error over a timestep. Berzins (1993) describes a similar strategy for hyperbolic equations, based on the local growth in the spatial error over each timestep. The local-in-time spatial error, $\hat{e}(t_{n+1})$, for the timestep from t_n to t_{n+1} is defined as the spatial error at time t_{n+1} given the assumption that the spatial error, $e(t_n)$, at time t_n is zero. A local-in-time error balancing approach is then given by

$$\parallel l_{n+1}(t_{n+1}) \parallel \; < \; \epsilon \parallel \hat{e}(t_{n+1}) \parallel, \; 0 < \epsilon < 1. \tag{20.7}$$

The error $\hat{e}(t_{n+1})$ is estimated by the difference between the computed second-order in space solution and the first-order piecewise constant solution which satisfies a modified o.d.e. system denoted by

$$\dot{v}_{n+1}(t) = G_N(t, v_{n+1}(t)), \tag{20.8}$$

where $v_{n+1}(t_n) = V(t_n)$, $\dot{v}_{n+1}(t_n) = G_N(t, V(t_n))$ and where $G_N(.,.)$ is obtained from $F_N(.,.)$ simply by setting the limiter function in the space discretisation to zero. The local-in-time space error is then given by

$$\hat{e}(t_{n+1}) = V(t_{n+1}) - v_{n+1}(t_{n+1}) \tag{20.9}$$

and is computed by applying the θ method of the previous section to equation (20.8). As only a rough estimate of the error is required it is sufficient to use only one functional iteration to compute v_{n+1}; combining this with the conditions on $v_{n+1}(t_n)$ gives

$$v_{n+1}(t_{n+1}) = V(t_n) + \theta k G_N(t_{n+1}, V(t_{n+1})) + (1 - \theta)k G_N(t_n, V(t_n)). \tag{20.10}$$

(It is also possible to ignore the $(1 - \theta)$ term as in Berzins *et al.*(1992).) Combining equation (20.10) with equations (20.5) and (20.9) gives

$$\begin{aligned}
\hat{e}(t_{n+1}) \;\; = \;\; & \theta k[F_N(t_{n+1}, V(t_{n+1})) - G_N(t_{n+1}, V(t_{n+1}))] \; + \\
& (1 - \theta)k[F_N(t_n, V(t_n)) - G_N(t_n, V(t_n))] \; .
\end{aligned} \tag{20.11}$$

This estimate will only approximate the error in the low-order solution and so local extrapolation in space is effectively being used when the high-order solution is used to move forward in time. As the right side of equation (20.11) has a factor of k the error control (20.7) for the step to t_{n+1} is of the LEPUS form given by

$$\parallel l_{n+1}(t_{n+1}) \parallel \; < \; k\, TOL \; \text{ where } TOL = \epsilon \parallel \hat{e}(t_{n+1})/k \parallel. \tag{20.12}$$

Berzins (1993) gives an analysis to show that although the method attempts to control accuracy a Courant-like stability condition is also satisfied. Although LEPUS control is generally thought to be inefficient for standard o.d.e.s, equation (20.7) may be also used directly as a LEPS control, with little difference in integration performance. The estimate of the spatial error used improves on that suggested by Berzins *et al.*(1991) using h-extrapolation as the present approach estimates the error in each triangle individually rather than on groups of four triangles merged into one large triangle.

20.5 NUMERICAL EXPERIMENTS

Consider the two dimensional Burgers' equation problem defined by :

$$u_t \; + \; u\,u_x \; + \; u\,u_y \; - \; \nu\,(u_{xx} \; + \; u_{yy}) \; = \; 0, \quad \nu = 0.0001\,, \qquad (20.13)$$

where $(x, y, t) \in (0,1) \times (0,1) \times (0.25, 1.25]$. The exact solution is given by $u(x, y, t) = (1 + e^{B})^{-1}$, where $B = (x + y - t)/(2\nu)$.

The problem was solved using fixed evenly-spaced square space meshes of 9x9 and 81x81 points and a cell-centered discretization used by Berzins (1993). The time integration was performed using the LEPUS strategy given by equation (20.12) with $\epsilon = 0.3$ and 0.5 , and with the standard absolute LEPS strategy given by equation (20.6) .

In Table 20.1 "NP" is the number of spatial mesh points. "TOL" is the time integration "LEPS" tolerance except that "New 0.5" refers to the new strategy, with $\epsilon = 0.5$. "L1 ERR" is the spatial error in the L1 norm at the specified output times. "CPU" is the time used on an Iris Indigo R4000. "NS" is the number of time steps used in the integration of the o.d.e.s. "NF" is the number of o.d.e. function evaluations and "EST 0.3" is the norm of the local-in-time space error estimator with $\epsilon = 0.3$.

Table 20.1 shows that as the local error tolerance is decreased the spatial error dominates. It can also be seen that the error balancing strategy with $\epsilon = 0.3$ or with $\epsilon = 0.5$ provides temporal integration results that are dominated by the spatial error and are computed efficiently. The Table also shows that the local-in-time spatial error estimates mimic the actual errors in Table 20.1 . A number of experiments using the new error control strategy are described by Berzins (1993).

Table 20.1 Fixed Mesh Burgers' Equation Results.

NP	TOL	L1 Error Norm at Time				NS	NF	CPU
		0.26	0.69	1.0	1.3			
	0.5D-1	1.1D-2	2.4D-2	3.5D-2	2.8D-2	32	91	0.2
9	0.5D-2	1.1D-2	2.4D-2	3.5D-2	2.8D-2	48	130	0.4
by	0.5D-3	2.6D-2	2.3D-2	3.3D-2	2.8D-2	73	189	0.6
9	New 0.3	5.5D-3	2.2D-2	3.3D-2	2.6D-2	60	138	0.4
	New 0.5	5.5D-3	2.2D-2	3.4D-2	2.6D-2	50	133	0.3
	Est 0.3	5.0D-3	4.0D-3	3.3D-3	4.9D-3			
	0.1D-1	2.4D-3	4.5D-3	5.9D-3	3.9D-3	244	711	142
81	0.1D-2	1.2D-3	2.8D-3	4.2D-3	3.1D-3	452	966	219
by	0.1D-3	9.5D-4	2.9D-3	4.0D-3	3.1D-3	538	1175	325
81	New 0.3	1.1D-3	2.8D-3	3.9D-3	3.0D-3	483	997	229
	New 0.5	1.1D-3	2.8D-3	3.9D-3	3.0D-3	468	982	225
	Est 0.3	4.7D-4	6.4D-4	8.7D-4	6.8D-4			

20.6   A PROTOTYPE AUTOMATIC ALGORITHM

The goal of the automatic algorithm described here is to ensure that the spatial mesh is fine or coarse enough so that the solution satisfies the users' accuracy and efficiency requirements. The adaptive algorithm was developed from that in Berzins *et al.*(1991), using the Shell Research mesh generator based on the ideas of George *et al.*(1991), and adopting a regular subdivision approach (e.g. Lohner (1987)), see Berzins *et al.*(1993). The strategies for deciding when to remesh are essentially those of Lawson and Berzins (1992). The input required from the user consists only of the problem specification, an initial spatial mesh from the mesh generator. and an error tolerance for the spatial discretization error, EPS. At each time step the estimate of $||\hat{e}(t)||$ is calculated, and if it is greater than 0.25 of EPS, then a new mesh is constructed that ensures that the subsequent error is less than $EPSDN$ where $EPSDN$ is a fraction of EPS. The underlying assumption in this agorithm is that the introduction of extra mesh points will cause the error to decrease. The selection of appropriate remeshing times is made by using a combination of present estimated errors and predicted future errors.
Once a new mesh has been found a "flying restart "is used. The computed solution and the time history array used by the time integrator are interpolated onto this mesh and the time integration is restarted with the same time step as used immediately before remeshing. Care must be taken to modify the accuracy tolerance for the time integration so that it reflects the expected reduction in the spatial discretization error. An illustration of how the solver works is provided by the Burgers' equation problem used above. The results are shown in Table 20.2.

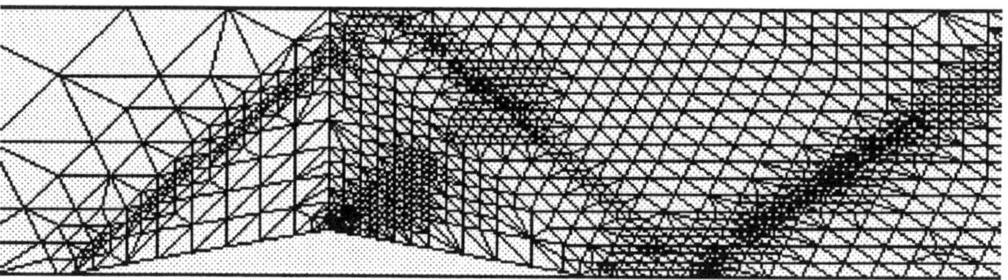

Figure 20.1   Final Mesh for Euler Equations Example.

Table 20.2   Adaptive Mesh Burgers' Equation Results.

	L1 Error Norm at Time							
TYPE	0.26	0.69	1.0	1.3	NS	NF	CPU	NRM
FIXD	1.0D-2	4.0D-2	3.8D-2	2.3D-2	720	1831	3602	0
ADAP	1.9D-2	4.0D-2	3.8D-2	2.4D-2	1458	3758	1361	160
NTRI	970	1378	931	251				
AUTO	1.9D-2	3.1D-2	2.6D-2	3.2D-2	1613	4156	1009	194
NTRI	762	856	322	217				

Three runs were done, the first (FIXD) used a fixed mesh of 8192 triangles and a time local error TOL of 1.0e-5 in an L1 vector norm. The second (ADAP) used the adaptive space strategy with a space local error control of 1.0d-5 and the same ordinary local error control and the third (AUTO) used the fully automatic code with error balancing and adaptivity. The cpu times are those on an Silicon Graphics R4000. The "NTRI" rows show the number of triangles used by the adaptive codes. The adaptive algorithm provides the same accuracy using less triangles and cpu. time than the fixed mesh code. "NRM" is the number of spatial remeshes automatically selected.
The final example is a well known Euler equations example of Mach 2 supersonic flow down a channel with a small triangular bump. The problem is solved as a time dependent problem by using initial conditions that suppose the bump is missing. The adaptive code solves the problem using 194 meshes ranging between 84 and 2899 triangles. A portion of the final triangulation is shown in Figure 20.1 with the reflecting shock patterns clearly illustrated by the position of the refined triangles.

20.7   CONCLUSIONS

This paper describes a prototype solver for time dependent p.d.e.s . The results obtained from preliminary experiments indicate that the algorithm is a promising start to developing codes which automatically control the error in the computed solution. The numerical results shows that this can be achieved for both simple convection problems and for Euler flows in two space dimensions.

ACKNOWLEDGEMENT

Justin Ware acknowledges the funding by SERC and Shell Research Ltd through a CASE studentship.

REFERENCES

Berzins M. (1993), Temporal Error Control in the Method of Lines for Convection Dominated Equations. *School of Computer Studies Report 93.18*, to appear in *SIAM J. Sci. Comp.*

Berzins M. , Baehmann P. , Flaherty J.E. and Lawson J. (1991), Towards Reliable Software for Time-Dependent Problems in CFD. *The Mathematics of Finite Elements and Applications VII* Mafelap 1990, pp 181-88. Ed J.R. Whiteman, Academic Press, London.

Berzins M. and Furzeland R.M. (1992), An Adaptive Theta Method for the Solution of Stiff and Non-Stiff Differential Equations . *Applied Num. Math.*, 7: 1–19.

Berzins M. , Lawson J. and Ware J. (1993), Spatial and Temporal Error Control in the Adaptive Solution of Systems of Conservations Laws. *Advances in Computer Methods for Partial Differential Equations VII* , pp. 60–66 Proc. of 1992 Conf., IMACS, New Jersey, USA.

Durlofsky L.J. , Engquist B. and Osher S. (1990), Triangle Based adaptive Stencils for the Solution of Hyperbolic Conservation Laws. *J. Comp. Physics* , 54: 545–581.

Georges P.L. Hecht F. and Saltel E. (1991), Automatic Mesh Generation with Specified Boundary. *Comp. Meth. Applied Mech. Eng.* , 92: 269–288.

Lawson J.L. and Berzins M. (1992), Towards an Automatic Algorithm for the Numerical Solution of Parabolic P.D.E.s using the Method of Lines. *Computational Ordinary Differential Equations* , pp 309–322. Eds. J.R. Cash and I.Gladwell, IMA Conference Series 1992, Oxford University Press.

Lohner R.L. (1987), An adaptive finite element scheme for transient problems in CFD. *Comp. Meths. in Applied Mech. and Eng.*, **61**: 323–338.

Spekreise S. (1987), Multigrid Solution of Monotone Second Order Discretizations of Hyperbolic Conservation Laws. *Mathematics of Computation* , **49**: 135–155.

Ware J. and Berzins M. (1993), Finite Volume Techniques for Time-Dependent Fluid-Flow Problems. *Advances in Computer Methods for Partial Differential Equations VII* , pp. 794–798 Proc. of 1992 Conf., IMACS, New Jersey, USA.

Chapter 21

Non-Reflecting Finite Elements

D. Givoli* and J.B. Keller**

* Department of Aerospace Engineering, Technion-Israel Inst. of Technology
Haifa 32000, Israel
** Departments of Mathematics and Mechanical Engineering,
Stanford University, California 94305, USA

21.1 INTRODUCTION

Wave problems in domains extending to infinity are very often encountered in various
fields of application such as acoustics and structural acoustics, geophysics, meteorology
and fluid dynamics. Standard solution procedures for such problems include the
boundary element method, the coupled finite element-boundary element method, and
the use of infinite elements. Another common procedure consists of the following three
steps: (a) first truncate the infinite domain by introducing an *artificial boundary $\mathcal{B}$*,
thus making the computational domain finite, (b) then choose a certain *boundary
condition* to be imposed on $\mathcal{B}$, (c) finally solve the problem in the finite computational
domain by the *finite element method*. This procedure as well as others are described
in a recent monograph (Givoli (1992(ii))).

It is well known that if the artificial boundary condition on $\mathcal{B}$ is poorly chosen,
then outgoing waves impinging on $\mathcal{B}$ may spuriously be reflected back into the
computational domain, and thus may give rise to large inaccuracies in the finite
element solution. An effective boundary condition must generate no (or very little)
spurious reflection of waves. Various such boundary conditions, which are called *non-
reflecting boundary conditions*, have been proposed in the last two decades. The subject
is reviewed in Givoli (1991). It is shown there that a robust non-reflecting boundary
condition, effective in the general case and not only in some specific cases, must be
either *nonlocal* or of sufficiently *high order*.

A finite element used in conjunction with a non-reflecting boundary condition on
an artificial boundary is called a *non-reflecting finite element (NRFE)*. We shall
distinguish between two types of NRFEs, namely nonlocal NRFEs which are based on
nonlocal boundary conditions, and local NRFEs, which are based on local boundary
conditions. In what follows we shall consider both types of NRFEs, discuss the
differences between the two approaches from various perspectives, and mention recent
and undergoing work.

The Mathematics of Finite Elements and Applications
Edited by J. R. Whiteman © 1994 John Wiley & Sons Ltd

21.2 NONLOCAL NON-REFLECTING FINITE ELEMENTS

In Keller and Givoli (1989) and Givoli and Keller (1989) we devised nonlocal boundary conditions which are *exact* and are thus perfectly non-reflecting. In the former paper we considered exterior time-harmonic acoustic waves, whereas in the latter we discussed nonlocal NRFEs in general. In Givoli and Keller (1990) we extended the ideas to time-harmonic elastic waves. In all cases, the artificial boundary $\mathcal{B}$ was chosen to be a circle in two dimensions and a sphere in three dimensions. The use of the nonlocal boundary conditions leads to a non-standard finite element formulation. The infinite domain outside the computational domain can be regarded as a "super finite element" with interacting degrees of freedom on $\mathcal{B}$.

21.2.1 ACOUSTIC TIME-HARMONIC WAVES

Exact Non-Reflecting Boundary Condition

Consider the reduced wave equation,

$$\nabla^2 u + k^2 u = 0 \, , \tag{21.1}$$

in the two-dimensional infinite domain exterior to a scatterer or an obstacle. The wave number k is given. A boundary condition is given on the boundary Γ of the scatterer, say the Neumann condition,

$$\frac{\partial u}{\partial \nu} = h \quad \text{on} \quad \Gamma \, . \tag{21.2}$$

Here $\partial u/\partial \nu$ is the normal derivative of u on Γ, and h is a given function. At infinity, the Sommerfeld condition holds:

$$\lim_{r \to \infty} r^{1/2} \left(\frac{\partial u}{\partial r} - iku \right) = 0 \, . \tag{21.3}$$

Here r is the distance from the origin which is fixed at a point inside the scatterer, and $i = \sqrt{-1}$. Condition (21.3) states that the waves are outgoing at infinity.

To make the computational domain finite, we introduce the artificial boundary $\mathcal{B}$. The computational domain, Ω, is now bounded internally by Γ and externally by $\mathcal{B}$. We choose $\mathcal{B}$ to be a *circle* with radius R. See Figure 21.1.

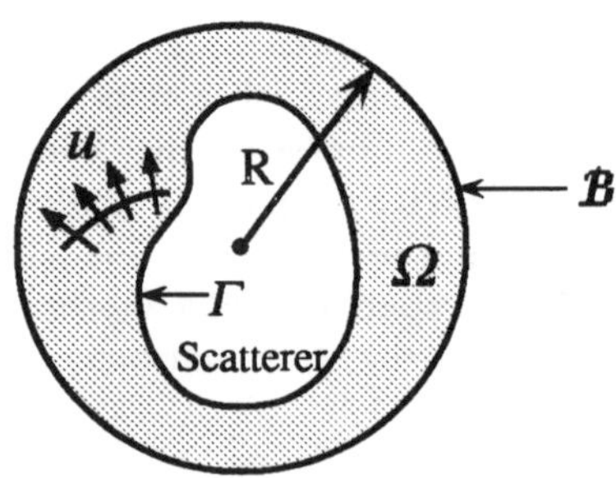

Figure 21.1 Setup of an exterior wave problem

On $\mathcal{B}$ we impose the boundary condition

$$\frac{\partial u}{\partial \nu} = -Mu \quad \text{on} \quad \mathcal{B} \; . \tag{21.4}$$

Here M is an operator called the Dirichlet to Neumann (DtN) map. In Keller and Givoli (1989) we find,

$$Mu(\theta) = \sum_{n=0}^{\infty}{}' \int_0^{2\pi} m_n(\theta - \theta')u(R, \theta')\, R\, d\theta' \; , \tag{21.5}$$

$$m_n(\theta - \theta') = -\frac{k}{\pi R} \frac{H_n^{(1)\prime}(kR)}{H_n^{(1)}(kR)} \cos n(\theta - \theta'). \tag{21.6}$$

In (21.5), the prime after the sum indicates that the $n = 0$ term is multiplied by a factor of $1/2$. In (21.6) the $H_n^{(1)}$ are the Hankel functions of the first kind. It can be shown that the boundary condition (21.4) is *exact*, and furthermore that the original problem (21.1)–(21.3) in the infinite domain is *equivalent* to the new problem in the finite domain Ω, consisting of equations (21.1), (21.2) and (21.4). See Givoli (1992(ii)) and Harari and Hughes (1992(ii)) for details. We note that the DtN map M in (21.5) is a *nonlocal* integral operator.

Finite Element Formulation

The finite element formulation corresponding to the problem consisting of (21.1), (21.2) and (21.4) can easily be obtained by formally following the standard Galerkin procedure. The final finite element matrix problem is,

$$\mathbf{Kd} = \mathbf{F} \tag{21.7}$$

$$\mathbf{K} = \mathbf{K}^a + \mathbf{K}^b \tag{21.8}$$

$$K_{AB}^a = \int_\Omega (\nabla \Phi_A \cdot \nabla \Phi_B - k^2 \Phi_A \Phi_B)\, d\Omega \tag{21.9}$$

$$K_{AB}^b = \int_B \Phi_A M \Phi_B\, d\mathcal{B} \tag{21.10}$$

$$F_A = \int_{\Gamma_h} \Phi_A h\, d\Gamma \; . \tag{21.11}$$

In (21.7), $\mathbf{K}$ is the stiffness matrix, $\mathbf{d}$ is the solution vector whose entries are the nodal values of the finite element solution, and $\mathbf{F}$ is the load vector. The matrix $\mathbf{K}$ is decomposed in (21.8) into the standard part $\mathbf{K}^a$ and the new part $\mathbf{K}^b$ which is contributed by the nonlocal boundary condition (21.4). In (21.9)–(21.11), Φ_A is the shape function associated with node A.

Owing to the locality property of the shape functions Φ_A, the entries K_{AB}^b of the matrix K^b are zero unless both nodes A and B are on the boundary $\mathcal{B}$. On the other hand, K_{AB}^b is in general non-zero for *any* pair of nodes A and B on $\mathcal{B}$. Let

N_B be the number of nodes on $\mathcal{B}$. Then there is a single $N_B \times N_B$ non-zero block in $\mathbf{K}^b$, corresponding to the degrees of freedom on $\mathcal{B}$. We denote this block $\mathbf{k}^{\mathcal{B}}$. Now, we may think of the infinite exterior domain which has been eliminated as a "super finite element" which has N_B degrees of freedom (the nodal values on $\mathcal{B}$) and whose stiffness matrix is $\mathbf{k}^{\mathcal{B}}$. As usual in the implementation of the finite element method, the assembly process is employed, in which the global matrix $\mathbf{K}$ is constructed by adding up element contributions. The stiffness matrix $\mathbf{k}^{\mathcal{B}}$ of the super element is included in the assembly process, as are the element stiffness matrices of all the other standard elements in the mesh.

Theoretical and Computational Aspects

Now we make some remarks on the nonlocal NRFE described above.

(a) The stiffness matrix $\mathbf{k}^{\mathcal{B}}$ of the super element is *symmetric* (Givoli and Keller (1989)).

(b) The calculation of $\mathbf{k}^{\mathcal{B}}$ may be carried out in an efficient way. In this way the added computational cost is marginal (Givoli (1992(ii)), Harari and Hughes (1992(i))).

(c) In practice, the infinite sum appearing in (21.5) must be truncated after a finite number of terms, N. Numerical experiments show that the error in the finite element solution due to this truncation becomes *rapidly* small as N becomes larger, even for very small computational domains. Thus, in general only the first few terms in the DtN map need to be computed (Givoli (1992(ii))).

(d) The method converges to the exact solution of the original problem (21.1)–(21.3) with the optimal rate of convergence (Givoli (1992(ii)), MacCamy and Marin (1980)). Harari and Hughes (1991, 1992(iii)) consider a similar finite element method based on the Galerkin Least Squares (GLS) formulation, and prove convergence in this case too. It turns out that it is easier to control the numerical stability of the scheme in the GLS method than in the standard Galerkin formulation.

(e) Numerical experiments (Keller and Givoli (1989), Givoli (1992(ii))) demonstrate that the nonlocal NRFEs are very effective. For example, for the reduced wave equation (21.1), the nonlocal NRFE performs better than the standard finite element method using the second-order Engquist-Majda boundary condition (Engquist and Majda (1979)) on the artificial boundary.

21.2.2 RECENT PROGRESS

Time-Dependent Waves

Nonlocal NRFEs have recently been constructed for the time-dependent case as well. In doing this the semi-discretization approach was adopted (Givoli (1992(i))), although other approaches are also possible (Givoli (1992(ii))). In the proposed scheme, the time-dependent wave equation is first discretized in time, leading to a sequence of boundary value problems in an unbounded domain. Then an artificial boundary is introduced, and a spatially exact boundary condition, somewhat more complicated than (21.4), is derived on it. This boundary condition is nonlocal in space but local in time. A NRFE, which is again a "super finite element," is constructed based on this boundary condition. It has an $N_B \times N_B$ constant stiffness matrix as well as an N_B-dimensional load vector which is re-calculated every time step.

Semi-infinite Geometry

Many problems in geophysics and meteorology are associated with semi-infinite domains. It turns out that in some of these problems the construction of a nonlocal NRFE is significantly more difficult than in the case of exterior problems of the type illustrated in Figure 21.1. Recently, such a NRFE has been constructed for the reduced wave equation and for classical elastostatics in two dimensions (Givoli and Vigdergauz (1993)). In both cases the artificial boundary $\mathcal{B}$ is chosen to be a semicircle. In the case of elastostatics the derivation of the nonlocal boundary condition on $\mathcal{B}$ is difficult since separation of variables fails. Instead, the derivation relies on complex analysis and on the use of Kolosoff-Muskhelishvili potentials. The stiffness matrix of the NRFE is obtained via the approximate solution of an infinite system of linear algebraic equations.

Elliptic Artificial Boundary

Sometimes it may be beneficial to use an artificial boundary $\mathcal{B}$ which is an ellipse (in two dimensions) or an ellipsoid (in three dimensions). Such an artificial boundary may be used to enclose slender obstacles in an efficient way, by adjusting the aspect ratio of the ellipse (Givoli (1992(ii))). Work is currently under way to construct nonlocal NRFEs for use with an elliptic artificial boundary in two dimensions. For Laplace's equation the computational cost is the same as with a circular artificial boundary. However, in the case of the reduced wave equation Mathieu functions are involved, and these are not so convenient to compute with. Careful implementation as well as the use of some special quadrature rules are needed to ensure the cost-effectiveness of the scheme. Preliminary results are encouraging.

21.3 LOCAL NON-REFLECTING FINITE ELEMENTS

Although nonlocal NRFEs are very effective and accurate, the fact that they behave like "super finite elements" and have many interacting degrees of freedom may be considered a disadvantage. Such elements cannot be incorporated in a finite element code using the standard finite element architecture. In addition, in certain cases they may require an equation-solving effort which is larger than is needed with a local non-reflecting boundary condition (Givoli (1992(ii))). On the other hand, low-order local non-reflecting boundary conditions cannot yield accurate results under *all* circumstances. Therefore we concentrate on the use of *high-order* local boundary conditions.

Local high-order non-reflecting boundary conditions were proposed by various authors; see Givoli (1991) for a survey. In general, such boundary conditions are not compatible with the standard symmetric finite element formulation. For the reduced wave equation (21.1), a boundary condition that leads to a symmetric finite element formulation has the form,

$$\frac{\partial u}{\partial r} + \sum_{n=0}^{N} \frac{\partial^n}{\partial \theta^n} \left(a_n(\theta) \frac{\partial^n u}{\partial \theta^n} \right) = 0 \quad \text{on} \quad \mathcal{B} . \tag{21.12}$$

It is sometimes possible, by making recursive use of the reduced wave equation (21.1) in polar coordinates, to reduce a high-order boundary condition to the form (21.12); see a similar reduction in Pinsky and Abboud (1990, 1991) for the three-dimensional time-dependent wave equation. The family of boundary conditions due to Feng (1983), and the family of localized boundary conditions discussed in the Appendix of Givoli and Keller (1990) (see next section) have the form (21.12). The first two Engquist-Majda conditions (1977) and the first two Bayliss-Turkel conditions (1980) can also be expressed as in (21.12). However, higher-order Bayliss-Turkel conditions have a non-symmetric form after reduction.

When a high-order boundary condition of the form (21.12) is used on the artificial boundary $\mathcal{B}$, standard C^0 finite elements *cannot* be used in the layer adjacent to $\mathcal{B}$, since (21.12) involves high-order spatial derivatives. A new finite element with high-order continuity on the boundary $\mathcal{B}$ is devised for this purpose. This element possesses C^1 or higher-order continuity on one of its sides (the side which lies on $\mathcal{B}$), and C^0 continuity elsewhere. It is fully compatible with standard finite element architecture. One layer of elements of this type is used near the artificial boundary $\mathcal{B}$.

21.3.1 A SEQUENCE OF LOCAL BOUNDARY CONDITIONS

A good way to obtain accurate local non-reflecting boundary conditions is to *localize* the nonlocal boundary condition (21.4). This localization yields a sequence of local boundary conditions, the Nth condition being *exact* for waves which consist of the first N harmonics. The Nth condition has the form (21.12), with

$$a_n \equiv (-1)^n \alpha_n . \tag{21.13}$$

The constants α_n are obtained by solving the linear system,

$$\begin{pmatrix} 0^0 & 0^2 & 0^4 & \cdots & \\ 1^0 & 1^2 & 1^4 & \cdots & \\ \cdot & & & & \\ \cdot & & & & \\ \cdot & & & & \\ N^0 & N^2 & N^4 & \cdots & N^{2N} \end{pmatrix} \begin{pmatrix} \alpha_0 \\ \alpha_1 \\ \cdot \\ \cdot \\ \cdot \\ \alpha_N \end{pmatrix} = \begin{pmatrix} Z_0 \\ Z_1 \\ \cdot \\ \cdot \\ \cdot \\ Z_N \end{pmatrix} . \tag{21.14}$$

Here 0^0 is defined to be 1, and

$$Z_n = -k \frac{H_n^{(1)\prime}(kR)}{H_n^{(1)}(kR)} . \tag{21.15}$$

The derivation may be found in Givoli and Keller (1990). Analogous boundary conditions may be derived in the three-dimensional case.

21.3.2 FINITE ELEMENT FORMULATION

Now we consider the finite element solution of the problem in Ω (see Figure 21.1) consisting of equations (21.1), (21.2) and (21.12). Formally, the finite element formulation is similar to the one considered previously. However, the boundary operator involved in the present case is local. The final finite element equations (21.7)–(21.11) remain unchanged except that the entries of the matrix $\mathbf{K}^b$ in (21.10) are replaced by,

$$K^b_{AB} = \sum_{n=0}^{N} (-1)^n \int_B \frac{\partial^n \Phi_A}{\partial \theta^n} a_n(\theta) \frac{\partial^n \Phi_B}{\partial \theta^n} \, dB \, . \tag{21.16}$$

It is important to note that the shape functions Φ_A are required to be in $H^1(\Omega) \cap H^N(B)$ (H^n being the Sobolev space), or, since the Φ_A are piecewise polynomials, they must be in $C^0(\Omega) \cap C^{N-1}(B)$. Thus, special finite elements with high-order regularity must be used in the layer near B. On the other hand, owing to the local character of (21.16), all the calculations may be carried out on the element level, and there is no interaction between remote degrees of freedom on B. Thus, standard finite element architecture is used, as opposed to the case of nonlocal NRFEs.

21.3.3 CONFORMING ELEMENTS WITH $C^{N-1}(B)$ REGULARITY

A hierarchy of conforming NRFEs with $C^{N-1}(B)$ regularity has recently been developed (Givoli and Keller (1993)). The first element in this hierarchy, with $C^1(B)$ regularity, is illustrated in Figure 21.2.

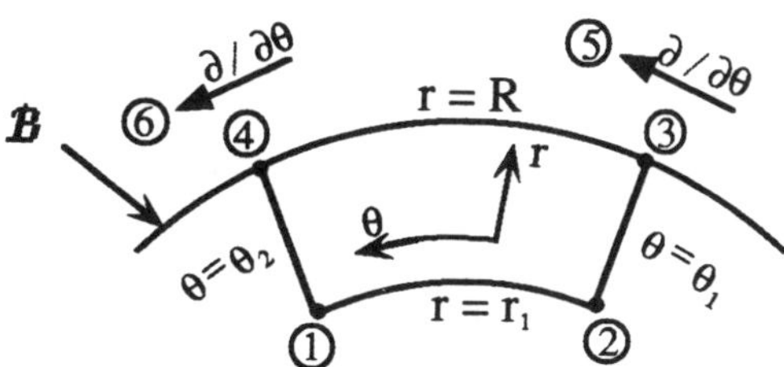

Figure 21.2 Local NRFE with $C^1(B)$ regularity

This NRFE has polar geometry, and is associated with four nodes and six degrees of freedom. The degrees of freedom are the function values at the four vertices and the values of the tangential derivative at the two nodes on B. Each degree of freedom is associated with a local shape function. The six shape functions are polynomials of first degree in r and third degree in θ, but they are linear in θ along the side $r = r_1$ (see Figure 21.2). It can be proved (Givoli and Keller (1993)) that this and the other NRFEs in the hierarchy are *conforming*, namely they fully satisfy the regularity requirements. Also, the construction of the shape functions is unique. A NRFE of this type connects along the two sides $\theta = \theta_1$ and $\theta = \theta_2$ (see Figure 21.2) to similar NRFEs. Along the side $r = r_1$, the NRFE connects to any C^0 element which has two nodes along that side. This includes the standard linear triangular element and the bilinear quadrilateral element. The implementation of these local high-order NRFEs is currently under way.

ACKNOWLEDGMENT

This work was supported in part by the L. Kraus Research Fund, VPRF no. 160-700, and by ONR.

REFERENCES

Bayliss A. and Turkel E. (1980), Radiation Boundary Conditions for Wave-Like Equations. *Comm. Pure Appl. Math.*, **33**, 707–725.

Engquist B. and Majda A. (1977), Absorbing Boundary Conditions for the Numerical Simulation of Waves. *Math. Comput.*, **31**, 629–652.

Engquist B. and Majda A. (1979), Radiation Boundary Conditions for Acoustic and Elastic Calculations. *Comm. Pure Appl. Math.*, **32**, 313–357.

Feng K. (1983), Finite Element Method and Natural Boundary Reduction. In: *Proc. of the Int. Congress of Mathematicians*, Warsaw, pp. 1439–1453.

Givoli D. (1991), Non-Reflecting Boundary Conditions: A Review. *J. Comput. Phys.*, **94**, 1–29.

Givoli D. (1992(i)), A Spatially Exact Non-Reflecting Boundary Condition for Time Dependent Problems. *Comp. Meth. Appl. Mech. Engng.*, **95**, 97–113.

Givoli D. (1992(ii)), *Numerical Methods for Problems in Infinite Domains*, Elsevier, Amsterdam.

Givoli D. and Keller J.B. (1989), A Finite Element Method for Large Domains. *Comp. Meth. Appl. Mech. Engng.*, **76**, 41–66.

Givoli D. and Keller J.B. (1990), Non-Reflecting Boundary Conditions for Elastic Waves. *Wave Motion*, **12**, 261–279.

Givoli D. and Keller J.B. (1993), Non-Reflecting Finite Elements. *TAE/BSF-1993 Report*, Technion, Haifa.

Givoli D. and Vigdergauz S. (1993), Artificial Boundary Conditions for 2D Problems in Geophysics. *Comp. Meth. Appl. Mech. Engng.*, to appear.

Harari I. and Hughes T.J.R. (1991), Finite Element Methods for the Helmholtz Equation in an Exterior Domain: Model Problems. *Comp. Meth. Appl. Mech. Engng.*, **87**, 59–96.

Harari I. and Hughes T.J.R. (1992(i)), A Cost Comparison of Boundary Element and Finite Element Methods for Problems of Time-Harmonic Acoustics. *Comp. Meth. Appl. Mech. Engng.*, **97**, 77–102.

Harari I. and Hughes T.J.R. (1992(ii)), Analysis of Continuous Formulations Underlying the Computation of Time-Harmonic Acoustics in Exterior Domains. *Comp. Meth. Appl. Mech. Engng.*, **97**, 103–124.

Harari I. and Hughes T.J.R. (1992(iii)), Galerkin/Least Squares Finite Element Methods for the Reduced Wave Equation with Non-Reflecting Boundary Conditions in Unbounded Domains. *Comp. Meth. Appl. Mech. Engng.*, **98**, 411–454.

Keller J.B. and Givoli D. (1989), Exact Non-Reflecting Boundary Conditions. *J. Comput. Phys.*, **82**, 172–192.

MacCamy R.C. and Marin S.P. (1980), A FInite Element Method for Exterior Interface Problems. *Int. J. Math. Math. Sci.*, **3**, 311–350.

Pinsky P.M. and Abboud N.N. (1990), Finite Element Solution and Dispersion Analysis for the Transient Structural Acoustics Problem. *Appl. Mech. Rev.*, **43** Suppl., 381–388.

Pinsky P.M. and Abboud N.N. (1991), Finite Element Solution of the Transient Exterior Structural Acoustics Problem Based on the Use of Radially Asymptotic Boundary Operators. *Comp. Meth. Appl. Mech. Engng*, **85**, 311–348.

Chapter 22

A Dispersion Analysis of Finite Element Methods on Triangular Grids for Maxwell's Equations

P.B. Monk*, A.K. Parrott** and P.J. Wesson**

* Dept of Mathematical Sciences, University of Delaware, Newark DE 19716, U.S.A.
** Oxford University Computing Laboratory, 11 Keble Road, Oxford,
OX1 3QD, U.K.

22.1   INTRODUCTION

Many important applications of Maxwell's equations in areas such as electromagnetic compatibility and bio-electromagnetism involve solving the equations in the presence of complex geometric interfaces. Automatic mesh generation using tetrahedra (in $\mathbb{R}^3$) or triangles (in $\mathbb{R}^2$) is increasingly able to provide reliable and good quality meshes in these circumstances. However the accuracy of finite element approximations should be carefully assessed on unstructured grids, particularly for transient problems where an important indicator of the merit of a given method is the accuracy with which it propagates electromagnetic waves. This can be characterised through the discrete dispersion relation of the method and in Monk and Parrott (1992) this is the basis for a comparison of a number of conforming and non-conforming piecewise polynomial representations of the fields (methods based on potentials are not considered here). Two methods, the linear-linear scheme and the edge(1)-constant scheme, have been shown to have significant advantages. The dispersion analysis for these two schemes is reviewed and extended and detailed wave propagation calculations for the edge(1)-constant case are presented. (See Lee and Madsen (1990), Monk (1992) for similar comparisons on quadrilateral elements.)

22.2   MAXWELL'S EQUATIONS

The two dimensional Maxwell's equations considered here govern the behaviour of the electric field $\boldsymbol{E} = (E_1(\boldsymbol{x}, t), E_2(\boldsymbol{x}, t))$ and the scalar magnetic field $H = H(\boldsymbol{x}, t)$ (where $\boldsymbol{x} = (x_1, x_2) \in \mathbb{R}^2$). The fields satisfy the following system of equations:

$$\frac{\partial \boldsymbol{E}}{\partial t} - \vec{\nabla} \times H = 0, \text{ and } \frac{\partial H}{\partial t} + \nabla \times \boldsymbol{E} = 0 \qquad \text{in } \mathbb{R}^2 \qquad (22.1)$$

The Mathematics of Finite Elements and Applications
Edited by J. R. Whiteman © 1994 John Wiley & Sons Ltd

where the two curls are defined by

$$\vec{\nabla} \times H = \left(\frac{\partial H}{\partial y}, -\frac{\partial H}{\partial x}\right)^T \quad \text{and} \quad \nabla \times E = \frac{\partial E_2}{\partial x} - \frac{\partial E_1}{\partial y}. \tag{22.2}$$

In equations (22.1), we have set the constitutive parameters ϵ and μ to one so that the speed of light for these equations is unity.

Equations (22.1) have solutions of the form

$$E = E_0 \exp(i(k \cdot x - \omega t)) \quad \text{and} \quad H = H_0 \exp(i(k \cdot x - \omega t)) \tag{22.3}$$

provided $\omega = \pm|k|$,   or $\omega = 0$, where ω, $k = (k_1, k_2)$, E_0 and H_0 are constants. The phase speed for the travelling waves ($\omega \neq 0$) is $c_\phi = \omega/|k| = \pm 1$.

22.3   FINITE ELEMENT APPROXIMATIONS

The electromagnetic fields E and H are approximated by

$$E^h(x, t) = \sum_{i=1}^{N_E} E_i(t)\phi_i^h(x) \quad \text{and} \quad H^h(x, t) = \sum_{i=1}^{N_H} H_i(t)\psi_i^h(x). \tag{22.4}$$

using piecewise polynomial vector basis functions $\{\phi_i^h\}_{i=1}^{N_E}$ and scalar basis functions $\{\psi_i^h\}_{i=1}^{N_H}$. Two types of piecewise polynomial basis functions are considered and their distribution of the degrees of freedom is shown in Figure 22.1. The first sets $E^h|_K \in (P_1)^2$ and $H^h|_K \in P_1$ on each triangle K, where P_1 is the set of linear polynomials in (x, y). The degrees of freedom on each triangle are then

$$\Sigma_K^E = \{E^h(a_i), 1 \leq i \leq 3\} \quad \text{and} \quad \Sigma_K^H = \{H^h(a_i), 1 \leq i \leq 3\}$$

where $a_i, i = 1, 2, 3$ are the vertex coordinates of K.

In the second "edge–constant" family, the electric field is discretized by using a set of vector elements with continuous tangential components (due to Nédélec (1980)), i.e.,

$$E^h|_K = \begin{pmatrix} a_K + b_K y \\ c_K - b_K x \end{pmatrix}$$

where a_K, b_K and c_K are constants. The degrees of freedom here are the tangential component of the field at the mid–point of each edge, i.e.

$$\Sigma_K^E = \{E^h(a_{i,j}) \cdot \tau_{ij} \quad 1 \leq i < j \leq 3\}$$

where $a_{i,j} = (a_i + a_j)/2$ and $\tau_{ij} = (a_j - a_i)/\|a_j - a_i\|$. The magnetic field is discretized by a simple piecewise constant basis with unknowns associated with the centroid of each element. Thus

$$H^h|_K = \text{constant} \quad \text{and} \quad \Sigma_K^H = \{H^h((a_1 + a_2 + a_3)/3)\} \ .$$

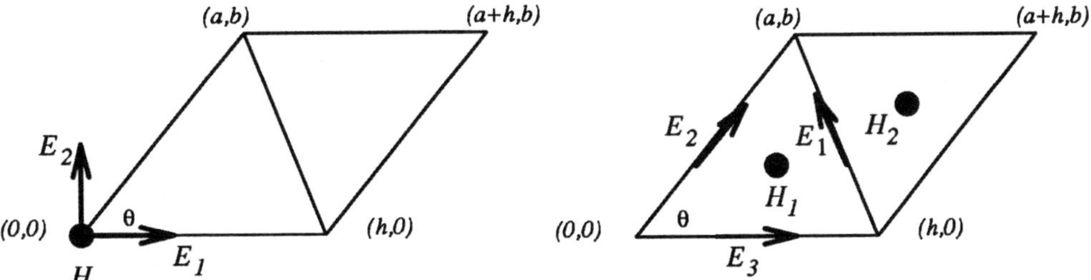

Figure 22.1   This figure shows the "unit cell" and degrees of freedom for the two methods discussed in this paper. The arrows represent electric degrees of freedom and the solid circle magnetic degrees.

22.4 DISPERSION ANALYSIS

To compute a dispersion relation for a finite element approximation on a triangular grid (see Monk and Parrott (1992) for more details) the mesh covering $\mathbb{R}^2$ has to be restricted to be translations of a "unit parallelogram" $\widehat{K}$ (see Figure 22.1). In this paper the parallelograms are additionally restricted to sides of length h but allow a variable angle θ so that the effects of element distortion can be examined .

For linear–linear elements we can compute difference equations satisfied by the degrees of freedom E_1, E_2 and H shown in Figure 22.1(a). Substituting (22.3) into these difference equations results in a 3×3 matrix eigenvalue problem for $\omega = \omega(k, h, \theta)$ giving three possible dispersion relations, $\omega = 0$ and

$$
\begin{aligned}
h^2\omega^2 \;=\; & \zeta_1^2 + \zeta_2^2 + \left(-\frac{1}{90} + \frac{\cos(\theta)}{36} - \frac{\cos(\theta)^2}{36}\right)\zeta_1^6 \\
& + \left(\frac{\sin(\theta)}{36} - \frac{\cos(\theta)^2\sin(\theta)}{36} - \frac{2\sin(\theta)\cos(\theta)}{45} + \frac{\cos(\theta)^3\sin(\theta)}{90}\right)\zeta_2\zeta_1^5 \\
& + \left(-\frac{\cos(\theta)}{36} - \frac{\cos(\theta)^4}{18} + \frac{\cos(\theta)^3}{9} - 1/36\right)\zeta_2^2\zeta_1^4 \\
& + \left(\frac{\cos(\theta)^2\sin(\theta)}{6} - \frac{\cos(\theta)^3\sin(\theta)}{9} - \frac{\sin(\theta)\cos(\theta)}{18}\right)\zeta_2^3\zeta_1^3 \\
& + \left(-\frac{\cos(\theta)^2}{12} - \frac{\cos(\theta)^3}{9} + \frac{\cos(\theta)^4}{9} - 1/36 + \frac{\cos(\theta)}{9}\right)\zeta_2^4\zeta_1^2 \\
& + \left(\frac{\cos(\theta)^3\sin(\theta)}{18} + \frac{\sin(\theta)}{36} - \frac{\cos(\theta)^2\sin(\theta)}{36} - \frac{\sin(\theta)\cos(\theta)}{18}\right)\zeta_2^5\zeta_1 \\
& + \left(-\frac{1}{90} + \frac{\cos(\theta)^2}{45} - \frac{\cos(\theta)^4}{90}\right)\zeta_2^6 + O(|\zeta|^8),
\end{aligned}
$$

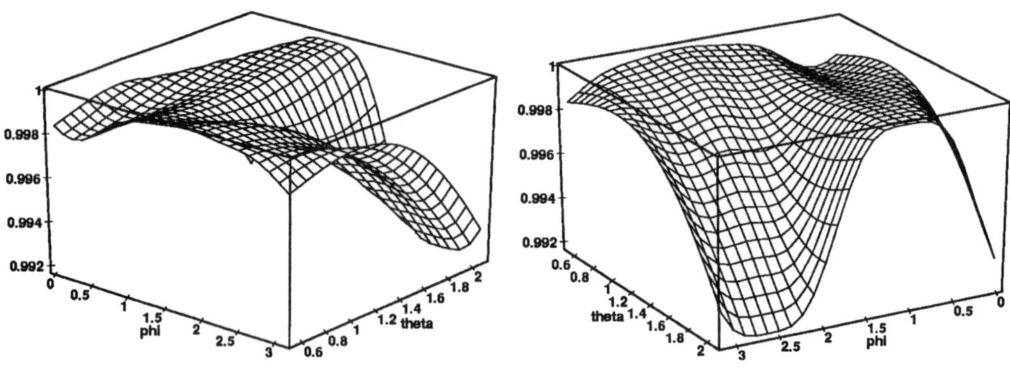

Figure 22.2 Surfaces of phase-velocity $(\omega/|\boldsymbol{k}|)$ against propagation angle ϕ $(\boldsymbol{k} = |\boldsymbol{k}|(\cos\phi, \sin\phi)\,)$ and grid angle θ for the linear-linear element family for eight triangles per wavelength.

where we define $\zeta_1 = k_1 h$ and $\zeta_2 = k_2 h$. This result extends the result in Monk and Parrott (1992) and shows that the linear–linear method has a fourth order convergent dispersion relation regardless of grid angle θ.

The accuracy of this dispersion relation clearly varies with propagation direction, mesh distortion (θ), and the number of triangle sides per wavelength. The phase velocity $\omega/|\boldsymbol{k}|$ is plotted in Figure 22.2 (of course this velocity should be exactly unity regardless of $\boldsymbol{k}$) and although the method deteriorates for $\theta > \pi/2$ it is otherwise remarkably accurate.

As yet it has not been possible to compute a Taylor series for ω as a function of ζ_1 and ζ_2 for variable θ for the edge–constant family . We have previously noted that the edge–constant scheme has a fourth order dispersion relation at $\theta = \pi/3$ (i.e., on equilateral triangles) but for $\theta = \pi/2$ only a second order dispersion relation. This can be seen in the plot of the phase velocity $w/|\boldsymbol{k}|$ against θ and ϕ in Figure 22.3 where the value $\theta = \pi/3$ is clearly associated with particularly high accuracy in the phase velocity. Away from $\theta = \pi/3$ the accuracy deteriorates, particularly for $\theta > \pi/2$ and is consistent with the proposition that this method fourth order convergent only for the single value $\theta = \pi/3$.

Note that, depending on the propagation angle ϕ, the numerical phase velocity can be either larger or smaller than the exact velocity. Thus compared to the linear–linear family discussed previously, the edge–linear family has poorer wave–propagation properties but it can be argued that these elements have the right degree of continuity for inhomogeneous media (see Bossavit (1988)).

A second aspect of the edge–constant scheme is the possibility of parasitic modes (Monk and Parrott (1992)). As can be seen from Figure 22.1(b) the unit cell for the edge–constant family contains 5 degrees of freedom (the remaining degrees of freedom

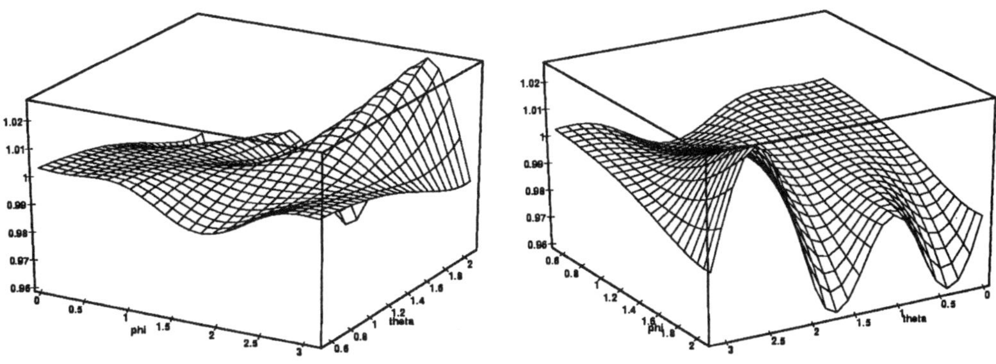

Figure 22.3   Surfaces of phase-velocity $(\omega/|\boldsymbol{k}|)$ against propagation angle ϕ and grid angle θ for the edge-constant element family for eight triangles per wavelength.

can be obtained by simple translation). Thus there are five dispersion relations: $\omega = 0, \omega^2 \simeq |\boldsymbol{k}|^2$ and $\omega^2 = 0(1/h^2)$. The modes corresponding to $\omega^2 = 0(1/h^2)$ are called parasitic. Using a perturbation expansion, we can show that

$$\omega = \frac{6\sqrt{2}}{h\sqrt{-2 + \cos(\theta)}} + O(1)$$

and the corresponding eigenmode, $(H_1, H_2, E_1, E_2, E_3)^T$, is

$$\left(-1, 1, -\frac{\sqrt{2}\sin(\theta)}{\sqrt{-2 + \cos(\theta)}}, \frac{\sqrt{2}\sin(\theta)}{\sqrt{-2 + \cos(\theta)}}, -\frac{\sqrt{2}\sin(\theta)}{\sqrt{-2 + \cos(\theta)}}\right)^T + O(h^2).$$

It is possible that these parasitic modes could give rise to high frequency in time oscillations on the solution. Notice also that the magnetic field oscillates very rapidly on the order of a grid cell.

22.5   NUMERICAL RESULTS

A number of test cases were run to evaluate the accuracy of the edge-constant method for wave propagation. For this case Maxwell's equations have the plane wave solution

$$\boldsymbol{E}(\boldsymbol{x}, t) = \begin{pmatrix} -k_2 \\ k_1 \end{pmatrix} g(t - \boldsymbol{k}.\boldsymbol{x}) \qquad \text{and} \qquad H(\boldsymbol{x}, t) = g(t - \boldsymbol{k}.\boldsymbol{x}). \tag{22.5}$$

A smooth Gaussian pulse was chosen, with

$$g(s) = \begin{cases} \frac{\exp(-10(s-1)^2) - \exp(-10)}{1 - \exp(-10)} & 0 \le s \le 2 \\ 0 & s > 2 \text{ or } s < 0. \end{cases}$$

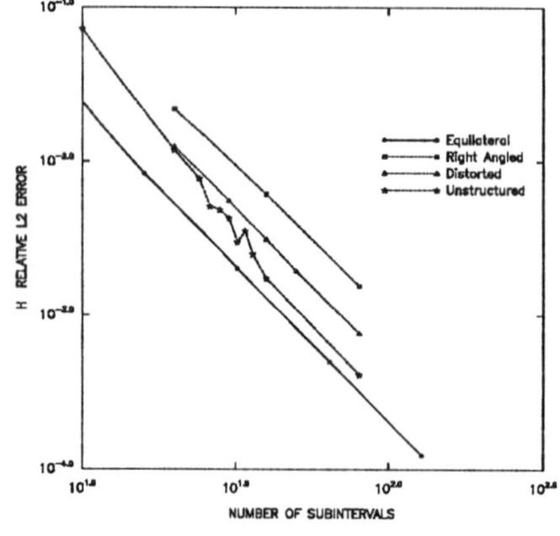

(a) Relative L2 errors in H

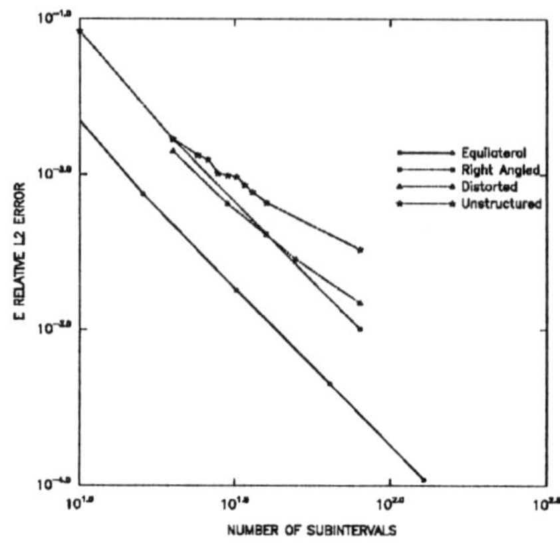

(b) Relative L2 errors in E

Figure 22.4   Errors for a series of regular and unstructured grids, plotted against the number of mesh intervals in the propagation direction

Leapfrog timestepping was selected for the fully discrete scheme, so that the fully discrete Galerkin equations are

$$\sum_{i=1}^{N_E} \left(\frac{E_i^{n+1} - E_i^n}{\Delta t} \right) (\phi_i, \phi_k) - \sum_{j=1}^{N_H} H_j^{n+\frac{1}{2}} (\psi_j, \nabla \times \phi_k) \; = \; 0 \qquad \forall k \geq 1 \qquad (22.6a)$$

$$\sum_{j=1}^{N_H} \left(\frac{H_j^{n+\frac{1}{2}} - H_j^{n-\frac{1}{2}}}{\Delta t} \right) (\psi_j, \psi_l) + \sum_{i=1}^{N_E} E_i^n (\nabla \times \phi_i, \psi_l) \; = \; 0 \qquad \forall l \geq 1. \qquad (22.6b)$$

where $(u, v) = \int_\Omega u \cdot v \, dx$. The additional error terms in the dispersion relations are easily shown to be isotropic and second order in Δt.

Four types of grids were used: equilateral, right-angle, distorted right-angle and randomised (created by randomly peturbing the positions of points in the right-triangle grids and then reconnecting). A computational domain of $[-1, 1] \times [-1, 1]$ was used in all cases except for the equilateral triangulation where a triangular region was more convenient. In each case the errors were calculated when the wave reached the middle of the domain, using a timestep corresponding to a CFL number of exactly 0.25 and for $k = (1, 0)$. The tangential components of the electric field were specified on the boundary of the domain using (22.5).

The errors in H were computed as follows; if

$$\|(H - H^h)(t)\|_{L^2,h} = \sqrt{\sum_{i=1}^{N_H} |H(\boldsymbol{x}_c, t) - H^h((\boldsymbol{x}_c, t)|^2 \mathcal{A}_i}, \qquad (22.7)$$

where $\boldsymbol{x}_c$ is the centroid of each triangle, and $\mathcal{A}_i$ is the triangle area, then the relative L^2 error is

$$\|(H - H^h)(t)\|_{rel} = \|(H - H^h)(t)\|_{L^2,h}/\|H^h(t)\|_{L^2,h}. \qquad (22.8)$$

Errors in the E were similarly defined in terms of the pointwise errors in the tangential components at mid-side nodes, weighted with the average of the two triangle areas adjacent to that edge.

The results for edge-constant elements, shown in Figure 22.4, show a number of interesting features. They confirm the prediction of the dispersion analysis that right triangle grids are less accurate then those on equilateral grids. The randomised grids typically lie somewhere in-between since the Delaunay method tends to generate a proportion of approximately equilateral triangles. For the series of refinements on regular grids both E and H converge like $O(h^2)$. These results are similar to those reported by Monk (1992) for rectangular grids. The results on irregular grids show an apparent disparity between E and H since the L_2 errors in H are clearly $O(h^2)$ while the errors in E are between first and second order. (Regression on the results of a large number of randomised grids gave slopes of -2.5 and -1.5, for H and E respectively.) This suggests that while H (and hence $\nabla \times E$) are superconvergent at the triangle centroids, E is not superconvergent at the mid-side nodes.

Further work is in progress to compare the accuracy of the edge-constant method in three dimensions on structured and unstructured tetrahedral grids.

REFERENCES

Bossavit A. (1988), A rationale for "edge elements" in 3-D fields computations. *IEEE Trans. Mag.*, **24**, *74-79*.

Lee R.L. and Madsen N.K. (1990), A mixed finite element formulation for Maxwell's equations in the time domain. *J. Comput. Phys.*, **88**, *284-304*.

Monk P.M. (1992), A comparison of three mixed methods for the time dependent Maxwell equations. *SIAM J. on Scientific and Statistical Computing* **13**, *1097-1122*.

Monk P.M. and Parrott A.K. (1992), A dispersion analysis of finite element methods for Maxwell's equations. *to appear in SIAM J. on Scientific and Statistical Computing*.

Nédélec J. (1980), Mixed finite elements in $\mathbb{R}^3$. *Numer. Math.*, **35**, *315-341*.

Chapter 23

Coupling of FEM and BEM for Transonic Flows

H. Berger , G. Warnecke and W.L. Wendland

Mathematisches Institut A, Universität Stuttgart,
D–70511 Stuttgart, Germany

23.1   FORMULATION OF THE PROBLEM

We consider a bounded domain $\Omega \subset \mathbb{R}^2$ with a slit Σ surrounding a given simply connected wing section $P \subset \mathbb{R}^2$. The boundary of Ω consists of three parts

$$\partial\Omega := \Gamma_\infty \cup \Gamma_P \cup \Sigma, \tag{23.1}$$

whose interiors are mutually disjoint. The curve Γ_∞ is the artificial exterior boundary of Ω which is taken in order to have a bounded computational domain for a transonic FEM approximation. The curve Γ_P is the common boundary between Ω and the profile, which has a corner at the trailing edge (TE), and is $\mathcal{C}^\infty$ otherwise. We denote by Σ a slit in Ω, joining the trailing edge with Γ_∞. The unbounded far field domain exterior to Γ_∞ will be denoted by

$$\Omega^c = \mathbb{R}^2 \backslash \overline{\Omega \cup P}. \tag{23.2}$$

The straight prolongation of the slit Σ in Ω^c to infinity will be denoted by Σ^c. Without loss of generality, we assume that the travelling velocity is given by a constant vector field $\boldsymbol{v}_\infty$ which is parallel to the x_1–axis, see Figure 23.1. In Ω we consider a stationary, compressible potential flow induced by a subsonic travelling velocity $\boldsymbol{v}_\infty$. A simple model of inviscid, steady, isoenergetic, homentropic, planar flow of an ideal gas is the full potential equation,

$$\operatorname{div}\left[\rho(|\nabla u|^2)\nabla u\right] = 0 \quad \text{in } \Omega \tag{23.3}$$

where $\boldsymbol{v} = \nabla u$ describes the velocity field. The generalized weak formulation for (23.3) admits transonic solutions with shock discontinuities in the velocity field. This equation can be derived from conservation of mass, momentum and energy, see Berger *et al.* (1990). The **density function** $\rho(s)$ is given as

$$\rho(|\nabla u|^2) = \rho_0 \left(1 - \frac{\kappa - 1}{2a_0^2}|\nabla u|^2\right)^{\frac{1}{\kappa - 1}}. \tag{23.4}$$

Here $\kappa > 1$ is the adiabatic gas constant, e.g. $\kappa = 1.4$ for dry air. The constants ρ_0, a_0 are the density and the local speed of sound, respectively, for the motionless gas. The

The Mathematics of Finite Elements and Applications
Edited by J. R. Whiteman © 1994 John Wiley & Sons Ltd

local speed of sound $a(|\nabla u|^2)$ is given by

$$\left[a(|\nabla u|^2)\right]^2 = a_0^2 - \frac{\kappa - 1}{2}|\nabla u|^2$$

and the **local Mach number** is $M := \frac{|\nabla u|}{a}$. Note that the density $\rho(s)$ is only defined for $|\nabla u|^2 \le \frac{2a_0^2}{k-1}$. This bound for the velocity will be assumed to hold in all further considerations. Equation (23.3) is elliptic in the subsonic ($M < 1$) and hyperbolic and in the supersonic ($M > 1$) region.

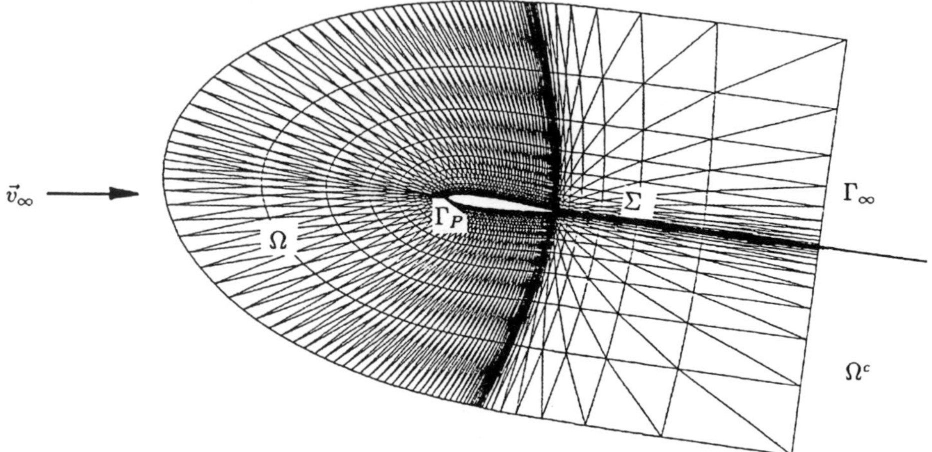

Figure 23.1: Computational domain for the coupling

In order to have a potential flow including circulation and lift we need the slit Σ across which we assume that ∇u is continuous, whereas the potential u may have a finite constant jump; i. e.

$$u^+ - u^- = \beta \quad \text{and} \quad \partial_n u^+ - \partial_n u^- = 0 \quad \text{on } \Sigma. \tag{23.5}$$

Here, n is an unit normal vector field and $\partial_n u = \nabla u \cdot n$. By u^+ and u^- we denote the one–sided boundary values on Σ. The constant β is an additional unknown and gives the circulation of the flow. To determine β, we need the additional **Kutta–Joukowski condition**

$$F(\beta) := \left|\nabla u^+\right|^2_{TE} - \left|\nabla u^-\right|^2_{TE} = 0 \tag{23.6}$$

at the trailing edge TE which follows from the requirement of continuous pressure there. Here we shall assume that the flow near TE is subsonic which implies (23.6), too. On the profile we impose the homogeneous Neumann boundary condition of non–penetration, i. e.

$$\partial_n u = 0 \text{ on } \Gamma_P. \tag{23.7}$$

It is generally assumed that the velocity v tends to v_∞ at infinity. A simple and widely used approximation of this condition is the mass flux condition of parallel flow v_∞ at Γ_∞ , i. e.

$$\rho(|\nabla u|^2)\partial_n u = \rho(|v_\infty|^2)v_\infty \cdot n_\infty \text{ on } \Gamma_\infty. \tag{23.8}$$

This condition was used by Berger *et al.* (1990). A slight improvement of (23.8) can be obtained by replacing $\boldsymbol{v}_\infty$ on the right–hand side by $\boldsymbol{v}_{0|_{\Gamma_\infty}}$, where $\boldsymbol{v}_0$ is the velocity of the incompressible potential flow in $\Omega \cup \Omega^c \cup \Gamma_\infty$, see Berger *et al.* (1988).

In this paper, however, we consider the coupling with Prandtl–Glauert flow exterior to Γ_∞ which is given by the equation

$$(1 - M_\infty^2)\varphi_{x_1 x_1} + \varphi_{x_2 x_2} = 0 \quad \text{for} \quad \boldsymbol{x} = (x_1, x_2) \in \Omega^c, \tag{23.9}$$

which is a linear approximation of (23.3). The **perturbation potential** φ is defined by

$$\varphi := u - \Psi \quad \text{with} \quad \Psi(x, \beta) := \boldsymbol{v}_\infty \cdot (x_1, x_2) + \frac{\beta}{2\pi} \arctan \frac{\sqrt{1 - M_\infty^2} \cdot x_2}{x_1}, \tag{23.10}$$

where M_∞ is the Mach number at infinity, see Zierep (1976) for details. The branch of the arctan is chosen such that both the perturbation velocity field $\nabla\varphi$ and also φ are continuous across the slit Σ^c. This gives the transmission conditions

$$\varphi^+ - \varphi^- = 0 \quad \text{and} \quad \partial_n \varphi^+ - \partial_n \varphi^- = 0 \quad \text{on } \Sigma^c. \tag{23.11}$$

Further, we require the radiation condition

$$\nabla\varphi(\boldsymbol{x}) \to 0 \quad \text{for } |\boldsymbol{x}| \to \infty. \tag{23.12}$$

In addition, we need two more transmission conditions for the coupling of u with φ at Γ_∞. The first transmission condition is obtained by definition (23.10) and continuity across Γ_∞. For the second transmission condition we require equality of the mass flux $\rho(|\nabla u|^2)\partial_n u$ to the corresponding expression defined by linearization about the Prandtl–Glauert solution. The latter leads for the density to $\rho \approx \rho_\infty := \rho(|\boldsymbol{v}_\infty|^2)$ in Ω^c. Then the coupling conditions become

$$u = \varphi + \Psi \quad \text{on } \Gamma_\infty \tag{23.13}$$

and

$$\rho(|\nabla u|^2)\nabla u \cdot \boldsymbol{n} = \rho_\infty \left\{ \nabla\Psi \cdot \boldsymbol{n} + (1 - M_\infty^2)\varphi_{x_1} n_1 + \varphi_{x_2} n_2 \right\} \quad \text{on } \Gamma_\infty \tag{23.14}$$

where $\boldsymbol{n} = (n_1, n_2)$ is the exterior unit normal to Γ_∞.

Collecting the equations (23.3), (23.5), (23.6), (23.7), (23.9), (23.11), (23.12) and the coupling conditions (23.13), (23.14), we get the following system of equations, boundary and transmission conditions:

Coupled boundary value problem

Find the functions u, φ in appropriate function spaces and the constant $\beta \in \mathbb{R}$ satisfying the

Interior full potential problem,

$$\begin{aligned}
\operatorname{div}\left[\rho(|\nabla u|^2)\nabla u\right] &= 0 & &\text{in } \Omega, \\
\partial_n u &= 0 & &\text{on } \Gamma_p, \\
u^+ - u^- &= \beta & &\text{on } \Sigma, \\
\partial_n u^+ - \partial_n u^- &= 0 & &\text{on } \Sigma, \\
F(\beta) = |\nabla u^+|_{TE}^2 - |\nabla u^-|_{TE}^2 &= 0;
\end{aligned} \tag{23.15}$$

Exterior Prandtl–Glauert problem,

$$
\begin{aligned}
(1 - M_\infty^2)\varphi_{x_1 x_1} + \varphi_{x_2 x_2} &= 0 && \text{in } \Omega^c, \\
\nabla\varphi &= o(1) && \text{for } |\boldsymbol{x}| \to \infty, \\
\varphi^+ - \varphi^- &= 0 && \text{on } \Sigma^c, \\
\partial_n\varphi^+ - \partial_n\varphi^- &= 0 && \text{on } \Sigma^c;
\end{aligned}
\tag{23.16}
$$

Coupling conditions,

$$
u = \varphi + \Psi \quad \text{on } \Gamma_\infty \quad \text{and} \tag{23.17}
$$

$$
\rho(|\nabla u|^2)\partial_n u = \rho(|\boldsymbol{v}_\infty|^2)\left\{\nabla\Psi \cdot \boldsymbol{n} + \left((1 - M_\infty^2)\varphi_{x_1}, \varphi_{x_2}\right) \cdot \boldsymbol{n}\right\} \quad \text{on } \Gamma_\infty \tag{23.18}
$$

Note that, for a solution (u, φ, β) of this coupled problem (23.15)–(23.18), one also has the solution $(u + c, \varphi + c, \beta)$ with an arbitrary constant c. We therefore fix this constant by the requirement

$$
\int_{\Gamma_\infty} \varphi\, ds = 0.
$$

With the Euler equations in Ω instead of the full potential model (23.15), Sofronov and Tscincov present a similar coupling formulation in Sofronov and Tscincov (1993).

23.1.1 THE WEAK FORMULATION

For $s \in \mathbb{R}$ and $p \in [1, \infty]$, let $W^{s,p}(\Omega)$ and $W^{s,p}(\Gamma_\infty)$ be the usual Sobolev spaces with the norms $\|\cdot\|_{W^{s,p}(\Omega)}$ and $\|\cdot\|_{W^{s,p}(\Gamma_\infty)}$, respectively. We define $\langle\cdot, \cdot\rangle$ to be the duality pairing between $H^s(\Gamma_\infty) := W^{s,2}(\Gamma_\infty)$ and the dual space $H^{-s}(\Gamma_\infty)$ with respect to the $L^2(\Gamma_\infty)$ inner product,

$$
\langle\chi, \psi\rangle := \int_{\Gamma_\infty} \chi(s)\psi(s)ds \quad \forall\ (\chi, \psi) \in H^s(\Gamma_\infty) \times H^{-s}(\Gamma_\infty). \tag{23.19}
$$

We further introduce the spaces

$$
\begin{aligned}
V &:= \left\{v \in W^{1,2}(\Omega) \mid v^+ - v^- = \beta \text{ on } \Sigma,\ \beta \in \mathbb{R}\right\}, \\
V^0 &:= \left\{v \in W^{1,2}(\Omega) \mid v^+ - v^- = 0 \text{ on } \Sigma\right\}, \\
H &:= \left\{\psi \in H^{-\frac{1}{2}}(\Gamma_\infty) \,\middle|\, \int_{\Gamma_\infty} \psi\, ds = 0\right\}
\end{aligned}
\tag{23.20}
$$

and, for fixed $s_0 \in\,]0, \frac{2a_0^2}{\kappa-1}[$, the closed, convex **set of admissible functions**

$$
K_{s_0} := \left\{v \in V \,\middle|\, |\nabla v|^2 \le s_0 \text{ a.e. in } \Omega\right\} \tag{23.21}
$$

in $W^{1,2}(\Omega)$. In order to simplify the notation we define the following nonlinear form

$$
a(u \mid v, w) := \int_\Omega \rho(|\nabla u|^2)\nabla v \cdot \nabla w\, d\boldsymbol{x} \quad \forall \text{ triplets } (u, v, w) \in K_{s_0} \times V \times V. \tag{23.22}
$$

For solving the exterior problem (23.16) we shall use a boundary potential formulation based on the Green representation theorem. For sufficiently smooth φ satisfying (23.16) having the behavior

$$\varphi(x) = \frac{\beta}{2} \log \left(\frac{x_1^2}{1 - M_\infty^2} + x_2^2 \right) + \varphi_\infty + o(1) \quad \text{for } |x| \to \infty,$$

with constant φ_∞, one obtains with the fundamental solution

$$G(x, y) \quad := \quad \frac{1}{2\pi} \log \left| \left(\frac{1}{\sqrt{1 - M_\infty^2}} [x_1 - y_1], [x_2 - y_2] \right) \right| \tag{23.23}$$

and the kernel of the double layer potential

$$K(x, y) \quad := \quad \frac{1}{2\pi} \frac{([x_1 - y_1], [x_2 - y_2]) \cdot n(y)}{\left| \left(\frac{1}{\sqrt{1-M_\infty^2}} [x_1 - y_1], [x_2 - y_2] \right) \right|^2} \tag{23.24}$$

the Green representation formula in the form

$$\varphi(x) \quad = \frac{\epsilon}{\sqrt{1 - M_\infty^2}} \int_{\Gamma_\infty} \{(1 - M_\infty^2)\varphi_{y_1} n_1(y) + \varphi_{y_2} n_2(y)\} G(x, y) \, ds_y$$

$$+ \frac{\epsilon}{\sqrt{1 - M_\infty^2}} \int_{\Gamma_\infty} \varphi(y) K(x, y) \, ds_y + \varphi_\infty \tag{23.25}$$

where $\epsilon = 1$ for $x = (x_1, x_2) \in \Omega^c$ and $\epsilon = 2$ for $x \in \Gamma_\infty$. In the latter case, (23.25) defines a boundary integral equation relating the Cauchy data $(1 - M_\infty^2)\varphi_{x_1} n_1(x) + \varphi_{x_2} n_2(x)$ and $\varphi(x)$ to each other. It can easily be shown that (23.25) corresponds to the choice of zero for the additional constant mentioned at the end of Section 23.1.1 and used in the definition of H in (23.20). We introduce the co–normal derivative of φ on Γ_∞ by

$$\lambda(x) := \rho(|v_\infty|^2)\{(1 - M_\infty^2)\varphi_{x_1} n_1(x) + \varphi_{x_2} n_2(x)\} \tag{23.26}$$

and define the boundary integral operators of single and double layer potentials

$$\mathcal{V}\lambda(x) \quad := -\int_{\Gamma_\infty} \frac{2G(x, y)}{\rho(|v_\infty|^2)\sqrt{1 - M_\infty^2}} \lambda(y) \, ds_y \, ,$$

$$\mathcal{K}\varphi(x) \quad := \int_{\Gamma_\infty} \frac{2K(x, y)}{\sqrt{1 - M_\infty^2}} \varphi(y) \, ds_y \, . \tag{23.27}$$

Thus, for $x \in \Gamma_\infty$, we may write (23.25) in short as

$$\varphi(x) + \mathcal{V}\lambda(x) - \mathcal{K}\varphi(x) = \varphi_\infty \, . \tag{23.28}$$

For smooth Γ_∞ we have the following well known result, see Hsiao and Wendland (1977).

Lemma 23.1.1 *For a C^∞-boundary Γ_∞, and any $\sigma \in \mathbb{R}$, the boundary integral operators*

$$\begin{aligned}
\mathcal{V} &: \quad H^{\sigma-\frac{1}{2}}(\Gamma_\infty) \quad \to \quad H^{\sigma+\frac{1}{2}}(\Gamma_\infty) \\
\mathcal{K} &: \quad H^{\sigma+\frac{1}{2}}(\Gamma_\infty) \quad \to \quad H^{\sigma+\frac{3}{2}}(\Gamma_\infty)
\end{aligned} \tag{23.29}$$

are continuous.

Further, we define the bilinear forms

$$b(\lambda, \psi) := \langle \mathcal{V}\lambda, \psi \rangle \quad \text{and} \quad d(\lambda, \psi) := \langle \mathcal{K}\lambda, \psi \rangle \quad \text{for } \lambda, \psi \in H^{-\frac{1}{2}}(\Gamma_\infty) \tag{23.30}$$

and, with the function $\Psi(x, \beta)$ given by (23.10), for any $\beta \in \mathbb{R}$ the functionals

$$\ell_1(v, \beta) := \rho_\infty \langle \nabla\Psi(\cdot, \beta) \cdot \boldsymbol{n}, v \rangle \quad \text{and} \quad \ell_2(\psi, \beta) \langle (\mathcal{I} - \mathcal{K})\Psi(\cdot, \beta), \psi \rangle$$

$\forall\, v \in W^{\sigma,p}(\Omega)$ and all $\psi \in H^{-\frac{1}{2}+\sigma}(\Gamma_\infty)$. For every fixed $\beta \in \mathbb{R}$, these are bounded linear functionals.

Lemma 23.1.2 *The nonlinear form $a(\cdot \mid \cdot, \cdot)$ satisfies the following estimates*

$$\begin{aligned}
|a(u \mid v, w)| \;&\leq\; \rho_0 \, \| v \|_{W^{1,2}(\Omega)} \| w \|_{W^{1,2}(\Omega)} \\
&\qquad \forall\, (u, v, w) \in K_{s_0} \times V \times V, \\
|a(u \mid u, w) - a(v \mid v, w)| \;&\leq\; C \, \| u - v \|_{W^{1,2}(\Omega)} \| w \|_{W^{1,2}(\Omega)} \\
&\qquad \forall\, (u, v, w) \in K_{s_0} \times K_{s_0} \times V.
\end{aligned} \tag{23.31}$$

Moreover, if we choose $s_0 < \dfrac{2a_0^2}{\kappa + 1}$, then we obtain uniform monotonicity for the corresponding purely subsonic flow fields, i.e.

$$a(u \mid u, u - v) - a(v \mid v, u - v) \geq \gamma_1 \, \| u - v \|_{W^{1,2}(\Omega)}^2 \;\; \forall\, (u, v) \in K_{s_0} \times K_{s_0} . \tag{23.32}$$

For a C^∞-boundary Γ_∞ and $\sigma \in \mathbb{R}$ there exists a constant C such that the form $b(\lambda, \psi)$ satisfies

$$|b(\lambda, \psi)| \leq C\|\lambda\|_{H^{\sigma-\frac{1}{2}}(\Gamma_\infty)}\|\psi\|_{H^{-\sigma-\frac{1}{2}}(\Gamma_\infty)} . \tag{23.33}$$

Moreover, there exists a constant $\gamma_2 > 0$ such that

$$b(\psi, \psi) \geq \gamma_2\|\psi\|_{H^{-\frac{1}{2}}(\Gamma_\infty)}^2 \quad \forall\, \psi \in H . \tag{23.34}$$

For the linear forms we have with some constant C the estimates

$$\begin{aligned}
|\ell_1(v, \beta)| \;&\leq\; \{|\varrho_\infty| \, |\boldsymbol{v}_\infty| + C|\beta|\} \, \|v\|_{H^{-1}(\Gamma_\infty)}, \\
|\ell_1(v, \beta_1) - \ell_1(v, \beta_2)| \;&\leq\; C|\beta_1 - \beta_2| \, \|v\|_{H^{-1}(\Gamma_\infty)},
\end{aligned} \tag{23.35}$$

and

$$\begin{aligned}
|\ell_2(\psi, \beta)| \;&\leq\; C \{|\boldsymbol{v}_\infty| + |\beta|\} \, \|\psi\|_{H^{-\frac{1}{2}}(\Gamma_\infty)}, \\
|\ell_2(\psi, \beta_1) - \ell_2(\psi, \beta_2)| \;&\leq\; \in C|\beta_1 - \beta_2| \, \|\psi\|_{H^{-\frac{1}{2}}(\Gamma_\infty)} .
\end{aligned} \tag{23.36}$$

Remark: The coerciveness inequality (23.34) and the mapping properties of $\mathcal{V}$ and $\mathcal{K} : H^{\sigma+\frac{1}{2}}(\Gamma_\infty) \to H^1(\Gamma_\infty)$ in (23.29) remain valid with $|\sigma| \leq \frac{1}{2}$ even for Lipschitz curves Γ_∞ due to Costabel (1988). $\qquad\square$

If we multiply equation (23.28) by a test function $\psi \in H$ and integrate over Γ_∞, we obtain the **weak formulation** for the boundary integral equation (23.28),

$$\langle \mathcal{V}\lambda, \psi \rangle + \langle \varphi, \psi \rangle - \langle \mathcal{K}\varphi, \psi \rangle = b(\lambda, \psi) + \langle \varphi, \psi \rangle - d(\varphi, \psi) = 0 \quad \forall\, \psi \in H \,. \quad (23.37)$$

The weak formulation for a solution of the full problem (23.15)–(23.18) can be obtained by the usual variational approach and coupling with Ω^c can be obtained with (23.17), (23.18) in variational form via the flux balance (23.14) through Γ_∞ and reads:

Variational problem
Find the three quantities $(u, \lambda, \beta) \in K_{s_0} \times H \times \mathbb{R}$ *such that*

$$\begin{cases} a(u \mid u, v) - \langle \lambda, v \rangle & = \ell_1(v, \beta) & \forall\, v \in V^0 \,, \\ b(\lambda, \psi) + \langle u, \psi \rangle - d(u, \psi) & = \ell_2(\psi, \beta) \ \forall\ \psi \in H^{-\frac{1}{2}}(\Gamma_\infty) \quad \text{and} \\ F(\beta) & = |\nabla u^+|^2_{TE} - |\nabla u^-|^2_{TE} = 0 \,. \end{cases} \quad (23.38)$$

We further assume that there is a vicinity of TE where the flow is purely subsonic which implies for solutions in K_{s_0} that ∇u is continuous at TE. For purely subsonic cascade flow, Feistauer *et al.* (1992) also use the Kutta–Joukowski condition for finding the appropriate finite element solution providing circulation.

To treat the full nonlinear problem (23.38) with an unknown circulation β we present a nonlinear iteraton in function spaces satisfying the Kutta–Joukowski condition by correcting the iterates with appropriate β–values enforcing the regularity of ∇u at TE in every step.

23.2   EXISTENCE, UNIQUENESS AND THE ENTROPY CONDITION

The coupled boundary value problem (23.15)–(23.18) is an approximation to the exterior boundary value problem without Γ_∞. For the solution to the full exterior **subsonic** problem see Bers (1958) and Bojarski (1967). For the coupled problem (23.15)–(23.18), however, we take the linearization on Ω^c and the corresponding coupling boundary conditions on Γ_∞. Under the restriction

$$\|\nabla u\|_{L^\infty(G)} < a^* \quad \text{in } G \subset \Omega \cap \Gamma_\infty \cup \Omega^c \,, \quad (23.39)$$

the equation (23.3) is elliptic there, i.e. the flow is subsonic in G. A unique solution exists for v_∞ given small enough to imply that (23.39) holds **everywhere**. In this case, the coupled problem corresponds to a purely elliptic nonlinear problem with discontinuous coefficients across Γ_∞. Using reflection at an appropriately large ellipse enclosing Γ_∞ which is mapped into a circle by the Prandtl–Glauert transformation, this problem is equivalent to one in a bounded domain. Based on (23.31) and strict monotonicity (23.32), existence and uniqueness then can be established in a function space satisfying the Kutta–Joukowski condition.

For the **transonic** case in which the geometry of the problem and the boundary conditions lead to a locally supersonic flow near the profile, the situation changes quite dramatically. In the supersonic regions, equation (23.3) becomes hyperbolic and the monotonicity property (23.32) is lost. In this case the existence of solutions to (23.3)–(23.8) is still open, even for problems in bounded domains, see Feistauer and Nečas (1985) and Morawetz (1964).

Further complication arises from the fact that mathematical analysis by Morawetz (1985), as well as physical and numerical experience, show that one generally has to expect solutions that contain shocks. Hence, one has to consider weak solutions which in turn lead to non–uniqueness and the existence of non–physical solutions to the variational formulation of the boundary value problem, see Nečas (1989). This fact, well known from the related theory of systems of conservation laws, is the reason why Bristeau *et al.* (1980) and Bristeau *et al.* (1985) need to supplement the problem with an additional admissibility condition for generalized solutions. We will concentrate here on the specific numerical entropy condition used for our numerical implementation which we adopted from Glowinski and Pironneau (1978). Further discussion on admissibility of transonic shocks may be found in Nečas (1989), Keyfitz and Warnecke (1991), and Warnecke (1990). Due to the assumption of isentropy, the shocks in transonic potential flow conserve entropy but not momentum. The selection condition we have used requires that the divergence of the flow field must be bounded from above, i.e.

$$\operatorname{div} \nabla u \leq B \quad \text{with an appropriate constant } B \in \mathbb{R}. \tag{23.40}$$

In Göhner and Warnecke (1993), it is shown that this inequality is equivalent to the solution with compressive shocks and is violated by the non–physical expansion shocks. In weak form this means

$$-\int_\Omega \nabla u \cdot \nabla \psi \, dx \leq B \int_\Omega \psi \, dx \quad \forall \, \psi \in C_0^\infty(\Omega) \ \text{ with } \ \psi \geq 0. \tag{23.41}$$

The implementation of (23.41) via penalization is presented in Section 23.4.1.

23.3   THE COUPLED FEM–BEM FORMULATION OF THE PROBLEM

In order to discretize the coupled problem (23.38), we approximate the domain Ω by a family of polygonal domains Ω_h where h denotes the parameter of meshwidth. The outer boundary Γ_∞^h of Ω_h is supposed to be a polygonal curve with nodes on Γ_∞. In the same way we define Γ_P^h as the approximation of the profile boundary Γ_P. Without loss of generality we may assume that the slit Σ is already a part of the boundary of Ω_h. Together with Ω_h we introduce a family of regular triangulations $\{\mathcal{T}_h\}_{h>0}$ with $\mathcal{T}_h := \bigcup_{i \in \mathbb{D}} T_i$ and finite $\mathbb{D} \subset \mathbb{N}$. The nodal points of the triangulation are denoted by $p_i, i = 1, \dots, N$. For convenience we assume that the nodes on the slit Σ are numbered as the first $2L$ points, i.e. p_j^+, p_j^- for $j = 1, \dots L$, with p_j^+ and

p_j^- having the same coordinates but characterizing the limits from above and below, respectively. By w_{h_i}, $i = 1, \ldots, N$ we denote the usual piecewise linear hat functions satisfying $w_{h_i}(p_j) = \delta_{ij}$, $i, j = 1, \ldots, N$, which form a basis of nonnegative functions (hat–functions) for the piecewise linear continuous Courant elements. By A_i we denote the areas of the corresponding supports, i.e. $A_i =$ meas (supp w_{h_i}), $i = 1, \ldots, N$. We say that a family of regular triangulations $\{T_h\}_{h>0}$ satisfies the **angle property** if there is a minimal angle $\alpha > 0$ such that for any $h > 0$ and any $T_i \in T_h$ one has $\alpha_i \geq \alpha$ where α_i denotes the smallest angle in T_i.

We denote by S_i the segments on Γ_∞^h given by the edges of the boundary triangles T_i of T_h. By $\{\mathcal{S}_h\}_{h>0}$ with $\mathcal{S}_h := \bigcup_{i \in \mathbb{B}} S_i$ we denote the corresponding induced family of polygonal approximations of the boundary Γ_∞^h, where $\mathbb{B} \subset \mathbb{D}$ is finite. Without loss of generality we assume that our triangulation Ω_h is chosen such that the corresponding family $\{\mathcal{S}_h\}_{h>0}$ admits an **inverse assumption**, see Ciarlet (1978, (3.2.28)) implying inverse estimates Ciarlet (1978, Theorem 3.2.6). The inverse estimates on $\{\mathcal{S}_h\}_{h>0}$ are not used in this paper. However, we implemented the boundary element coupling with piecewise constant boundary trial functions, using point collocation. Then the asymptotic error estimates used in Chapter 5 remain valid if the inverse estimates hold. The family $\{T_h\}_{h>0}$ itself does not need to be quasiuniform. For all further considerations, the parameter h will stand for the maximum diameter of all triangles $T_i \in T_h$. Let $C^0(\overline{\Omega}_h)$ denote the set of all continuous functions on Ω_h, having one–sided limits on the slit Σ and $\partial\Omega_h$, respectively. For the discretization of (23.38) we introduce the following finite–dimensional spaces on Ω_h,

$$\widetilde{V}_h \quad := \quad \left\{ \widetilde{v}_h \in C^0(\overline{\Omega}_h) \,\middle|\, \widetilde{v}_{h|_{T_i}} \text{ is linear on every } T_i \in T_h \text{ and} \right.$$

$$\left. \widetilde{v}_h^+ - \widetilde{v}_h^- = \beta \text{ on } \Sigma \text{ with } \beta \in \mathbb{R} \right\}, \tag{23.42}$$

$$\widetilde{V}_h^0 \quad := \quad \left\{ \widetilde{v}_h \in \widetilde{V}_h \,\middle|\, \widetilde{v}_h^+ - \widetilde{v}_h^- = 0 \text{ on } \Sigma \right\}, \tag{23.43}$$

$$\widetilde{H}_h \quad := \quad \left\{ \widetilde{\psi}_h \in L_2(\Gamma_\infty^h) \,\middle|\, \widetilde{\psi}_{h|_{S_i}} \text{ is constant on every } S_i \in \mathcal{S}_h \text{ and } \langle \widetilde{\psi}_h, 1 \rangle_h = 0 \right\} \tag{23.44}$$

and the set of admissible finite elements

$$\widetilde{K}_{s_0} := \left\{ \widetilde{v}_h \in \widetilde{V}_h \,\middle|\, |\nabla \widetilde{v}_h|^2 \leq s_0 \right\}. \tag{23.45}$$

Here we use the notation

$$\langle \phi, \psi \rangle_h := \int_{\Gamma_\infty^h} \phi(s)\psi(s)\, ds \tag{23.46}$$

for the L_2–scalar product on Γ_∞^h. For the following, we need an approximate kernel K_h which is defined as in (23.24) where $n(y)$ is to be replaced by $n_h(y)$, the linear interpolant of the normal vectors to Γ_∞ at the vertices of Γ_∞^h. The associated operator will be denoted by $\mathcal{K}_h$. We further define the discrete forms

$$a_h(\widetilde{u}|\widetilde{v}, \widetilde{w}) \quad := \quad \int_{\Omega_h} \rho(|\nabla \widetilde{u}|^2) \nabla \widetilde{v} \cdot \nabla \widetilde{w}\, d\boldsymbol{x},$$

$$
\begin{aligned}
\widetilde{b}_h(\widetilde{\lambda}, \widetilde{\psi}) &:= \langle \mathcal{V}\widetilde{\lambda}, \widetilde{\psi}\rangle_h \,, \\
\widetilde{d}_h(\widetilde{u}, \widetilde{\psi}) &:= \langle \mathcal{K}_h \widetilde{u}, \widetilde{\psi}\rangle_h \,, \\
\widetilde{\ell}_1^h(\widetilde{v}, \beta) &:= \rho_\infty \langle \nabla\Psi(\cdot,\beta)\cdot \boldsymbol{n}_h, \widetilde{v}\rangle_h \quad \text{cf.(23.10)}\,, \\
\widetilde{\ell}_2^h(\widetilde{\psi}, \beta) &:= \langle (\mathcal{I} - \mathcal{K}_h)\Psi(\cdot,\beta), \widetilde{\psi}\rangle_h \,.
\end{aligned}
\tag{23.47}
$$

Now the discrete analogue of problem (23.38) reads as follows:

The discrete variational problem

Find $(\widetilde{u}_h, \widetilde{\lambda}_h, \beta_h) \in (\widetilde{V}_h \cap \widetilde{K}_{s_0}) \times \widetilde{H}_h \times \mathbb{R}$ *such that*

$$
\begin{aligned}
a_h(\widetilde{u}_h|\widetilde{u}_h, \widetilde{v}_h) - \langle \widetilde{\lambda}_h, \widetilde{v}_h\rangle_h &= \widetilde{\ell}_1^h(\widetilde{v}_h, \beta_h) \quad \forall\ \widetilde{v}_h \in \widetilde{V}_h\,, \\
\widetilde{b}_h(\widetilde{\lambda}_h, \widetilde{\psi}_h) + \langle \widetilde{u}_h, \widetilde{\psi}_h\rangle_h - \widetilde{d}_h(\widetilde{u}_h, \widetilde{\psi}_h) &= \widetilde{\ell}_2^h(\widetilde{\psi}_h, \beta_h) \quad \forall\ \widetilde{\psi}_h \in \widetilde{H}_h
\end{aligned}
\tag{23.48}
$$

subject to $F(\beta_h) = 0$.

Problem (23.48) leads to a system of nonlinear equations, where we have one unknown per node in the triangulation $\mathcal{T}_h$ of Ω_h, one unknown per segment S_i of the polygonal boundary Γ_∞^h and the unknown circulation β_h.

For the error analysis we convert problem (23.48) into a form which allows the use of subspaces to the admissible function spaces used in (23.38). We will define subspaces V_h, V_h^0 and H_h satisfying the conformity inclusions $V_h \subset V$, $V_h^0 \subset V^0$ and $H_h \subset H$. This reformulation of problem (23.48) allows an analysis similar to that for conforming finite and boundary elements, see Johnson and Nedelec (1980). We introduce a mapping $\Phi_h : \Gamma_\infty^h \to \Gamma_\infty$, where $\Phi_h(\boldsymbol{x})$ is the point on Γ_∞ closest to the point $\boldsymbol{x} \in \Gamma_\infty^h$. For h sufficiently small, the mapping Φ_h becomes a bijection, see Le Roux (1974). Hence, the inverse mapping Φ_h^{-1} exists. It transforms integrals along Γ_∞^h into integrals along Γ_∞ by

$$
\int\limits_{\Gamma_\infty^h} \widetilde{\psi}(s)\,ds = \int\limits_{\Gamma_\infty} \widetilde{\psi}\circ\Phi_h^{-1}(s) J(\Phi_h^{-1}(s))\,ds\,,
\tag{23.49}
$$

where $J(\Phi_h^{-1}) = \left|\frac{\partial\Phi_h^{-1}}{\partial s}\right|$ is the one–dimensional Jacobian. Here $\frac{\partial}{\partial s}$ denotes the differentiation in the tangential direction. We now define the conforming boundary space

$$
H_h := \left\{ \psi = J(\Phi_h^{-1})\widetilde{\psi}_h \circ \Phi_h^{-1} \,\middle|\, \widetilde{\psi}_h \in \widetilde{H}_h \right\} \subset H\,.
\tag{23.50}
$$

We also need an appropriate subspace $V_h \subset V$ based on the definition of $\widetilde{V}_h$. Let $\widetilde{v}_h \in \widetilde{V}_h$ be an arbitrarily given function. Then $\widetilde{v}_h$ is well–defined in $\Omega \cap \Omega_h$ by taking the restriction. In the **skin**

$$
\omega_\infty^h := (\Omega\backslash\Omega_h) \cap \{x \mid \operatorname{dist}(x, \Gamma_\infty) \leq h\}\,,
$$

however, we need an extension of $\widetilde{v}_h$ which we define by setting

$$
v_h(\boldsymbol{y}) := \widetilde{v}_h(\boldsymbol{x})
$$

with $x \in \Gamma^h_\infty$ and any given y lying on the line segment between the two points $x \in \Gamma^h_\infty$ and $\Phi_h(x) \in \Gamma_\infty$. In the same way as described above, we define the function $\tilde{v}_h$ in the skin

$$\omega^h_P = (\Omega \backslash \Omega_h) \cap \{x \mid \text{dist}\,(x, \Gamma_P) \leq h\}\,.$$

This defines an extension operator which is only slightly different from Zlamal's operator used by Feistauer and Ženíšek (1987) and Gatica and Hsiao (1992). Our version is chosen in accordance with the boundary element approximation defined by Le Roux (1974) and used by Johnson and Nedelec (1980). Note that for the piecewise $\mathcal{C}^\infty$–curve Γ_P we have

$$\text{meas } \{(\text{supp } w_{h_i}) \cap \omega^h_P\} \leq c h_i^3 \leq h_i A_i\,.$$

We now denote by V_h the space of all functions v_h defined from $\tilde{v}_h \in \tilde{V}_h$ in the above way. Since on $\Omega_h \backslash \Omega$ the functions v_h are well defined, V_h consists of functions given on $\Omega \cup \Omega_h$ while their restrictions to Ω define a subspace of V. By V^0_h we denote the set of all functions $v_h \in V_h$, which are continuous across the slit Σ. By changing the integrations from Γ^h_∞ to Γ_∞ and using the definitions of V_h, V^0_h and H_h, we can reformulate problem (23.48) as follows. We note that

$$a_h(u_h | v_h, w_h) := a_h(\tilde{u}_h | \tilde{v}_h, \tilde{w}_h)\,,$$

and define

$$b_h(\lambda, \psi) \quad := \quad \frac{2}{\rho(|v_\infty|^2)\sqrt{1 - M^2_\infty}} \int\!\!\!\int_{\Gamma_\infty \Gamma_\infty} \lambda(y)\psi(x) G\left(\Phi^{-1}_h(x), \Phi^{-1}_h(y)\right)\, ds_x ds_y\,,$$

$$d_h(u, \psi) \quad := \quad \frac{2}{\sqrt{1 - M^2_\infty}} \int\!\!\!\int_{\Gamma_\infty \Gamma_\infty} u(y)\psi(x) \cdot K\left(\Phi^{-1}_h(x), \Phi^{-1}_h(y)\right) [J \circ (\Phi^{-1}_h(y)]ds_x ds_y\,,$$

$$\text{(23.51)}$$

$$\ell^h_1(v, \beta) \quad := \quad \rho_\infty \langle (\nabla\Psi(\cdot, \beta) \cdot n_h) \circ \Phi^{-1}_h, v \rangle\,,$$

$$\ell^h_2(\psi, \beta) \quad := \quad \langle \Psi(\cdot, \beta) \circ \Phi^{-1}_h, \psi \rangle - d_h(\Psi(\cdot, \beta) \circ \Phi^{-1}_h, \psi)\,.$$

Note that the transformations Φ_h do not always map all functions of $\tilde{K}_{s_0}$ into K_{s_0}. But taking $s'_0 < s_0$, the functions on $\tilde{K}_{s'_0}$ are mapped into K_{s_0} provided $h > 0$ is small enough. This distinction of s'_0 and s_0, however, is not significant for the error analysis. Therefore, in the following we consider the analysis for the admissible approximations in $V_h \cap K_{s_0}$. Using these definitions we obtain:

The conforming discrete variational problem

Find $(u_h, \lambda_h, \beta_h) \in (V_h \cap K_{s_0}) \times H_h \times \mathbb{R}$ satisfying $F(\beta_h) = 0$ such that

$$\begin{aligned}
a_h(u_h | u_h, v_h) - \langle \lambda_h, v_h \rangle &= \ell^h_1(v_h, \beta_h) \quad \forall\, v_h \in V^0_h \quad \text{and} \\
b_h(\lambda_h, \psi_h) + \langle u_h, \psi_h \rangle - d_h(u_h, \psi_h) &= \ell^h_2(\psi_h, \beta_h) \quad \forall\, \psi_h \in H_h\,.
\end{aligned} \qquad \text{(23.52)}$$

The following consistency estimates essentially follow from the corresponding results by Feistauer and Ženíšek (1987) and by Johnson and Nedelec (1980), respectively.

Lemma 23.3.1 *Provided the polygonal approximation S_h of Γ_∞ satisfies an inverse assumption, we have the following estimates for the bilinear and linear forms in* (23.51)*:*

$$
\begin{aligned}
|a(u_h|u_h, v_h) - a_h(u_h|u_h, v_h)| \leq\ & ch\|u_h\|_{W^{1,2}(\Omega)}\|v_h\|_{W^{1,2}(\Omega)} \\
& \forall\ (u_h, v_h) \in (V_h \cap K_{s_0}) \times V_h ,
\end{aligned}
\tag{23.53}
$$

$$
\begin{aligned}
|b(\lambda_h, \psi_h) - b_h(\lambda_h, \psi_h)| \leq\ & ch\|\lambda_h\|_{H^{-\frac{1}{2}}(\Gamma_\infty)}\|\psi_h\|_{H^{\frac{1}{2}}(\Gamma_\infty)} \\
& \forall\ (\lambda_h, \psi_h) \in H_h \times H_h ,
\end{aligned}
\tag{23.54}
$$

$$
\begin{aligned}
|d(v, \psi_h) - d_h(v, \psi_h)| \leq\ & ch^{3/2}\|v\|_{W^{1,2}(\Omega)}\|\psi_h\|_{H^{-\frac{1}{2}}(\Gamma_\infty)} \\
& \forall\ (v, \psi) \in V \times H_h ,
\end{aligned}
\tag{23.55}
$$

$$
\begin{aligned}
|\ell_1^h(v, \beta) - \ell_1(v, \beta)| &\leq (c_1 + c_2|\beta|)h\|v\|_{W^{1,2}(\Omega)} &&\forall\ v \in V , \\
|\ell_2^h(\psi, \beta) - \ell_2(\psi, \beta)| &\leq (c_1 + c_2|\beta|)h\|\psi\|_{H^{-\frac{1}{2}}(\Gamma_\infty)} &&\forall\ \psi \in H .
\end{aligned}
\tag{23.56}
$$

For later use we collect the following results on finite element approximation.

Lemma 23.3.2
(a) *Let us denote by $P_h : H \to H_h$ the L^2-projection defined by*

$$
\langle P_h\varphi, \psi_h \rangle := \langle \varphi, \psi_h \rangle \quad \forall\ \psi_h \in H_h.
\tag{23.57}
$$

Then we have

$$
\|\varphi - P_h\varphi\|_{H^{-\frac{1}{2}}(\Gamma_\infty)} \to 0 \ \ for\ h \to 0 .
\tag{23.58}
$$

(b) *Further, we define the* **Ritz projections** *$\widetilde{R}_h : V \to \widetilde{V}_h$, respectively, $R_h : V \to V_h$ by*

$$
\int_{\Omega_h} \nabla \widetilde{R}_h v \cdot \nabla \widetilde{v}_h \, dx = \int_\Omega \nabla v \cdot \nabla v_h \, dx \quad for\ all\ \widetilde{v}_h \in \widetilde{V}_h
\tag{23.59}
$$

where $v_h \in V_h$ is the extension of $\widetilde{v}_h$ to Ω. Then

$$
\lim_{h \to 0} \|v - R_h v\|_{W^{1,2}(\Omega)} = 0 \quad \forall\ v \in V
$$

and

$$
\left\|\widetilde{R}_h v\right\|_{W^{1,2}(\Omega_h)} \leq c \|v\|_{W^{1,2}(\Omega)} \quad \forall\ v \in V .
$$

(c) *There exist families $\{\mathcal{T}_h\}_{h>0}$ of triangulations having additional properties which imply that the Ritz projection defined in* (23.59) *satisfies the stability estimate*

$$
\left\|\nabla \widetilde{R}_h v\right\|_{L^\infty(\Omega_h)} \leq c^* \|\nabla v\|_{L^\infty(\Omega)} \quad \forall\ v \in W^{1,\infty}(\Omega)
\tag{23.60}
$$

where the constant $c^* \geq 1$ is independent of v, v_h and h.

(a) and (b) are well known properties, see Babuška and Aziz (1972). The estimate (23.60) in (c) is e. g. a consequence of the quasiuniformity assumption as was proved by Rannacher and Scott (1982). To us it is not clear how far these assumptions can be relaxed such that (23.60) still holds.

23.4 THE SOLUTION OF THE DISCRETE PROBLEM

23.4.1 THE MINIMIZATION PROBLEM

The goal of this Section is the formulation of problem (23.48) as a discrete minimization problem. The underlying idea goes back to Glowinski and Pironneau (1978), and has since been further developed by Bristeau *et al.* (1980) and Berger *et al.* (1990). For transonic flow, the hyperbolic character of the supersonic region creates additional difficulties since then we must take into account an additional selection principle. This will be done by a penalization. To this end, we define the following family of functionals $\mathcal{J}_h : \widetilde{V}_h \cap \widetilde{K}_{s_0} \to \mathbb{R}$ by

$$\mathcal{J}_h(\phi_h) := \begin{cases} \dfrac{1}{2} \displaystyle\int_{\Omega_h} |\nabla \xi_h(\phi_h)|^2 \, d\boldsymbol{x} & \text{if globally } s_0 < \dfrac{2a_0^2}{\kappa+1} \text{ (subsonic flow)}, \\[2ex] \dfrac{1}{2} \displaystyle\int_{\Omega_h} |\nabla \xi_h(\phi_h)|^2 \, d\boldsymbol{x} + \mathcal{P}_h(\phi_h) & \text{for } s_0 < \dfrac{2a_0^2}{\kappa-1} \text{ (transonic flow)}, \end{cases}$$

(23.61)

where the penalty functionals $\mathcal{P}_h : \widetilde{V}_h \cap \widetilde{K}_{s_0} \to \mathbb{R}$ are given by

$$\mathcal{P}_h(\phi_h) := \frac{\mu}{2} \sum_{\substack{i=1 \\ P_i \notin \Sigma \cup \Gamma_\infty}}^{N} \frac{1}{A_i^\epsilon} \left(\left[-\int_{\Omega_h} \nabla\phi_h \cdot \nabla w_{h_i} \, d\boldsymbol{x} - B \int_{\Omega_h} w_{h_i} \, d\boldsymbol{x} \right]^+ \right)^2 . \quad (23.62)$$

By $[\cdot]^+$ we denote the nonnegative part of the quantity in brackets. Here, $\mu > 0$, $B > 0$ and $2 > \epsilon > 1$ are constants, which do not depend on h. These constants can be chosen according to numerical experiments depending on the specific profile, the travelling velocity $\boldsymbol{v}_\infty$ etc., but then they are fixed for mesh refinement. The function $\xi_h(\phi_h) \in \widetilde{V}_h^0$ is the solution of the state equation,

$$\int_{\Omega_h} \nabla\xi_h(\phi_h) \cdot \nabla v_h dx = a_h(\phi_h|\phi_h, v_h) - \langle \lambda_h(\phi_h, \beta_h), v_h \rangle_h - \widetilde{\ell}_1^h(v_h, \beta_h) \ \forall \ v_h \in \widetilde{V}_h^0 .$$

(23.63)

This state equation is the finite element approximation of the Neumann problem for the Poisson equation with given Neumann data on Γ_P, see (23.15), and on Γ_∞ where λ_h is given. On the other hand, $\lambda_h(\varphi_h) \in \widetilde{H}_h$ is to be determined by the Galerkin discretization of the boundary integral equation (23.28), i.e.

$$\widetilde{b}_h(\lambda_h(\phi_h, \beta_h), \psi_h) = \widetilde{\ell}_2^h(\psi_h, \beta_h) - \langle \phi_h, \psi_h \rangle_h + \widetilde{d}_h(\phi_h, \psi_h) \ \forall \psi_h \in \widetilde{H}_h . \quad (23.64)$$

Instead of solving discrete equations (23.48), we will now solve the following:

Discrete minimization problem

Find $(u_h, \lambda_h, \beta_h) \in (\widetilde{V}_h \cap \widetilde{K}_{s_0}) \times \widetilde{H}_h \times I$ such that

$$\mathcal{J}_h(u_h) := \min_{\phi_h \in \widetilde{V}_h \cap \widetilde{K}_{s_0}} \mathcal{J}_h(\phi_h) \quad (23.65)$$

where $I \subset \mathbb{R}$ is an appropriately fixed finite interval.

To simplify the notation, we are not using the $\tilde{\ }$-sign for finite element functions as previously. Since β_h is the circulation of u_h, it is sought only in a bounded interval I. The solution of problem (23.65) subject to $F(\beta_h) = 0$, to (23.63) and to (23.64) exists, since we minimize a differentiable functional over a bounded, convex and nonempty subset of a finite–dimensional space. The nonemptyness is due to the fact that the zero function lies in $\widetilde{V}_h \cap \widetilde{K}_{s_0}$ and satisfies $F(0) = 0$. Note that the minimization of the first term of (23.61) in view of (23.63) is equivalent to the minimization of the $W^{1,2}(\Omega)$–seminorm of the Riesz representation of the residual of (23.3) in the least squares sense. The above method, hence, can be considered as a least squares method.

The solution $u_h \in \widetilde{V}_h \cap \widetilde{K}_{s_0}$ is not necessarily unique; nevertheless, any solution u_h defines a corresponding flux $\lambda_h(u_h) \in \widetilde{H}_h$ as the unique solution of (23.64). Thus we may assume for a given sequence of meshsizes the existence of a sequence $\{(u_h, \lambda_h, \beta_h)\}_{h>0}$, where $u_h \in \widetilde{V}_h \cap \widetilde{K}_{s_0}$ is a solution of (23.65) and $\lambda_h \in \widetilde{H}_h$ is the corresponding solution of (23.64) with the corresponding β_h. If $\mathcal{J}_h(u_h) = 0$, then $(u_h, \lambda_h, \beta_h)$ is a solution of (23.48).

23.4.2 A CONJUGATE GRADIENT ALGORITHM FOR THE MINIMIZATION PROBLEM

In this Section we present a Polak-Ribière type conjugate gradient algorithm, which takes into account the constraint $F(\beta) = 0$ and the weak coupling equation (23.64). The treatment of the Kutta–Joukowski condition $F(\beta) = 0$ is similar to the method described by Berger *et al.* (1990). The conjugate gradient algorithm reads as follows:

Starting Value:

As an initial approximation we choose the solution of the incompressible irrotational flow problem:

Find $u_h^0 \in \widetilde{V}_h$ and $\beta^0 \in \mathbb{R}$ such that

$$\int_{\Omega_h} \nabla u_h^0 \cdot \nabla v_h \, d\boldsymbol{x} = \int_{\Gamma_\infty^h} \boldsymbol{v}_\infty \cdot \boldsymbol{n} \, v_h \, d\boldsymbol{x} \quad \forall \ v_h \in V_h^0 \tag{23.66}$$

subject to $F(\beta^0) = 0$.

The corresponding variational problem has a unique solution, as was shown by Kirchhoff via conformal mapping. The finite element approximation (23.66) is also uniquely solvable for h sufficiently small, see Becker *et al.* (1984, Section 1.7 ff.). Then we define $\lambda_h^0 \in \widetilde{H}_h$ as the solution of

$$\widetilde{b}_h(\lambda_h^0, \psi_h) = \widetilde{\ell}_2^h(\psi_h, \beta^0) - \langle u_h^0, \psi_h \rangle_h + \widetilde{d}_h(u_h^0, \psi_h) \ \forall \ \psi_h \in \widetilde{H}_h \,. \tag{23.67}$$

Equations (23.67) are uniquely solvable for any given u_h^0 as follows from the results for boundary integral equations of the first kind by Hsiao and Wendland (1977) and Le Roux (1974).

Outer iteration for λ_h^m:
Starting value: For $m = 0$ take λ_h^0.

Inner iteration:
1. Initial Step:

For the descent direction $g_h^0 \in \widetilde{V}_h^0$ we solve the following variational equation:
Find g_h^0 such that

$$\int_{\Omega_h} \nabla g_h^0 \cdot \nabla v_h \, d\boldsymbol{x} = [\mathcal{J}_h'(u_h^0), v_h] \ \ \forall \, v_h \in \widetilde{V}_h^0 . \tag{23.68}$$

Here the Fréchet derivative $\mathcal{J}_h'$ of $\mathcal{J}_h$ at u_h^0 operating on v_h is given by

$$
\begin{aligned}
[\mathcal{J}_h'(u_h^0), v_h] \ := \ & \int_{\Omega_h} \rho\left(|\nabla u_h^0|^2\right)\left(\nabla \xi_h^0 \cdot \nabla v_h\right) \, d\boldsymbol{x} \\
& + \ \int_{\Omega_h} \rho'\left(|\nabla u_h^0|^2\right)\left(\nabla \xi_h^0 \cdot \nabla u_h^0\right)\left(\nabla u_h^0 \cdot \nabla v_h\right) \, d\boldsymbol{x} \\
& + \ \mu \sum_{i=1}^{N} \frac{1}{A_i^\epsilon}\left[-\int_{\Omega_h} \nabla u_h^0 \cdot \nabla w_{h_i} \, d\boldsymbol{x} - B \int_{\Omega_h} w_{h_i} \, d\boldsymbol{x}\right]^{+} \cdot \\
& \qquad\qquad \cdot \left[-\int_{\Omega_h} \nabla v_h \cdot \nabla w_{h_i} \, d\boldsymbol{x}\right]
\end{aligned}
\tag{23.69}
$$

where $\xi_h^0 := \xi_h^0(u_h^0, \lambda_h^m, \beta^0) \in \widetilde{V}_h^0$ is defined as the FEM–solution of the variational Poisson equation

$$\int_{\Omega_h} \nabla \xi_h^0 \cdot \nabla v_h \, d\boldsymbol{x} = a_h(u_h^0 \mid u_h^0, v_h) - \langle \lambda_h^m, v_h \rangle_h - \widetilde{\ell}_1^h(v_h, \beta^0) \ \ \forall \, v_h \in \widetilde{V}_h^0 . \tag{23.70}$$

For $s_0 < \dfrac{2a_0^2}{\kappa + 1}$ we take $\mu = 0$ in (23.69). Finally, we set for the first descent direction

$$z_h^0 := g_h^0 . \tag{23.71}$$

2. Inner Iteration Loop:

Assume now that $u_h^n, \lambda_h^m, \beta^n, g_h^n, z_h^n$ are known for $n \geq 0$. Then the computation of the new values $u_h^{n+1} \lambda_h^m, \beta^{n+1}, g_h^{n+1}$ and z_h^{n+1} proceeds as follows:

Step 1:

Define the value of the functional $\mathcal{J}_h^{n+1}(u_h^n)$ by

$$
\mathcal{J}_h^{n+1}(u_h^n) := \begin{cases} \frac{1}{2} \int\limits_{\Omega^h} |\nabla \xi_h^n|^2 \, d\boldsymbol{x} & \text{for } s_0 < \frac{2a_0^2}{\kappa+1} , \\[2ex] \frac{1}{2} \int\limits_{\Omega^h} |\nabla \xi_h^n|^2 \, d\boldsymbol{x} + \mathcal{P}_h(u_h^n) & \text{for } \frac{2a_0^2}{\kappa+1} \leq s_0 < \frac{2a_0^2}{\kappa-1}, \end{cases}
\tag{23.72}
$$

where $\xi_h^n := \xi_h^n(u_h^n, \lambda_h^m, \beta^n) \in \widetilde{V}_h^0$, is the solution of the Poisson problem

$$\int_{\Omega_h} \nabla \xi_h^n \cdot \nabla v_h \, d\boldsymbol{x} = a_h(u_h^n \mid u_h^n, v_h) - \langle \lambda_h^m, v_h \rangle_h \quad \forall \, v_h \in \widetilde{V}_h^0 \,. \tag{23.73}$$

Then we solve the one–dimensional minimization problem (line search) for α^n,

$$\mathcal{J}_h^{n+1}(u_h^n - \alpha^n z_h^n) \leq \mathcal{J}_h^{n+1}(u_h^n - \alpha z_h^n) \quad \forall \, \alpha \leq 0 \tag{23.74}$$

and set

$$\widetilde{u}_h^{n+1} := u_h^n - \alpha^n z_h^n \,. \tag{23.75}$$

Note that $\lambda_h^m \in \widetilde{H}_h$ will be fixed for the solution of (23.75). This saves some computational effort, since for the evaluation of $\mathcal{J}_h^{n+1}$ we do not need to solve always the boundary integral equation.

Step 2:

In this step we incorporate the constraint $F(\beta) = 0$ and compute the new u_h^{n+1} and β_h^{n+1} by a procedure very similar to the one in Berger *et al.* (1990). We modify $\widetilde{u}_h^{n+1}$ by

$$\widetilde{u}_h^{n+1} \mapsto \bar{u}_h^{n+1} \in \widetilde{V}_h^n \tag{23.76}$$

so that the modified function $\bar{u}_h^{n+1}$ satisfies the Kutta-Joukowski condition at TE in the form

$$F(\beta^{n+1}) = \left| \nabla \bar{u}_h^{n+1} \big|_{T_+} \cdot \boldsymbol{t}_+ \right|^2 - \left| \nabla \bar{u}_h^{n+1} \big|_{T_-} \cdot \boldsymbol{t}_- \right|^2 = 0 \,, \tag{23.77}$$

where $\boldsymbol{t}_+, \boldsymbol{t}_-$ are the unit tangential vectors at the trailing edge pointing in the direction of TE. Let T_+ and T_- denote the respective triangles adjacent to TE as indicated in Figure 23.2.

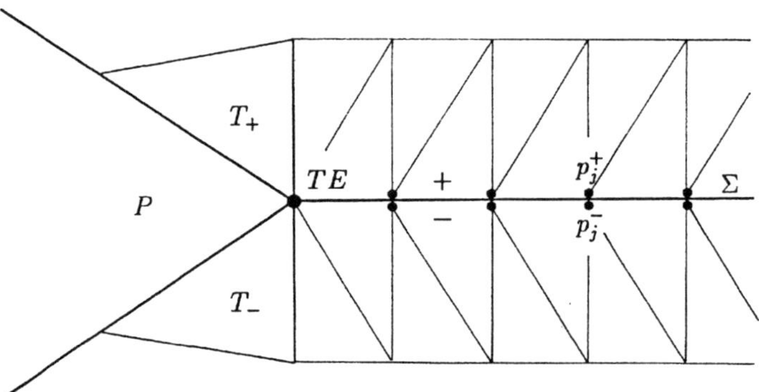

Figure 23.2: The triangular grid involving the Kutta–Joukowski condition

The main idea of the modification is that we change $\widetilde{u}_h^{n+1}$ only in the $2L$ points p_j^+, p_j^- along the slit Σ where the new function $\bar{u}_h^{n+1}$ differs from $\widetilde{u}_h^{n+1}$. For points p_j, which do not lie on Σ, i.e. $p_j \in \Omega\backslash\Sigma$ we define

$$\bar{u}_h^{n+1}(p_j) := \widetilde{u}_h^{n+1}(p_j) \quad \text{for} \quad p_j \in \Omega\backslash\Sigma \,. \tag{23.78}$$

Then we compute the tangential speeds

$$s_+ := \nabla \widetilde{u}_h^{n+1}\big|_{T_+} \cdot t_+ \quad \text{and} \quad s_- := \nabla \widetilde{u}_h^{n+1}\big|_{T_-} \cdot t_- \tag{23.79}$$

at the trailing edge from above and below, respectively. At the points $p_1^+, p_1^- = TE \in \Sigma$ we define the modified potential values $\bar{u}_h^{n+1}(p_1^+)$ and $\bar{u}_h^{n+1}(p_1^-)$ such that the following equations hold,

$$\begin{aligned}
\nabla \bar{u}_h^{n+1}\big|_{T_+} \cdot t_+ &= \tfrac{1}{2}(s_+ + s_-), \\
\nabla \bar{u}_h^{n+1}\big|_{T_-} \cdot t_+ &= \tfrac{1}{2}(s_+ + s_-).
\end{aligned} \tag{23.80}$$

This uniquely defines the jump

$$\bar{\beta}^{n+1} := \bar{u}_h^{n+1}(p_1^+) - \bar{u}_h^{n+1}(p_1^-) \tag{23.81}$$

and the Kutta-Joukowski condition in the form (23.77) is satisfied. The potential values $\bar{u}_h^{n+1}(p_j^+)$, $\bar{u}_h^{n+1}(p_j^-)$ for $j = 2, \ldots, L$ at the remaining points on Σ will be determined by the condition that $\bar{u}_h^{n+1}$ will have a constant jump of strength $\bar{\beta}_h^{n+1}$ and, if possible, continuous normal derivatives on Σ. This yields the set of equations

$$\bar{u}_h^{n+1}(p_j^+) - \bar{u}_h^{n+1}(p_j^-) = \bar{\beta}^{n+1} \tag{23.82}$$

and

$$\int_{\Omega_h} \nabla \bar{u}_h^{n+1} \cdot \nabla w_{h_j}^+ \, dx + \int_{\Omega_h} \nabla \bar{u}_h^{n+1} \cdot \nabla w_{h_j}^- \, dx = 0 \quad \text{for} \quad j = 2, \ldots, L. \tag{23.83}$$

Here $w_{h_j}^+$, $w_{h_j}^-$ are the one–sided basis functions corresponding to the points p_j^+, p_j^- for $j = 2, \cdots, L$. The equations (23.82), (23.83) give a system of $2(L-1)$ linear equations for the $2(L-1)$ unknowns $\bar{u}_h^{n+1}(p_j^+), \bar{u}_h^{n+1}(p_j^-)$ for $j = 2, \cdots, L$.

Numerical experiments did not always show satisfactory convergence; in particular, at the very beginning of the iteration. Hence, for stabilizing, we used a relaxation technique,

$$\begin{aligned}
u_h^{n+1} &:= \omega \bar{u}_h^{n+1} + (1 - \omega)\widetilde{u}_h^{n+1} \\
\beta^{n+1} &:= \omega \bar{\beta}^{n+1} + (1 - \omega)\beta_h^n
\end{aligned} \quad \text{with some } \omega \in (0, 1] . \tag{23.84}$$

In particular, we chose $\omega = \tfrac{1}{2}$ as the starting value, then increased $\omega < 1$ and, finally, set $\omega = 1$ when $|\beta_h^n - \bar{\beta}^{n+1}|$ was smaller than a given tolerance.

Remark: This step of the algorithm based on (23.84) converges for angles of attack in a computationally determined interval. In Göhner *et al.* (1994) we describe an alternative procedure for enforcing (23.77) by a global correction of the potential where vanishing of the stress intensity factor at the trailing edge TE is enforced, which turns out to converge for a larger variety of angles of attack. $\quad\square$

Step 3:

We compute the new gradient $g_h^{n+1} \in \widetilde{V}_h^0$ as the FEM–solution of the variational Poisson problem,

$$\int_{\Omega_h} \nabla g_h^{n+1} \cdot \nabla v_h \, dx = [\mathcal{J}_h'(u_h^{n+1}), v_h] \quad \forall \, v_h \in \widetilde{V}_h^0. \tag{23.85}$$

Here $[\mathcal{J}_h'(u_h^{n+1}), v_h]$ is defined as in (23.69) with u_h^{n+1} instead of u_h^0 and with λ_h^m. The new descent direction $z_h^{n+1} \in \widetilde{V}_h^0$ is now given by

$$z_h^{n+1} := g_h^{n+1} + \gamma^{n+1} z_h^n \quad \text{with} \quad \gamma^{n+1} := \frac{\int_{\Omega_h} \nabla g_h^{n+1} \cdot \nabla \left(g_h^{n+1} - g_h^n\right) d\boldsymbol{x}}{\int_{\Omega_h} \left|\nabla g_h^n\right|^2 d\boldsymbol{x}} . \tag{23.86}$$

The inner iteration procedure, returning to *Step 1*, will be repeated until the minimum of the functional $\mathcal{J}_h$ subject to $F(\beta) = 0$ is obtained within a given accuracy.

The final solution we denote by $(\widetilde{u}_h^m, \beta_h^m)$.
End of inner iteration.

Outer iteration step:

This step is devoted to the computation of $\lambda_h^{m+1} \in \widetilde{H}_h$ by solving the discrete boundary integral equation

$$\widetilde{b}_h\left(\lambda_h^{m+1}, \psi_h\right) = \widetilde{\ell}_2^h(\psi_h, \beta_h^m) - \langle \widetilde{u}_h^m, \psi_h \rangle_h + \widetilde{d}_h\left(\widetilde{u}_h^m, \psi_h\right) \quad \forall \, \psi_h \in \widetilde{H}_h . \tag{23.87}$$

With λ_h^{m+1}, repeat the outer iteration till $\|\lambda_h^{m+1} - \lambda_h^m\|_{L^2(\Gamma_\infty)}$ is smaller than a given tolerance.
End of outer iteration.

For practical computations we executed numerical experiments with different choices for the number of inner iterations per outer iteration. Mostly we used 10 inner iterations.

23.5 ON THE CONVERGENCE OF THE MINIMIZATION METHOD

In the paper Berger *et al.* (1993) we present a complete proof for the convergence of the sequence $\{(\widetilde{u}_h, \widetilde{\lambda}_h, \beta_h)\}_{h>0}$ of solutions of the minimization problems (23.65) which will be sketched here. Under the assumption of existence and uniqueness for the solution to problem (23.38) we will assume that the sequence

$$\{(\widetilde{u}_h, \widetilde{\lambda}_h, \beta_h)\}_{h>0} \in (\widetilde{V}_h \cap \widetilde{K}_{s_0}) \times \widetilde{H}_h \times I$$

converges to this solution. For subsonic flow, the proof is rather straightforward, whereas the case of transonic flow is difficult since for the proof we need a compactness result by Mandel and Nečas (1987) and Murat (1981) provided by a discrete entropy condition which enforces convergence.

23.5.1 THE CASE OF SUBSONIC FLOW

For this case, we need a preliminary result which involves standard results of approximation theory.

Lemma 23.5.1 *Let $u \in V \cap K_{s_0}$ with $s_0 < \frac{2a_0^2}{\kappa+1}$ a given function which has the additional regularity*

$$u \in W^{2+\varepsilon,2}(\Omega) \text{ for some } \varepsilon > 0. \tag{23.88}$$

To this function u there corresponds $\beta \in \mathbb{R}$ providing $F(\beta) = 0$. Then for sufficiently small mesh sizes h, there exist $\widetilde{s}_0$ and a sequence of $\{\widetilde{u}_h^I\}_{h>0}$ with corresponding $\{\widetilde{\beta}_h\}$ satisfying

$$\widetilde{u}_h^I \in \widetilde{V}_h \cap \widetilde{K}_{\widetilde{s}_0} \text{ with } s_0 < \widetilde{s}_0 < \frac{2a_0^2}{\kappa+1},$$

$$F(\widetilde{\beta}_h) := \left|\nabla\widetilde{u}_h^I\big|_{T_+} \cdot t_+\right|^2 - \left|\nabla\widetilde{u}_h^I\big|_{T_-} \cdot t_-\right|^2 = 0, \tag{23.89}$$

$$\left\|u - u_h^I\right\|_{W^{1,p}(\Omega)} \xrightarrow{h\to 0} 0 \quad \text{for every } p \in [1,\infty]$$

where $u_h^I \in V_h$ is the extension of $\widetilde{u}_h^I \in \widetilde{V}_h$ to all of Ω as defined in Section 23.3.

For the proof we combine standard approximation results for $\varepsilon \in \mathbb{N}_0$, see Ciarlet (1978, p. 123), interpolate these inequalities for noninteger ε and finally use the property that the width of the skin is of order h^2 in connection with the Kutta–Joukowski condition as a point condition.

Theorem 23.5.2 *Let $u \in V \cap K_{s_0}$ with $s_0 < \frac{2a_0^2}{\kappa+1}$, $\lambda \in H$ and $\beta \in I$ be the unique solution of the coupled flow problem (23.38). Moreover let $u \in W^{2+\varepsilon,2}(\Omega)$ for some $\varepsilon > 0$.*

Then there exists a constant $\widetilde{s}_0 < \frac{2a_0^2}{\kappa+1}$ such that the sequence $\{(\widetilde{u}_h, \widetilde{\lambda}_h, \beta_h)\}_{h>0}$ corresponding to $\widetilde{u}_h \in \widetilde{V}_h \cap \widetilde{K}_{\widetilde{s}_0}$, $\widetilde{\lambda}_h \in \widetilde{H}_h, \beta_h \in I$, the sequence of solutions to the minimization problems (23.65) converges in $V \times H \times I$ to the solution (u, λ, β) of problem (23.38).

In Berger *et al.* (1993), the proof of Theorem 23.5.2 is split into four steps.

Step 1: Lemma 23.5.1 implies the existence of a constant $\widetilde{s}_0 < \frac{2a_0^2}{\kappa+1}$ and a function $\widetilde{u}_h^I \in \widetilde{V}_h \cap \widetilde{K}_{\widetilde{s}_0}$ interpolationg u and satisfying (23.89). To this interpolant $\widetilde{u}_h^I$ we take $\beta \in \mathbb{R}$ corresponding to the exact solution and determine λ_h^I by solving Equation (23.64) with these $\widetilde{u}_h^I$ and β. By (23.63) we then determine the corresponding $\widetilde{\xi}_h$. Thus, using the definition of $\{(\widetilde{u}_h, \widetilde{\lambda}_h, \beta_h)\}_{h>0}$ as the solution of minimization problem (23.65) subject to $F(\beta_h) = 0$, we obtain

$$0 \le \mathcal{J}_h(\widetilde{u}_h) = \min_{\widetilde{\phi}_h \in \widetilde{V}_h \cap \widetilde{K}_{\widetilde{s}_0}} \mathcal{J}_h(\widetilde{\phi}_h) \le \mathcal{J}_h(\widetilde{u}_h^I). \tag{23.90}$$

We then consider the expression $\mathcal{J}_h(\tilde{u}_h^I)$ as h tends to zero. With Definition (23.63), properties (23.31), (23.88) and Lemma 23.5.1 we finally obtain an estimate

$$
\begin{aligned}
\mathcal{J}_h(\tilde{u}_h^I) \;\leq\;& c\Big\{ h(\|u_h^I\|_{W^{1,2}(\Omega)} + 1 + |\beta|) \\
& + \|u - u_h^I\|_{W^{1,2}(\Omega)} + \|\lambda - \lambda_h^I\|_{H^{-\frac{1}{2}}(\Gamma_\infty)} \Big\} \|\xi_h\|_{W^{1,2}(\Omega)} \,.
\end{aligned}
\tag{23.91}
$$

Since λ_h is not the interpolation of λ, (23.91) needs to be improved.

Using the L^2–projection P_h defined in Lemma 23.3.2, inequality (23.34), Equations (23.38) and (23.52), the continuity of the form d which follows from (23.29), together with inequalities (23.33), (23.35), (23.36) and (23.53)–(23.56), we finally obtain an estimate of the form

$$
\begin{aligned}
\mathcal{J}_h(\tilde{u}_h^I) \;\leq\;& c\Big\{ h \left(\|u_h^I\|_{W^{1,2}(\Omega)} + \|\lambda_h^I\|_{H^{-\frac{1}{2}}(\Gamma_\infty)} + 1 + |\beta| \right) \\
& + \|\lambda - P_h\lambda\|_{H^{-\frac{1}{2}}(\Gamma_\infty)} + \|u - u_h^I\|_{W^{1,2}(\Omega)} \Big\} \|\xi_h\|_{W^{1,2}(\Omega)} \,.
\end{aligned}
\tag{23.92}
$$

The sequence $\{u_h^I\}_{h>0}$ is uniformly bounded in $W^{1,2}(\Omega)$ and, hence, by (23.52) and (23.34), we can show that $\{\lambda_h^I\}_{h>0}$ is also uniformly bounded in H. The construction of $\xi_h \in V_h^0$ from $\tilde{\xi}_h \in \tilde{V}_h^0$ implies together with (23.61) the estimate

$$
\|\xi_h\|_{W^{1,2}(\Omega)} \leq c \left\| \tilde{\xi}_h \right\|_{W^{1,2}(\Omega_h)} \leq c \left\| \nabla\tilde{\xi}_h \right\|_{L^2(\Omega_h)} \leq c\sqrt{\mathcal{J}_h(\tilde{u}_h^I)} \,.
\tag{23.93}
$$

Inserting (23.93) into (23.92) with the approximation properties of u_h^I and $P_h\lambda$, we finally obtain

$$
0 \leq \mathcal{J}_h(\tilde{u}_h) \leq \mathcal{J}_h(\tilde{u}_h^I) \to 0 \quad \text{for } h \to 0 \,.
\tag{23.94}
$$

Step 2: Since $\{u_h\}_{h>0} \subset K_{\tilde{s}_0}$, it follows that the sequence $\{u_h\}_{h>0}$ is bounded in V. Again we can use (23.52) and the coercivity (23.34) to show that the sequence $\{\lambda_h\}_{h>0}$ is bounded in H. Therefore, the sequence $\{(u_h, \lambda_h, \beta_h)\}_{h>0}$ is bounded in $(V \cap K_{\tilde{s}_0}) \times H \times I$. Since $V \cap K_{\tilde{s}_0}$ is a closed convex subset of V with respect to its topology, we may extract a subsequence $\{(u_h, \lambda_h, \beta_h)\}_{h>0}$ and find an element $(\hat{u}, \hat{\lambda}, \hat{\beta}) \in \left(V \cap K_{\tilde{s}_0} \right) \times H \times I$ such that

$$
\begin{aligned}
u_h \;&\rightharpoonup\; \hat{u} \quad \text{weakly in } W^{1,p}(\Omega) \text{ for every } p \in (2, \infty) \,, \\
\lambda_h \;&\rightharpoonup\; \hat{\lambda} \quad \text{weakly in } H \,, \\
\beta_h \;&\rightarrow\; \hat{\beta} \quad \text{in } I \subset \mathbb{R} \,.
\end{aligned}
\tag{23.95}
$$

Step 3: We show that $\{(u_h, \lambda_h, \beta_h)\}_{h>0}$ converges **strongly** to $(\hat{u}, \hat{\lambda}, \hat{\beta})$. With a very tedious analysis of all bilinear forms involved, we can show the inequality

$$
\int_{\Omega_h} \nabla\tilde{\xi}_h(\tilde{u}_h) \cdot \nabla(\hat{u}_h - \tilde{u}_h)\, dx \leq c \left\{ \frac{1}{2} \int_{\Omega_h} \nabla\xi_h(\tilde{u}_h)|^2\, dx \right\}^{\frac{1}{2}} = c\sqrt{\mathcal{J}_h(\tilde{u}_h)} \to 0 \text{ for } h \to 0
\tag{23.96}
$$

with $2 > \varepsilon > 1$ *as in* (23.62) *where*

$$\widetilde{E}_h := \left\{ \widetilde{v}_h \in \widetilde{V}_h \,\middle|\, \widetilde{v}_h \geq 0 \text{ and } \widetilde{v}_h = 0 \text{ on } \overline{\Gamma_\infty^h \cup \Sigma} \right\}.$$

Lemma 23.5.4 *Under the assumptions of Lemma 23.5.3 let* $\{\widetilde{u}_h\}_{h>0}$ *be a sequence of solutions to Problem* (23.65) *with* $h \to 0$*. Then there exists a subsequence* $\{\widetilde{u}_h\}_{h>0}$ *and a limit function* $u \in V \cap K_{s_0}$ *with strong convergence*

$$\|u - u_h\|_{W^{1,p}(\Omega)} \to 0 \text{ for } h \to 0 \text{ and for every } p \in [1, \infty) \tag{23.101}$$

and u *satisfies the entropy inequality*

$$-\int_\Omega \nabla u \cdot \nabla v \, dx \leq B \int_\Omega v \, dx \quad \forall\, v \in E := \left\{ v \in W^{1,2}(\Omega) \,\middle|\, v \geq 0 \text{ and } v = 0 \text{ on } \overline{\Gamma_\infty \cup \Sigma} \right\}.$$

$$\tag{23.102}$$

The proof of this lemma is based on the work by Mandel and Nečas (1987) and Feistauer and Nečas (1985, Theorem 4.23) .

For the convergence proof of the following theorem we have to modify the density function ρ to a function $\widetilde{\rho} : [0, \infty) \to [0, \infty)$ such that $\widetilde{\rho}(s) \geq \rho_c = $ constant> 0. We choose $s^* \in \left(s_0, \frac{2a_0^2}{\kappa - 1} \right)$, set $\widetilde{\rho}(s) = \rho(s)$ for $s \in [0, s^*]$ and use the extension described in Feistauer and Nečas (1985) whereby $\widetilde{\rho}(s) = \rho_c$ for large s. Note that the set of solutions to problem (23.38) is not changed by this modification.

Theorem 23.5.5 *Suppose that the following assumptions hold:*

(a) *Let the family of triangulations be quasiuniform.*
(b) *The problem* (23.38) *has exactly one solution* $(u, \lambda, \beta) \in (V \cap K_{s_0}) \times H \times I$*, which satisfies the entropy condition* (23.102)*.*
(c) *The solution of* (23.38) *is subsonic near the trailing edge.*

Then there exists a constant $\widetilde{s}_0 \geq s_0$ *such that the solutions* $\{(\widetilde{u}_h, \widetilde{\lambda}_h, \beta_h)\}_{h>0}$ *of the modified minimization problem*

$$\mathcal{J}_h(\widetilde{u}_h) = \min_{\widetilde{\varphi}_h \in \widetilde{V}_h \cap K_{\widetilde{s}_0}} \mathcal{J}_h(\widetilde{\varphi}_h) \tag{23.103}$$

subject to (23.63)*,* (23.64) *and* $F(\beta_h) = 0$ *converge to the unique solution* (u, λ, β) *of the variational problem* (23.38)*.*

The proof rests on the following three steps.

Step 1: The assumption (23.60) implies that $\widetilde{R}_h u \in \widetilde{V}_h \cap K_{c^* \cdot s_0}$. We modify $\widetilde{R}_h u$ along the slit such that the modified function $\widetilde{u}_h^I$ satisfies the Kutta–Joukowski condition and that $\widetilde{u}_h^I$ is uniformly bounded in $W^{1,\infty}(\Omega_h)$. Therefore, we can find a constant $\widetilde{s}_0 \geq c^* \cdot s_0$, such that

$$\left\| \nabla \widetilde{u}_h^I \right\|_\infty \leq \widetilde{s}_0 . \tag{23.104}$$

Moreover, the modified family $\{\widetilde{u}_h^I\}$ has the same approximation properties as $\widetilde{R}_h u$ implying

$$\|u - u_h^I\|_{W^{1,2}(\Omega)} \to 0 \text{ for } h \to 0 . \tag{23.105}$$

Step 2: Since the density ρ is now modified, the functional $\mathcal{J}_h$ may be applied to $\widetilde{u}_h^I \in \widetilde{V}_h \cap K_{\widetilde{s}_0}$; and we obtain from (23.104) the inequality

$$0 \le \mathcal{J}_h(\widetilde{u}_h) \le \mathcal{J}_h(\widetilde{u}_h^I) = \frac{1}{2} \int\limits_{\Omega_h} \left| \nabla \widetilde{\xi}_h(\widetilde{u}_h^I) \right|^2 \, d\boldsymbol{x} + \mathcal{P}_h(\widetilde{u}_h^I) \, . \tag{23.106}$$

By using the same arguments for the subsonic case we get with the help of (23.105)

$$\frac{1}{2} \int\limits_{\Omega_h} \left| \nabla \widetilde{\xi}_h(\widetilde{u}_h^I) \right|^2 \, d\boldsymbol{x} \to 0 \ \text{ for } \ h \to 0 \, . \tag{23.107}$$

Note that the function $\widetilde{u}_h^I$ coincides with $\widetilde{R}_h u$ in the points p_i for $i = 2L + 1, \ldots, N$. Since our penalty functional has no contributions from these nodes we have $\mathcal{P}_h(\widetilde{u}_h^I) = \mathcal{P}_h(\widetilde{R}_h u)$. Further, by using the definition of the Ritz projection (23.59), we can finally show

$$0 \le \mathcal{P}_h(\widetilde{R}_h u) \le c \sum_{\substack{i=1 \\ p_i \notin \Sigma \cup \Gamma_\infty}}^{N} h_i^2 h_i^{4-2\varepsilon} \le c h^{2(2-\varepsilon)} \, .$$

If we combine this with (23.106) and (23.107) we obtain

$$\mathcal{J}_h(\widetilde{u}_h) \to 0 \ \text{ for } \ h \to 0 \, . \tag{23.108}$$

Step 3: We now apply Lemmas 23.5.3 and 23.5.4 and get the existence of a subsequence $\{\widetilde{u}_h\}_{h>0}$ and a function $\widehat{u} \in K_{\widetilde{s}_0}$ such that

$$\|\widehat{u} - u_h\|_{W^{1,p}(\Omega)} \xrightarrow{h \to 0} 0 \ \forall p \in [1, \infty) \ \text{ and } \ -\int\limits_{\Omega} \nabla \widehat{u} \cdot \nabla v \, d\boldsymbol{x} \le B \int\limits_{\Omega} v \, d\boldsymbol{x} \ \forall v \in E \, . \tag{23.109}$$

The convergent subsequence $\{u_h\}_{h>0}$ is bounded. Then (23.52) holds. Taking $\psi_h = \lambda_h$ and using the coercivity (23.34) gives the boundedness of the corresponding sequence $\{\lambda_h\}_{h>0}$ in H. We may derive an estimate of $\widehat{\lambda} - \lambda_h$ in terms of $\lambda - P_h \lambda$ and u_h^I, λ_h^I. Then the strong convergence (23.109) together with the boundedness of $\{\lambda_h\}_{h>0}$ in H implies

$$\left\| \widehat{\lambda} - \lambda_h \right\|_{H^{-\frac{1}{2}}(\Gamma_\infty)} \to 0 \ \text{ for } \ h \to 0 \, , \tag{23.110}$$

$$b(\widehat{\lambda}, \psi) + \langle \widehat{u}, \psi \rangle - d(\widehat{u}, \psi) = \ell_2(\psi, \widehat{\beta}) \ \forall \, \psi \in H \, . \tag{23.111}$$

Using the same arguments as in *Step 4* of Theorem 23.5.1 together with (23.109)–(23.110) and assumptions (a), (b) we obtain the result. $\qquad \square$

23.6 NUMERICAL RESULTS

We present some results of our numerical computations. Here, we compare three different treatments of the farfield boundary condition.

The first condition corresponds to the condition (23.8) which is just the parallel flow at infinity.

In the second case we first compute the FEM–BEM approximation to the harmonic solution U of the exterior incompressible flow problem.

With U, we used as a second boundary condition

$$\rho(|\nabla u|^2)\partial_n u = \rho(|v_\infty|^2)\partial_n U \quad \text{on } \Gamma_\infty .$$

The third case shows the results of the complete coupling described in Section 23.4.

For better comparison we give results for two standard test cases of flows around the NACA–0012 profile. Two different sized C–grids were used. Here Γ_∞ has two corner points where the C^∞ assumption used in the foregoing analysis is violated; however, due to the Remark in Section 23.1.2, our convergence result can be extended to this simple Lipschitz curve. A large grid with 115 by 15 nodes, and the outer boundary 6 chord lengths distance from the profile; and a smaller grid with 111 by 13 nodes, and the outer boundary 3 chord lengths distance from the profile. We give the lift coefficients c_a calculated from the pressure distributions along the profile.

Test Case 1: (purely subsonic flow): $M_\infty = 0.63$, $\alpha = 2^0$

	c_a large grid	c_a small grid
1. case	0.3470	0.3684
2. case	0.3395	0.3481
full coupling	0.3371	0.3391

In a research report Kroll and Jain (1987) give a lift coefficient of $c_a = 0.3333$. They used an O–grid with 256 by 64 nodes, outer boundary 50 chord lengths from the profile. Their calculation was done with a potential as well as with an Euler code.

Test Case 2: (transonic flow): $M_\infty = 0.8$, $\alpha = 1.25^0$

	c_a large grid	c_a small grid
1. case	0.4853	0.5852
2. case	0.4705	0.5257
full coupling	0.4209	0.4459

In Rizzi and Viviant (1986) a number of solutions for this test case are given. These were obtained by finite difference and finite volume discretizations of the full potential equation. The calculated lift coefficients vary between 0.5 and 1.1. In AGARD (1985) newer results for the Euler equations were published. There the lift coefficients vary between 0.35 and 0.37.

ACKNOWLEDGMENT

The research reported in this paper was supported by the "Stiftung Volkswagenwerk".

where $\widehat{u}_h$ is an appropriate approximation of $\widehat{u}$.

The continuity of the operator $\mathcal{K}$ in (23.29) implies the compactness of the bilinear form d on V; hence the weak convergence (23.95) for the sequence $\{\lambda_h\}_{h>0}$ yields

$$d(u_h, \widehat{\lambda} - \lambda_h) = \langle u_h, \mathcal{K}'(\widehat{\lambda} - \lambda_h) \rangle \to 0 \quad \text{for} \quad h \to 0. \tag{23.97}$$

Thus, (23.96), (23.97) imply together with (23.53)–(23.56) the weak convergence (23.95). Then the approximation property of P_h and a suitable modification of u_h along Σ imply that $\widetilde{u}_h^I - P_h u$ can be used as a test function. This implies with (23.96) that

$$(u_h, \lambda_h, \beta_h) \to (\widehat{u}, \widehat{\lambda}, \widehat{\beta}) \quad \text{strongly in} \quad V \times H \times \mathbb{R} \quad \text{for} \quad h \to 0. \tag{23.98}$$

Step 4: In the last step we show that $(\widehat{u}, \widehat{\lambda}, \widehat{\beta})$ must coincide with the solution (u, λ, β) of (23.38) by using the consistency properties collected in Lemma 23.3.1.

It still remains to prove that $\widehat{u} \in V \cap K_{\widetilde{s}_0}$ satisfies the Kutta–Joukowski condition (23.6). However, $\widehat{u}$ is a weak solution of (23.38) except the Kutta–Joukowski condition; and $\widehat{u}$ is uniformly bounded in $W^{1,2}(\Omega)$. However, as explained in Berger *et al.* (1993), any solution u which lies in V either satisfies the Kutta–Joukowski condition or has an arbitrary growth of the gradient in a neighbourhood of the trailing edge TE. Hence, it follows that $\widehat{u}$ must satisfy the Kutta–Joukowski condition at TE. Thus, $(\widehat{u}, \widehat{\lambda}, \widehat{\beta})$ is a subsonic flow solution of (23.38). The assumption on the uniqueness of such a solution implies that $(\widehat{u}, \widehat{\lambda}, \widehat{\beta}) = (u, \lambda, \beta)$ holds and, moreover, that the whole sequence of solutions $\{(u_h, \lambda_h, \beta_h)\}_{h>0}$ of the minimization problem (23.65) converges to the given solution (u, λ, β).

Note that due to the uniform L^∞-bound for the gradients of the sequence $\{u_h\}_{h>0}$, the latter converges in $W^{1,p}(\Omega)$ for every $p \in [1, \infty)$.

23.5.2   THE CASE OF TRANSONIC FLOW

The main underlying ideas of the convergence proof for the sequence $\{(\widetilde{u}_h, \widetilde{\lambda}_h, \beta_h)\}_{h>0}$ of solutions to (23.65) go back to Berger (1989), who has shown the convergence for **interior** problems on polygonal domains with a simplified boundary condition. An extension of this work to domains with arbitrary curved boundaries can be found in Berger and Feistauer (1993). In Berger *et al.* (1993), we prove the following slightly stronger result than Berger (1989, Theorem 4.1).

Lemma 23.5.3 *Let $\{\mathcal{T}_h\}_{h>0}$ be a family of triangulations satisfying the uniform angle property. This implies that the supports of the hat functions w_{h_i} contain at most a certain finite number of triangles independently of h. This in turn implies that there is a constant $c > 0$ such that*

$$A_i \leq ch^2 \quad \text{for} \quad i = 1, \ldots, N \text{ and all } h > 0. \tag{23.99}$$

Then a family of solutions to (23.65) satisfies the estimate

$$-\int_{\Omega_h} \nabla \widetilde{u}_h \cdot \nabla \widetilde{v}_h \, d\boldsymbol{x} \leq B \int_{\Omega_h} \widetilde{v}_h \, d\boldsymbol{x} + ch^{\varepsilon - 1} \|\widetilde{v}_h\|_{L^\infty(\Omega)} \; \forall \; \widetilde{v}_h \in \widetilde{E}_h \tag{23.100}$$

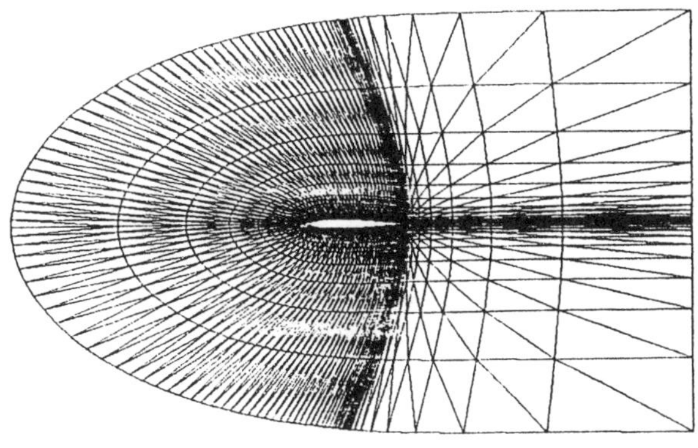

Figure 23.3: Small grid

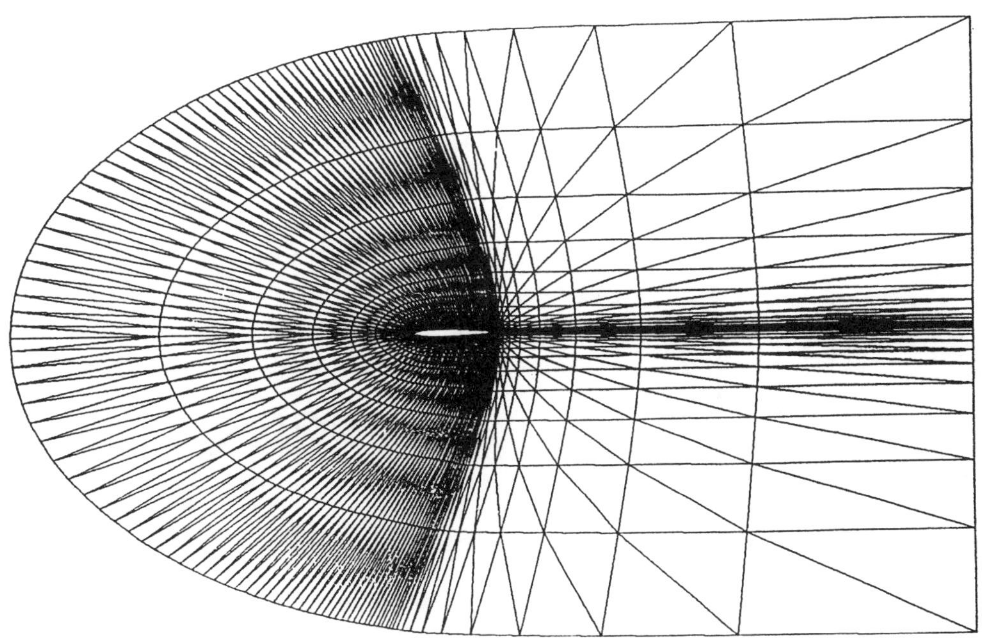

Figure 23.4: Large grid, containing the small grid

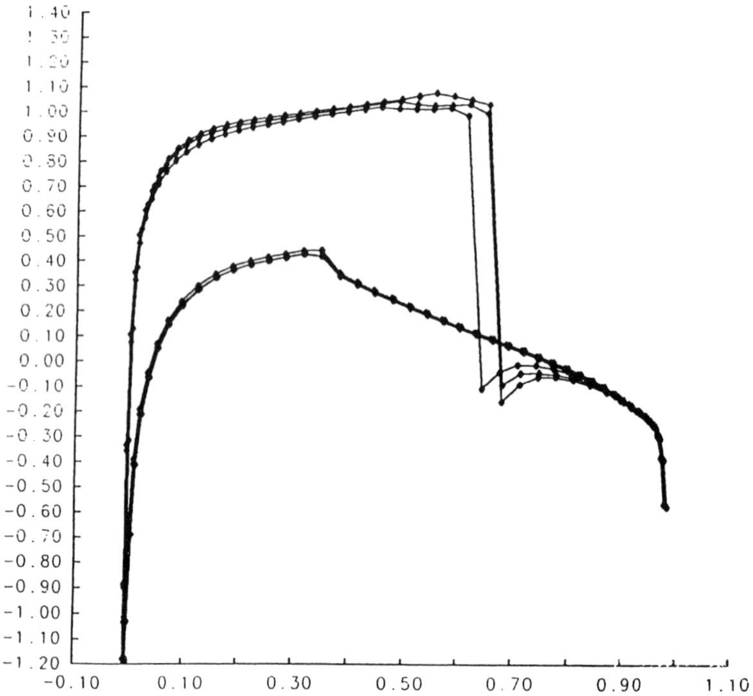

Figure 23.5: Pressure distributions for the transonic test case 2 on the large grid

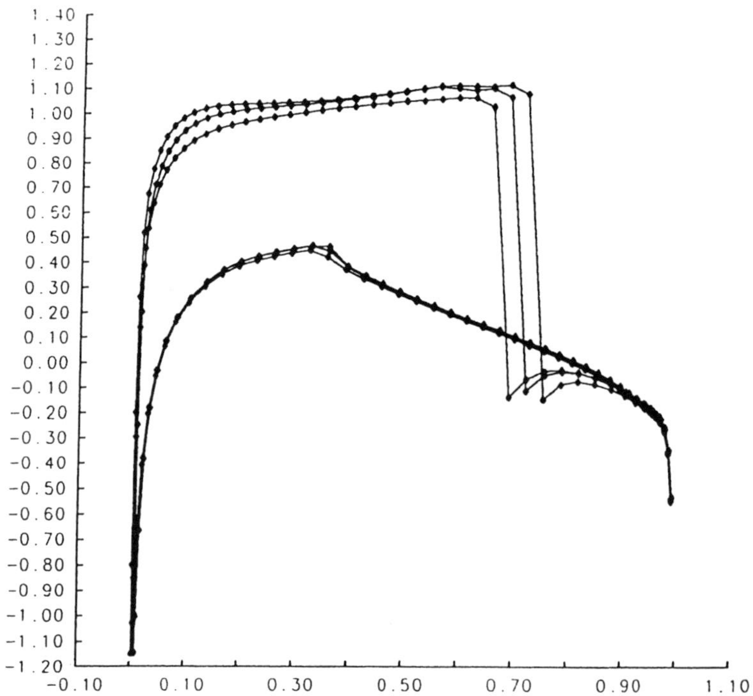

Figure 23.6: Pressure distributions for the transonic test case 2 on the small grid

REFERENCES

AGARD, 1985, Advisory Report No 211, Test Cases for Inviscid Flow Field Methods, Report of Fluid Dynamics Panel Working 07.

Babuška, I. and Aziz, A. K. (1972), Survey lectures on the mathematical foundations of the finite element method, in *The Mathematical Foundation of the Finite Element Method with Applications to Partial Differential Equations*, A. K. Aziz (ed.), Academic Press, New York 3-359.

Becker, E. B., Carey, G. F. and Oden, J. T. (1984), *Finite Elements, Vol. VI: Fluid Mechanics*, Prentice-Hall Inc., Englewood Cliffs, New Jersey.

Berger, H. (1989), A convergent finite element formulation for transonic flows, Numer. Math. **56**, 425-447.

Berger, H. and Feistauer, M. (1993), Analysis of the finite element variational crimes in the numerical approximation of transonic flow. To appear in Math. Comp.

Berger, H., Warnecke, G. and Wendland, W. (1988), Finite Element-Berechnungen fur transsonische Stromungen unter Berucksichtigung verschiedener Fernfeldrandbedingungen. In: *Stromungen mit Ablosungen*, DGLR-Bericht 88-05, Bonn, 233-242.

Berger, H., Warnecke, G. and Wendland, W. (1990), Finite elements for transonic potential flows, *Numer. Meth. Part. Diff. Eqns.* **6**, 17-42.

Berger, H., Warnecke, G. and Wendland, W. (1993), Analysis of a FEM/BEM coupling method for transonic flow computations, in preparation (Preprint 93-9, Mathematisches Institut A, University Stuttgart).

Bers, L. (1958), Mathematical aspects of subsonic and transonic gas dynamics, in *Surveys in Applied Mathematics III*, Wiley, New York.

Bojarski, B. (1967), Subsonic flow of compressible fluid, in *Mathematical Problems in Fluid Mechanics*, Polish Academy of Sciences, Warsaw.

Bristeau, M. O., Glowinski, R., Periaux, J., Perrier, P., Pironneau, O. and Poirier, G. (1980), Application of optimal control and finite element methods to the calculation of transonic flows and incompressible flows, in *Numerical Methods in Applied Fluid Dynamics*, B. Hunt (ed.), Academic Press, New York, 203-312.

Bristeau, M. O., Glowinski, R., Periaux, J., Perrier, P., Pironneau, O. and Poirier, G. (1985), On the numerical solution of nonlinear problems in fluid dynamics by least squares and finite element methods II. Application to transonic flow simulations, *Comput. Methods Appl. Mech. Engrg.* **51**, 363-394.

Ciarlet, P. G. (1978), *The Finite Element Method for Elliptic Problems*, North-Holland, Amsterdam.

Costabel, M. (1988), Boundary integral operators on Lipschitz domains: Elementary results, *SIAM J. Math. Anal.* **19**, 613-626.

Feistauer, M., Felcman, J., Rokyta, M. and Vlasek, Z. (1992), Finite-element solution of flow problems with trailing conditions, *J. Comp. Appl. Math.* **44**, 131-165.

Feistauer, M. and Nečas, J. (1985), On the solvability of transonic potential flow problems, *Z. Anal. Anw.* **4**, 305-329.

Feistauer, M. and Ženíšek, A. (1987), Finite element solution of nonlinear elliptic problems, *Numer. Math.* **50**, 451-475.

Gatica, G. N. and Hsiao, G. C. (1992), The coupling of boundary element and finite element methods for a nonlinear exterior boundary value problem, *Numer. Math.* **61**, 171-214.

Glowinski, R. and Pironneau, O. (1978), On the computation of transonic flows, in *Funct. Anal. and Num. Anal.*, H. Fujita (ed.), Jap. Soc. Prom. Sci., Tokyo-Kyoto, 143-173.

Göhner, U. and Warnecke, G. (1993), A shock indicator for adaptive transonic flow computations. To appear in *Numer. Math.*

Göhner, U., Warnecke, G. and Wendland, W. (1994), The Kutta-Joukowski condition for transonic airfoil conditions. In preparation.

Hsiao, G. C. and Wendland, W. L. (1977), A finite element method for an integral equation of the first kind, *J. Math. Anal. Appl.* **58**, 449-481.

Johnson, C. and Nedelec, J. C. (1980), On the coupling of boundary integral and finite element methods, *Math. Comp.* **35**, 1063-1073.

Keyfitz, B. L. and Warnecke, G. (1991), The existence of viscous profiles and admissibility for transonic shocks, *Commun. Part. Diff. Eqns.* **16**, 1197-1221.

Kroll, N. and Jain, R. K. (1987), *Solution of Two-Dimensional Euler Equation Experience with a Finite Volume Code*, DFVLR-Forschungsbericht 87-41, DFVLR Institut fur Entwurfsaerodynamik, Braunschweig.

Le Roux, M. N. (1974), *Résolution numérique du problème du potential dans le plan par une méthode variationelle d'éléments finis*, Doctoral Thesis, University Rennes.

Mandel, J. and Nečas, J. (1987), Convergence of finite elements for transonic potential flows, *SIAM J. Numer. Anal.* **24**, 985-996.

Morawetz, C. S. (1964), Non-Existence of Transonic Flow Past a Profile III, *Commun. Pure Appl. Math.* **17**, 357-367.

Morawetz, C. S. (1985), On a Weak Solution for a Transonic Flow Problem, *Commun. Pure Appl. Math.* **38**, 797-818.

Murat, F. (1981), L'injection du cône positiv de H^{-1} dans $W^{-1,q}$ est compact pour tout $q < 2$, *Math. Pure Appl.* **60**, 309-322.

Nečas, J. (1989), *Composite par Entropie et Evulements de Fluides*, Lecture Notes Université de Charles et E.N-S., Masson, Paris.

Rannacher, R. and Scott, R. (1982), Some optimal error estimates for piecewise linear finite element approximations, *Math. Comp.* **38**, 437-445.

Rizzi, A. and Viviant, H., Eds. (1986), *Numerical Methods for the Computation of Inviscid Transonic Flows and Shock Waves - Notes on Numerical Fluid Mechanics III*, Vieweg, Braunschweig.

Sofronov, I. L. and Tscincov, S. V. (1993), Application of the linear potential flow model to the artificial boundary conditions construction for the Euler equations. (Preprint Nat. Center of Math. Simulation, Russian Acad. Sci. Moscow, 1991).

Warnecke, G. (1990), Admissibility of solutions to the Riemann problems for systems of mixed type - transonic small disturbance theory. In: *Nonlinear Evolution Equations that Change Type*, B. L. Keyfitz and M. Shearer (eds.), IMA-Series Volume 27, Springer-Verlag, New York, 258-284.

Zierep, J. (1976), *Theoretische Gasdynamik*, Braun, Karlsruhe.

ABSTRACTS OF OTHER LECTURES

ON GRADIENT SMOOTHING IN GRIDS WITH SINGULARITY ELEMENT PATCHES

J. Aalto*, J. Louhivirta* and J.R. Whiteman**

*Department of Civil Engineering,
University of Oulu, Kasarmintie 8, 90100 Oulu, Finland

**BICOM, The Institute of Computational Mathematics,
Brunel University, Uxbridge, UB8 3PH, U.K.

One possibility for dealing with corner singularities in the finite element analysis of boundary value problems is to use a special subgrid of curved elements, called a singularity element patch, around each corner point. This is done by introducing a suitable geometrical mapping between the parametric u, v-plane and the physical x, y-plane. In the u, v-plane the geometry of the elements is linear. In the x, y-plane the elements are distorted according to the equations of mapping.

This paper shows how typical gradient smoothing techniques can be modified so that they work properly in grids with singularity element patches. The idea of the modification is based on reasonable smoothed approximation for the gradient $\vec{g} = \nabla\phi$. Within the elements outside the singularity element patches the components $g_x \equiv \partial\phi/\partial x$ and $g_y \equiv \partial\phi/\partial y$ of the gradient are represented typically as

$$\bar{g}_x = \sum_{i=1}^{m} N_i a_{xi}, \quad \bar{g}_y = \sum_{i=1}^{m} N_i a_{yi}. \tag{1}$$

In the vicinity of the singularity point, however, the gradient components g_x and g_y are singular. Therefore it is not reasonable to use approximations (1), as such, within the elements inside a singularity element patch. Instead of these, the components $g_u \equiv \partial\phi/\partial u$ and $g_v \equiv \partial\phi/\partial v$, which are regular, are represented as

$$\bar{g}_u = \sum_{i=1}^{m} N_i a_{ui}, \quad \bar{g}_v = \sum_{i=1}^{m} N_i a_{vi}. \tag{2}$$

In terms of these the components $\bar{g}_x$ and $\bar{g}_y$ are obtained using known relations between the gradient components g_x, g_y and g_u, g_v.

Such modifications for three well known smoothing procedures: Nodal Averaging (NA), Global Least Squares smoothing (GLS) and the Zienkiewicz-Zhu Superconvergent Patch Recovery technique (SPR) are dealt with. Numerical examples of both (a) the Poisson's equation and (b) the plane elasticity problem are given.

The Mathematics of Finite Elements and Applications
Edited by J. R. Whiteman © 1994 John Wiley & Sons Ltd

TIME-DEPENDENT FINITE ELEMENT SOLUTION OF BOLTZMANN EQUATION FOR PHOTON TRANSPORT

R.T. Ackroyd, C.R.E. de Oliveira and A.J.H. Goddard

*Energy Systems Section, Department of Mechanical Engineering,
Imperial College of Science, Technology and Medicine, SW7 2BX, U.K.*

A method of solving the time-dependent Boltzmann equation for particle transport in the case of material-radiation interaction is described. Spatial dependence of the solution is treated by the finite element method. A discontinuous variational formulation is employed in order to cope with the abrupt changes of the physical characteristics of the system with time. Particle scattering is treated by means of a spherical harmonic expansion for particle direction. The method is illustrated by the solution of the nonlinear equation for photon transport in a reactive atmosphere by presenting solutions for the classical Marshak thermal wave problem. The results show that the method is capable of accurately predicting the correct speed of the thermal wave even in the case of very coarse meshes.

STRESSES COMPUTED FROM HIERARCHIC PLATE MODELS

R.L. Actis* and B.A. Szabó**

**Center for Computational Mechanics,
Washington Uiversity, St Louis, Missouri, U.S.A.*

***University of Miskolc, Department of Machine Elements,
H3515 Miskolc, Egyetemváros, Hungary*

Stress distributions near boundaries, discontinuities and, in the case of laminated plates, at interfaces are generally very different from the stress distribution in the interior regions of structural plates. Boundary layer effects are normally present, and the problem in those regions is essentially three-dimensional. Hiearchic models for laminated plates make it possible to approximate the three-dimensional problem through the solution of two-dimensional problems without the expense of a fully three-dimensional analysis.

Higher order models can be selected in regions of local discontinuities or near boundaries, while low-order models are used in the smooth interior regions of the domain. Improved transverse stresses can be obtained for any model of the hierarchy by using an extraction technique. In particular, the integration of the equilibrium equations has shown substantial improvement over the direct computation for the transverse shear and normal stresses. Examples are presented which show that hierarchic models are capable of approximating the fully three-dimensional solution.

BOUNDARY INTEGRAL EQUATIONS FOR THE EXTERIOR ACOUSTIC PROBLEM

S. Amini and N.D. Maines

*Department of Mathematics and Computer Science,
University of Salford, Salford, M5 4WT, U.K.*

The boundary integral solution of the Helmholtz equation

$$(\nabla^2 + k^2)\phi = 0, \quad p \in D \tag{1}$$

in an unbounded domain, D_+, exterior to a smooth C^∞ boundary Γ, satisfying a general boundary condition on Γ and an appropriate radiation condition at infinity is considered. By a careful application of the Green's second theorem the unique solution of this problem can be obtained in the form

$$\phi(p) = \int_\Gamma \phi(q)\frac{\partial G_k}{\partial n_q}(p,q)d\Gamma_q - \int_\Gamma G_k(p,q)\frac{\partial \phi}{\partial n_q}d\Gamma_q, \quad p \in D_+ \tag{2}$$

In 2-D, the Green's function $G_k(p,q) = i/4H_0^{(1)}(k|p-q|)$. The Helmholtz integral representation formula (2) requires the Cauchy data $(\phi, \partial\phi/\partial n)$. The boundary condition on Γ gives either ϕ or $\partial\phi/\partial n$ or a relationship between the two in the case of "spring like" scatterer. We need another boundary relationship between ϕ and $\partial\phi/\partial n$. This is usually obtained by taking the limit as $p_+ \in D_+ \to p \in \Gamma$ in (2). We obtain the so-called "Surface Helmholtz Equation"

$$-\frac{1}{2}\phi(p) + \int_\Gamma \phi(q)\frac{\partial G_k}{\partial n_q}(p,q)d\Gamma_q = \int_\Gamma G_k(p,q)\frac{\partial \phi}{\partial n_q}d\Gamma_q. \tag{3}$$

The operators on either side of equation (3) can be shown to be singular for a countable set of real positive values of k which correspond to the standing wave solutions of respective interior problems. The following formulation due to Burton and Miller overcomes this non-uniqueness problem and is based on coupling (3) with its derivative in the direction of the normal to Γ at p,

$$\left(\frac{-1}{2}I + M_k + i\eta N_k\right)\phi(p) = \left(L_k + i\eta(\frac{1}{2}I + M_k^t)\right)\frac{\partial\phi}{\partial n}(p), \quad p \in \Gamma. \tag{4}$$

In (4), L_k and M_k are the single and double layer potentials with M_k^t and N_k as their respective derivatives. In appropriate Sobolev spaces $H^r(\Gamma)$ it can be shown that L_k, M_k, $M_k^t : H^r \to H^{r+1}$ are smoothing operators of order -1 whilst $N_k : H^{r+1} \to H^r$ is principally an order +1 differentiation operator.

In this paper we study the spectral properties of the four Helmholtz integral operators. The choice of the coupling parameter η and also the convergence analysis of collocation and discrete collocation solutions of the strongly elliptic pseudo–differential operator equation (4) will be discussed. We also report on several 2-grid iterative methods for our boundary element equations.

ON THE MODELLING ERRORS IN FRACTURE
ANALYSIS OF DELAMINATED PLATES

B. Andersson

The Aeronautical Research Institute of Sweden,
Box 11021, S-161 11 Bromma, Sweden

Composite materials are frequently used in engineering applications. One critical failure mechanism for this class of materials is related to the delamination of sets of laminate layers. For safe design using these materials, reliable methods for prediction of growth and arrest of delamination damage are needed. Delaminations have been observed to propagate in very complex patterns. Initially circular delamination fronts may under compressive loading (so-called buckling-driven delamination) form "worm"-like shapes or "telephone cord"-shapes. The numerical models that attempt to simulate this type of delamination growth are generally based on nonlinear plate theory and fracture mechanics. The reasons for using plate models are that the delaminations appear to be "thin" and that a three-dimensional analysis is considered to be too costly. A critical question in this context is, under what circumstances do plate models yield accurate solutions?

The growth of delamination fronts is assumed to be governed by a local growth criterion of the type $f(K_I(y), K_{II}(y), K_{III}(y)) = 0$, where $K_\alpha(y)$ are the stress intensity functions and y is a coordinate along the delamination front. It is of special importance to separate the individual fracture modes α, since experiments indicate significant differences in energy release rates in opening and sliding/tearing modes. The reliability of solutions obtained from this type of numerical model depends critically on the modelling error, the discretisation error, and the extraction procedures used to derive the stress intensity functions $K_\alpha(y)$. For plate models, bending moments and shear forces are determined. With known moments and forces, the stress intensity functions are estimated using analytical solutions (split-beam solutions) for a two-dimensional domain representing the delamination-tip-region.

In the present paper, the modelling errors obtained using the Kirchoff and the Reissner-Mindlin plate models are studied. Laminates of isotropic, bi-material isotropic and orthotropic materials having quadratic, elliptic, or (some) real-life shaped delamination fronts are considered. Three-dimensional almost exact reference solutions (i.e. maximum relative pointwise errors in $K_\alpha(y)$ of the order 10^{-3}) are derived using the $hp-$ version of FEM and a novel computational procedure for extracting edge stress intensity functions at 3D interfacial delaminations in orthotropic laminates. The solutions for the Kirchoff and the Reissner-Mindlin plate models are derived using the p version of FEM. Modelling errors, thus isolated from discretization errors, are determined virtually exactly for different plate models, fracture mechanics models, material combinations, delamination shapes, and delamination/thicknesses ratios for a simple loading case.

Since boundary layers are very different for different models, so are calculated intensity functions. For delaminations with a diameter/thickness ratio of $\kappa{=}20$ and simple shapes of the delamination front (i.e. quadratic or elliptic shape), the modelling errors $\left(K_\alpha^{3D} - K_\alpha^{Plate}\right)/K_\alpha^{3D}$ are of the order 5-15%. For near-circular delaminations with $\kappa{=}20$ and a "wavy" boundary (as sometimes observed in practice), the modelling error may be as large as 25-40%. The conclusion is that since energy release rates G_α are proportional to $K_\alpha{}^2$, plate model solutions often do not provide reliable quantitative results when applied to real-life delaminations with complex boundaries.

A FINITE DIFFERENCE METHOD OF DETERMINATION OF LOCAL COEFFICIENTS IN DENTAL ENAMEL AND OTHER INHOMOGENEOUS PERMEABLE MEDIA

P. Anderson, J.C. Elliott and S.E.P. Dowker

Department of Child Dental Health,
The London Hospital Medical College, Turner Street, London , E1 2AD, U.K.

Methods to study transport processes in teeth (composed of inhomogeneous, permeable materials) are important in the understanding of various, destructive and repair processes. In normal and diseased enamel, and many other inhomogeneous materials, the local porosity and tortuosity, and therefore the diffusion coefficient vary with position. To measure diffusion coefficients in homogeneous permeable materials (e.g. a glass frit), a solute with high X-ray absorbance (e.g. 1 mole L^{-1} KI) is allowed to diffuse from a constant source at one surface into water contained within its pores. The concentration of KI at any point along the 1-dimensional diffusion path can be calculated from X-ray attenuation measurements. Repeated measurements along the diffusion path yield a time series of concentration-distance profiles from which the diffusion coefficient for KI can be calculated using the Crank-Nicolson (CN) algorithm for Fick's 2nd equation. The aim of this study is to extend this method to measure the local diffusion coefficient (D_{local}) in inhomogeneous materials. Experimental data was simulated for the diffusion of a solute within a permeable material of predetermined inhomogeneity using the CN algorithm with very small distance and time steps. Using this data, D_{local} was calculated at many positions using a modification of the CN algorithm. Systematic errors due to the approximations required for the finite difference method were investigated. Errors associated with intrinsic uncertainties due to X-ray photon counting statistics were studied using Monte-Carlo simulations. Improvement in accuracy by increasing count time is constrained by the change in concentration with time. However, KI can be repeatedly diffused in and out of the permeable solid and an averaging algorithm used which effectively increases the count-time for any particular concentration determination without introducing systematic errors.

ANISOTROPIC FINITE ELEMENTS AND APPLICATION TO EDGE SINGULARITIES

T. Apel

Technische Universität Chemnitz-Zwickau, Fachbereich Mathematik,
Postschileßfach 964, 0-9010 Chemnitz, Germany

The classical local interpolation error estimates (see e.g. Ciarlet 1978) were derived under an assumption which is in two dimensions known as Zlámal's minimal angle condition. This condition was weakened by different authors (Jamet 1976, Babŭska/Aziz 1976, Křižek 1989) to a maximal angle condition. But the possible advantage of using mesh sizes with different asymptotics in different directions which leads to small angles, was not extracted. In this paper, anisotropic interpolation error estimates in two and three dimensions, proved by Dobrowolski and Apel (1992) are reviewed. Here, one derives benefit from the different asymptotics of the mesh sizes. The results are applied to the finite element approximation of elliptic equations on domains with edges. Practical computations show that the proposed finite element meshes are well suited for elliptic problems with solution of r^λ-type.

THE BOUNDARY LAYER FOR THE REISSNER-MINDLIN PLATE MODEL

D.N. Arnold

*Department of Mathematics,
Pennsylvania State University, University Park, PA 16802, U.S.A.*

The solution to the Reissner-Mindlin plate model, in contrast to that of the biharmonic model, exhibits a complex dependence on the thickness of the plate. For thin plates there may be a boundary layer, the existence and structure of which depends on the boundary conditions, the plate geometry, and the solution component considered. For example, the transverse displacement variable does not exhibit any edge effect, but the rotation vector exhibits a boundary layer except for very special data or plate geometry. The bending moment tensor and shear force vector have more pronounced boundary layers. The boundary layer is stronger for a free or simply-supported plate than for a clamped one. In this talk we describe recent results which furnish a complete asymptotic description of the dependence of the solution on plate thickness. The techniques used to obtain the asymptotic expansions and rigorous error bounds for them are described and illustrated by concrete examples. Applications to finite element analysis are presented.

A STABLIZED GALERKIN FINITE ELEMENT METHOD FOR SOLVING THE BOUSSINESQ EQUATIONS

A. Auge and G. Lube

*Technische Universität Magdeburg, Fakultät für Mathematik,
PF 4120, D-3010 Magdeburg, Germany*

We consider the Boussinesq approximation of the stationary, incompressible and nonisothermal Navier-Stokes equations. Spurious numerical solutions of standard mixed Galerkin finite element formulations may be generated by

(a) inappropriate combinations of velocity/pressure interpolation functions and/or

(b) the presence of (locally) dominating convective terms.

As a remedy we extend the least-squares stabilization approach Franca *et al.* (1992) to the Boussinesq equations. In particular, we add weighted residuals of the basic equations to the Galerkin discretization in order to stabilize the method and to allow arbitrary finite element approximations. For a linearized problem of Oseen type, we prove asymptotic error estimates on a non-uniform mesh and give an explanation of the parameter design which is frequently used in CFD. As a basic tool we take advantage of a modified Babŭska-Brezzi condition. Furthermore, we extend existence and convergence results to the case of regular solutions of the original problem using Banach's fixed point theorem (Lube and Auge (1992)).

We conclude with numerical results for some benchmark problems.

Franca, L.P., Frey, S.L., Hughes, T.J.R. (1992), *Comp. Meths. Appl. Mech. Engrg.* **95** (1992).
Lube, G., Auge, A. (1992), Regularized mixed finite element approximations of incompressible flow problems, Preprint TU Magdeburg.

A NEW ELEMENTAL FORMULATION FOR THE DYNAMIC ANALYSIS OF STRUCTURES

S. Azimi

*IIEES, International Institute of Earthquake Engineering and Seismology,
PO Box 19395, 3913 Tehran, Iran*

Use of a new "Dynamic Mode Shape Function" (DMSF) based on the dynamic behaviour of every structural element is under study. Using this DMSF the new elemental stiffness and mass matrices are obtained. The solution of the problem for the free vibration results in exact frequencies and mode shapes of the structures. As an application the method has been applied to a bar finite element. Some examples of vibration of bar systems have been considered for clarification and the results are compared with the previous published results.

A FINITE VOLUME PROCEDURE TO PREDICT DEFORMATION AND RESIDUAL STRESS IN CASTINGS

C. Bailey, M. Cross and P. Chow

*Centre for Numerical Modelling and Process Analysis,
Department of Mathematics, University of Greenwich, London SE18, U.K.*

The foundry/casting industry plays a major part within the manufacturing base of the United Kingdom. For a new cast design the present practice is to undertake numerous trials and tests to find the optimum design options. This practice is very wasteful in both materials and lost man hours. During the casting process hot liquid metal is poured into a mould. Once the mould is filled with liquid metal, the casting together with the running system and feeders will cool and solidify. The development of a code to completely simulate all the macroscopic processes involved in castings will greatly aid the foundry engineer in reducing lead times and producing sound castings. The developed code must include the following types of analysis:

1) Fluid flow analysis for mould filling.
2) Thermal flow analysis of solidification and the evolution of latent heat.
3) Fluid Flow analysis to represent thermal convection.
4) Deformation, hence stress analysis to predict geometrical soundness of cast.
5) Macroscopic porosity formation.

There is a high degree of coupling between the above phenomena, for example the convective currents are dependent on the temperature changes. These in turn are dependent on the geometric deformation as heat losses are restricted across the cast/mould interface due to gap formation. Obviously it would be of great benefit to solve all the above using a single code. The route taken in this research is to discretise the conservation equations, representing the above phenomena, using Finite Volume procedures. This presentation concentrates on the methods used for deformation and stress and how these methods have been coupled to the flow and solidification calculations to simulate some simple castings.

ADAPTIVE METHODS FOR GRID GENERATION AND FINITE ELEMENT SOLUTIONS OF TIME-DEPENDENT PARTIAL DIFFERENTIAL EQUATIONS

M.J. Baines

Department of Mathematics,
University of Reading, PO Box 220, Reading, RG6 2AX, U.K.

A direct variational approach has been used to generate algorithms which determine optimal discontinuous piecewise linear and constant L_2 fits with adjustable nodes to a continuous function, see Baines (1993(i)). The process generates data-dependent grids. The algorithms are fast and robust, the mesh cannot tangle, and convergence can be proved under certain conditions. It can be shown that one iterate of such an algorithm corresponds to an explicit time step of the Moving Finite Element (MFE) procedure for the solution of a particular time-dependent differential equation , see Baines (1993(ii)). When the MFE procedure is run to steady state for this equation it follows a path leading to the best fit, with adjustable nodes, of the steady state operator to the zero function. The result generalises to a wider class of differential equations and operators. For first order partial differential equations it is known, see Baines (1991), that the MFE procedure approximates the method of characteristic strips. The above result shows that there also exists a superimposed speed component arising from the L_2 projection which is "best fit seeking". An alternative moving grid strategy is proposed based on a two pass method, see Baines (1992). The first pass is a fixed grid predictor: this is followed by one or two iterations of the best fit algorithm with adjustable nodes. A second pass of the solver is then made on the moving grid, the motion being provided by the above node adjustment.

Baines, M.J. (1993(i)), On algorithms for best L_2 fits to continuous functions with variable nodes, Numerical Analysis Report 1/93, Department of Mathematics, University of Reading, UK.
Baines, M.J. (1993(ii)), On the relationship between the Moving Finite Element method and best fits to continuous functions with adjustable nodes. Journal of Numerical Methods for

Partial Differential Equations (to appear).

Baines, M.J. (1991), An analysis of the Moving Finite Element method. *SIAM J. Numer. Anal.* **28**, 1323-1349.

Baines, M.J. (1992), MASTERFUL. Monotonic Adaptive Solutions of Transient Equations using Recovery, Fitting, Upwinding and Limiters. Numerical Analysis Report 5/92, Department of Mathematics, University of Reading, UK.

INFLUENCE OF EXTENSIONAL VISCOSITY ON VORTEX DEVELOPMENT IN COMPLEX GEOMETRIES

A. Baloch, P. Townsend and M. F. Webster

Department of Computer Science,
University of Wales at Swansea, Singleton Park, Swansea SA2 8PP, Wales, U.K.

In this study we are concerned with the influence of strain thickening and inertia upon the steady laminar incompressible flow of non-Newtonian fluids. We concentrate on flows through circular contraction geometries and chose to work at contraction ratios of four to one. A generalised Newtonian material description is adopted with an extensional viscosity depending on second and third invariants of the rate of deformation tensor. Particular attention is paid to the effect of the abrupt re-entrant corner by introducing in contrast a rounded corner geometry. Under such circumstances this work casts light on the formation of vortices, their origin and subsequent growth with increase in flow rate and elasticity number. Both salient and lip vortices are observed under different conditions.

The numerical scheme employed in the simulations is a semi-implicit time-stepping procedure that combines the merits of a Taylor-Galerkin weighted residual method with those of a pressure-correction method. The resulting procedure is a fractional staged scheme taken over each time step, that has already proved its flexibility in solving a range of complex incompressible viscous flows. The required extensional behaviour is incorporated through an appropriate functional form for the viscosity. Our numerical solutions are compared with other numerical predictions in the literature, and with some experimental data.

ON THE APPROXIMATION OF JUNCTIONS BETWEEN THIN SHELLS BY CURVED C^1-FINITE ELEMENT METHODS

M Bernadou

INRIA,
Domainede Voluceau, Rocquencourt 78153 Le Chesnay Cedex, France

Many thin plate problems are formulated on curved boundary plane domains. Likewise, by using a mapping technique, thin shell problems can be formulated on such curved boundary plane domains. Such mapping techniques can be extended to the junction between thin shells. This means that for all these cases, we need to approximate a fourth order problem set on curved boundary plane domains. These approximations by conforming finite element methods require the use of curved C^1-finite elements.

In this talk, we will present some results obtained in the analysis of curved C^1-finite elements and on their applications to solve the above mentioned problems.

Bernadou, M. (1991), On the approximation of junctions between plates or between shells, Computational Mechanics, Cheun, Lee and Leung (eds.), 35-45, Balkema, Rotterdam.

Bernadou, M. (1993(i)), C^1-curved finite elements with numerical integration for thin plate and thin shell problems. Part 1: Construction and interpolation properties of curved C^1 finite elements (to appear in *Comp. Meth. Appl. Mech. Engng.*).

Bernadou, M. (1993(ii)), C^1-curved finite elements with numerical integration for thin plate and thin shell problems. Part 2: Approximation of thin plate and thin shell problems, (to appear in *Comp. Meth. Appl. Mech. Engng.*).

Bernadou, M. and Boisserie, J.M. (1993(iii)), Curved finite elements of class C^1: implementation and numerical experiments. Part 1: Construction and numerical tests on the interpolation properties, (to appear in *Comp. Meth. Appl. Mech. Engng.*).

Bernadou, M. and Boisserie, J.M. (1993(iv)), Curved finite elements of class C^1: implementation and numerical experiments. Part 2: Applications to thin plate and thin shell problems. (for submission to *Comp. Meth. Appl. Mech. Engng.*).

ON THE DISCRETIZATION OF AN OPTIMAL DESIGN PROBLEM FOR AN UNILATERALLY SUPPORTED VISCOELASTIC PLATE

I. Bock and J. Lovisek

Department of Mathematics, Faculty of Electrical Engineering, Slovak Technical University, 81219 Bratislava, Slovakia

The discretization of an optimal design problem for a viscoelastic plate with respect to its variable thickness is solved. The middle surface of the plate is identified with an open bounded plane domain G with a Lipschitz boundary $\partial G = \Gamma_u \cup \Gamma_K$. The plate is clamped on Γ_u and unilaterally supported on Γ_K. Let $K \subset H^2(\Omega)$ be the closed convex set of admissible deflections. The deflection $w(e, t, x)$ is a solution of the initial-boundary value problem for a pseudoparabolic variational inequality.

$$w(e, t) \in K, \quad t \in [0, T], \tag{1}$$

$$\int \int_G e(x)^3 \Big(A_{ijkl}^{(0)} \frac{\partial w_{,ij}}{\partial t} + A_{ijkl}^{(1)} w_{,ij} \Big) (v - w(e, t, x)) \, dx \tag{2}$$
$$\geq \langle f(e, t), v - w(e, t) \rangle \quad \forall v \in K, \quad t \in [0, T]$$

$$w(e, 0) = w_0(e) \in K. \tag{3}$$

We connect with (1), (2), (3) the design problem

$$\|Dw(\bar{e}, T) - z_d\|_X^2 + j(\bar{e}) = \min_{e \in U_{ad}} \big\{ \|Dw(e, T) - z_d\|_X^2 + j(e) \big\}, \tag{4}$$

where $U_{ad} \subset C^{0,1}(\bar{G})$ is the admissible set of thickness-functions, X is any Hilbert space, $D : X \to H^2(G)$ is a linear bounded operator. Let K_{2h} and U_a^{hd} be finite element approximations of the sets K and U_{ad} respectively, $d = T/N$. Using finite differences with the step d with respect to t we approximate the problem (1) - (4) by

N elliptic optimal control problems with admissible deflections and thickness-functions from K_h and U_{ad} respectively.

A HYBRID DIRECT-ITERATIVE METHOD FOR THE SOLUTION OF FINITE ELEMENT LINEAR SYSTEMS WITH POSITIVE DEFINITE MATRICES

L. Brusa and F. Riccio

CISE spa, Casella postale 12081 1-20134 Milano, Italy

The major cost of many finite element analyses arises from the solution of large sparse sets of linear equations which can often be performed only by using vector and parallel computers. Direct methods, based on Gaussian elimination, are efficient but they can easily require prohibitively large amounts of storage, which hamper their practical application for complex engineering analyses. This drawback is avoided by using iterative schemes, mainly the conjugate gradient method, but their performance is often unsatisfactory because of the large number of iterations required for convergence, due to the ill-conditioning of the matrices. Hybrid methods, based on the use both of direct and iterative techniques, have been proposed in an attempt to exploit the advantages of the two approaches and to mitigate their drawbacks. The starting point of the methods considered in the paper is the substructure technique which arranges the elements in groups (or substructures) so that two classes of variables can be defined, i.e. the internal variables relevant to nodes within substructures, and the interface variables relevant to nodes belonging to two or more substructures. The internal variables are decoupled and are eliminated by applying a direct method to factorize the related matrices. The reduced system obtained after this step involves only the interface variables and is solved by means of the preconditioned conjugate gradient method (PCGM). In this way the efficiency of Gaussian elimination is exploited in the solution of medium-sized problems, while PCGM, applied to solve the reduced system, deals with a matrix which is better conditioned than the original one. This feature of the reduced system, as far as we know, has been experimentally verified and demonstrated only for special matrices. The paper presents a proof that the condition number of the Schur complement (the matrix of the reduced system) of a positive definite matrix is lower than the one relevant to the global matrix. On the basis of this result, indications are given to build preconditioners for the reduced system, starting from those defined for the original problem. Some of these preconditioners are tested for the solution of structural mechanics problems and the performances are compared with those obtained by applying PCGM to the complete problem. The numerical experimentation has been performed on the parallel shared memory multiprocessor ALLIAN FX/80 and future work will concern the implementation of the method on a distributed memory hardware platform.

HYBRID ARNOLDI-NEWTON ALGORITHM

T. Bulenda

Zeitlbergstraße 29, 8411 Zeitlarn, Germany

Applying a predictor-corrector scheme to the augmented set of nonlinear equilibrium equations in statics $G(u, \lambda) = 0$, we get for an inexact Newton-Raphson iteration in the corrector phase (superscripts stand for the iteration number):

$$\begin{pmatrix} {}^{i}K_T & -P \\ {}^{i}f_u{}^{T} & {}^{i}f_\lambda \end{pmatrix} \begin{pmatrix} {}^{i+1}\delta u \\ {}^{i+1}\delta_\lambda \end{pmatrix} = \begin{pmatrix} -{}^{i}R = -F({}^{i}u) + {}^{i}\lambda P \\ -f({}^{i}u, {}^{i}\lambda) \end{pmatrix} + {}^{i}r = -G + {}^{i}r \text{ with } \frac{\|{}^{i}r\|}{\|G\|} \leq {}^{i}\delta$$

$$(1)$$

with the tangential stiffness matrix ${}^{i}K_T = {}^{i}(\partial R/\partial u)$, the residual vector R, the vector of inner forces F (dependent on the vector of displacements u), the reference load vector P and the scaling factor for the loads λ. The equation $f = 0$ describes the surface of iteration and ${}^{i}f_u = {}^{i}(\partial f/\partial u)$, ${}^{i}f_\lambda = (\partial f/\partial \lambda)$. The inexact Newton method requires some coupling of the linear equation solver and the Newton iteration. This is achieved via the term ${}^{i}\delta$ which sets the residuum ${}^{i}r$ of the linear equation in relation to that of the Newton process ${}^{i}G$. The value of ${}^{i}\delta$ is chosen as a forcing sequence. In the present work we use: ${}^{i}\delta = \min(10^{-i}, 0.1\|{}^{i}G\|/{}^{i}\lambda)$.

To get an approximate solution of the linear equations, we need a proper equation solver. In this work the Arnoldi algorithm is used. It can be regarded as a generalization of the well known Lanczos algorithm to unsymmetric system matrices. The solution of our linear system is thus reduced to the repeated computation of (system-matrix)-vector products. Via preconditioning and re-orthogonalization respectively the algorithm's performance and stability is improved. As the Arnoldi algorithm can handle unsymmetric matrices, we solve directly the augmented system (1) instead of using the widely known partitioning method of Batoz and Dhatt (1979). The results of our computations show great savings in the total number of Arnoldi iterations we get by working with the augmented system instead of the partitioning method. Whether this is a saving in computer time also, depends on the structure of K_T and the cost relation for building the matrix-vector product in each step of the Arnoldi process and performing one Arnoldi step. If K_T is unsymmetric, the augmented system is favourable anyway. Then the additional effort for solving an $n + 1$-dimensional system can be neglected compared with solving a second n-dimensional system when using the partitioning method. If K_T is symmetric, the decision, which way to choose, depends on the above mentioned cost relation. The higher the cost for the matrix-vector product, the more favourable is the augmented system. For any strategy the inexact Newton method additionally produces great savings.

A FEM APPROACH FOR THE ANALYSIS OF COMPOSITE ELEMENTS

M. Buonsanti

Istituto di Ingegneria Civile ed Energetica,
Faculty di Ingenaria Univ di Reggio Calabria, 89100 Reggio Calabria, Italy

The study of the behaviour under load of heterogeneously built solids is treated by analysing the contact area of two different constituents in detail. The problem of unilateral contact with or without friction as well as the deformative adaptation that results from the application of the loads is studied. The disjunction among elements, both in the particular and the bi-dimensional case is probed. In each case we suggest a finite element model for numerical applications.

CUBIC FINITE ELEMENTS IN ELLIPTIC PROBLEMS WITH INTERFACES AND SINGULARITIES

P. Burda

*Department of Mathematics,
Czech Technical University, Karlovo n 13, 121 35 Prague 2, Czech Republic*

In applications of the finite element method to singular problems, two approaches are most widely used. In the first one makes use of some kind of refinement of the mesh near the singularity. In the second one uses additional trial functions in the neighbourhood of the singularity. In our paper we give an alternative to these techniques, based on the use of cubic polynomials in the finite element method.

COMPUTATION OF THE HOMOGENIZED BEHAVIOUR OF A GASKET

P. Cartraud, C. Wielgosz and O. Debordes

Ecole Centrale de Nantes, 1 Rue de la Noe, 44072 - Nantes Cedex, France

The present study is devoted to the determination of mechanical behaviour of a cylinderhead gasket. We consider here a gasket made of a heterogeneous material, with an iron core (perforated sheet) and a paper (non asbestos material) which is clamped by rolling on both sides of the sheet. This gasket exhibits an elastoplastic behaviour and its effective properties are computed by a homogenization technique.

The homogenization method implies a good description of the microstructure, and especially the characteristics of the gasket's components. At this stage, the main difficulty is to model the paper behaviour. This behaviour, displayed with compression tests, looks like that of soil.

We therefore use an elastoplastic model, including nonlinear and anisotropic elasticity, plasticity with two yield surfaces and isotropic hardening. One homogeneous test is used to determine the material parameters based on an analytical solution. We then compute the numerical solution of another compression test, with the finite element code SIC (Université Technologique de Compiègne), which incorporates specific developments for the numerical integration of elastoplastic constitutive laws for geomaterials. We obtain good agreement between the model and experimental results.

One can then compute the three dimensional homogenized behaviour of the gasket. We first make use of the planar periodicity of the gasket material (due to the sheet perforation) to reduce the study to that of a representative volume element (r.v.e.).

The homogenized behaviour of the gasket is then obtained from numerical simulation of elementary mechanical tests on a finite element model of this r.v.e. Here again we use the finite element code, SIC, in which a specific module of homogenization has been developed. In practice, the r.v.e. is submitted to an average strain, increasing in time, and we compute the associated average stresses. The effective properties of the gasket define the relations between the averages of strains and stresses. One can thus construct the elastoplastic evolution law of the gasket, by considering successive strains in different directions. These computations compare well with experimental tests on the gasket.

These results will enable us to incorporate the gasket behaviour more effectively into the finite element computation of an engine.

ADAPTIVE MESH REFINEMENT FOR THE
3D MAGNETIC CODE TRIFOU

P. Chaussecourte, B. Métivet and G. Nicolas

Electricité de France, Etudes et Recherches, MMN,
1 av du Général de Gaulle, 92140 Clamart, France

For some years, we have been working to develop TRIFOU, a finite element code devoted to 3D magnetic field problems. One application is the modelling of eddy current non destructive testing. In this problem, we must build a mesh for cracked bodies. This should be done carefully: an accurate representation of the induced current around the crack is necessary to achieve a good representation of the control signals. But forecasting the shape of the currents and building an appropriate mesh are often difficult. Therefore, we decided to implement an adaptive mesh refinement strategy in TRIFOU.

The Maxwell equations for time periodic situations lead to solving the following problem: find the complex magnetic field h satisfying $i\omega\mu h + \mathrm{curl}(\mathrm{curl}(h)/\sigma) = 0$, where ω, μ and σ are calculation parameters. The variational formulation of this problem can be seen as: find h in $H = \{h \in \mathrm{IL}^2(E), \mathrm{curl}(h) \in IL^2(E)\}$ such that $a(h, h') = \langle b, h' \rangle$, for all h' in H, where b belongs to the dual of H. The approximate solution, h_k, is sought in the finite dimensional space, H_k, which is connected to the mesh, M_k. We can define the residual of the discrete formulation, R_k, by: $\langle R_k, h' \rangle = \langle b, h' \rangle - a(h_k, h')$, for all h' in H. The natural norm of R_k is given by the maximum of $|\langle R_k, h' \rangle|$ for all h' in the unit sphere of H. This value is a norm of the global error, but its computation is difficult. Moreover, this value does not allow us to say where and how the mesh should be refined. Nevertheless, we shall use this idea for our adaptive mesh refinement, which can be seen as an optimisation algorithm: the successive meshes are built up in order to minimise this residual.

The mesh M_{k+1} is obtained by refining some carefully chosen areas of the mesh M_k. We define a unit refinement as the lowest increase in the approximation space H_k. An adaptive mesh iteration is divided into two stages. First, we test all the possible unit refinements of the mesh: for each of these, we enrich the approximation space H_k temporarily with the few new functions related to this virtual refinement; we evaluate the efficiency of this. In the second stage of the process, we begin by selecting the most efficient unit refinements. We then build up the H_{k+1} space, adding all the selected unit refinements to H_k.

To evaluate the efficiency of a virtual unit refinement, we consider the highest value of $|\langle R_k, h' \rangle|/\|h'\|$ for h' in the set of the new functions. We have to solve a small linear system; its rank is approximately the number of degrees of freedom which are introduced by the unit refinement.

The results of the first applications of this technique are promising, and the refinement occurs in the immediate surroundings of the cracks or the singularities.

EXTRAPOLATION METHODS FOR THE STREAMLINE DIFFUSION FINITE ELEMENT METHOD

H. Chen and R. Rannacher

*Institute of Applied Mathematics, University of Heidelberg,
Im Neuenheimer Feld 293, D-6900 Heidelberg, Germany*

The Richardson extrapolation is a classical technique for increasing the accuracy in numerical analysis and has been applied to the standard finite element method, c.f. Blum, Lin and Rannacher (1986). The purpose of this paper is the study of extrapolation methods for the streamline diffusion finite element method (the SD method) that has been developed by Hughes and Brooks (1979) for solving convection-dominated convection diffusion problems numerically. The theoretical analysis of the SD method has been started by Johnson and Nävert in (1981). Global and local error estimates in L^2-norm of order $O(h^{k+1/2})$ have been derived. These estimates are optimal for general quasi-uniform meshes. In Johnson, Schatz and Wahlbin (1987) the pointwise error behaviour was analysed for the linear finite element and an error estimate of order $O(h^{5/4}|\log(h)|^{3/2})$ was proved. An improved error estimate of order $O(h^{11/8}|\log(h)|)|$ has been obtained in Niijima (1990). Up to now, this estimate is the best pointwise error estimate. In this paper we derive error asymptotic expansions with remainders of order up to $O(h^{3+1/2}|\log(h)|^{1/2})$ for the SD method. These are the theoretical approach of the extrapolation methods for increasing accuracy. Our methods require that the exact solution is sufficiently locally smooth. In the meantime, a quasi-optimal pointwise error estimate of order $O(h^{2/3}|\log(h)|^{1/2})$ will be derived for locally uniform meshes by making use of the known superconvergence results for the finite element interpolation.

As a model problem, let us consider the following boundary value problem:

$$-\rho u_{xx} - \epsilon u_{yy} + u + x + u = f \quad \text{in } \Omega, \quad u = 0 \quad \text{on } \partial\Omega \tag{1}$$

where $\Omega \in \mathbb{R}^2$ is a bounded smooth domain and ρ, ϵ are small positive parameters. Let $S_h \subset H_0^1(\Omega)$ be a piecewise linear finite element space constructed on a quasi-uniform mesh T_h. The so-called streamline diffusion finite element approximation $u^h \in S_h$ to the problem (1) is determined through the relation

$$B_h(u^h, \chi) = (f, \chi + \rho_h \chi_x), \quad \forall \chi \in S_h, \tag{2}$$

where

$$B_h(u^h, \chi) = \rho(u_x^h, \chi_x) + \epsilon_h(u_y^h, \chi_y) + (u_x^h + u^h + \rho_h\chi_x)$$
$$= (\rho_h + \rho)(u_x^h, \rho_x) + \epsilon_h(u_y^h, \chi_y) + (1 - \rho_h)(u_x^h, \chi) + (u^h, \chi)$$

and $\rho_h = ch$, $\epsilon_h = ch^{3/2}$. For simplicity, we assume that $\epsilon \le \epsilon_h$, $\rho \le \rho_h$. Then we will derive the following error asymptotic expansions

$$u^h = u + \epsilon_{hE_1} + \rho_h\rho E_2 + h^2 E_3 + \epsilon_h\rho_{hE_4} + \epsilon_h{}^2 E_5 + \rho_h{}^2\rho E_6$$
$$+ h^2\rho_{hE_7} + O(\epsilon + h^{3+1/2}|\log(h)|^{1/2})$$

Based on this expansion, we design some extrapolation methods to gain higher accuracy.

Blum, H., Lin, Q. and Rannacher, R. (1986), Asymptotic error expansion and Richardson extrapolation for linear finite elements, *Numer. Math.* **49**, 11-37.

Hughes, T.J.R. and Brooks, A.N. (1979), A multidimensional upwind scheme with no crosswind diffusion, in Finite Element Methods for Convection Dominated Flows (T J R Hughes, ed.), AMD, vol.34. ASME, New York, pp. 19-35.

Johnson, C. and Nävert, U. (1981), An analysis of some finite element methods for advection-diffusion problems in Analytical and Numerical Approaches to Asymptotic Problems in Analysis (O Axelsson, L S Frank and A van der Sluis eds.), North Holland, Amsterdam, pp. 99-116.

Johnson, C., Schatz, A.H. and Wahlbin, L.B. (1987), Crosswind smear and pointwise errors in streamline diffusion finite element methods, *Math. Comp.* **49**(179), 25-38.

Nävert, U. (1982), A finite element method for convection diffusion problems, thesis, Chalmers University of Technology and University of Gothenburg.

Niijima, K. (1990), Pointwise error estimates for a streamline diffusion finite element scheme, *Numer. Math.* **56**, 707-719.

Rannacher, R. and Zhou, G. (1993), An adaptive mesh and pointwise error analysis for hyperbolic equations in the streamline diffusion methods, to appear.

Zhu, Q. and Chen, H. (1988), High accuracy error analysis of the linear finite element method for general boundary value problems, *Natural Sci. J. of Xiangtan Univ.*, **10**(4), 1-10.

A NEW MODEL OF JOINTED ROCK MASSES REINFORCED BY PASSIVE, FULLY-GROUTED BOLTS

S.H. Chen

*Department of Civil Engineering,
University College of Swansea, Swansea, SA2 8PP, Wales, U.K.*

Rock bolts have been used as a general reinforcement measure to improve the strength of jointed rock masses in recent decades. Several useful models have been proposed by some researchers.

This paper presents a new model for the jointed rock masses reinforced by passive, fully-grouted bolts. The model is based on an "equivalent continuum" approach. On one hand, the increment of load is shared among the bolts, rock materials, and joints respectively. On the other hand, the strain increments of the reinforced jointed rock masses are given as the sum of the increment strains of the reinforced rock materials and reinforced joints. The proposed model has an advantage in comparison to previous "equivalent continuum" models - it can calculate the bolt stresses separately in joints and rock materials. Because their deformations are different, this is a more reasonable approach to the real work state of bolts in the jointed rock masses. A computer program of the Finite Element Method has been compiled, and a typical case study of the rock slope stability problem has been made.

STRUCTURAL DYNAMIC FINITE ELEMENT ANALYSES OF SPACE SHUTTLE PROPULSION COMPONENTS

E.R. Christensen

*Sverdrup Corporation, Sverdrup Technology Inc, MSFC Group,
620 Discovery Drive, Huntsville AL35806, U.S.A.*

This paper describes three finite element structural dynamic analyses of the space shuttle main engine and solid rocket boosters which have recently been conducted for the NASA Marshall Space Flight Center in Huntsville, Alabama, USA, in support of the Advanced Turbopump Development (ATD) and Advanced Solid Rocket Motor (ASRM) programs. These analyses were performed using the Engineering Analysis Language (EAL) and COSMIC NASTRAN finite element codes. In the first analysis, three dimensional, solid finite element models (FEM) of the ATD fuel pump and LOX pump first stage turbine blades were constructed and analyzed using EAL. Since modal analysis results indicated possible resonance conditions, the transient response of the blades due to the operating pressure loads was calculated. Several transient analyses were conducted for various operating conditions with the results indicating the possibility of fatigue life problems. The second analysis involved an investigation of the fluid structure interaction between liquid hydrogen fuel and toroidal shaped inlet structure on the high pressure fuel turbopump. A FEM representing both the fluid and the structure was constructed and a modal analysis was performed. The results indicated the presence of a large number of "spurious" fluid modes occurring at nonzero frequencies. These spurious modes are a direct result of the displacement formulation of the EAL fluid elements which does not impose the constraint of irrotationality on the fluid. When these constraints were enforced through the use of a global penalty function, however, it was found that the spurious modes could be removed from the solution. The final analysis involved determining the effect of internal pressure on the dynamic characteristics of the ASTM which is currently being developed as a replacement for the solid rocket motors presently used on the shuttle. Several new NASTRAN elements have been developed to model the pressure stiffness which is a nonlinear effect due to the change in direction of the pressure force as the structure deforms. Neglecting this effect can lead to errors of up to 48% in the computed frequency of the pitch/roll mode. Prediced frequencies obtained using the new NASTRAN elements, however, agree with modal test results to within 5%.

DOMAIN DECOMPOSITION TECHNIQUES FOR MASSIVELY PARALLEL MACHINES

R.K. Coomer and I.G. Graham

School of Mathematics, University of Bath, Bath BA27 7AY, U.K.

In this talk we describe massively parallel domain decomposition methods for solving large ill-conditioned linear systems arising from discretisations of symmetric elliptic PDEs. The method has particular power when the ill-conditioning is made worse by the presence of discontinuous coefficients in the PDE. Methods based on mapping subdomains to an array of processors and local elimination of interior unknowns are reviewed. The resulting Schur complement system is solved by the preconditioned conjugate gradient method with preconditioner (in 2D) based on local solves in vertex and/or edge spaces, plus a coarse grid solve to simulate global interaction of nodes. This is an "additive Schwarz" type method, and its analysis has been recently investigated by M Dryja, O Widlund and B Smith. It is explained why this method yields an optimal algorithm (i.e. its convergence rate is independent of the number of degrees of freedom), but in general its convergence rate is affected by the jumps of the coefficients of the PDE across subdomain boundaries. On the other hand, neglecting the vertex solves yields a weakly sub-optimal algorithm but with convergence rate unaffected by these jumps. The method is implemented for a class of model problems on the MasPar MP-1, a SIMD machine with 1024 processors using inner iterations to solve the preconditioning problems. Numerical results which support the theoretical properties of the algorithms are reported.

p-VERSION PRECONDITIONING FOR THE MASS MATRIX

A. Craig

*Department of Mathematical Sciences,
University of Durham, South Road, Durham, DH1 3LE, U.K.*

The p-version of the finite element method has become more popular over the last few years. The increase in popularity is due to several factors; including the greater return in accuracy for problems with singularities on the mesh and the ease of design of p-adaptive codes. Here, instead of refining the mesh, we increase the accuracy by using higher degree basis functions. There is an observed problem which causes computational difficulties: the discrete operator from the finite element method is highly ill-conditioned. Widlund, Pasciak *et al.* . and Babuška *et al.* ., to name but a few, have all given experimental results which reduce the order of the condition number for h and p-versions of the finite element stiffness matrix from polynomial to log-squared in the number of degrees of freedom; also, numerical results suggest a similar method for the hybrid hp-version. An additional problem arises in the p-version. We observe that hierarchical bases, while being very natural bases for the p-version stiffness matrix, are very unnatural bases for the mass matrix. This can be seen by noting that the growth in the condition number is exponential in p, the degree of the polynomial on each element. We often need to solve problems using the mass matrix, in particular, time-dependent problems, but with our adaptive codes and the use of hierarchical bases, we observe that we cannot use mass-lumping to remove our ill-conditioning. Using methods derived from those for the stiffness matrix, we shall show that numerically, and in one case analytically, we can control this ill-conditioning at little expense.

NONHOMGENEOUS DEGENERATE EIGENVALUE PROBLEM

P. Drábek

Department of Mathematics, Faculty of Applied Sciences,
University of West Bohemia, Americka 42, 306 14 Plzen, Czech Republic

Let us consider the nonhomogeneous eigenvalue problem

$$\operatorname{div}(a(x,u)|\nabla u|^{p-2}\Delta u) = \lambda b(x,u)|u|^{p-2}u, \quad \text{in } \Omega, \quad u = 0 \quad \text{on } \partial\Omega. \tag{1}$$

We assume that $p > 1$ is the arbitrary real number and the Carathéodory functions $a(x,s)$, $b(x,s)$ satisfy the following hypotheses. Let $w \in L^1_{loc}(\Omega)$, $1/w \in L^{1/(p-1)}_{loc}(\Omega)$ and

$$\frac{w(x)}{c} \le a(x,s) \le cw(x)g(|s|), \quad 0 < b(x,s) \le a(x) \tag{2}$$

for a.e. $x \in \Omega$ and for all $s \in \mathbb{R}$, where $c > 0$ is a constant, $a(x) \in L^\infty(\Omega)$ and $g : \mathbb{R}^+ \to [1,\infty)$ is a nondecreasing function. We say that λ is the *eigenvalue* and $u \in W_0^{1,p}(\omega,\Omega)$ is the corresponding *eigenfunction* of the eigenvalue problem (1) if the identity

$$\int_\Omega a(x,u)|\nabla u|^{p-2}\nabla u\nabla\phi dx = \int_\Omega b(x,u)|\nabla u|^{p-2}u\phi dx$$

holds for any $\phi \in W_0^{1,p}(\omega,\Omega)$ (here $W_0^{1,p}(\omega,\Omega)$ is the *weighted Sobolev space*).

THEOREM *For a given real number $r > 0$ there exists the least eigenvalue $\lambda_r > 0$ and a corresponding eigenfunction $u_r \ge 0$ (a.e. in Ω) of the eigenvalue problem (1) such that $\|u_r\|_{L^p(\Omega)} = r$.*

Let us note that neither global results for nonlinear eigenvalue problems, nor Ljusternik-Schnirelmann theory, can be used, since the operators are not, in general, potential operators.

Let us note also that the principal part of (1) may contain the *degeneration* (or *singularity*) and it may depend on the modulus of the function u in a very general way (see (2)).

A FLUX CONTINUOUS APPROXIMATION OF THE PRESSURE EQUATION FOR H-ADAPTIVE GRIDS

M.G. Edwards

BP Research Centre,
Chertsey Road, Sunbury on Thames, Middlesex TW16 7LN, U.K.

A new flux continuous discretisation of the reservoir simulation pressure equation at an h-adaptive grid interface is presented. For regular grids it is well known that a use of a harmonic mean of the permeabilities to define the discrete pressure equation coefficients preserves flux continuity and is important in the case of discontinuous coefficients. This property is generalised to h-adaptive grid interfaces. The standard cell centred finite volume discretisation of the pressure equation leads to an $O(1/h)$ pointwise local truncation error at the adaptive grid interface. For arbitrary grid aspect ratio Ar, and interface ratio Gr, it has been shown that the coefficient of this error is proportional to the product Ar^*Gr, and this error has a severe impact on results for large Ar, see Edwards (1992).

A correction implemented in Edwards (1992) leads to a non-symmetric matrix, conditional diagonal dominance and increased support of the scheme resulting in increased complexity in the connectivity of the pressure matrix. These difficulties are handled by implementing the correction explicitly following Quandalle and Besset (1985).

In this paper a new correction is presented which removes the local leading $O(1/h)$ error. The new correction uses the same support as the standard scheme and symmetry of the pressure matrix is maintained. The correction is constructed such that the flux is continuous at all adaptive grid interfaces, which in turn results in an unconditionally diagonally dominant pressure matrix for isotropic coefficients. The simplicity of the new correction enables a fully implicit and consistent implementation of the scheme.

Finally the benefits of the new correction are demonstrated with results for the general case of a dynamically varying h-adaptive grid which tracks a moving high resolution shock front.

Edwards M.G. (1992), A dynamically adaptive Godunov scheme for reservoir simulation on large aspect ratio grids. Conference on Numerical Methods for Fluid Dynamics, 7-10 April, Reading University.

Quandalle P. and Besset P. (1985), Reduction of grid effects due to local sub-gridding in simulations using a composite grid, SPE 13527.

PARALLEL IMPLEMENTATION OF THE hp-VERSION OF THE FINITE ELEMENT METHOD ON A SHARED-MEMORY COMPUTER

H.C. Elman

Department of Mathematics, UMIST, Manchester M60 1QD, U.K.

We study the costs of solving two-dimensional elliptic partial differential equations discretized by the hp-version of the finite element method, on a shared memory parallel computer. The computational algorithm consists of elimination of "high order" unknowns associated with element interiors, followed by preconditioned conjugate gradient solution of the global problem on element boundaries, using the submatrix associated with nodal unknowns as a preconditioner. We systematically examine costs in CPU time of the individual steps of the computation, including construction of the local stiffness matrices. Our general observations are that costs of the "naturally" parallel computations associated with local elements are significantly higher than any global computations, so that the latter do not represent a significant bottleneck to parallel efficiency. However, in order to avoid inefficiency caused by hierarchical computer memories, it is necessary to organize the local operations into blocks of computations. We show how to do this effectively using the Level 3 BLAS linear algebra kernels.

NUMERICAL APPROXIMATION AND ASYMPTOTIC PROPERTIES OF SOME TWO DIMENSIONAL PLATE MODELS

R. Falk

Department of Mathematics, Rutgers University, New Brunswick, New Jersey, U.S.A.

The finite element approximation of some two dimensional plate models derived from mixed variational principles is studied. A crucial element in the derivation of error estimates is to first understand the asymptotic properties of the models. In particular, we consider the boundary layer behaviour and the dependence of the regularity of the solution on the plate thickness. Differences between minimum energy models and complementary energy models are highlighted.

THE UNCOUPLING OF BOUNDARY INTEGRAL AND FINITE ELEMENT METHODS FOR NONLINEAR BOUNDARY VALUE PROBLEMS

G.N. Gatica* and G.C. Hsiao**

**Departamento de Matematica,*
Universidad de Concepcion, Casilla 3-C, Concepcion, Chile

***Department of Mathematical Sciences, University of Delaware,*
501 Ewing Hall, Newark, Delaware 19716, U.S.A.

Recently, the usual coupling of boundary integral and finite element methods has been modified to yield simpler variational formulations for linear exterior boundary value problems. This simplifed procedure is based on choosing the artificial coupling boundary as a circle or a sphere, which allows us to invert the boundary integral operators exactly. This leads to the so called *uncoupling technique*, by means of which the resulting weak formulations reduce to almost the same as those arising from Neumann boundary value problems on bounded domains. In fact, the only boundary integral operator appearing in the formulations is the one provided by the *simple layer potential*. The purpose of this paper is to apply the uncoupling procedure to study the weak solvability of certain nonlinear exterior boundary value problems. As a model we consider a nonlinear second order elliptic equation in divergence form in a bounded inner region, which becomes the Laplace equation in the corresponding unbounded exterior region. We provide sufficient conditions for the nonlinear coefficients from which existence, uniqueness and approximation results are established. In particular, nonlinear equations yielding both monotone and non-monotone operators are analyzed. Also, the effect of numerical integration, and the polygonal approximation of the coupling boundary, are taken into account to perform the error analysis of the discrete scheme.

NEW HIGH ORDER INFINITE ELEMENTS WITH DIFFERENT DECAY TYPES

L. Gavete, C. Manzano and A. Ruiz

Catedra de Calc Num e Informatica, Escuela Tec Superior de Ing de Minas, Universidad Politecnica de Madrid, Rios Rosas 21, 28003 Madrid, Spain

A class of problems of interest in mathematical physics and engineering is characterized by partial differential equations on unbounded domains. For example, the irrotational flow of an incompressible fluid exterior to a body in $\mathbb{R}^3$ is described by Laplace's equation in the exterior domain, Ω, with no flow through the body boundary, Γ, and uniform flow at infinity. Similarly, in acoustics and electromagnetism, Helmholtz's equation describes the exterior scattering from a body in $\mathbb{R}^3$. In this case, the "boundary" condition designates the decay far field rate of the solution at infinity. Existence and uniqueness of solutions have been obtained by several authors using various techniques, but the treatment of unbounded domains has presented problems for the finite element analyst. Discretizations are typically extended to large distances in attempts to minimize inaccuracies. Unfortunately, coverage of this extensive domain results in CPU costs and storage penalties.

The focus of the present work is the numerical solution of partial differential equations by using infinite elements instead of finite domain truncation. Infinite elements are becoming increasingly popular as an economical means of extending the finite element method to deal with unbounded domains. Various approaches have been adopted in the extension of the method, and these techniques tend to merge into each other. In this paper, new high order mapped infinite elements are investigated.

We conclude that in order to solve efficiently a partial differential equation with an unbounded exterior domain, it is necessary to consider the following:

(a) The type of infinite element to be used and its control.
(b) The location of the interface between finite and infinite elements.
(c) Alternative approaches to Finite Element Formulation (only in some cases).

These questions can be solved by using high order mapped infinite elements with the appropriate decay type. Also we consider the case that it is necessary to employ a Petrov-Galerkin method. The advantages of using the new approach proposed in this paper are shown with different examples of solutions of partial differential equations on unbounded domains.

USING EXPLICIT PRECONDITIONED SCHEMES FOR
SOLVING INITIAL/BOUNDARY VALUE PROBLEMS

G.A. Gravvanis and E.A. Lipitakis

Airforce Academy, Decelia Atticis, Greece

A new class of hybrid time implicit-explicit approximating schemes combined with Approximate Inverse Finite Element Matrix (AIFEM) techniques and explicit preconditioning semi-direct methods for the numerical solution of initial/boundary value problems is presented. The AIFEM techniques based on the concept of adaptable LDL^T factorization procedures have been recently introduced for computing explicit pseudoinverses of large sparse symmetric matrices of irregular structure, derived from the FE discretization of elliptic and parabolic partial differential equations without inverting the corresponding decomposition factors. This category of equations represents a large class of commonly occurring problems in Mathematical Physics and Engineering.

The AIFEM techniques originated from the known algorithmic procedures of inverting a real $(n \times n)$ matrix A, which can be decomposed by adaptive approximate factorization methods (i.e. $A = L_r D_r L_r^T$, where D_r is a diagonal matrix and L_r is a sparse lower triangular matrix of the same profile as A), by computing explicitly the elements of the inverse A^{-1}, without inverting the decomposition factors. The effectiveness of the Explicit Preconditioned iterative methods using these AIFEM techniques for solving numerically initial/boundary value problems is related to the fact that the exact inverse of the original sparse coefficient matrix A (although full) exhibits a similar "fuzzy" structure to A, i.e. the largest elements are clustered around the principal diagonal and m-diagonal where m is the semi-bandwidth of A.

It should be noted that an important feature of the AIFEM algorithmic techniques is the provision of both explicit preconditioning direct and iterative methods for solving partial differential equations in two and three space variables. Additional facilities are provided by the choice of "fill-in" (factorization) and "retention" (pseudoinverse) parameters that allow the best method for a given problem to be selected. Additionally a class of composite Picard/Newton methods combined with Approximate Inverse Finite Element Matrix (AIFEM) techniques and explicit preconditioning semi-direct methods for the numerical solution of non-linear elliptic partial differential equations is presented.

USE OF COMPUTER ALGEBRA SYSTEMS TO
GENERATE ELEMENT STIFFNESS MATRICES

D.V. Griffiths

*Department of Engineering,
University of Manchester, Oxford Road, Manchester, M13 9PL, U.K.*

Finite element matrices such those for stiffness and mass are usually generated using Gaussian quadrature because it leads to convenient formulations in terms of local coordinates. This paper describes how element matrices can be generated in *closed form* with the help of Computer Algebra Systems (CAS) and how this can lead to improvements in run-times. The CAS approach is also shown to be able to generate matrices which would normally be obtained using reduced or selective reduced integration (SRI).

Consider a square 4-node elastic element in plane strain, fixed on two adjacent sides, with a horizontal force P applied to the free corner. Computer Algebra can be used to compute the horizontal deflection of the corner for various element stiffness integration schemes. The expressions show why locking occurs as the material approaches incompressibility with full integration.

Integration	δ_H	$\delta_H(\nu \approx 0.5)$
FULL	$\dfrac{96(3-4\nu)(1+\nu)(1-2\nu)P}{(9-16\nu)(15-16\nu)E}$	0
SRI(λ,μ)	$\dfrac{16(2-3\nu)(1+\nu)P}{3(5-6\nu)E}$	$\dfrac{2P}{E}$
SRI(G,K)	$\dfrac{144(17-25\nu)(1+\nu)P}{25(43-50\nu)E}$	$\dfrac{2.16P}{E}$
REDUCED	$\dfrac{(3-4\nu)(1+\nu)P}{(1-\nu)E}$	$\dfrac{3P}{E}$

NEWTON'S METHOD FOR OBSTACLE PROBLEMS

C. Grossmann

Department of Mathematics, Kuwait University, PO Box 5969, Safat 13060, Kuwait

The obstacle problem can be considered as a variational problem with inequality constraints. The discretization by piecewise linear finite elements leads to large scale optimization problems with a special structured objective functional and with simple bounds for the variable as constraints. In this investigation we include the constraints into an auxiliary objective functional by means of a penalty technique. The unconstrained problems obtained are nonlinear variational equations. These can be solved by Newton's method. As a well known fact the auxiliary problems generated in penalty techniques are ill conditioned in the limit. Thus an important question is to estimate the area of contraction of Newton's method independent of the penalty and of the discretization parameter. In the first part we summarize some basic facts on obstacle problems, their discretization and on an adapted penalty technique. Later we derive optimal parameter selection rules to adjust the error caused by the penalty terms to the same magnitude as the discretization error of the finite element approximation. Finally, we investigate the convergence behaviour of Newton's method applied to the variational equality obtained. Here the area of contraction of Newton's method is estimated independent of the discretization parameter and the penalty parameters by means of maximum principles for elliptic problems. Finally, we adjust the iterations on each level of discretization and investigate the method on families of grids.

LAPACK: HIGH PERFORMANCE SOFTWARE FOR LINEAR ALGEBRA

S. Hammarling

NAG Ltd, Wilkinson House, Jordan Hill Road, Oxford, OX2 8DR, U.K.

NAG Ltd and US Scientists have collaborated in the development of LAPACK, a high performance linear algebra package, already proven for shared memory parallel machines, now being extended to distributed memory machines.

FINITE ELEMENT METHODS FOR GENERAL SECOND-ORDER ELLIPTIC PARTIAL DIFFERENTIAL EQUATIONS: MODEL PROBLEMS

I. Harari* and T.J.R. Hughes**

**Department of Solid Mechanics, Materials and Structures,*
Faculty of Engineering, Tel Aviv University, Ramat Aviv 69978, Israel

***Division of Applied Mechanics, Durand Building, Stanford University,*
Stanford, California 94305, U.S.A.

Finite element methods for second-order elliptic partial differential equations were initially applied to problems which contain only derivatives of the highest order. In such cases finite elements based on the Galerkin method possess several advantageous attributes, with guaranteed performance on configurations of practical interest. The addition of other terms such as *first-order* derivatives in advective-diffusive problems and *undifferentiated* terms in the Helmholtz equation has potentially destabilizing effects. The concept of Galerkin/least squares (GLS) obtained by appending terms in *least squares form* to the standard Galerkin formulation enables the design of stable methods for such applications, retaining the higher-order accuracy of the Galerkin method for problems governed by the Laplacian, at relatively low added computational cost.

General second-order equations, which are studied in this work, have exact solutions which may differ widely in nature depending on values of the coefficients, with rapidly varying values in domain interiors as well as close to boundaries. An analysis of numerical solutions obtained by the Galerkin method employing linear finite elements indicates that relatively fine meshes may be required to retain an acceptable degree of accuracy for certain ranges of values of the physical coefficients. Numerous criteria for improving accuracy by employing GLS technology are examined. Of these, several are found to provide the lowest error in nodal amplification factors, each in a certain range of ratios of physical coefficients. Guidelines for required mesh resolutions are presented for Galerkin and GLS, showing that in large parts of the parameter space substantial savings may be obtained by employing GLS. Numerical computations bear out these observations.

THE FOURIER-FINITE ELEMENT METHOD FOR ELLIPTIC PROBLEMS WITH AXISYMMETRIC EDGES

B. Heinrich

*Fachbereich Mathematik, Technische Universität Chemnitz,
Postschließfach 964, 9010 Chemnitz, Germany*

The Fourier-finite element method is a semi-analytic technique for the approximate solution of three-dimensional elliptic problems in axisymmetric domains with nonaxisymmetric data. The method employs truncated Fourier series with coefficients which are the solutions of a finite number of two-dimensional elliptic problems posed on the meridian plane of the domain and approximated by the finite element method. This approach is often used by engineers for thermal and stress analysis of axisymmetric bodies and can be implemented on a parallel computer.

In this paper we present results on the mathematical justification of the Fourier-finite element approach for solving problems in 3D, especially for Poisson's equation, the description of the properties of the approximation, local mesh refinement in the meridian plane for an appropriate treatment of singularities near re-entrant edges and error estimates in the H^1- and L_2-norms, with respect to the discretization parameters N and h (Fourier and finite element approximation) for solutions belonging to Sobolev spaces. Further, algorithmic aspects and the computer implementation of the method are discussed.

NODAL FINITE ELEMENT METHODS FOR PARTIAL DIFFERENTIAL EQUATIONS

J.P. Hennart

IIMAS-UNAM (Mexico), Apdo Postal 20-726, 01000 Mexico DF, Mexico

Modern nodal methods were developed in the 1970's in numerical reactor calculation. Roughly speaking, nodal methods are fast and accurate methods which try to combine the attractive features of the finite difference method (FDM) and the finite element method (FEM). From the FEM, they borrow a piecewise continuous, usually (but not always) polynomial, behaviour over a given coarse mesh. With the FDM, they have in common the fact that the final algebraic systems are usually quite sparse and well structured, *at least over domains which are not too irregular such as unions of rectangles*. Nodal methods can thus be viewed as fast solvers, to which techniques of vectorization and (or) parallelism are applicable. They are especially suited to all physical situations which have been traditionally modelled by finite differences, and for which the domain as well as the coefficients of the equation(s) are not known with sufficient accuracy. It is then tempting to use a rectangular grid to discretize the domain and to assume that over each cell of the grid the physical parameters are not known with great accuracy, so that a constant or mean value only is available. This situation is prevailing for instance in groundwater hydrology, oil reservoir simulation, and air pollution modelling problems.

In this paper, we show that most nodal schemes can be viewed as finite element schemes provided the FEM is viewed in a fairly liberal way with moments (and not point values) as principal unknowns, with piecewise polynomials (or quasi-polynomials) as basic functions and with a systematic use of variational crimes like numerical quadrature and nonconformity.

These ideas were applied to the three basic types of PDE's, elliptic, parabolic, and hyperbolic, with concrete applications (and numerical results) in multidimensional diffusion, space-time kinetics and 2D transport of neutrons in nuclear reactors, as well as for benchmark problems of the types mentioned above, including the Poisson equation and the heat diffusion equation.

FINITE ELEMENTS FOR EXTENDED VARIATIONAL FORMULATION IN NONLINEAR PROBLEMS

Z. Kestřánek

ČKD Intech, Freyova 27, Prague 9, Czech Republic

Nonlinear problems such as the elastohydrodynamic lubrication problem, involving nonconservative mechanical systems, are of considerable importance in the field of hydro- and aero-elasticity and space mechanics. These problems involve nonsymmetric operators, and therefore variational formulations in a strict sense are not available. The variational formulation presented in Tonti (1984) and Kestřánek (1990) is called extended in the following sense: Given a problem $N(u) = o_v$ with a nonlinear operator $N : D(N) \subset U \to R(N) \subset V = U^*$, find a functional $\hat{F}$, if any, whose critical points are solutions to the problem and vice versa. This implies that for a given operator N an operator $\hat{N}$ exists such that $\delta\hat{F} = \left\langle \hat{N}(u), \delta u \right\rangle$ and the problems $N(u) = o_v$ and $\hat{N}(u) = o_v$ have the same solution. This statement requires only that the critical points should coincide with the solutions, without posing the additional requirement that N must be the gradient of the functional. The problem can be rewritten as $\hat{N}(u) = N_u'^*(u; KN(u)) = o$, provided that $D(K) \subset R(N)$, $R(K) \subset D(N_u'^*)$, where K is a linear, invertible and symmetrical operator. The modified operator $\hat{N}$ satisfies the symmetry condition for the Gâteaux derivative, $\hat{N}_u' = \hat{N}_u'^*$ and it follows that there exists a potential functional $\hat{F}[u] = 1/2 \left\langle N(u), KN(u) \right\rangle$ whose gradient is the operator $\hat{N}$. It is worth noting that the operator K is not uniquely determined and the practical construction of K for the general case presents some difficulties. In numerical applications, however, K can easily be defined. The functional vanishes when the solution is reached. Moreover if K is positive definite, then $\hat{F}[u]$ is minimum at the critical point. The functional $\hat{F}$ can be defined over finite-dimensional subspaces S. The kernel $k(s,t)$ of the linear operator K in integral form is naturally defined over $S \times S$ and may be expressed as $k(s; t) = \{L(s)\}^T [A]\{L(t)\}$, where $[A]$ is a symmetric constant matrix and $\{L\}$ is a vector of the shape functions. Here the choice of the kernel in a finite subspace, generated by means of functions with local compact support, makes it possible to provide finite element models in a straightforward way. The main novelty in this standard procedure is a complication arising from the integro-differential nature of the basic equation. To find the minimum of the functional $\hat{F}$ we can use a Newton-like type algorithm. This method for the solution of the inverse problem has been successfully developed for special numerical applications.

Tonti, E. (1984), Variational formulation for every nonlinear problem, *Int. J. Eng. Sci.* **22**, 1343-1371.

Kestřánek, Z. (1990), Finite elements in the extended variational formulations for nonlinear problems, Proceedings, Congress WCCM II, Stuttgart.

A PARALLEL SEMI-DISCRETE GALERKIN SPLITTING METHOD FOR SECOND-ORDER HYPERBOLIC EQUATIONS

A.Q.M. Khaliq

Department of Mathematics, College of Arts and Sciences, Western Illinois University, 476 Morgan Hall, Macomb, Illinois 61455, U.S.A.

Second-order hyperbolic partial differential equations (PDE's) in multi-space dimensions occur frequently in viscoelastic theory Larsson *et al.* (1991), acoustics Lee *et al.* (1988), and in many geoscience applications, for example seismology Fitzgibbon and Wheeler (1992 (i) and (ii)) and Larsson (1991). Computational techniques are proposed by several authors, see for example Fitzgibbon and Wheeler (1992 (i) and (ii)), Mullen and Belytshko (1982) and Seron *et al.* (1990) and references therein, for the numerical solution of the wave equation by finite-element methods. Their investigations have been restricted to mainly, two-dimensional wave equations and implementation on serial machines. Recently an explicit finite-difference scheme has been proposed by Mufti (1990), for the numerical solution of large-scale three-dimensional seismic models. Since analytical methods also are not effective because they are restricted to simple geometries with homogeneous structures like the finite-difference methods, there is an ever growing need to develop efficient and stable finite-element methods which may be implemented effectively on modern parallel computers.

A new parallel method for solving hyperbolic PDE's in many space dimensions is presented by applying the finite-element technique, and is unconditionally stable. The proposed method is based on the semi-discrete Galerkin approach utilizing Strang-type splitting techniques in which a multi-dimensional problem is reduced to a series of one-dimensional problems. This approach avoids the difficulty of constructing a finite element space for multi-dimensional problems. A large scale problem requiring large storage and CPU times is thus solved using only a one-dimensional finite-element method an MIMD parallel machine. This approach is seen to be highly competitive to those presented in Seron *et al.* (1990). Second order convergence and error estimates of the scheme are given using piecewise linear basis functions. However, it is anticipated that cubic B-splines may also be used as basis functions and their convergence may be proved by semi-group theory. Numerical results are demonstrated on problems from the literature. The proposed parallel algorithm for the numerical solution of the wave equation will provide geophysicists with a very useful and efficient tool for predicting and understanding seismic wave propagation. It will also be useful to engineers because of its efficiency for solving large scale problems in acoustics and in numerical approximation to the solution of the linear elastodynamic equations.

Fitzgibbon, W.E. and Wheeler, M.F. (eds) (1992(i)), *Modeling and Analysis of Diffusive and Advective Processes in Geosciences*, SIAM, Philadelphia.

Fitzgibbon, W.E. and Wheeler, M.F. (eds) (1992(ii)), *Computational Methods in Geosciences*, SIAM, Philadelphia.

Larsson, S., Thomee, V. and Wahlbin, L.B. (1991), Finite-element methods for a strongly damped wave equation, *IMA J. Num. Anal.* **11**, 115-142.

Lee, D., Sternberg, R.L. and Schultz, M.H. (eds) (1988), *Computational Acoustincs (wave propagation)*, North-Holland, New York.

Mufti, R.I. (1990), Large-scale three-dimensional seismic models and their interpretive significance, *Geophysics*, **55**, 1166-1182.

Mullen, R. and Belytshko, T. (1982), Two dimensional wave equation, *Int. J. Num. Meth. Engng.* **18**, 11-29.

Seron, F.J., Sanz, F.J., Kindelan, M. and Badal, J.I. (1990), Finite element methods for elastic wave propagation, *Comm. Appl. Num. Meth.* **6**, 359-368.

ON THE STABILITY OF LAYERED CIRCULAR RINGS

B. Kovács* and F.J. Szabö**

**University of Miskolc, Institute of Mathematics,*
***University of Miskolc, Department of Machine Elements,*
H3515 Miskolc, Egyetemváros, Hungary

The purpose of this paper is to find the critical external pressure of layered circular rings buckling in their planes. The composite ring is subjected to normal gas pressure and constant-directional pressure. The stability of an arch is analysed by formulating the governing differential equations and the associated boundary conditions using a variational principle. The cross-sections in individual layers remain planes after deformation.

In the case of free rings, the critical external pressures are investigated, using a theoretical method, the COSMOS/M and the SYSTUS finite element software packages.

As a theoretical investigation the eigenproblem is solved by a successive approximation based on the Schwarz-method and by the Ritz-method, reducing the problem to a matrix-eigenvalue problem.

For the finite element solution the model of the ring is built up using four-node composite shell elements and the numerical results are obtained by the Inverse Power iteration technique in case of COSMOS/M and by the Lanczos algorithm in SYSTUS.

The analysis of the numerical results shows that in the case of free rings the constant directional pressure gives higher critical values than the hydrostatic pressure and one can find good agreement between the theoretical and finite element solutions.

FREE VIBRATIONS OF LAYERED CIRCULAR RING SEGMENTS

B. Kovács and F.J. Szabö

Address as above.

The free vibration eigenvalue problem of circular arc shaped three layered sandwich ring segments was investigated theoretically and numerically in the case of linear elastic layered materials. The comparison of the different solution methods makes it possible to study the plane bending free vibration behaviour of the curved layered beams more accurately.

The free vibration eigenvalue problem of the ring segments was solved theoretically by using the Schwarz-method, which is a fast and flexible, easy-to-program algorithm. This algorithm can be easily completed by an optimization method and can be applied for some special optimization problems, which cannot be handled using the built-in optimization of the known finite element programs (multiobjective optimization, discrete variables, complicated constraints, etc.).

The results of the theoretical solution were compared with those obtained by a finite element analysis made with the SYSTUS finite element program which runs on a VAX 3100 workstation under VMS. For the eigenvalue solution the LANCZOS method was used. Two different finite element models were built up: -The first model contains eight-node hexahedron brick elements and this makes it possible to apply several elements along the thickness, which increases the accuracy of the model. -The second model consists of four-node composite shell elements, which results in a very easy-to-solve finite element model which needs much less storage and computation capacity than the previous one.

Numerical results for the theoretical and finite element analysis were made for the circular arc with radius varied in the ranges between $100mm$ to $200mm$, with the same thickness and width data, in the case of two different core materials. (Although both of the first four bending eigenvalues show very close agreement between the theoretical and finite element results, which is better in case of long (length of $150mm$ or more) beams. In the case of very short (length of $50mm$ or less) beams there may be a considerable difference between the results, so for these structures it could be necessary to investigate the problem experimentally, too. Comparing the two different finite element models, one can say that the model containing hexahedron elements needs more computing and storage capacity but gives more accurate results than composite shell elements. Because of the possibly long calculation time in the case of complicated, long solution procedures (nonlinear or transient dynamic analysis containing many time cards) the composite shell elements are more suitable.

FINITE ELEMENT APPROXIMATION OF A NONLINEAR ANISOTROPIC HEAT CONDUCTION PROBLEM

M. Křížek

Matematicky Üstav CSAV, Zitna 25, 115 67 Prague 1, Czech Republic

We prove the existence and uniqueness of a solution of the following second order quasilinear elliptic problem of a divergence form with nonlinear mixed boundary conditions:

$$\mathrm{div}(A(\boldsymbol{x}, u)\mathrm{grad}(u)) = \boldsymbol{F}(\boldsymbol{x}, u) \quad \text{in } \Omega,$$

$$u = \bar{u} \quad \text{on } \Gamma_1,$$

$$\boldsymbol{n}^t A(s, u)\mathrm{grad}(u) + \boldsymbol{G}(s, u) = 0 \quad \text{on } \Gamma_2$$

where $\Omega \subset \mathbb{R}^d$, $d \in \{1, 2, ...\}$, is a bounded domain with a Lipschitz boundary $\partial\Omega$, n is the outward unit normal to the boundary, Γ_1, Γ_2 are respectively open and disjoint subsets of $\partial\Omega$ and $\bar{\Gamma}_1 \cup \bar{\Gamma}_2 = \partial\Omega$, $\bar{u} \in W_2^1(\Omega)$ and $\boldsymbol{A} = (a_{ij})_{i,j=1}^d$ is a uniformly positive definite matrix function. We further assume that $F(x, .)$ and $G(s, .)$ are nondecreasing functions for almost every $x \in \Omega$ and $s \in \Gamma_2$ (in particular, they can be independent of u). We show that the problem is nonpotential in general and also that the theory of monotone operators cannot be applied. Sufficient conditions from Hlaváček, Křížek and Malý (1993) which guarantee the uniqueness of the classical and weak solutions are introduced. Note that there exist examples of non-unique solutions if the equation is not in divergence form (see e.g. Meyers (1963)). We also give a finite element approximation of the problem and present some results concerning the existence and uniqueness of the Galerkin solution. We then introduce two convergence theorems of Hlaváček, Křížek and Malý (1993). A practical example of computing a temperature field in the magnetic cores of transformers is shown.

Hlaváček, I., Křížek, M. and Malý, J. (1993), On Galerkin approximations of a quasilinear nonpotential elliptic problem of a nonmonotone type, (submitted to *J. Math. Anal. Appl.*).
Meyers, N.G. (1963), An example of non-uniqueness in the theory of quasi-linear elliptic equations of second order, *Arch. Rational Mech. Anal.* 14, 177-179 (1963).

A VARIATIONAL INEQUALITY FORMULATION FOR FLOW THEORY PLASTICITY

M.S. Kuczma* and J.R. Whiteman**

Institute of Building Structures,
Technical University of Poznań, 60-965 Poznań, Poland

**BICOM, The Institute of Computational Mathematics,*
Brunel University, Uxbridge, UB8 3PH, U.K.

We consider the quasi-static deformation process of an elasto-plastic body, under the assumption of small strains. The elasto-plastic behaviour of the material is assumed to be governed by the von Mises yield function with linear strain hardening (softening). The yield function is written in terms of strains, and we define the incremental problem as the system that consists of (a) a variational equation being the equilibrium condition for the body and (b) a variational inequality expressing the unilateral character of the plastic strain-rate multiplier. An iterative method for solving this non-linear system is proposed, and results of some numerical experiments based on the FEM-approximation are provided.

A BE-FORMULATION FOR FINITE STRAIN ELASTOPLASTICITY

G. Kuhn

Lehrstuhl für Technische Mechanik,
Universität Erlangen-Nurnberg, Egerlandstre 5, W-8520 Erlangen, Germany

An extension of the boundary-element-method to finite strains and to constitutive relations based on the concept of an intermediate configuration is presented. Within this concept material behaviour is described in terms of a hyperelastic relation relative to the intermediate configuration together with evolution equations for this configuration and further internal variables. In the context of numerical solution techniques this enables an operator split leading to a global hyperelastic problem and local time integration schemes for the internal variables. The proposed BEM-approach allows for arbitrary hyperelastic relations relative to the intermediate configuration and leads to a nonlinear set of equations with displacement gradients as basic unknowns. As a Total-Lagrange scheme is adopted the expensive computation of the system matrices has to be done only once in the initial configuration. Nevertheless the nonlinear set of equations can be consistently linearised with respect to the local integration schemes. Constitutive equations, integration schemes and linearisation techniques used in FEM can be directly applied. The BEM-formulation is presented for associated rate independent elastoplasticity using a yield condition in strain space. Some numerical results will be shown.

ADAPTIVE MESH REFINEMENT IN THE FINITE ELEMENT METHOD

G. Kunert

Humboldtstr 31, 0-9200 Freiberg, Germany

The rate of convergence of the finite element method can be improved by applying adaptive mesh refinement. This has been investigated for different kinds of error indicators, and numerical experiments performed for the Laplace operator are described.

THE FACTORIZATION OF UNSTRUCTURED SPARSE SYMMETRIC POSITIVE DEFINITE MATRICES ON DISTRIBUTED MEMORY MIMD PARALLEL COMPUTERS

R. Levkovitz and G. Mitra

*Department of Mathematics and Statistics,
Brunel University, Uxbridge, UB3 3PH, U.K.*

The efficient solution of unstructured large sparse symmetric positive definite equations on a distributed memory computer using direct methods is an important and challenging area of parallel computing research. We concentrate on the design and implementation of a parallel Cholesky factorization algorithm and focus the discussion on the distribution of data, efficient distribution of the computation and the aggregation of the communication. We show that the accumulation of non-zeros in certain areas of the Cholesky factor lends itself to a hybrid sparse-dense computation paradigm and demonstrate how to extend this approach to a distributed memory parallel MIMD computer.

THE APPLICATION OF THE FINITE ELEMENT METHOD TO THE SOLUTION OF NONLINEAR PARTIAL DIFFERENTIAL EQUATIONS AND A COMPARISON TO THE FINITE DIFFERENCE METHOD

H. Maisch, H. Karl and G. Lehner

*Institut für Theorie der Elektrotechnik, Universität Stuttgart,
Pfaffenwaldring 47, 7000 Stuttgart 80, Germany*

Solitons are special solutions of certain nonlinear partial differential equations with increasing importance in plasma physics, hydrodynamics, electrical engineering and other fields. In this paper we have studied the applicability of the finite element method to solve soliton equations. As the most important and famous example we have solved the Korteweg-de Vries equation (KdV) using Hermite elements. Here we have studied the interaction of solitons among themselves and also the generation of solitons from arbitrary initial conditions. The results agree very well with theoretical predictions. In the case of soliton generation there are often no analytical results available, therefore we have checked the three lowest order conservations laws. While for the Korteweg-de Vries equation the continuity of the first derivative is absolutely necessary, for most other equations continuity conditions are less stringent and Lagrange type elements are sufficient. Some numerical results for the regularized long wave equation (RLW), the Sine-Gordon equation (SG), nonlinear Schrödinger equation (NLS) and the Zakharov equations are also presented briefly. The numerical solution of two dimensional soliton equations is now also possible due to increasing computer power. We have chosen the two dimensional Sine-Gordon equation as an example. For comparison we solved the Korteweg-de Vries equation and the two-dimensional Sine-Gordon equation also by using finite differences, especially the method of lines. It turned out, that the more flexible finite element method definitively performed better in both cases.

BALANCING DOMAIN DECOMPOSITION PRECONDITIONERS AND DISCONTINUOUS COEFFICIENTS

J. Mandel

University of Colorado at Denver, Denver, 80217-3364, U.S.A.

This talk presents the formulation and experience with several preconditioners based on solving, in each iteration, a local problem on each subdomain coupled with a coarse level problem. This provides for an efficient global propagation of the error and guarantees that the possibly singular local problems, solved in every iteration, are consistent.

The preconditioner is based on solving in each iteration possibly singular problems based on the local stiffness matrices of the subdomains. Is it proved that the condition number remains bounded for arbitrary size jumps of coefficients between subdomains in any number of dimensions, is bounded independently of the number of subdomains, and grows only as the logarithm of subdomain size. Computational experiments confirm the theory and show that the method is remarkably robust and performs very well for general strongly discontinuous coefficients as well as unstructured subdomains. Possible modifications for shells and plates are also discussed.

FAST ITERATIVE SOLVERS FOR p-VERSION SOLIDS, PLATES, AND SHELLS

J. Mandel

University of Colorado at Denver, Denver, 80217-3364, U.S.A.

We present the formulation of and experience with several methods for the iterative solution of large-scale systems of algebraic linear equations arising from the p-version finite element method in three dimensions. The basic idea of the method is to use domain decomposition approaches with each element considered to be a subdomain. The resulting iterative methods have been shown to be superior to state-of-the-art direct solvers for real-world problems with severely distorted solid elements, such as engine crankshaft, complete aircraft fuselage, and stiffened shell panel with a window rim. The data have been generated by the packages MSC/PROBE and STRIPE.

The method is essentially block preconditioning, with one block based on selected lower degree functions and other blocks associated with vertices, edges, and faces of the elements.

A bound on the condition number independent of the number of elements is proved, and reasonably low growth of the condition number has been observed in practice, including thin elements, i.e., plates and shells. The selection of the specific preconditioner adapts itself to the local characteristics of the problem with no user intervention required.

FINITE ELEMENT ANALYSIS OF TURBULENT FLOW THROUGH AN ORIFICE METER WITH A CONCENTRIC OBSTRUCTION OF CONICAL SHAPE AT ITS CENTRE

M.R. Mokhtarzadeh-Dehghan, D.J. Stephens and L. Yu

Department of Mechanical Engineering, Brunel University, Uxbridge, UB8 3PH, U.K.

Orifice meters provide a simple and cheap method of measuring flow rates. The problem of flow through an annular constriction, namely a simple orifice plate modified by a concentric plug, arises in flow meters which are designed to adjust the flow area with variations in the flow rate. Such adjustment is done by allowing the plug to move in the flow direction and therefore change the available area through which the fluid flows. In a simple orifice plate the relationship between the flow rate and the pressure difference across the orifice is non-linear and therefore the measured pressure difference, and the accuracy of the meter, decreases rapidly as the flow rate decreases. If the flow area is allowed to change, by careful design of the plug profile, it is possible to obtain a linear relationship between the flow rate and the pressure drop and maintain the accuracy over a wider range of flow rates. The disadvantage in this case is the increase in cost and higher permanent pressure loss.

The paper describes a finite element study of turbulent flow through a simplified model, consisting of an orifice plate with a concentric conical plug at its centre. The mathematical model is based on the time-averaged equations of the continuity and momentum in a cylindrical polar coordinate system and the k - ϵ turbulence model. The computer program FIDAP has been used as the computational tool.

The velocity and pressure fields, the pressure variation along the pipe wall and the pressure losses across the device and the corresponding discharge coefficient are presented and discussed. Comparisons are made with the results obtained for a simple orifice meter.

REMARKS ON A-POSTERIORI ERROR ESTIMATION FOR FINITE ELASTICITY

R. Mücke* and J.R. Whiteman**

*Institut für Festkörpermechanik,
Technische Universität Otto von Guericke, 0-3010 Magdeburg, PSF 4120, Germany

**BICOM, The Institute of Computational Mathematics,
Brunel University, Uxbridge, UB8 3PH, U.K.

Several methods of *a posteriori* estimation of the discretization error in finite element solutions are well established and widely used in engineering practice for linear boundary value problems. In contrast we are concerned here with finite elasticity. The main features of finite elasticity are first summarized. Using the residuals in the equilibrium conditions (Babuška approach) an error measure in the energy norm is proposed and physically interpreted. The effects arising from the nonlinear displacement-strain mapping and the accuracy of the proposed error measure are demonstrated by a numerical example. The estimated error distribution is then used to control an h adaptive finite element procedure.

FINITE ELEMENT ANALYSIS IN NUCLEAR SAFETY

J. Nedoma

8-Kobylisy, Frýdlantská 1313/17, 182 00 Prague, Czech Republic

In order to produce criteria for secure siting, location and protection of operation of high level radioactive waste repositories (HLRWRS) and nuclear power plants (NPPS) mathematical simulations must be undertaken of geodynamic and geomechanic processess in both the global and the local areas, where HLRWRS or NPPS, RESP, will be situated. In the contribution the finite element analyses of mathematical models based on the contact problems of Signorini type in thermoelasticity and on the nonstationary incompressible thermo-Bingham problem are discussed. The methods for monitoring the stress-strain field in situ, of the determination of the optimal design, of the measuring tensometric probes and of the post-processing of measured data based on the FEM are also briefly discussed. Existence and convergence theorems are proved, and the algorithms are presented.

FEM ANALYSIS OF ARTIFICIAL SUBSTITUTES
OF HUMAN JOINTS AND THEIR OPTIMAL DESIGN

J. Nedoma

8-Kobylisy, Frýdlantská 1313/17, 182 00 Prague, Czech Republic

The actual situation in human joints under loading is much more complicated than that described with classical methods of biomechanics such as Bombelli (1983). Some new ideas of biomechanics of human joints and of their artificial substitutes were given in Nedoma and Stehlik (1989), Nedoma (1991). These are based on contact problems in elasticity and thermoelasticity (plasticity).

In the contribution mechanical processes which take place during static burdening in the contact area of the acetabulum and the head of the hip and of their artificial substitutes - the artificial acetabulum and the head of the endoprosthesis - are discussed. The mathematical problem introduced in the contribution represents a simulation of the function of human joints and their total substitutes by total endoprostheses (TEP). The model problems lead to coercive and semi-coercive contact problems in quasi-coupled thermoelasticity. The FEM is used for numerical analyses of such model problems. The optimal design of the TEP's based on the FEM analyses of the contact problems in thermoelasticity is also briefly discussed and existence and uniqueness of the solutions of the model problems and a convergence theorem are proved.

Bombelli, R., (1983) Osteoarthritis of the Hip. Springer Verlag.

Nedoma, J. and Stehlik, J., (1989) Mathematical simulation of function of great human joints and optimal design of their substitutes. Part I: Biomechanics, construction of the TEP and tribology; Part II: Mathematical analysis of the problem. Technical Notes V-406, V-407, Institute of Information and Computer Science, Czech Academy of Science, Prague (in Czech).

Nedoma, J. (1991) Biomechanics of static and dynamic joints and of human motion. Technical Report No.507, Institute of Information and Computer Science, Czech Academy of Science, Prague.

NUMERICAL SIMULATION OF FLOW CIRCULATION
IN THE GOLDEN HORN

H. Örs

*Department of Mechanical Engineering,
Bogaziçi University, 80815 Babek, Instanbul, Turkey*

A numerical model of the flow in the Golden Horn is presented. The model is based on shallow water theory. Considering the complex geometry of the domain of interest, the governing set of partial differential equations, with appropriate boundary conditions, is solved by the finite element method. Given the inner sea nature of the Marmara Sea with minimal tidal effects, the steady state case is considered. The computational domain is discretized into 140 triangular elements with 111 node points. Different interpolation functions are tried in the context of the Galerkin approach. The final system of equations is solved by a Cholesky factorization method. Flow field values thus obtained are compared with on site measurement values. The ultimate goal of the project, supported by Boğaziçi University Research Fund, being to study the pollution mechanisms of the Golden Horn and ways of cleaning it. A second code is written to compute the dispersion rates of the pollutants. Values of the velocity field previously obtained are used in incorporating convection terms. Different wind force distributions and pollution sources are also tried. Computations are carried out on a HP 720 workstation and results are presented graphically.

ERROR ESTIMATES FOR DISCRETIZED BOUNDARY ELEMENT METHODS FOR THREE-DIMENSIONAL CRACK PROBLEMS

F. Penzel

*Fachberiech Mathematik, Technische Hochschule Darmstadt,
Schlossgartenstrasse 7, D6100 Darmstadt, Germany*

The displacement in an unbounded elastic solid with a given internal crack under normal traction can be calculated with the aid of boundary element methods. Let us assume that the crack surface $\Gamma \subset \{(x_1, x_2, 0) : (x_1, x_2) \in \mathbb{R}^2\}$ is a polygon. The crack opening displacement $u \in \left(H^{1/2}(\Gamma)\right)^3$ is known to be a solution of the hypersingular boundary integral equation

$$Du(x) = F^{-1}\sigma Fu(x) = v(x) \quad (x \in \Gamma), \tag{1}$$

with the two-dimensional Fourier-transform denoted by F, $v \in \left[H^{-1/2}(\Gamma)\right]^3$ is the known traction vector and σ is the operator of multiplication with the matrix function

$$\sigma(\xi) = 2\pi \begin{pmatrix} |\xi| - \nu\xi^2/|\xi| & \nu\xi_1\xi_2/|\xi| & 0 \\ \nu\xi_1\xi_2/|\xi| & |\xi| - \nu\xi^2/|\xi| & 0 \\ 0 & 0 & |\xi| \end{pmatrix} \tag{2}$$

The displacement u is approximated by a Galerkin scheme using a regular triangulation of Γ and piecewise-linear finite elements. Let h be the supremum of the diameters of the elements in a fixed triangulation. If the traction data v is a sufficiently smooth function, then it is known that the solution u_h of the Galerkin scheme converges to u with an order of $h^{1/2-\eta}$, for $\eta > 0$ chosen arbitrarily small.

Here we present a fully discretized Galerkin scheme obtained by use of a quadrature formula in the definition of the scalar products. We reduce the analysis of convergence of the fully discretized Galerkin method for the hypersingular integral equation (1) to the analysis of the weakly singular integral operator

$$Vw(y) = \int_\Gamma \frac{w(x)}{|x - y|} dx$$

Under the assumption that the quadrature error is sufficiently small we prove convergence of order $h^{1/2-\eta}$, of the solution of the fully discretized Galerkin method. By use of explicitly determined error bounds we formulate restrictions on the quadrature formulas which are sufficient to ensure this convergence result.

A SIMPLE AND EFFICIENT SOLUTION FOR THE
FACET APPROXIMATION OF SHELLS

E.D. Providas

101 Epeirou, Larisa 412 22, Greece

A method for adding drilling rotation degrees of freedom to the constant strain triangle is presented. The method is free from any deficiencies and does not alter the strain energy of the constant strain triangle. Superposition with the displacement version of the hybrid bending element yields the required six degrees of freedom per vertex. Several standard shell problems are solved. Attention is focussed on testing the capacity of the element to simulate inextensional bending of real arbitrary shells. For this purpose a displacement prescribed patch test and a matrix procedure are employed. The element, in contrast to other similar elements, is proven to respond satisfactorily to this type of deformation mode.

ANISOTROPIC h-ADAPTIVE FINITE ELEMENT
METHOD FOR HYPERSONIC VISCOUS FLOWS
WITH SHOCK-BOUNDARY LAYER INTERACTION

W. Rachowicz

*Applied Mathematics Section, Institute of Mathematics,
Technical University of Cracow, ul. Warszawska 24, 31-155 Krakow, Poland*

Viscous hypersonic flows are analysed using second order Taylor-Galerkin schemes. An h-adaptive version of the finite element method is used to approximate the solutions. The method supports anisotropic h-refinements, i.e. the possibility of breaking of quadrilateral elements in one of the two possible directions. The strategy of selecting elements to be refined and the direction of the refinement is based on minimization of the interpolation error and it involves estimation of second order derivatives of the solution. The anisotropic strategy is especially effective in boundary layers where solutions are almost one-dimensional. The method allows for accurate resolution of aerothermal loads with considerably smaller computational effort than with the standard adaptivity. Numerical simulation of a shock-boundary layer interaction problem illustrates the approach.

ON PRECONDITIONING FOR FINITE ELEMENT
EQUATIONS ON IRREGULAR GRIDS

A. Ramage* and A.J. Wathen**

**School of Mathematics, University of Strathclyde, Glasgow G1 1XH, U.K.*
***School of Mathematics,
University of Bristol, University Walk, Bristol, BS8 1TW, U.K.*

One efficient algorithm for solving Galerkin finite element equations is the preconditioned conjugate gradient method. A large number of preconditioning strategies have been proposed and these are generally analysed and compared using model problems: simple discretisations of Laplacian operators on regular computational grids, generally in two space dimensions. Many of these techniques have attractive theoretical convergence estimates on such model problems and it is principally for geometrically irregular (non-model) problems that the applicability and economy of preconditioned conjugate gradient methods is most useful. This is particularly true for problems on irregular unstructured three-dimensional grids.

Developing rigorous theory for irregular problems has proved to be more difficult than in analogous regular cases and, as a result, many aspects of working with unstuctured finite element grids are a lot less well understood. Here we present some theoretical results which apply to any finite element grid regardless of irregularity. We extend the work of the second author to find easily computed *eigenvalue* bounds for a class of *finite element* matrices on irregular grids. In addition, we obtain (weak) bounds on the interior *eigenvalues* of a general *finite element* matrix. These are useful in the analysis of various preconditioners for irregular finite element problems. We also present a result concerning the effect of a certain class of element-based preconditioners on the matrix condition number, which is again relevant to the rate of PCG convergence.

A PRECIS OF DEVELOPMENTS IN MODELLING EMBEDDED REINFORCEMENT FOR THE FINITE ELEMENT ANALYSIS OF REINFORCED CONCRETE STRUCTURES

A. Ranjbaran

*Department of Civil and Structural Engineering,
UMIST PO Box 88, Manchester, M60 1QD, U.K.*

To date three alternatives are available for modelling of reinforcement in reinforced concrete elements, i.e. discrete, smeared and embedded.

In this paper various techniques of modelling reinforcement are critically reviewed and their merits and shortcomings are highlighted. The embedded models are put into two general categories, i.e. locally embedded model (LEM) and globally embedded model (GEM). When the trajectory of a reinforcement in isoparametric coordinates of concrete is assumed as *a priori*, the model is called LEM. On the other hand, if the trajectory in global coordinates is known the term GEM is used. A totally general GEM is proposed. The relevant mathematical formulation for computation of stiffness and load contribution of a reinforcing bar to those of concrete are derived. The two- and three-dimensional problems are treated in a unified manner. Moreover, it is shown that all previous embedded models, available in the literature, can be considered as special cases of the proposed model. The proposed model is implemented in a finite element program. Numerical examples are included to verify the efficacy of the model.

IMPROVING THE ACCURACY OF FINITE-ELEMENT SOLUTIONS TO TWO-DIMENSIONAL ELLIPTIC PROBLEMS

S. Rippa and B. Schiff

School of Mathematical Sciences,
Tel Aviv University, Ramat Aviv 69978, Tel Aviv, Israel

We present a method for improving the accuracy of a finite-element solution of a two-dimensional elliptic problem over a triangular mesh by modifying the triangulation whilst keeping the mesh points unaltered. The triangulation is modified by a process of "local optimisation" in which each quadrilateral formed by two triangles sharing a common edge is considered in turn. The value of the energy integral over the two triangles is compared with the integral over the two triangles which would be obtained had the given quadrilateral been divided by the opposite diagonal. If the latter integral is less than the former, we modify the triangulation by replacing the diagonal of the given quadrilateral by the opposite diagonal. As the remainder of the triangulation is unaltered, the value of the energy integral over the whole domain has decreased. After completing a sweep of local optimisation over all of the quadrilaterals, the finite element equations may be solved again over the new triangulation to obtain new nodal values. Alternatively, before re-solving, additional sweeps of local optimisation can be carried out using the same nodal values. In either case, once the equations have been solved over the new triangulation, local optimisation is found to lead to no significant improvement in the solution. A third possibility is to terminate the procedure without re-solving at all.

The method has been tested for the Poisson equation using linear elements over each triangle, and for the biharmonic equation using (C^1) cubic macro-elements of Clough-Tocher and reduced Clough-Tocher (normal derivative linear over the external edges of the macro-element) type. Various problems with known solutions were solved, and the process often led to considerable improvement in the results, particularly when the solution has a marked directional dependence. For the biharmonic equation, two problems with a boundary singularity were also solved, the determination of the Airy stress-function for a rectangular elastic plate containing an edge crack under uniform tension, and the Stokes flow in a square domain with one of the four walls moving with constant velocity (the "driven cavity" problem). Comparison with existing solutions obtained for these two problems by a variety of other methods shows that the "local optimisation" technique results again in a considerable improvement of the original finite-element solution.

THE NUMERICAL RESOLUTION OF BOUNDARY LAYERS AND LOCKING EFFECTS IN THE REISSNER-MINDLIN PLATE MODEL

C. Schwab and M. Suri

*Department of Mathematics and Statistics,
University of Maryland Baltimore County, Baltimore, MD 21228, U.S.A.*

The Reissner-Mindlin plate model is characterized by both boundary layers and locking as the thickness t tends to zero. The boundary layers have the form $\exp(-\lambda p/t)$, where $\lambda > 0$ is a known constant, and p is the distance to the boundary. For t small, the approximation of such layers can be quite unsatisfactory unless the discretization parameter is sufficiently small. Locking effects can also cause the approximations to deteriorate when t is small. These occur due to the *Kirchhoff constraint* being imposed on both the true and the approximate solutions in the limiting case, as t tends to zero. In order to accurately approximate such models numerically, it is essential that the numerical method under consideration be designed in such a way that both these effects are resolved uniformly, i.e. with errors independent of the thickness.

In this talk, we present various results for the approximation of boundary layers by the h, p and hp finite element method. For the h version we show how exponentially graded meshes towards the boundary yield an optimal convergence rate independent of t. We also show that by a suitable combination of mesh refinement and degree selection (the hp method), one can obtain uniform exponential rates of convergence.

Next, we discuss the locking and robustness properties of various h and p type finite element methods. We quantify the exact degree of locking that can be expected if various triangular and rectangular meshes are used. We show that the p version is free of locking for this problem.

NUMERICAL EXPERIENCE WITH GRID ADJUSTMENT BASED ON A POSTERIORI ERROR ESTIMATORS

K. Segeth

*Mathematical Institute of the Czech Republic, Academy of Sciences,
zitná 25, 115 67 Prague 1, Czech Republic*

Recently, a variety of techniques for a posteriori error estimation have been theoretically developed and practically applied. A posteriori estimates can serve as a means for grid adjustment ensuring the optimal number and optimal distribution of grid points in the finite element method. We, however, need a solution of the problem on some grid to construct a new, optimal grid. This approach, therefore, is very suitable e.g. for solving parabolic partial differential equations by the method of lines. The analysis of the approximate solution at a fixed time level then yields a new grid to be used for the time step leading to the next time level.

In the paper, some approaches using the idea of monitor equidistribution, with monitors based on a posteriori error bounds as well as with those independent of error estimators, are presented together with numerical experience.

EFFICIENT PRECONDITIONERS FOR INCOMPRESSIBLE FLOW

D. Silvester* and A.J. Wathen**

*Mathematics Department, UMIST, PO Box 88, Manchester, M60 1QD, U.K.
**School of Mathematics,
University of Bristol, University Walk, Bristol, BS8 1TW, U.K.

Having an efficient Stokes solver is an important requisite of many Navier-Stokes solution algorithms. The Stokes operator is self-adjoint so iterative solution methods are an attractive approach for large problems. The only impediment to efficiency is the indefinite nature of the discrete system.

In this talk, some new theoretical results will be presented showing the effect of preconditioning discrete Stokes systems. Our analysis covers a variety of preconditioners, including simple diagonal scaling, incomplete factorisations and multilevel preconditioning. Our results apply to stable approximations of the Stokes operator, typically based on staggered grids, and also to recently developed *stabilised* approximations: equal order finite element methods or finite difference methods on unstaggered grids.

NUMERICAL SOLUTION OF A PARABOLIC EQUATION WITH A WEAKLY SINGULAR POSITIVE TYPE MEMORY TERM

M. Slodička

Institute of Applied Mathematics, Faculty of Mathematics and Physics,
Comenius University, 842 15 Bratislava, Slovakia

There exist many models from heat conduction in materials with memory, population dynamics and visco-elasticity, which can be described by integro-differential equations. Usually, the memory term depends on the solution u or on its spatial derivatives. But there exist cases where this memory term can depend on the time derivative of u. Integro-differential equations of this nature appear in describing diffusion models for fractured media (c.f. Hornung-Showalter (1990)). We discuss the numerical solution of the following problem:

$$u_t(t) - \Delta u(t) + \int_0^t a(t-s)u_s(s)ds = f(t, u(t)) \quad \text{in } \Omega, \quad t > 0,$$

$$u = 0 \quad \text{on } \partial\Omega, \quad t > 0,$$

$$u(0) = v \quad \text{in } \Omega,$$

where Ω is a bounded convex polyhedral domain in $\mathbb{R}^d(d \geq 1)$, $v \in H_0^1(\Omega)$. Our integral kernel in the memory term is supposed to be weakly singular and positive type. We use the backward Euler method for discretization in time and the Galerkin finite element method for discretization in space. We prove the convergence of our approximation scheme in some functional spaces, the existence and uniqueness of the solution, and produce some regularity results for exact solution.

Hornung, U. and Showalter, R.E. (1990), Journal of Mathematical Analysis and Applications **147**, 69-80.

DESIGN OPTIMIZATION OF RUBBER ELASTIC STRUCTURES

E. Stein and F-J. Barthold

*Institut für Baumechanik und Numerische Mechanik,
Universität Hannover, Appelstraße 9A, 3000 Hannover 1, Germany*

Sensitivity analysis of physical and geometrical nonlinear behaviour is an outstanding research field within design optimization. In this lecture sensitivity analysis of rubber elastic structures and its efficient implementation into a design optimization procedure is outlined.

Based on continuum mechanics at finite strains, essential remarks on the finite element method for nearly incompressible rubber elastic materials and on the efficient use of adaptive mesh refinement are discussed. Then the sensitivity analysis for rubber elastic materials with respect to changes of geometry and material parameters is described using a parametric model consisting of space coordinates, time and design variables. Due to large strains occurring in technical applications of rubber materials a current configuration approach is used both in structural and sensitivity analysis. The concept of a design optimization procedure and integrated algorithms for Computer Aided Design (CAD), Finite Element Method (FEM), Mathematical Programming and Sensitivity Analysis is explained. A highly modular subprogram and data base concept for these methods is used to implement efficiently the analytical derived sensitivities into the FE research program INA-OPT (INelastic Analysis and OPTimization) developed at our institute. Applications of the optimization procedure to material parameter identification problems in tyre layout are shown.

The methodology for the approximate solution of the inverse problem was used for the identification of material parameters in the Ogden material equation which can also be understood as a validation of the constitutive model for the tyre material investigated.

An optimal design of steel cords in a tyre layout based on a minimal internal energy equilibrium state is derived using the optimization procedure described.

ON LINEAR AND BILINEAR ELEMENTS FOR REISSNER-MINDLIN PLATES

R. Stenberg

*Faculty of Mechanical Engineering,
Helsinki University of Technology, 02150 Espoo, Finland*

We consider the following elements:

1. The MITC4 analyzed by Brezzi and Bathe in the MAFELAP 1984 Conference for the case of rectangular elements. We give the error analysis for the more general class of meshes for which the "$Q_1 - P_0$" Stokes element was analyzed by Pitkäranta and Stenberg in the same conference.

2. We consider the linear triangular MITC element in which the deflection and rotation are linear and the reduction operator is given by the lowest order rotated Raviart Thomas element. We give an error analysis for the case when the triangular

mesh is obtained from a quadrilateral one by drawing the two diagonals in each quadrilateral. The error analysis can now be performed for the same quadrilateral meshes as for the MITC4. Our analysis seems to explain the good behaviour reported by Hughes and Taylor in MAFELAP 1981 for a very similar element. As a byproduct we also give an error analysis of the well known linear constant strain triangle with a "criss-cross" mesh for an incompressible material.

3. We show that both of the above elements can be modified so that they are optimally convergent without any restrictions on the meshes.

FINITE ELEMENT SOLUTION OF UNSTEADY-STATE MASS TRANSFER THROUGH A STATIONARY LIQUID

J.L. Torres

Universidad de las Americas - Puebla, Puebla, Mexico 72820

Mass transfer by diffusion of one liquid component through another can be modelled by the expression:

$$-\nabla \cdot cD_{AB}\nabla y_A + \nabla \cdot c_A v + \frac{\partial c_A}{\partial t} - R_A = 0. \tag{1}$$

If constant liquid density and constant diffusivity are assumed, and if the diffusion occurs through a stationary layer, the expression reduces to the so-called Fick's Second Law of Diffusion for non-reacting systems:

$$\frac{\partial c_A}{\partial t} = D_{AB}\nabla^2 c_A. \tag{2}$$

The calculation of a concentration profile as a function of time and position in such systems is usually carried out using equation (2), mainly because the rigorous application of (1) can lead to computational complications.

The assumptions that lead to equation (2) are usually appropriate, with one important exception. When the diffusion takes place in a stationary liquid layer, in many cases the diffusivity is a strong function of concentration; in these cases equation (1) would have to be used, with its non-linear terms. The calculations cannot be carried out using a conventional finite-difference formulation, because, in essence, the system becomes nonisotropic almost immediately, due to the rapidly varying diffusivity coefficient.

In this paper, an alternative finite element method for solving problems with high concentration dependence in D_{AB} is described. The problem requires a re-statement of the domain properties at every time step. The calculation of concentration-dependent diffusivity was carried out using a cubic spline model. The paper discusses accuracy vs. computational effort, and also a possible application in the experimental measurement of diffusivities.

A FAMILY OF TRIANGULAR PLATE BENDING ELEMENTS, AS A BASIS FOR GEOMETRICALLY NONLINEAR SHALLOW SHELL ELEMENTS

F. van Keulen

Laboratory for Engineering Mechanics,
Faculty of Mechanical Engineering and Marine Technology,
Delft Univ of Technology, PO Box 5033,2600 GA Delft, The Netherlands

A class of shell elements can be formulated as a combination of plate bending and membrane elements. Generally, this results in facet elements. Some well-known examples are the facet elements based on the constant bending moment triangle, the Discrete Kirchhoff Triangle and on the Discrete Shear Triangle. In order to account for initially curved geometries and to improve numerical efficiency, shallow shell terms can be considered. This results in shallow shell elements, which were originally facet elements.

Here our attention is concentrated mainly on a family of plate bending elements which is very suitable as a basis of the formulation of shallow shells. This family consists of low-order as well as higher-order plate bending elements. The family has members that are based on the Kirchhoff-Love assumptions and members that are based on first order shear deformation theory. The independent kinematic degrees of freedom are the nodal displacements, the rotations about the element sides and additionally, the transverse shear strains along the element sides.

Some numerical examples are shown, which demonstrate the features of the plate bending elements under consideration.

Finally, the way towards geometrically nonlinear shallow shell elements based on this family of plate bending elements is outlined. The formulation of efficient shallow shell elements is not completely straightforward. In particular, the higher-order elements will require more attention, in order to avoid membrane locking and to achieve a correct description of rigid body motions.

ON AN EXTERNAL FINITE ELEMENT METHOD FOR A SECOND-ORDER EIGENVALUE PROBLEM ON A CONCAVE 2D-DOMAIN WITH DIRICHLET BOUNDARY CONDITIONS

M. Vanmaele

Seminar Mathematical Analysis, Galglaan 2, B-9000 Gent, Belgium

We consider a finite element method for a second-order elliptic eigenvalue problem in a concave bounded domain $\Omega \subset \mathbb{R}^2$ with sufficiently regular boundary $\partial\Omega$, where homogeneous Dirichlet boundary conditions are imposed. In the method a 'variational crime' is committed, the approximate eigenvalue problem being formulated on a domain $\Omega_h \not\subset \Omega$. Hence the finite element approximation space $V_h \not\subset V, V = H_0^1(\Omega)$.

We restrict ourselves to a linear triangular finite element mesh (h mesh parameter). The convergence of the method and optimal error estimates for both the eigenvalues and the eigenfunctions are obtained in Vanmaele and Ženíšek (1993). The proofs of the estimates in that paper rest heavily upon the $0(h^2)$-estimate for the error of the elliptic projection in the $L_2(\Omega_h)$ norm, see Vanmaele and Ženíšek (1993) , Theorem 4.8. In the present paper we prove this estimate in an alternative way, proceeding similarly as in the proof of the well-known Aubin-Nitsche trick. Moreover in this paper we obtain an optimal, (Oh^2),estimate (from above) for the eigenvalues under weaker conditions on $\partial\Omega$ than in Vanmaele and Ženíšek (1993), Theorem 5.5.

Vanmaele, M. and Ženíšek, A. (1993), External finite element approximations of eigenvalue problems, *RAIRO $M^2 AN$* (to appear)

THE COMPUTATIONAL MODELLING OF THE THERMOFORMING PROCESS FOR THE CREATION OF AXISYMMETRIC CONTAINER STRUCTURES

M.K. Warby and J.R. Whiteman

*BICOM, Institute of Computational Mathematics,
Brunel University, Uxbridge UB8 3PH, U.K.*

This paper is concerned with the computational modelling of the thermoforming process which is a process used in the packaging industry to create polymeric container structures by forcing thin sheets into moulds. In our model only axisymmetric geometries are considered and the sheet, although generally multilayered, deforms to a reasonable approximation like a membrane. When the containers to be formed are "shallow" it is usually satisfactory to use pressure alone as the forcing action and this case has been considered by several authors in the recent literature. In this context, the term satisfactory usually refers to the thickness distribution of the formed container (pot). In our model the process starts with a sheet of uniform thickness and as the deformation proceeds, the thickness distribution becomes non uniform. When the thickness distribution becomes highly non uniform, or more specifically when any part of the sheet becomes too thin, then we would describe the container as being unsatisfactory. This is the situation that occurs when pressure alone is used and the pot is deep. In the manufacturing process deep pots are instead created by the combined actions of a rigid plug and an applied pressure. The use of the plug leads to a different deformation history before contact is made with the mould than is the case when pressure alone is used and, since the material is viscoelastic, this has the potential of leading to a thickness distribution which is different from that obtained with pressure alone. However the main reason for the more satisfactory distribution of the thickness being obtained is as a result of the slight cooling effect that the plug has on the sheet. Specifically the plug, which is cooler than the sheet, has the effect of slightly cooling and hence slightly stiffening the part of the sheet to which it makes contact. The purpose of constructing a computational model is to be able to quantify this effect for different plug shapes, depths and temperatures. This requires a model which takes account of the following effects.

The model must take account of the large deformation and hence the equations are nonlinear. There is a contact problem as the sheet slides on the plug and for the part of the sheet in contact with the plug there is the nonisothermal cooling effect. When the sheet is inflated against the mould there is a simpler contact problem as the sheet sticks to the mould. For the constitutive model for the polymers only standard Mooney Rivlin elastic and simple viscoelastic relations have been considered as no more better experimentally based information was available. Fortunately the model involved only one space dimension and thus the finite element discretization is straightforward. In our model cubic Hermite elements using a carefully chosen moving mesh is used in order to capture the main features at relatively low cost. A form of Newton's method is used at each stage to solve the nonlinear system of equations. Results will be presented for the intermediate membrane profiles and the final thickness distribution predicted using the model.

ELEMENT-BY-ELEMENT PRECONDITIONING

A.J. Wathen

School of Mathematics,
University of Bristol, University Walk, Bristol, BS8 1TW, U.K.

Element-by-element preconditioning methods were introduced by T J R Hughes and co-workers in 1983 as efficient and highly vectorisable techniques which arise out of and are highly consistent with natural finite element data structures.

In this talk we present a description and analysis of such 'EBE' preconditioning methods for self-adjoint elliptic partial differential equations in 2- and 3-dimensional domains. As well as analysis for model Poisson problems, we present results for problems with discontinuous variable coefficients and for constant coefficient problems discretised using elements with large aspect ratios.

Some of this work is joint with Han-Chow Lee.

A POSTERIORI ERROR INDICATOR AND ESTIMATOR FOR TIME-SPACE FE DISCRETIZATION OF PARABOLIC PROBLEMS

J. Weisz

Institute of Applied Mathematics, Faculty of Maths and Physics,
Comenius University, Mlynska Dolina, 84215 Bratislava, Slovakia

In the area of adaptive numerical methods for parabolic partial differential equations there are two traditional approaches: In the first one the equation is discretized in space and then the corresponding system of ordinary differential equations is solved by usual methods using the adaptive control of the time step. The second possibility is to discretize the equation in time and then to solve the corresponding elliptic equations adaptively.

The simultaneous time-space discretization Axelsson and Maubach (1990) is a generalization of the above methods. The equation is discretized in time and space simultaneously using time-space finite element spaces. We obtain an equation which is "elliptic" and finite-dimensional. Our aim is to derive a-posteriori error indicators and estimators for such discretization of the linear parabolic initial-boundary value problem. Using the technique of Baranger, Amri and Estimateurs (1991) the error of the finite element approximation is estimated in terms of the discretization parameter, data of the problem and the approximate solution.

Axelsson, O., Maubach, J. (1990), A time-space finite element method for nonlinear convection diffusion problems, NMFM 30, Vieweg Braunschweig, 6-23.
Baranger, J., Amri El., Estimateurs, H. (1991), a posteriori d'erreur pour le calcul adaptif d'ecoulements quasi-newtoniens, M²AN 25, 31-48.

FINITE-ELEMENT SOLUTIONS FOR MECHANICAL DRILLING METHODS

H. Wern

Hochschule fuer Technik und Wirtschaftes Saarlandes, Goebenstr 40, D-W-6000 Saarbruecken, Germany

The principle for evaluating the unknown depth distributions of residual stresses from drilling methods relaxed strain data is viewed as an Integral Method, see Schajer (1988) and Wern and Peiter (1988(i-ii)). The mechanical hole drilling, see Mathar (1934) and Peiter *et al.* (1970), as well as the ring-core-method Wolf and Böhm (1971) and Hofer and Schmidt (1988), are analyzed using finite elements. The influence of the notch sensitivity is discussed for different hole-, slot- and sample geometries. Temperature effects caused by the drilling procedure are considered. From the calculated strain relaxation data, possible integral operators for the relaxation process are discussed.

Schajer, G.S. (1988), Transactions of the ASME, 344, 349.
Wern, H. and Peiter, A. (1988(i)), Steel Research 59, 115, 120.
Wern, H. and Peiter, A. (1988(ii)), Materialprüfung 30, 99, 101.
Mathar, J. (1934), ASME 58, 249, 254.
Peiter, A., Grieger, N., and Klein, H. (1970), Materialprüfung 12, 262, 268.
Wolf, H. and Böhm, W. (1971), Arch. Eisenhüttenwesen, 195, 200.
Hofer, G. and Schmidt J. (1988), DVS, 125, 128.

IMPROVED PATCH STRESS RECOVERY PROCEDURE BY EQUILIBRIUM AND BOUNDARY CONDITIONS

N-E. Wiberg and F. Abdulwahab

Department of Structural Mechanics, Chalmers University of Technology, Sven Hultins gata 8, S-412 96, Goteborg, Sweden

The cornerstone of an adaptive FE-analysis is a cheap and accurate calculation of the local and global distribution of the error, so that successively improved solutions can be obtained. The Zienkiewicz-Zhu error estimate is undoubtedly the most practical *a posteriori* error estimator capable of estimating both local and global error of discretization. This error estimator, for instance in the energy norm, is asymptotically exact if the recovered derivatives are superconvergent. Very recently two important recovery procedures that recover superconvergent gradients have been developed.

Namely the Superconvergent Patch Recovery (developed by Zienkiewicz and Zhu) and patch recovery based on superconvergence and equilibrium (developed by the authors).

Both methods are based on local patches of elements. Their performance is excellent for recovering derivatives for points in inner patches. Generally speaking the quality of recovered derivatives at or near boundaries is however inferior compared to that at interior points. Both methods do not take into consideration the boundary conditions which can be either prescribed displacements or prescribed tractions. In this paper, a post-processing technique for points on or near boundaries is described. We introduce a weighted least squares polynomial fit in an attempt to 'force' the recovered derivative field to satisfy the prescribed boundary conditions. In the literature, the least square methods are deemed to be insensitive to different weighting functions over a wide range. In spite of this, we assume that higher values of weighting functions for prescribed boundary conditions would improve the recovered derivatives near boundaries. Several numerical examples illustrate the very good performance of the proposed recovery technique at or near boundaries.

MATRIX AND IMAGE DECOMPOSITION USING WAVELETS

J.R. Williams and K. Amaratunga

*Department of Civil Engineering, Massachusetts Institute of Technology,
Room 1-272 MIT, Cambridge, MA 02139, U.S.A.*

This paper demonstrates how the wavelet transform may be advantageously applied to matrix and image data. The resulting decomposition yields a reduced set of data which retains the essential properties of the original data. As a result, the wavelet decomposition enables us to capture the behaviour of a physical system in a greatly simplified mathematical model.

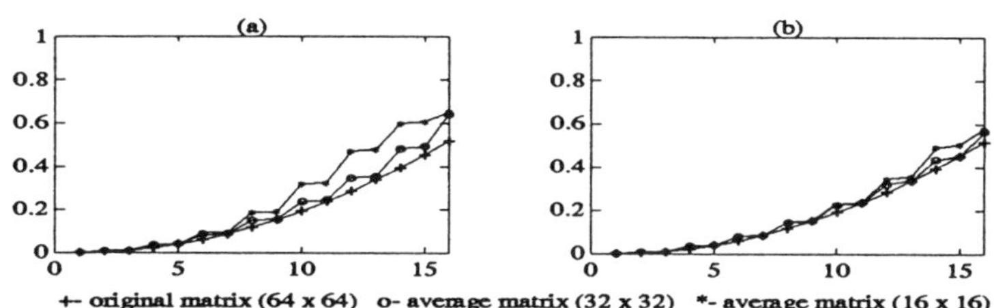

Figure 1: Eigenvalues of a stiffness matrix and wavelet reduced matrices using (a) D6 and (b) D20

Figure 1 compares the eigenvalues of the stiffness matrix for a spring-mass system with those of the average matrices of 1/4 and 1/16 the size derived using the Daubechies D6 and D20 wavelet systems. The figure shows that good estimates of the lower eigenvalues of the stiffness matrix may be obtained from the averaged matrices. This result has significant implications in modal analysis, where the lower eigenfrequencies are often the most critical of all.

Similarly, in the wavelet-Galerkin method for solving PDE's, the resulting wavelet differential operator matrix may be decomposed using the wavelet transform, leading to an averaged differential operator matrix and a reduced set of equations. By solving the reduced set of equations, a good estimate of the solution may be obtained with relatively little work.

The averaging property of wavelets is attractive in itself for the rapid solution of engineering problems. The appeal of wavelets is greatly enhanced by their hierarchical properties, which permits the subsequent addition of detail to the solution, thereby providing the capability for progressive refinement of the solution.

A GENERAL BOND-SLIP FORMULATION FOR EMBEDDED REINFORCING BAR ELEMENTS

Z.P. Wu and D. Phillips

Faculty of Civil Engineering, Delft University of Technology, PO Box 5048, 2600 GA Delft, The Netherlands

This is a generalized formulation developed for isoparametric elements including embedded reinforcing bars and bond-slip effects, in which the representation of the bond-slip behaviour is evaluated by prescribing bond characteristics at a set of artificial "nodes" introduced along the bar. The embedded bar element and the parent concrete element are now general finite elements removing all the restraints from the conventional approaches. The condition of compatibility for the system is obtained by only condensing the middle nodes of the bond-slip while the equilibrium condition is satisfied by assembling the stiffness matrix of the reinforcing bar and bond slip into that of the parent concrete element at element level. Several examples are employed to illustrate the performance of the formulation, i.e. pull-out test, transfer test and beamcolumn joints. The numerical results are compared with experimental ones where good agreements have been achieved.

Index

The index is compiled alphabetically word-by-word.

Italic numbers denote pages on which diagrams appear.

a-posteriori error analysis 154-8, 267-86

a-posteriori error bounds 278-86

a-posteriori error estimates 36-42, 108-10, 112-20, 279, 286, 292-3, 385

a-posteriori error estimators 391, 398

a-posteriori error indicator 33, 397-8

a-priori error analysis 267-86

a-priori error bound 269

a-priori error estimates 12, 21, 25, 108-10, 112-14, 117

acetate 206, *210*

acoustic problem, exterior 353

acoustic time-harmonic waves 308

actual error 294

actual error norm 123

adaptive algorithm 105, 111-12, 114, 118, 120, 289, 304, 305

adaptive error control 105

adaptive mesh algorithm 299

adaptive mesh finite element analysis 179-80, 183-5

adaptive mesh refinement (AMR) 364-5

adaptive mesh refinement procedures 124-6, 269

adaptive mesh refinement strategies 127, *130*

adaptive space strategy 305

adaptivity 110-12

admissible function spaces 332

admissible functions, set of 326

advancing front method 216

advection-diffusion problems 273

advection-diffusion reaction system 205

advection equations, two-dimensional linear 290

advection problem, linear scalar 292

aerodynamic flows 215, 216

ageing effect 50

aircraft, model of 226, *227*, *228*

aircraft wing, oscillating, flow over 232-9

airfoil 294

airfoil, Korn 126, 129-30

algorithm
adaptive 105, 111-14, 118, 120, 304, 305
Arnoldi-Newton 362
characteristics-mixed 203
Cholesky factorization 382
conjugate gradient 336-40
Crank-Nicolson 355
direct-solution 249
frontal solution 172
grid solution 219-25
iterative 271
iterative solution 245
Jacobi preconditioned conjugate gradient 206
Lanczos 362
Newton-Raphson 8
nonoverlapping parallel domain decomposition 204

The Mathematics of Finite Elements and Applications
Edited by J. R. Whiteman © 1994 John Wiley & Sons Ltd

parallel 185, 187
Polak-Ribiére type conjugate
    gradient 336
remeshing 179
shock fitting 270
spatial discretisation 300
algorithms, time-marching 89-97, 99
AMR *see* adaptive mesh refinement
angle property 331
anisoparametric interpolations 192
anisotropic interpolation error 356
approximate approximations 77-104
approximate approximations, for
    Navier-Stokes equations 97-100
approximate partitions of unity 78
approximate velocity 99
approximation
    asymptotically stable 147
    fourth order 79
    sixth order 80
approximation errors 10, 147
approximation estimates 81-4
approximation space, global 160
aquifer 243
Arnoldi iterations 362
Arnoldi-Newton algorithm, 362
artificial boundary 307, 310
artificial viscosity 107, 114, 271-3
asymptotic convergence 146-8
asymptotic exactness 38
attenuation factor 61
automatic mesh generation 315
automobile catalytic converter 256
axisymmetric container structures
    396-7
axisymmetric contraction 69-76

B-L law 28-31
    peak strain *32*
    peak stresses for *30*
Babuška-Brezzi condition 274
backstep channel problem 16-21
Baiocchi transformation 182
balancing procedure 205
bandwidth reduction process 184
bar elements, embedded reinforcing
    400

barycenters 291, 292
basis functions 77, 84, 101, 102, 139,
    140, 316
global 6
integral operators on 84-9
scalar 316
Bayliss-Turkel conditions 312
beam, two-dimensional 65
bending moments 131
bending specimen *173*
Bernoulli's equation 267
Bessel functions 88, 101, 104, 164
bilinear elements, for
    Reissner-Mindlin plates 394
bilinear form 3, 152, 162
bilinear functions 283, 293
bio-electromagnetism 315
biodegradation kinetics 199-211
bioremediation results 206-9
biorestoration techniques 199
block tri-diagonal system matrix
    245
body forces 3
bolts, passive, fully-grouted 366-7
Boltzmann equation,
    time-dependent 352
Boltzmann superposition 51
bond-slip formulation 400
boundary
    piecewise smooth 31
    wavy 354
boundary and initial conditions 299
boundary condition 7, 46, 72, *73* ,
    168 , 257, 307
    Engquist-Majda, second-order
        310
    exact 281
    farfield 345
    high-order 312
    homogeneous 4
    inflow 276
    local
        high-order 311
        sequence of 312
    localised 312
    natural 164
    non-reflecting 307

nonlocal 308
spatially exact 310
traction 64
boundary element formulation 381-2
boundary element method 307
boundary element method
  coupling of 323-48
  discretized 387
boundary and initial conditions 299
boundary integral equation 101,
  103, 329, 353
  Burton-Miller 138
  discretization of 77
boundary integral methods,
  uncoupling of 371-2
boundary integral operators 327,
  328
boundary layers 105
boundary layers, numerical
  resolution of 391
boundary node 271
boundary operator 313
Boundary Point Method (BPM) 77,
  100-4
boundary trial functions, piecewise
  constant 331
boundary value problem 10, 33, 275,
  282, 290, 371-2
  local 281, 285
  mixed initial 286
  model hyperbolic 279
  multidimensional initial 268
  stationary 100-11
  in unbounded domain 310
bounded domain 137, 151, 255, 257,
  323
bounded open set 268
bounds, upper & lower 39
Boussinesq equations 356-7
box scheme 267, 268, 271
BPM *see* Boundary Point Method
Bramble-Hilbert lemma 278
bubble functions 5
Burgers' equation problem 304
  two-dimensional 303
Burgers' equation results
  adaptive mesh *304*

fixed mesh *303*
Burton-Miller boundary integral
  equation 138

C^1-finite element methods 359-60
canonical norm 152
capillary pressure 244
Carathéodory functions 369
carbon tetrachloride (CCl$_4$) 206,
  209
cartesian co-ordinates 217, 299
cartesian directions 290
casting, deformation and stress in
  357-8
Cauchy data 61, 63, 327, 353
Cauchy problems 77, 88-9, 96, 97,
  100
Cauchy-Schwarz inequality 157
Cauchy stress tensor 46
cell centred finite volume approach
  300
cell centre schemes 267
cell error 281, 282
cell error indicator
  lower 282
  upper 282
cell-piecewise invariant 38
cell residuals 269-71
cell-translation invariant 38
cell vertex equations 268
cell vertex finite volume
  approximation 291
cell vertex finite volume method 279
cell vertex method 291, 294
cell vertex scheme 267, 277
  for compressible flow problems
  269-73
centred schemes 220
central difference/trapezoidal
  approximation 58
centroid 301
CFD *see* computational fluid
  mechanics
CFL stability control 301
Chaboche law 28, *29* , *30*
  peak strain for *32*
  peak stresses for *34*

characteristics-mixed algorithm 203
Chevron mesh 39, *40*
Cholesky factorization algorithm
    382
Chorin scheme 8
circular contraction *73*
Clough-Tocher, cubic
    macro-elements 390
coarse mesh 159, 164
coerciveness inequality 329
collocation finite element mesh *246* ,
    247
collocation system of equations 248
column-wise subdividing 181
compatibility conditions 138
compressible flow 217-19, 269-73
    high speed 215-39
compression stress 29
compression wave problem 273
computational biology 106
computational chemistry 106
computational fluid mechanics 106,
    119-20, 289
computational mechanics, reliability
    of 25-42
computational physics 106
computational solid mechanics 106
computer algebra systems 373-4
computers
    distributed memory
        multi-processors 185
    multi-processor 179
    parallel 179, 183
    parallel MIMD 382
    shared-memory 370-1
concrete 50, 51
concrete structures, finite element
    analysis of 389
conductivity matrix 38
conformal mapping 336
conforming discrete variational
    problem 333
conjugate gradient, method of 175
conjugate gradient algorithm 336-40
conjugate gradient method,
    preconditioned 361
conservation law 270

consistency 108
constant linear advection equation
    273
constant vector field 323
constitutive matrix 122
constitutive model 54-9
constitutive relations 47
constrained nodes 6, 143
constraint conditions 6
contaminant transport 243, 248
    subsurface 199-211
    with biodegradation 200-1
continuity constant 147
continuity equation 200, 257
continuous bilinear interpolant 268
continuous bilinear vector functions
    270
continuous dual problem 113, 115
continuous function 81, 270, 280
continuous piecewise bilinear
    functions 268
continuous positive weight function
    280, 282, 283
continuum problem 257
contour and surface plots *295*
convection matrix 71
convection terms, explicit 70
convective fluxes 300
convective term 260
convective velocity fields 294
convergence 108, 249
    second-order 278
convergence acceleration techniques
    226-9
convex set 183
coordinate system, local *263*
corner
    abrupt 69, 72, *73*
    re-entrant 69
    rounded *73*
coupled boundary value problem
    325-6
coupled nonlinear equations 244
coupling 325
    computational domain for 324
Courant elements, piecewise linear
    continuous 331

Courant-like stability condition 302
Courant number 229
Courant number, global cell-based 272
crack problems, three-dimensional 387
Crank-Nicolson algorithm 355
Crank-Nicolson coefficient 72
Crank-Nicolson method 66, 67, 70
creep 45, *52*
creep curve coefficients *56*
creep data 54-9
creep function 50, 55
creep response 55
creep test *40* , 50
cubic displacement element 195
cylinder
    infinite circular, scattering of plane wave by *234*
    thick 126-9

D'Alembert solution 59
Darcy velocity, mass average 244
Darcy's Law 200, 201, 244
deflection of waves, spurious 307
deformation 46
deformation tensor, rate of 70
degrees of freedom 6, 114, 140, 164, 245, 311, 317
    nodal 192
    remote 313
Delaunay mesh generation, automatic 265
Delaunay method 321
Delaunay tessellation 258, 263
Delaunay tetrahedra 265
Delaunay tetrahedral meshes 255, 256, *257*
Delaunay/Voronoi tessellations, dual 255-65
density function 323
dental enamel 355
derefinement 123, 125
design optimization, of rubber elastic structures 393
difference equations 317
diffraction potential 86-7

diffusion coefficient 164
diffusion coefficient, scalar 221
diffusion-dominated convection-diffusion problems, stationary 107
diffusion/dispersion step 205-6
dilatation 54
Dirac delta function 245
direct-iterative method, hybrid 361
direct-solution algorithm 249
Dirichlet boundary conditions 164, 182, 245-8, 395-6
    homogeneous 151
Dirichlet boundary data 59
Dirichlet interface boundary data 204
Dirichlet map 309
Dirichlet problem, for the Laplacian 101
Dirichlet series, finite 49
discontinuous coefficients 329, 383
discrete continuity equations 258
discrete divergence operator 258
discrete LBB-constant 147
discrete minimization problem 335-6
discrete problem, conditioning 147
discrete stress 191
discrete time level 113
discrete variational problem 332
discretisation error 105, 109, 354
discretisation method 299, 300
discretisation scheme, spatial 300
dispersion, convergent 318
dispersion analysis 317-19
dispersion relation 317
dispersion relation
    fourth order 318
    second order 318
dispersive flux 202
displacement 46
displacement
    irrotational & solenoidal components 60
    peak of 29
displacement errors *64*
displacement field 46
displacement field

dilatational 60
equivoluminal 60
displacement field wave speed,
    irrotational 61
displacement vector 137, 140
distorted right-angle grids 320
distribution matrices 271-3
divergence theorem 300
divergence theorem, 3D 258
divergence/pressure gradient matrix
    71
Dolittle's equation 54
domain decomposer 181
domain decomposition 14-15
domain decomposition
    automatic 176-8
    hp 160-4
domain decomposition algorithm
    204
domain decomposition method,
    parallel 243
domain decomposition
    preconditioners 383
domain decomposition scheme 245
domain decomposition techniques
    368
domain integral 219
domain subdividing *187*
drilling methods, mechanical 398
dual problem 115, 116, 117, 118
dual variable method 263-5
duality pairing 284
Duvaut-Lions space 279, 284
dynamic analysis of structures 357
dynamic mode shape function 357

edge-constant 316
edge-constant elements 321
edge-constant method *317*, 321
edge-constant scheme 315
edge functions 5, 6
edge gradients 221
edge limiters 223
edge nodes 5
edge singularities 355-6
effectivity index 16, *18*, 34-9, 41,
    42, 283, 284

efficiency index 294
eigenmode 319
eigenvalue, maximum 221
eigenvalue problem 395-6
eigenvalue problem,
    nonhomogeneous degenerate 369
eigenvalues 147, 225
eigenvectors 225
elastic foundation 194
elastic-plastic fracture analysis
    program 171-8
elastic problem 38, 60
elastic rod 194
elastic scattering 137-48
elasticity 76
elasticity
    linear 153
    modulae of 27
elasticity equations 138
elasto-plasticity 109
elastodynamic equations 59, 60
elastostatics, classical, in 2D 311
electric field 315
electromagnetic compatibility 315
electromagnetic fields 316
electromagnetic scattering 232-3
electromagnetic waves 232, 315
electron acceptors 200, 201
electron donors 200, 201
element(s) *see* finite elements
elementary cycles 264
ellipse artificial boundary 311
ellipsoid artificial boundary 311
elliptic pressure equation 249
elliptic problems 112-13, 118, 122,
    124, 363, 376
energy norm 35, 37, 118, 122, 124,
    159
energy norm control 116
energy norm errors 36
Engquist-Majda boundary
    condition, second-order 310
entropy problem, existence,
    uniqueness 329-30
entropy inequality 344
equation
    Bernoulli's 267

boundary integral, Burton-Miller
138
conservation of momentum,
vector-valued, 257
constant linear advection 273
continuity 200, 257
differential algebraic (DAE) 262
discrete 292
elasticity 138
elliptic-like pressure 243
elliptic pressure 249
Fick's second 355
heat 88-9
Helmholtz 137, 353, 375
integral 387
Laplace 101, 311
linear hyperbolic 273
momentum 257, 258
neutron diffusion 267
perturbation 109
Poisson 70, 335, 337, 390
pressure 370
pressure stiffness matrix 71
scalar linear hyperbolic 272
semi-discrete momentum 262
time-dependent wave 310
wave 59, 88-9, 107, 109
reduced 311, 312
equation solver, linear 362
equations
boundary integral 353
Boussinesq 356-7
cell vertex 268
collocation system of 248
coupled nonlinear 244
difference 317
discrete continuity 258
elasticity 138
elastodynamic 59, 60
Euler 217, 233, 279, 292, 293,
300, 301, 305, 326, 346
Euler
3D unsteady compressible 217
compressible 220, 223, 225,
226, 279, 289
of compressible flow 217-19
of compressible gas dynamics
283
numerical 299
one-dimensional 274
subsonic 273
two-dimensional compressible
290-1
hyperbolic 273, 302
hyperbolic advection 268
integral 106
linear advection 268
linear elliptic 289
Maxwell's 232, 315-21, 364
Navier-Stokes 4, 8, 21, 89, 151-8,
279, 356
Navier-Stokes
3D, modeling of 225-65
compressible 279
incompressible 265
of mapping 351
non-stationary 97-100
steady-state 4
nonlinear convection-diffusion
244
nonlinear parabolic 89-97
ordinary differential 26, 109, 265,
299, 301
parabolic 89-97, 245, 392
parabolic-like contaminant
species-transport 243
parabolic-like saturation 243
partial differential 46, 106, 243,
268, 289, 375-6
partial differential
nonlinear 383
time-dependent 299-305,
358-9
phase mass balance 244
predictor-corrector 70
reaction diffusion 159, 160
St. Venant 271
scalar hyperbolic 286
scalar linear hyperbolic 272
second-order hyperbolic 378
shallow water 267
Stokes 151-8
transonic gas dynamics 267
transport 202, *252*

equilateral grids 320
equilibrium 398-9
equilibrium error, residual *197*
equilibrium technique 10
error
    discretisation 105, 109, 354
    global 13-16, 122-9, 153, 279,
       281, 301
    intermediate 13
    local 13
    on macroscopic scale *166*
    on microscopic scale *167*
    permissible 124
    quadrature 160
    relative *36*
    spatial 299, 302, 303
    target 13, 14
    temporal 299
    transmitted 281
    truncation 108, 109, 116, 117,
       270, 272
error bound 63
error control 114
    choice of norm for 115
error energy norm 122, 129
error estimate 23, 34, 80
    a-posteriori 12, 13, 25, 36-42,
      108-10, 112-17, 120, 279, 286,
      292-3
    a-priori 12, 21, 25, 108-110,
      112-14, 117
    for boundary element methods
      387
    global 12
error estimation 189-98
error estimation
    in computational fluid dynamics
      1-23
    residual 196-7
    residual equilibrium 189
error estimator 37, 269, *295*
    definition of 121-2
    performance of 38
    post-processing 190
    residual based 190
    robustness of 38
    Zienkiewicz-Zhu 121

error function 103
error index 12, 14
    target 12
error indicator 33, 36, 269, 294
    local 123, 156-7
error level, intermediate 12
error norm 196
    actual 123
    global 130
    required 123, 125
error tolerance 304
errors, approximation 10
Euler constant 85
Euler equations 233, 293, 300, 301,
    304 , 305, 326, 346
    3D unsteady compressible 217
    compressible 220, 223, 225, 226,
      279, 289
    of compressible gas dynamics 283
    one-dimensional 274
    subsonic 273
    two-dimensional compressible
      290-1
Euler flows, compressible transonic
    221
Euler method, backward 119, 392
expansion wave problem 273
explicit multi-stage time stepping
    scheme 233
explicit time stepping 232
extrapolation approaches 38
extrapolation methods 365-6

face nodes 5
facet approximation, of shells 388
fading memory hypothesis 48
far field scattering data 233
FCT *see* flux corrected transport
Fick's second equation 355
Fick's second law of diffusion 394
fictitious frequencies 138
field vectors, electric and magnetic
    232
finite (control) volume 267
finite difference approximations 268
finite difference equation generation
    265

finite difference method 355, 376, 383
finite difference operators 268
finite dimensional spaces 146, 331
finite dimensional test space 106
finite dimensional trial spaces 106, 292
finite elasticity 385
finite element
  3-noded linear triangular 127
  anisoparametric one-dimensional 194
  anisotropic 355-6
  bisection 6
  collocation 243-53
  composite 362-3
  conforming, with regularity 313-14
  cubic 363
  cubic displacement 195
  curved, subgrid of 351
  h refinement 140
  isoceles right triangle 180
  linear displacement 195, *196*
  linear tetrahedral or triangular 219
  moving 358
  neighbouring 143, 180
  non-reflecting 307-14
  with overlay *162*
  prismatic body 140
  quadratic 16, 129
  quadratic displacement 195, *196*
  quadratic smoothing 194
  shell 395
  single 8-noded quadrilateral creep results for *52*
  smoothing 191
  super 308, 310, 311
  target 180
  Timoshenko beam 194
  triangular plate bending 395
  uniform quadratic 16
  vector 316
  Zienkiewicz-Zhu 124
finite element analysis *197*
  adaptive mesh 179-80

in nuclear safety 385
of space shuttle propulsion components 367
finite element approximations 316-17, 380-1
finite element-boundary element method, 307, 330-4
finite element-by-element preconditioning 397
finite element computations, 121-34
finite element discretization, 153-4, 397-8
finite element equations 313
  preconditioning 388-9
finite element error 13, 124
finite element formulation 313
  symmetric 312
finite element Lax-Wendroff method 220-3
finite element level error indicator 13
finite element matrix, approximate inverse 373
finite element matrix problem 309
finite element method 106, 307, 375, 383
  adaptive 105-20
  anisotropic 388
  C^1 359-60
  coupling of 323-48
  dispersion analysis of 315-21
  external 395-6
  hierarchical *hp*-version of 159-69
  hp-version 370-1
  nodal 376
  s-version of 159
  streamline diffusion 113, 365-6
  uncoupling of 371-2
finite element method *see also* method
finite element refinement parameter 123, 124, 126
finite element shape functions 5
finite element space 115, 275
  standard 160
finite element stiffness matrix 173, 373

finite element vertex node 143
finite strain elastoplasticity 381-2
finite volume approximation 270,
    276, 279
finite volume method 267-86, 291-2,
    299-305
    adaptive 289-97
finite volume procedure, for castings
    357-8
finite volume solution 292
first-order piecewise constant
    solution 302
first-order scheme 300
flow
    compressible 109, 215-39
    incompressible 70, 109
    inviscid *230* , *231*
    inviscid, steady, isoenergetic,
        homentropic, planar 323
    laminar incompressible 359
    non-Newtonian 69-76
    of semi-miscible incompressible
        fluid phases 244
    over a shuttle *224*
    stationary compressible potential
        323
    steady compressible 271
    Stokes 390
    subsonic 341-3, *346*
    supersonic 330
    turbulent 384-5
flow circulation, in Golden Horn
    386-7
flow model, incompressible potential
    129
flow problem
    co-volume discretization of
        258-62
    exterior incompressible 346
    incompressible 257-8
flow theory plasticity 381
fluid, viscous incompressible 3
fluid density 70, 138
fluid dynamics, computational 1-23
fluid dynamics system 262
fluid flow 3
    dense non-aqueous phase 243-53

inviscid compressible 290
fluid flow problems 129
    multi-dimensional 278
fluid-variable methodology 140
fluids, non-Newtonian 359
flux 300
    approximate 11
    averages 156
flux balancing technique 10
flux continuous approximation 370
flux corrected transport (FCT)
    222-3
flux evaluations, diffusive 300
flux function 221, 223, 290
    numerical 292
flux projection method 39, *41*
flux projections estimators 37
flux vectors 232, 269, 272
    inviscid 217
forcing function 294
forcing function vector 72
form
    bilinear 3
    compact sesquilinear 146
    linear 3
    trilinear 3
    V-coercive sesquilinear 146
    weak primitive-variable 3
FORTRAN, ACE 173, 174, 175
Fourier analysis 272
Fourier components 61
Fourier-finite element methods 376
Fourier transform 78, 83
Fredholm alternative 276
free boundary seepage problem
    182 , 183
Friedrichs systems 286, 293
frontal solution algorithm 172
frontal solution method 173-5
functional, local 155
functional iteration 301
functions
    bubble 5
    edge 5, 6
    Holder class of 80
    local spaces of 153
    non-smooth 80

symmetric linearly independent
33
trilinear 5

Galerkin approximation 139, 146
   mixed 71
Galerkin basis functions 140
Galerkin discretization error 110
Galerkin formulations 356
Galerkin method 106, 107, 392
   discontinuous 113
   stabilized 356-7
   standard 219, 220, 309
Galerkin methodology 49
Galerkin orthogonality 105, 109,
   110, 115-18, 120
Galerkin problem, discrete symbolic
   115
Galerkin procedure, standard 309
Galerkin scheme, fully discretized
   387
Galerkin spatial discretization 70
Galerkin splitting methods, parallel
   semi-discrete 378
Galerkin/least squares 375
Gårding inequality, discrete 267,
   268, 277
gas, ideal 290, 323
gas constant, adiabatic 323
gas dynamics 267
gasket, homogenized behaviour 363
Gateaux derivative 377
Gauss-Green formula 269
Gauss quadrature points 245
Gauss-Seidel scheme 181
Gauss' theorem 267
Gaussian elimination 172
Gaussian pulse, smooth 319
Gaussian quadrature, one-point 294,
   295
Gibb's free energy 53
global basis functions 6
global effectivity indices 16
global error 13-16, 122, 126, 153,
   279, 281, 301
   and mesh optimality criterion
   124-5

permissible 122, 127, 129
global error norm 124, 130
global error parameter 123, 127
global least squares smoothing 122
global matrix 310
GLS method 310
Gram matrix 147
gravity vector 244
Green representation theorem 327
Green's function 353
Green's identity 156
Green's second theorem 353
grid
   coarse 226, *227* , *236*
   distorted right-angle 320
   equilateral 320
   fine 226, *227*
   gradient smoothing in 351
   h-adaptive 370
   intermediate *236*
   irregular 321, 388-9
   quadrilateral 268
   randomised 320, 321
   rectangular 321
   regular and unstructured *320*
   right triangle 321
   tetrahedral 220
   triangular 138, 140, 315-21
   twisted 205
   uniform 294
   unstructured tetrahedral 216,
      321
   unstructured triangular 216
grid adjustment, on *a-posteriori*
   error estimators 391
grid data structure 225
grid generation
   adaptive methods for 358-9
   unstructured 216-17, 255-7
grid solution algorithms,
   unstructured 219-25
groundwater 243

H^{-1} 289, 294
H^{-1} norm 284
H_2 234
h-adaptive finite element method,

anisotropic 388
h-adaptive grids 370
h-adaptive mesh 13
h-approximation 159
h-extrapolation 302
h-methods 111
h-refined elements 140
h-refinement
    anisotropic 143-6
    isotropic 143
h-refinement derefinement 6
h-version function 160
Hankel functions 309
Hausdorff measure 269
heat conduction problem 38
    anisotropic 380-1
heat equations 88-9
heat source 166, *168*
Heaviside function 283
HELIOS 173, 174
Helmholtz equations 137, 139, 353,
    375
Helmholtz representation formula
    140, 353
Hermite interpolation 245, 248
Hermite polynomial basis functions,
    bicubic 245
Hermite polynomials 81
heuristic analysis 248, 249
hierarchic modelling, problem of
    31-6
hierarchic plate models, stresses
    computed from 352
hierarchical FE-approximation 160
hierarchical hp-version 159-69
hierarchy of plate models 33
Hilbert space 146, 284, 360
    with scalar product 115
Hilbert transform 84-5
Holder class, of functions 80
Holder space 82
hook, optimization of 126, 130, *133*
Hooke's law 59
hp-adaptive boundary element/FE
    methods 137-48
hp-adaptive strategy 21, 140
hp-domain decomposition 160-4

hp-finite element scheme 21
hp-interpolant 6
hp-mesh 5
hp-mesh
    fixed 9
    nonuniform 7
    optimal 140
    sequence of 16
hp-spaces, general 6
hp-version 159-69
hydraulic conductivity 201
Hyper Frontal Solver 173
hyperbolic equation 268, 273, 302
hyperbolic flow problems, steady
    289
hyperbolic operator 275, 279, 281,
    284, 290
hyperbolic problems 113-14, 278,
    289-97
    linear and nonlinear 284
    two-dimensional 279

identity matrix 272
implicit backward differencing
    scheme 244
implicit estimators 37, *40*
implicit schemes 226-9
implicit time stepping 216
incidence matrix 263
incompressibility constraint 70
incompressible flow 70
    preconditioners for 392
incompressible flow problem 257-8
inf-sup condition 152, 153
infinite dimensional problem 157
infinite domain 307, 309, 310
infinite elements 307
    high order 372
inflow, parabolic *18*
inhomogeneous permeable media
    355
initial conditions 27, 28, 46, 63-5,
    72, 257, 305
initial data 3, 7
initial error index 13
initial strain 58
initial value problem 3, 301, 373

integral equation 106
    hypersingular 387
integral operators 50
    on basic functions 84-9
integration rule, composed 163
interface continuity 245
interface relaxation 245, 248
internal variable 26, 27
interpolation constant 113, 116, 117
interpolation estimates 115, 117
interpolation projector 275
inverse assumption 331
invertible map 5, 6
inviscid flow *228*, *230*, *231*, 323
irregular grids 321, 388-9
iso-smoothing 193, 195, 197
isothermal conditions 46
isotropic hardening law 27-8, *29*, *30*

isotropic law, peak stresses for *30*, *32*
isotropic point insertion technique 294
iteration
    explicit 221
    outer and inner 337
iteration loop, inner 337-40
iteration scheme 292
iteration step, outer 340
iterative algorithm 245, 271
iterative interface relaxation
    solution technique 245
iterative parallel FE computational
    method 179-88
iterative solvers, fast 384

Jacobi matrices 272
Jacobi preconditioned conjugate
    gradient algorithm 206
Jacobian, one-dimensional 332
Jacobian matrix 220, 221
joints, human 386

kernel function 83
kinematic hardening law 27, *29*, *30*
kinematic law, peak stresses for *30*, *32*

kinematic viscosity, of fluid 257
Kirchoff constraint 391
Kirchoff-Love assumptions 395
Kirchoff plate models 354
Kirchoff triangle, discrete 395
Knauss model 54, *57*, 58
Kolosoff-Muskhelishvili potentials 311
Korn airfoil 126, 129-30
Kronecker delta 290
Kutta-Joukowski condition 324, 329, 330, 336-41, 343, 344

L_2 approximation 157
L_2 errors 321
L_2 inner product 290
L_2 norm 116, 119, *197*, 283, 284, 289, 294
Lagrange basis functions, bilinear 245
Lagrangian functional 155
Laguerre polynomials 80
Lamé coefficients 47, 62
Lanczos algorithm 362
LAPACK 375
Laplace equation 101, 311
Laplace operator 81
Laplace transform 50
Laplacian 375
    Dirichlet problem for 101
    square root of 88
Lax-Richmyer stability 268
Lax-Wendroff flux function 220
Lax-Wendroff method 70, 220-3
Lax-Wendroff pseudo-timestepping procedure 272
Lax-Wendroff type scheme 222, 229
Layton-Rabier scheme 15
LBB condition 7
least-squares functional, discrete 191
Lebesgue space 151
Legendre polynomial 5, 33
LEPS *see* local time error per step
LEPUS *see* local time error per unit step
line relaxation 216, 229
linear advection *295*

linear advection equations 268
linear displacement element 195,
    196
linear elastic behaviour 126
linear elastic limit 61
linear elastic responses, Boltzmann
    type 47
linear elasticity 47, 153
    stiffness matrix of 62
linear elasticity problem 31
linear elements, for
    Reissner-Mindlin plates 394
linear flow fields 262
linear form 3
linear hyperbolic equations 274-8
linear hyperbolic problems 270
linear-linear method *317*, 318
linear operator 115
linear scalar convection-diffusion
    problems 109
link characteristics 264
lip vortex 69
Lipschitz boundary 269, 279, 284,
    360, 380
Lipschitz curve, simple 346
Lipschitz domain 8, 31
Ljusternik-Schnirelmann theory 369
load balance 15, 175, 181
load functional 162
load vector 140, 162, 309
loads, concentrated 160
local error indicator 123, 294
local error parameter 125, 127
local error residual problem 158
local error tolerance 303
local least squares smoothing 122
local time error per step (LEPS)
    301, 303
local time error per unit step
    (LEPUS) 301, 302, 303
local-in-time spatial error 302
locking 191, 192
logarithmic potential 85-6
LU-factorization 15

Mach number 223-*231*, 294, *296*,
    325

local 324
magnetic field 316
map, invertible 5, 6
mapping 273, 292, 329, 332, 351, 359
Marshak thermal wave problem 352
mass conservation property 109
mass density 51
mass matrix 62, 71, 368-9
    diagonal lumped 221
    standard consistent 219
master element 5
master mesh 39
material
    internal 53
    non-aging 48
    synchronous & asynchronous 48
    synchronous non-aging 50
matrices, positive definite 361, 382
matrix eigenvalue problem 317
matrix-valued piecewise constant
    test function 272
Maxwell's equations 232, 315-21,
    364
memory
    distributed 199
    of viscoelastic materials 46
mensional 300
mesh *168*
    1-irregular refinements of 6
    2-D rectangular 143
    Chevron 39, *40*
    coarse 159, 164
    composed *161*
    composed, and global solution
        165
    criss-cross *40*
    fine 159, 164
    intermediate 13, *19*, *20*
    intial, estimated error for *19*
    m-irregular 143
    optimal 111, 127, 140, *141*
    quadrilateral 300
    refinement-path dependent 143
    regular *40*
    sequence of *42*
    spatial 299, 303
    tetrahedral 258

triangular 300
Union Jack 39, *40* , *41*
mesh-dependent seminorms 276
mesh distortion 318
mesh elements, initial 143
mesh function 111, 112, 270
mesh generation 180, 229
  automatic 315
mesh generator 42, 304
  multiblock algebraic 140
mesh optimality criteria 121-34
  & specific error 125-6
mesh parameter 189
mesh points 301
  interface 181
  interior 181
mesh refinement 123, 283
  adaptive, in finite element
    method 382
  non-consistent oscillatory 125
mesh size 5
  local 111
mesh strategy, one-irregular 143-6
method
  advancing front 216
  boundary element 307, 323-48,
    387
  Boundary Point (BPM) 77
  of conjugate gradients 175
  Crank-Nicolson 70
  Delaunay 321
  direct-iterative 361
  domain decomposition 243
  dual variable 263-5
  edge-constant *317* , 321
  Euler, backward 119
  extrapolation 365-6
  finite difference 355, 376, 383
  finite volume 267-86, 289-97,
    299-305
  flux projection 39, *41*
  Fourier finite element 376
  frontal solution 173-5
  Galerkin splitting 378
  grid 215-19
  h− 111
  iterative parallel 179-88

Lax-Wendroff 70, 229
linear-linear *317* , 318
of lines 299
mechanical drilling 398
mesh generation 229
minimisation 340-5
multi-stage 221-2
multigrid 226
Newton inexact 362
Newton's 271, 374
of weighted residuals 245
orthogonal collocation 245
p− 111
parallel domain-decomposition
    243
parallel iterative 181
Schwarz 380
see also finite element method
SOR iterative 184
spatial discretisation 233, 299
Texas 3-step 11-15, 16
upwind 223-5, 293, 300
upwind vertex-centred 291-2
variational 189-98
of weighted residuals 245
microbial mass 200, 201, *209*
mid-point solution values 300
minimisation method 340-5
minimization principle 154
minimization problem 335-6
  conjugate gradient algorithm for
    336-40
  one-dimensional 338
MMOC-Galerkin *see* modified
  method of characteristics
modelling errors 355
modified method of characteristics
  (MMOC-Galerkin) 202, 206
Moire measurements 190
momentum diffusion matrix 71
momentum equation 257, 258
momentum step 7
Monte-Carlo approach 29
Monte-Carlo simulations 355
Mooney-Rivlin relations 397
multi-processing systems 171
multi-stage method 221-2

multigrid acceleration 216, 233
multigrid method 226
MUSCL technique 223, 292

Navier-Stokes equations 4, 8, 21, 89,
    151-8, 356
    3D, modelling of 225-65
    compressible 279
    incompressible 265
    non-stationary 97-100
    steady-state 4
Navier-Stokes problem 15-21
network, primary and secondary 291
Neumann condition 207, 245-8, 308,
    324
Neumann data 335
Neumann map 309
neutron diffusion equation 267
Newton iteration 362
Newton-Raphson algorithm 8
Newton-Raphson iteration 13
    inexact 362
Newton's law of viscosity 46
Newton's method 271, 362, 374
Newton's second law 46
Newtonian potential 86
nitrate 206, *210*
nitrite 206
no-slip 72
nodal residuals 271-3
nodal smoothing technique 129
nodal spacing 216
nodal stress recovery techniques 122
nodal stress values 122
node
    constrained 6, 143
    edge 5
    element vertex 143
    external and internal *173*
    face 5
    interior 5
    irregular 6, 143
    multiple 143
    multiple constrained *144*
    parent 143
    vertex 5
node-based iteration scheme 273

non-dimensional variables 71
non-dissipative difference scheme
    268
non-Newtonian flow 69-76
non-Newtonian fluids 359
non-Newtonian inelastic materials
    70
non-reflecting boundary condition,
    exact 308-9
non-reflecting finite element 307-14
non-reflecting finite element
    (NRFE) 307-14
    conforming 313
    high-order 313
    local *313*
    nonlocal 310
non-wetting 244
nonlinear convection diffusion 244
nonlinear equations 292
nonlinear minimization problem 114
nonlinear monotone elliptic
    problems 109
nonlinear parabolic equations 89-97
nonlinear viscoelasticity 51-9
nonlocal boundary condition 309
nonlocal integral operator 309
nonreactive trace *208*
nonsmooth problem 284
norm 111, 115
    canonical 152
    choice of, for error control 115
    energy 159
    energy-like 9
    error energy 122
    L_2 119, *197*, 283, 284, 289
    Sobolev 4, 82, 284
    stress error *195*
    strong 116, 117
    symmetriser 153
    weak 117
    weighted graph 279
NRFE *see* non-reflecting finite
    element 307
nuclear safety 385
null space 157
numerical flux function 221
numerical model, parallel 199-211

numerical quadrature 110
Nylon 45, 53, 54, 58, 66
Nylon 66 creep curves *56*

ONERA M6 wing *236* , *237* , *238*
operator 147, 148
    pseudodifferential 88
operator splitting technique 70
optimal design problem,
    discretization of 360-1
optimal error bounds 267
optimal mesh 111, 125, 127
optimization problem 111
ordinary differential equations 26,
    109, 265, 299
    non-stiff 301
orifice meter 384-5
orthogonal collocation method 245
orthotropic material 39
Osher scheme 300
outflow *18*

p-adaptivity 6
p-approximation, hierarchical sum
    of 159
p-methods 111
p-version 159, 160, 384
p-version preconditioning 368-9
parabolic equations 89-97, 245, 392
parabolic formulation 67
parabolic-like contaminant
    species-transport equations 243
parabolic-like saturation equation
    243
parabolic problems 113, 397-8
parabolic velocity profile 72
paradigm for adaptivity 108-10
parallel backward substitution 173
parallel domain-decomposition
    method 243
parallel iterative main net solver
    175-6
parallel iterative method,
    formulation of 181
parallel machines, domain
    decomposition techniques for 368
parallel processing 14-15, 171-8,

185-7
parallel solution collocation (PSC)
    245-9
parallel substructure technique
    172-6
parasitic modes 318, 319
parent node 143
PARIX 175
partial differential equations 46,
    106, 107, 268, 376
    coupled 243
    linear elliptic 289, 375
    nonlinear 383
    time-dependent 299-305, 358-9
partial differential operator 111
partitions, family of 5
patch of elements 189, 190
patch stress recovery procedure
    398-9
Peace-Rachford iteration 15
Peclet numbers 202
penalty constraint functional 191
penalty functionals 335
permeability, relative 244
permeability field *207*
perturbation equation 109
perturbation expansion 319
perturbation potential 325
Petrov-Galerkin formulation 272
Petrov-Galerkin method 270, 372
    nonconforming 267
phase mass 244
phase velocity 61
phase velocity, surfaces of *318* , *319*
phase viscosity 244
photon transport 352
Picard iteration technique 245
piecewise constant boundary trial
    functions 331
piecewise constant functions 268,
    292
piecewise constant values 300
piecewise linear functions 10, 39, 71,
    72
    discontinuous 203
piecewise linear hat functions 331
piecewise polynomial approximation

106, 111, 156
piecewise polynomial functions 106
piecewise polynomial subspaces 153
piecewise polynomial vector basis
    functions 316
piecewise polynomials 49, 313, 315
piecewise quadratic functions 71, 72
piecewise smooth boundary 31
plane strain conditions 126
plane stress 51
plane stress model 131
plane wave *141*, *142*
plane wave solution 319
plastic region 27
plasticity
    1-D problem of 26
    modulae of 27
plasticity problem 26-31
PLASTPARA 171, 173, 175
plate models, 2-dimensional 371
plates, delaminated 354-5
point collocation 331
pointwise error 164, 320
Poisson's equation 70, 335, 390
Poisson's kernel 90
Poisson's problem 8, 338
Poisson's ratio 48, 51, 54, 87, 131
Poisson's summation formula 78
Polak-Ribiére type conjugate
    gradient algorithm 336
polygonal boundary 268
polymers 45
polynomial 9, 11, 13
polynomial interpolation 118
polytope, convex 255, *256*
polyvinylacetate 54
porosity 201
porous media 243-53
post-processing 37
power law 50
Prandtl-Glauert flow 325
Prandtl-Glauert transformation 329
preconditioning, element-by-element
    397
predicator-corrector 70, 92-4
PREDIVIDER 177, 178
pressure 3, 22, 100, 151, 259, 348

computed, for mesh *22*
pressure-based scheme 7
pressure equation, for h-adaptive
    grids 370
pressure stiffness matrix 71
pressure vectors 71
primitive discrete model 265
prismatic elements 140
problem
    2D elliptic 390
    3D 34
    of *a-posteriori* error estimation
        36-42
    advection 292
    advection-diffusion 273
    backstep channel 16-21
    Burgers' equation 303, 304
    Cauchy 77, 88-9, 96, 97, 100
    compressible flow 269-73
    compressible wave 273
    contaminant transport 248
    continuous dual 115
    continuum 257
    convection-dominated hyperbolic
        109
    coupled, elastic scattering 146
    coupled boundary value 325-6
    diffusion-dominated
        convection-diffusion 107
    discrete 147, 335-40
    discrete variational 332
    dual 115, 116, 117, 118
    dual continuous 118
    eigenvalue 369, 395-6
    elasticity 38, 60
    elliptic 11, 112-13, 118, 122, 124,
        363, 376
    entropy 329-30
    expansion wave *273*, 274
    flow 257-62, 346
    fluid flow 129
    forced vibrations 146
    free boundary seepage *182*, 183
    Galerkin 115
    heat conduction 38, 380-1
    of hierarchic modelling 31-6
    hyperbolic 113-14, 278, 279, 284,

289-97
hyperbolic flow 289
incompressible flow 257-8
infinite dimensional 148, 157
initial value 3, 301, 373
interior 325, 343
of linear elasticity 31
linear hyperbolic 270
linear scalar convection-diffusion 109
local error residual 158
Marshak thermal wave 352
matrix eigenvalue 317
minimisation 335-40
model, for heat conduction in a plate 31
model elastic scattering 146
Navier-Stokes 15-21
nonlinear minimization 114
nonlinear monotone elliptic 109
nonsmooth 284
optimal design 360-1
optimization 111
parabolic 113, 397-8
plasticity 26-31
Poisson 8, 338, 339
Prandtl-Glauert 326
quasistatic non-ageing 48
Riemann 300
scalar advection model 273-4
semi-discrete approximating 49, 66
shell, elastic spherical 140, *241* , *142*
Stokes 151, 153, 274
structural 122
subsonic 329
transient 4, 229-39
transversal flow 273
U-notch *175* , *176*
variational 329
viscous flow 15-16
Volterra 63
wave *308*
procedure, line relaxation 229
product space 153, 154
propagation direction 318

PSC *see* parallel solution collocation
pseudodifferential operator 88, 268

quadratic elements 129
uniform 16
quadratic smoothing element 194
quadratic triangles *130*
quadrature error 160
quadrature rule, one point 300
quadrilateral, convex 275
quadrilateral mesh 300
structured 268
quasistatic creep response 51
quasistatic formulation 48
quasistatic non-aging problem 48

radial velocity 72
radiation condition 325
Radon measures 269
random vectors 29
randomised grids 320, 321
rate of convergence, exponential 159
Raviart-Thomas spaces 201, 204
re-entrant corner 69
reaction coefficient 164
reaction diffusion equations 159, 160
reaction zones 105
reactions step 205
rectangular grids 321
refinement process, 2 step 125
refinement strategy 180
region, open bounded 3
regular mesh *40*
Reissner-Mindlin plate model 354, 391, 394
boundary layer for 356
relative error *36*
relative permeability 244
relaxation factor 125, 183
relaxation function 51, 54, 55
relaxation parameter 272
remeshing algorithm 179
residual estimators 37
residuals 10, 113
retardation factor 201
Reynolds number 74, 76, 120

Rice 3D Parallel Groundwater Reactive Flow and Transport Code (RPGW) 207, 209, 211
Richardson extrapolation 365
Riemann problems, one-dimensional 300
Riesz operator, discrete 147, 148
Riesz-Schauder theory 156
right triangle grids 321
ring segments, vibrations of layered 379-80
rings, layered circular 379
Ritz projection 334, 345
Ritz-Volterra projections 63
rock masses, jointed 366-7
Roe matrix, 222, 225
row-wise subdividing 181
RPGW *see* Rice 3D Parallel Groundwater Reactive Flow and Transport Code
rubber elastic structures 392
Runge-Kutta scheme 7-8, 205

s-version, of finite element method 159
St. Venant equations 271
saturation assumption 156
scalar advection model problems 273-4
scalar magnetic field 315
scalar potential 60
scalar relaxation parameter 272
Schapery model *57*
Schur complement, condition number of 361
Schur complement system 368
Schwarz method 380
SEA *see* smoothing element analysis
second-order hyperbolic equations 378
second-order in space solution 302
selective reduced integration 374
semi-discrete momentum equation 262
semi-discrete problem 49, 66
semi-discretisation approach 310

semi-infinite geometry 311
sequential main net solver 173
sequential work *176*
sesquilinear form 268
shallow water equations 267
shape function 159, 160, 194, 313
    element 5
    local 313
    local spectral order of 5
    piecewise linear 71, 72, 219
    piecewise quadratic 71, 72
    standard bilinear 5
shape function polynomials, degree of 124
sharp reaction 160
shear modulus 87
shear triangle, discrete 395
shell elements, shallow 395
shell problem
    elastic 140, *141*
    spherical *141*, *142*
shells
    approximation of 388
    thin, approximation of junctions 359-60
shock 105, 271, 272, 278
shock-boundary layer interaction 388
shock capturing 222
shock fitting algorithm 270
shock fronts 160
shock waves 283
simple layer potential 371
singularity element patches 351
six triangle stencil 300
skin 332
slit 344
smooth boundary 151
smooth flow 119, 120
smooth functions 83, 270
smooth solution 38, 63
smooth stresses 191, 194-7
    variational method for 189-98
smoothing element analysis (SEA) 194, 195
    optimal 193
smoothing procedures 351

Sobolev norms 4, 82, 284
Sobolev space 139, 151, 313, 326,
    353, 376
    complex 275
solid density 138
solution
    on macroscopic scale *169*
    near origin *169*
    unique 4
solution methodology 138-40
solution vector 309
Sommerfeld condition 138, 308
SOR iterative method 181, 184
SOR relaxation factor 183
sound speed, in fluids 138
space discretization 302
space grid step 96
space local error control 305
space shuttle propulsion
    components 367
space and time errors, balancing of
    301-2
spatial approximations, high-order 7
spatial discretisation 49, 233, 299,
    300
spatial discretisation error 301, 304
spatial mesh 299
spatial mesh points 303
SPD system 15
specific heats, ratio of 290
speedups 172, 177, 181, 183, 185
    theoretical 172
spline approximations 84
splines, cubic B- 378
split-beam solutions 354
stability 108
stability constant 113, 116, 118, 148
standing wave 65
state vector 290
stationary liquid 394
steady compressible flow 271
step function 283
stiffness matrix 140, 185, 309, 310
    constant 311
    symmetric 310
Stokes equations 151-8
Stokes flow 390

Stokes problem 16, 151, 153, 274
Stokes solver, efficient 392
Stokes theorem 259, 260
stopping criterion 112
stopping strategy 23
strain 46, 47, *52*
    final 58
    initial 58
strain energy 122
strain operator 124
strain tensor 46
    infinitesimal 138
strain thickening 69, 74
stream function *22*
streamline diffusion finite element
    method 113
streamline projections *75*
stress 45, 46, 47, 138
    compression 29
    computed 28
    exact 122
    measured 28
    nonproportional or decaying 48
    yield, for aluminium 29
stress error 65, *196*
stress error norm *195*
stress intensity functions 354
stress interpolating functions 122
stress relaxation 46
stress relaxation matrix 47
stress tensor 7, 151, 192
strong stability 105, 109, 110,
    115-18, 120
subdomain 153
subdomain estimator *40*
subsonic flow 328, 341-3, *346*
subsonic problem 329
super finite element 308, 310, 311
super-smoothing 193, 195, 197
supercomputers, parallel 200
superconvergent patch recovery 399
supersonic flow 330
surface tractions 3, 7, 46
symmetrizer norm 153
synchronous material 48
system matrices 72

T-junction, speedup and efficiency
for *174*
tangential components 7
tangential traction 72
target element 180
target error 13, 14
Taylor-Galerkin
    pseudo-timestepping procedure
    272
Taylor-Galerkin type scheme,
    explicit 233
Taylor-Galerkin/pressure-correction
    scheme 70
Taylor series 70, 78, 318
technique
    balancing 10
    biorestoration 199
    convergence acceleration 226-9
    domain decomposition 368
    flux balancing 10
    isotropic point insertion 294
    iterative interface relaxation
        solution 245
    mapping 359
    MUSCL 223, 292
    nodal smoothing 129
    nodal stress recovery 122
    operator splitting 70
    parallel substructure 172-6
    Picard iteration 245
    zooming 159-69
ten triangle stencil 300
tensor
    deformation 70
    diffusion/dispersion 201
    intrinsic permeability 244
tensor of elasticities 138
tensor product non-uniform meshes
    278
Tessler's smoothing finite element
    method 190
test functions, 140, 162, 274
test space 106, 268
tetrahedra *263* , 315
tetrahedral control volume *258*
tetrahedral edges 255
tetrahedral face *261*

tetrahedral grid generation,
    unstructured 215
tetrahedral grids 220, 321
tetrahedral mesh 258
tetrahedron *263*
Texas 3-step method 11-15, 16
theorem
    divergence 258, 300
    Gauss' 267
    Green's second 353
    Stokes' 259, 260
thermoforming process 396-7
Theta method 301
three-dimensional divergence
    theorem 258
three-dimensional-elastic potential
    87-8
three-dimensional Navier-Stokes
    equations 255-65
three-dimensional tetrahedral
    meshes 256
three-noded linear triangular
    elements 127
three phase flow code 210
three-step *hp*-adaptive strategy
    11-12
time-dependent finite element
    solution 352
time-dependent partial differential
    equations 299-305, 358-9
time-dependent wave equation, 3D
    312
time-dependent waves 310-11
time frequency 138
time-harmonic elastic waves 308
time-harmonic waves, acoustic 308
time integration 299, 300-1, 303
time-marching algorithms 89-97, 99
    one step 89-92
time-marching scheme, second order
    semi-implicit 70
time-step increments 202
time stepping scheme 320
time stepping scheme
    explicit multi-stage 221, 223, 226
    implicit 216, 226
    leapfrog 320

time steps 58, *66* , 70, 299, 301, 303
  constant 67
Timoshenko beam elements 194
tolerance 111
tolerance level 105
trace operator 280
  inward 270
trace theorem, 157
traction 49, *52*
trailing edge 323, 324, 339
transfer operators, intergrid 226
transient problem 4, 229-39
transmission condition, second 325
transmitted error 281
transonic flow 294, 323-48, *346*
transonic gas dynamics equations
  267
transport equation iterations *252*
transport equations 202
transputer networks 171-8
transversal flow problem 273
trapezoidal rule 49, 58, 62, 66
travelling velocity 323
travelling waves 316
trial space 106, 268
triangle inequality 269, 278
triangular bump 305
triangular element
  plate bending 395
  standard linear 313
triangular grid 140, 315-21, *338*
triangular meshes 300
triangulation
  families of 334
  family of 344
  regular 160
    family of 331
trilinear form 3
trilinear functions 5
truncation error 108, 109, 116, 117,
  270, 272
twisted grid 205
two-dimensional domain 395-6
two-sided bound 282
two space dimensions 94-7

U-notch problem

FE mesh for *175*
  speedup and efficiency of *176*
uniaxial tensile test 53, 54
uniform grid 294
uniform refinement 123
Union Jack mesh 39, *40* , *41*
unit normal vector field 324
unit outward directed normal 46
unit outward normal vector 270
unit parallelogram 317
unit vector 11, 257, 258
unsteady-state mass transfer 394
update global solution 248, 249
upstream & downstream length 74
upwind differencing, on tetrahedral
  mesh *261*
upwind methods 223-5, 293, 300
upwind vertex-centred method
  291-2
user-specified constants 216, 221
user supplied accuracy tolerance
  (TOL) 301, 303

van Leer limiter 300
variable, internal 26
variables 172
  non-dimensional 71
variational inequality formulation
  182, 381
variational principle 191
variational problem 329
vector 33
  3-D parameter 27
  5-D parameter 28
  6-D parameter 28
  of conservation variables 217
  consistent load 192
  displacement 140
  forcing function 72
  outer iteration 337
  outward normal 258
  pressure 71
  solution 309
  state 290
  unit 11, 257, 258
  unit outward normal 270
  velocity 70, 71, *110* , 129, 257,

260
vector elements 316
vector function 269, 273
  two-component 274, 279
vector potential 60
velocity 3, 66
  nodal 71
velocity components 99
velocity dependent hydrodynamic
  dispersion tensor 244
velocity field 129
  nondivergence-free 70
  reconstruction of 262-3
velocity nodal points 72
velocity vector 70, 71, *110*, 129,
  257, 260
vertex-centred method, upwind
  291-2
vertex nodes 5, 263
vertex pressure nodal points 72
vibration 59
virgin material 27, *29*
viscodynomics 48, 59-67
viscoelastic body, isotropic,
  compressible, solid 46
viscoelastic decay, swamped 66
viscoelastic functions 48, 50
viscoelastic materials, memory in 46
viscoelastic plate, unilaterally
  supported 360-1
viscoelastic relations 397
viscoelastic standing wave 59
viscoelastic travelling wave 59
viscoelasticity
  dynamic 45-67
  nonlinear 51-4
  quasistatic 45-67
viscosity 152
  artificial 114, 271-3
  extensional 359
viscosity constant 70
viscous flow problem 15-16
viscous term 259
Vol'pert's generalisation 269
Volterra problem, elliptic-, 63
Von Mises stresses 131
Voronoi face *261*
Voronoi polygon 264
Voronoi polyhedron *256*

Voronoi polytope 255, 256, 260,
  261, 263, 264
Voronoi tessellation 255, 256, *257*
vortex
  abrupt corner 74
  lip 69, 74, 76
  salient corner 74, 76
vortex development, in complex
  geometries 359

wave, plane, scattering of 140
wave equation 59, 88-9, 107
  linear 109
  reduced 311, 312
  time-dependent 312
wave equation Mathieu functions,
  reduced 311
wave number 138
wave problem, exterior *308*
wavelets 399
waves, travelling 316
weak formulation 329
weak stability 109
weakly singular positive type
  memory term 392
weight function 217, 219, 245
  positive 279
weighted graph norm 279
weighted least squares 107
weighted residuals method 245
Weissenburg number 71
wetting 244
wing, discretisation of surface of *218*

wing section 323
worst brackets approach 29

yield stress 27
Young's modulus of relaxation 53,
  60, 131

Zienkiewicz lecture, second 1-23
Zienkiewicz-Zhu (ZZ)
  element 124
  estimator 41, 179
  stress recovery procedure 189
Zlamal's minimal angle condition
  356
Zlamal's operator 333
zooming technique 159-69